10th International Symposium on NDT in Aerospace 2018

Berichtsband BB 168

Dresden, Germany
24-26 October 2018

ISBN: 978-1-5108-9404-4

Printed from e-media with permission by:

Curran Associates, Inc.
57 Morehouse Lane
Red Hook, NY 12571

Some format issues inherent in the e-media version may also appear in this print version.

Deutsche Gesellschaft Fuer Zerstoerungsfreie Pruefung (DGZfP)
Max-Planck-Str. 6
12489 Berlin
Germany

Phone: +49 30 67807-0
Fax: +49 30 67807-109

mail@dgzfp.de

Additional copies of this publication are available from:

Curran Associates, Inc.
57 Morehouse Lane
Red Hook, NY 12571 USA
Phone: 845-758-0400
Fax: 845-758-2633
Email: curran@proceedings.com
Web: www.proceedings.com

TABLE OF CONTENTS

COMPOSITES (TECHNICAL SESSION)

EDDY CURRENT

PLENARY TALK ROBOTICS IN NDT

ROBOTICS IN NDT (SHORT PRESENTATIONS)

PLENARY TALK NDT AND ADDITIVE MANUFACTURING

APPLICATIONS (TECHNICAL SESSION)

STRUCTURAL HEALTH MONITORING (TECHNICAL SESSION)

POSTER

Simulation of a fiber-optic based SHM system. Application to a bonded repair patch

Alfredo GÜEMES [1], Maria Eugenia REYES [1], Jaime GARCIA-RAMIREZ [1], C. AGUILAR[1], Antonio FERNANDEZ-LOPEZ [1]
[1] Universidad Politecnica de Madrid (UPM), Madrid, Spain

Contact e-mail: alfredo.guemes@upm.es

Abstract. Simulation tools are very useful to understand and optimize NDT systems, and particularly the SHM systems. An algorithm has been proposed to quantify the damage detection capability of a fiber optic sensor network, with a finite number of strain measurements. A damage index is obtained by simulation, dependent of the damage size and position, and the number and distribution of sensors, and also of the noise of the signals; by repeating the simulation under random noise, a POD (Probability of Detection) could be calculated. The analysis is applied to a simple representative experiment, a bonded repair patch, under mechanical and thermal loads.

Introduction

Structural Health Monitoring (SHM) is defined as the process of acquiring and analyzing data from on-board sensors to evaluate the health of the structures [1]. SHM shares with NDT the objective of detecting local damages, but while NDT techniques rely on external instruments, like X-Ray sources and detectors, or ultrasonic probes, which imply that the structural element needs to be accessed and probably dismounted for inspection, for SHM the sensors responsible of getting information must be permanently attached to the structure at fixed positions; the structure can be remotely inspected, without disassembly, without specialized personnel, which imply significant savings in maintenance. Sensors have to be distributed through the structure, because the location for damage occurrence is usually unknown; consequently the sensors need to be small size and cheap, to allow for tens to hundreds of sensors for each structural element. PZT and FOS sensors are the best candidates, and the more widely studied, but alternative solutions like MEMS are also feasible.

Fiber-optic sensors (FOS) are recognized as excellent strain/temperature sensors. It can be said that their technology readiness level (TRL) is about 8–9, they have already been demonstrated in real aircrafts, and they are routinely used in many other industrial applications, like monitoring oil wells. But as said at [2], to get information about damage from strain measurements, additional strategies are needed. Damage information can only be got by processing and comparing the raw signals received from the sensors before and after damage occurrence, in an attempt to identify the "features", or parameters that are sensitive to minor damages and that can be distinguished from the response to loads and environmental disturbances.

To validate an SHM system and enable their large-scale applicability, the effectiveness and damage detection sensitivity of the SHM technique must be quantified with respect to its ability to accurately determine the actual health condition, as it is done for any other NDE method. The effectiveness of the NDE method is quantified by the term NDE reliability, which is defined as "the probability of detecting a crack in a given size group under specified procedures and inspection conditions" [3]. NDE reliability is expressed in terms of the flaw size having a detection probability of 90%, a90. Nevertheless, there is certain statistical uncertainty associated with the value of a90, which is represented by stating a 95% confidence interval. This crack size is represented as a90/95 and serves as an important parameter to quantify NDE reliability. The NASA POD developments for quantifying the NDE capabilities were proposed in the '70s and very soon accepted as a standard method. MIL-HDBK- 1823A [4] is currently considered as the state-of-the-art guidance for conducting POD studies by the USAF and other industries. Very succinctly, what is needed is to get, for the given inspection system, a plot similar to the drawing given at Figure 1

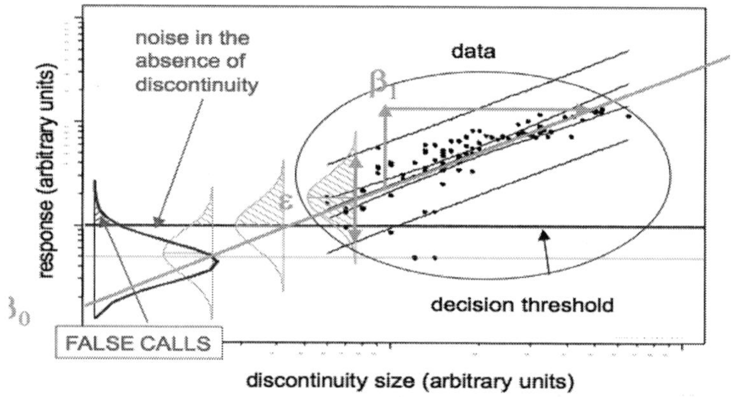

Figure 1. Reliability chart requirements for SHM systems (from Ref 5)

To build such a plot requires a lot of experimental data, which are costly and time consuming; the experimental effort can be reduced with an adequate numerical model of the NDT process. Simulation allows for a deeper understanding of the inspection method, allowing for a larger number of 'virtual experiments' conditions and the optimization of the system. And this is particularly true for the case of SHM system, because sensors are at fixed positions, and running real experiments changing the position of the damage respective to the sensors would be costly prohibited.

Monitoring damage in bonded repair patches, instrumented either with PZTs or with FOS, has been a classical exercise for SHM, because certification of bonded structures poses many difficulties, and SHM could be a way to overcome this issue [6]. It has usually been done under a pure experimental approach [7], and because of the limited number of experiments, the conclusions have mainly a qualitative nature, hardly to extrapolate to other cases.

1. Damage detection from strain measurements

Measuring strains is not the same as detecting damage. It is important to recognize that damage is not a physical parameter, it is just a local change in the materials properties or at the structure connectivities (a crack is simply a new boundary), which degrades their structural performances. Damage can only be detected by comparing the response of the structure to some external estimulae (forces, elastic waves, thermal heating, …), before and after damage occurrence.

There are several procedures for doing it, as mentioned at [2], which may be classified as local or global methods. At this paper we deal only with the global methods, able to detect the damage anywhere at the structure.

1.1 Damage detection algorithm

The changes in the global strain field promoted by a local damage are very small a few centimeters away from the crack, and may be easily faded out by natural disturbances or by the noise of the measurement system. Data analysis procedures are needed to identify the systematic deviation caused by the damage. We are using one of the simplest and more robust procedures, the Principal Component Analysis (PCA). The original data are re-expressed in a new orthogonal basis where the data are arranged along directions of maximal variance and minimal redundancy, called the principal components. Damage indicator for this technique is the Q-index. Verified software for doing PCA analysis is available as Matlab tools. Full details are given at [8].

1.2 Simulation algorithm for a fiber-optic based SHM system

The former algorithm may work with experimental data from strain measurements, obtained from mechanical tests on the structure, before and after damage occurrence. A large number of tests are needed in order to get statistical significance. To run several experiments on the pristine structure is quite feasible, but only one case of damage location may be tested, so a limited number of experimental cases can be done.

Strain data can also be generated by a Finite Element Model (FEM). As sketched at figure 2, a conventional FEM analysis can be applied to the structure, and once the sensors position are defined, it is straightforward to get the array of strain data for these positions. Similarly, after the damage position and size is defined, a similar FEM model may be applied, and again the strain data at the sensors position can be produced. It has to be noted that essentially the same FEM model may be used, the crack is simulated just by disconnecting the nodes of the elements around the crack; a refined mesh at the crack tip is not needed, because only the global strain field is needed. It has to be pointed out that everytime something is changed at the FE model, some differences will appear at the numerical results; but to mimic the reality, it is needed to add to the numerical data a random value representing the noise of the measurement system.

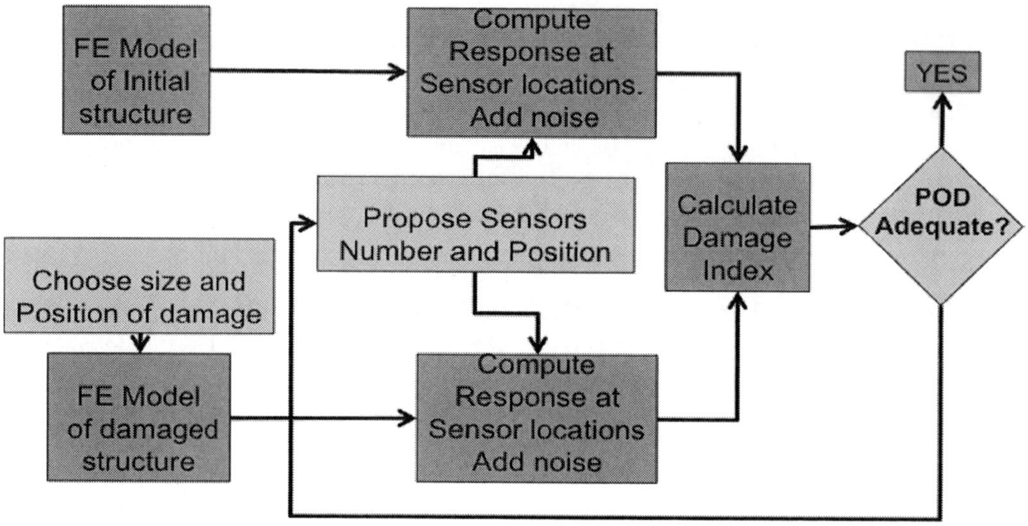

Figure 2. Algorithm for simulation of a strain-based SHM system

1.3 FEM model description

The above-mentioned algorithm is applied to a simple exercise, simulating a repair patch, as sketched at fig 3. The finite element analysis is done with NASTRAN. A linear static analysis of a rectangular aluminium plate is represented with continuous shell elements (CQUAD and CTRIA). The plate is uniformly loaded in one of the edges of the panel and clamped at the opposite side. The length of the panel is large enough to obtain a homogeneous stress distribution in the far field. A single horizontal slit was placed in the middle of the panel. The slit is simulated by releasing the node connectivity throughout the slit length.

An aluminium square cover plate is simulated with the same material as the main plate (Al 7075: E = 71.7 GPa and e = 0.33) and the same thickness (2 mm). This patch is placed on top of the slit and 45 degrees reoriented respect to the principal direction of the panel.
The linear static analysis was repeated after detaching different percentages of the cover plate (0%, 5%, 10%, 15%, 25% and 50%). The detached percentage is simulated by disconnecting the common nodes between the panel and the cover plate from the left side to the right side of the plate. Strain measurements were obtained at the four corners of the crack cover plate in the principal direction of the panel.

Figure 3. Plate with a bonded repair patch. Red arrow are sensors position

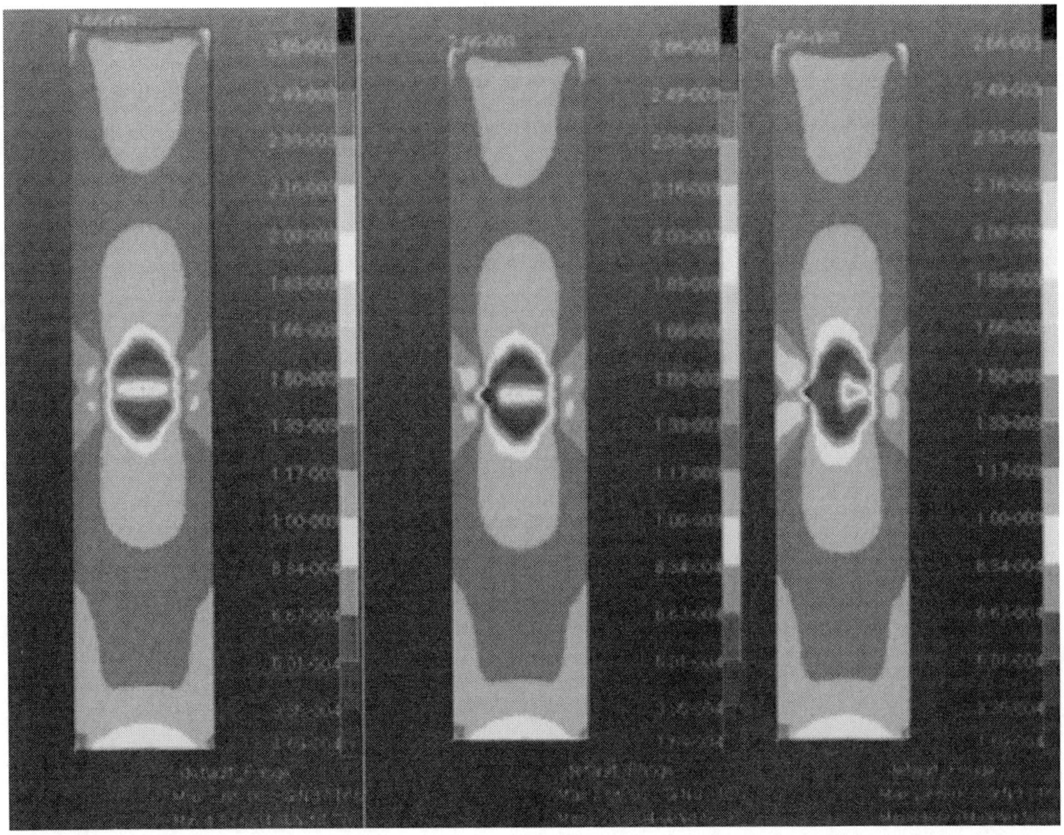

Figure 4. Strain field (longitudinal component) at the pristine structure, and after detaching 25% and 50% , respectively

With clean numerical results, all cases could be distinguished, but after adding a random noise to the data (assuming that noise of the interrogation system has a normal distribution with a standard deviation of 5 microstrains), the differences were not so evident. After applying the PCA, a damage index is obtained for each simulated measurement data, as represented at figure 5. Results are not fully satisfactory for this example; there is a nearly linear dependence of the damage index and damage size, as it should be, but the dispersion of data are growing with the damage size, which impedes the linear regression analysis as explained at [4], which require uniform variance.

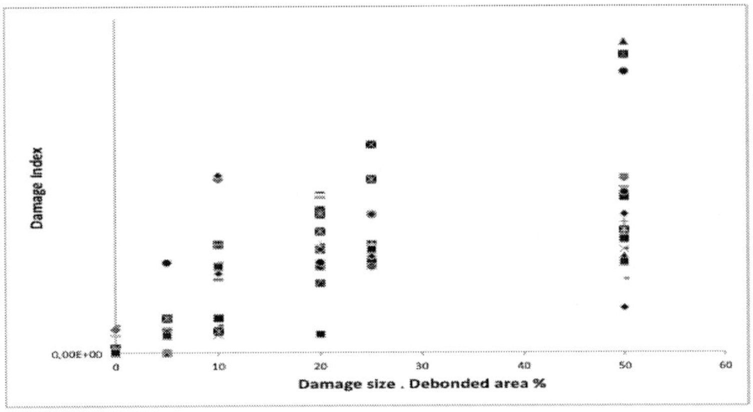

Figure 5. Damage Index .VS. increasing percentage of debonding

2. Conclusions

To validate an SHM system and enable their large-scale applicability, the reliability of the SHM technique must be quantified, with procedures similar to those used for NDT and described at MIL HDBK1823. It required a large number of test data, obtained experimentally or with the help of numerical simulation.

A fiber optic based SHM system is relatively easy to simulate, because optical sensors are just strain sensors; Finite element models allow to reproduce the strain field, with/without damage. At the example worked at this article, a repair patch instrumented with four sensors, it has been found that the dispersion of results grows with damage size, which will do the approach non valid.

A careful analysis of the results shows that it is not due to the modeling itself, but to the data comparison algorithm. PCA is useful to handle large multivariate data sets, but it implies a data compression to the main principal components. When the number of variables is not so large (4 for this analysis), this data reduction introduces significant errors. Other procedures, like the Mahalanobis distance, which does not simplify the data set, would be more efficient for this analysis.

Acknowledgements

This work has been partially sponsored by the DACOMAT Project. The DACOMAT project has received funding from the European Union's Horizon 2020 research and innovation programme under GA No. 761072.

References

[1] Guidelines for Implementation of Structural Health Monitoring on Fixed Wing Aircraft. SAE ARP6461 (2013)

[2] Güemes, A. Fernández-López, A, et al., "Structural Health Monitoring in Composite Structures by Fiber-Optic Sensors" Sensors 2018, 18(4), 1094.

[3] A.P. Berens and P.W.Hovey, 1988. Statistical evaluation of NDE reliability in the aerospace industry. Review of Progress in Quantitative Nondestructive Evaluation, Vol. 7B, 1761-1768, Springer, USA.

[4] MIL-HDBK-1823-A, 2009. Nondestructive evaluation system reliability assessment. US Department of Defense.

[5] D. Forsyth, et al., "Development and Validation of the Model Assisted Probability of Detection Method," QNDE 2009, Providence RI

[6] Alan Baker, Andrew J. Gunnion & John Wang (2015) On the Certification of Bonded Repairs to Primary Composite Aircraft Components, The Journal of Adhesion, 91:1-2

[7] M. Frovel, et al., "Repair Patch Monitoring With Embedded Optical Sensors by a Residual Strain Release Technique", EWSHM 2014 Proceedings, Nantes

[8] J. Sierra, A. Güemes and L. Mujica. "Damage Detection by Using {FBG}s and Strain Field Pattern Recognition Techniques." Smart Materials and Structures. 25011-25020. (2013)

Development of multi-physics multi-scale modelling platform for CFRP composites using inductive thermography techniques.

Abdoulaye BA [1], Bengisu YILMAZ [2], Huu Kien BUI [1] ,Elena JASIUNIENE [2], Gerard BERTHIAU [1]
[1] IREENA and NDTonAIR, Saint Nazaire, France
[2] Kaunas University of Technology and NDTonAIR, Kaunas, Lithuania
Contact e-mail: abdoulaye.ba@etu.univ-nantes.fr

Abstract. Due to their excellent mechanical performance, the use of carbon fiber composites has been growing in recent decades. However, the large-scale development of these materials depends on the improvements of the processes during the various stages of their whole life cycle (producing, forming, assembly, inspection, recycling). At various stages of the life cycle of the material, non-destructive testing (NDT) methods can be used to characterize the health state of the material. They play a vital role in the quality control and risk management. The Induction Thermography is significantly promising NDT technique for inspection of composites and it based on the measurement of eddy current thermal effects. The development of these methods requires multiphysics electromagnetic – thermal modelling. The developed models will deal with some numerical issues concerning thin regions of strong anisotropy and the multiscale geometries. In this paper, we develop a FEM coupled electromagnetic-thermal model in order to investigate the detectability of metal foil incorporate on the interface of three layer composite-adhesive bond.

Introduction

During the manufacturing process or operation time, several flaws can occur in the composites materials. These defects can reduce the mechanical strength and increase the failure risk[1]. In aerospace, composites materials replace metals in structural components due to their light weight and excellent configurable mechanical properties. Adhesive bonding of composites is a new and efficient technology with weight and homogenous load advantages compared to other joining techniques for composites. The imperfections such as porosity, delamination, fiber rupture and contamination are the most encountered flaw types in adhesive bonds. These flaws are considered in mesoscopic scale, i.e. the elementary ply scale. In this work, we use non-destructive testing to investigate the detectability of metal foil incorporate on the interface of three layer composite-adhesive bond numerically[2]. 3D multi-physics and multi-scale modeling approach with Finite Element Method is used.

1. Induction Thermography Technique

Fig. 1 shows our experimental device of the induction thermography techniques. The specimen is heated due to joule effect of eddy-current by a circular coil which is fed by an induction generator via an impedance adaptation circuit. The temperature exposed to the surface of the specimen is then measured with an infrared thermal camera. The presence of defects can disturb the induced currents flow and generate intensified current areas or inhibit the heat diffusion in the composite.

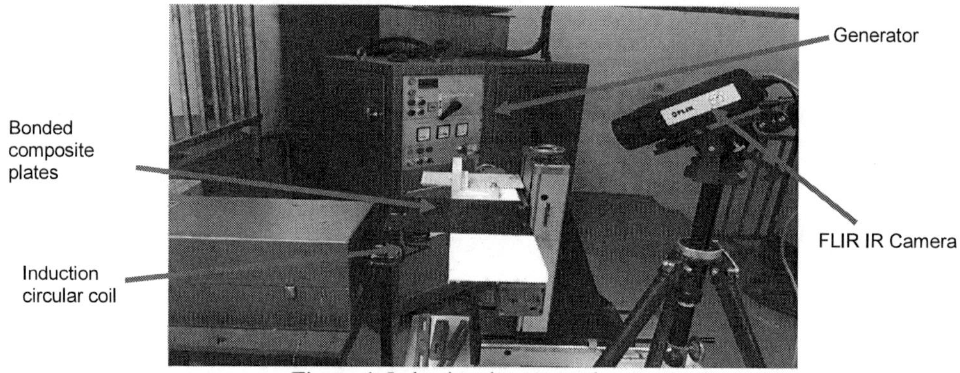

Figure 1: Induction thermography setup

In order to detect and characterize the defects, the temporal evolution of the temperature on the surface of the specimen is processed and analysed. The informations of the coil which we use is shown in Tab. 1.

Table 1: Dimensions of the coil

External Radius	Rext (mm)	12.5
Internal Radius	Rint (mm)	7.5
Number of Turns	N_Turn	5
Height	H (mm)	30

Fig. 3 shows the description of the model that we simulate and experiment with its dimensions.

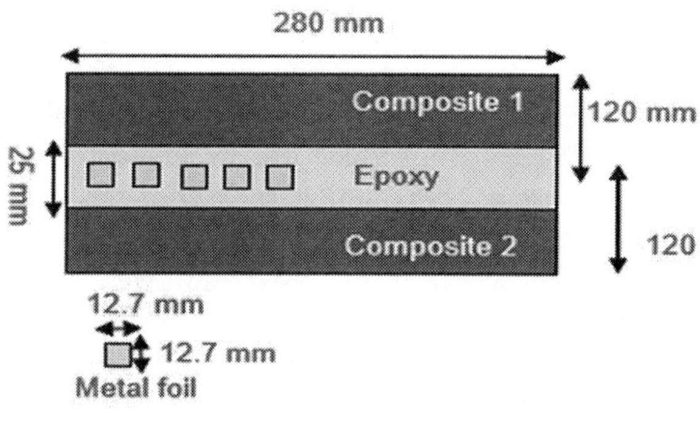

Figure 2: Description of the model. Defects are Metal foils

In this study, we simulate two composite rectangular plates 120mm*280mm bonded together with 25mm bondline with epoxy AF163. The thickness of metal foil is about 0.2413mm. the distance between two metal foil is 15mm. Each composite plate has a thickness of 2.2 mm. The two composite plates have 6 plies of 5H Satin (090) textile. The purpose of the simulation is to detect defects artificial metal foil defects on the interface of the bonded structure by using induction thermography.

Numerical models permit us a better understanding of defects behaviors and are a very useful for a parametric study. In this study, the goal is to focus much power over the metal foil. To do this, we will make an optimization of the electromagnetic frequency in order to choose the frequency which gives the best ratio power over the metal foil by total power injected (composite plates, metal foil). As the infrared camera records the temporal evolution of the temperature at the upper face, we do also the ratio of power in the metal foil by total power on the top composite plate.

2. Numerical modelling

Induction thermography simulation model involves a coupled electromagnetic and thermal problem[3][4]. Weak electromagnetic $\mathbf{A}-\varphi$ and nodal thermal formulations are used to calculate the eddy currents and temperature distribution on the surface of the specimen.

2.1 Electromagnetic model

The electromagnetic problem can be described as shown in fig. 3. We consider a bounded domain $D=D_{nc} \cup D_c$ with boundary $\Gamma=\Gamma_B \cup \Gamma_H$. The conducting and non-conducting parts of D are denoted by D_c and D_{nc}. The conducting domain is composed of inductor domain and loads domain.

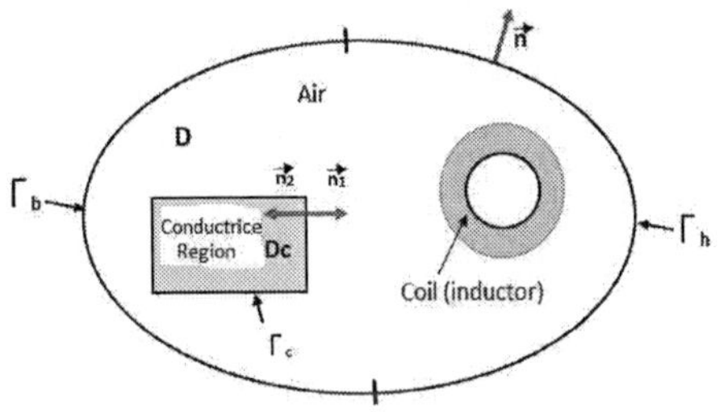

Figure 3: Studied domain

Weak electromagnetic formulation is given by [3][5]:

$$\int_D \left(\mathbf{curl}\left(w_a \frac{1}{[\mu]} \mathbf{curl}(\mathbf{A}) \right) + jw[\sigma] w_a (\mathbf{A} + \mathbf{grad}\varphi) \right) d\Omega$$
$$- \int_\Gamma w_a \left(n \wedge \frac{1}{[\mu]} \mathbf{curl}(A) \right) = \int_D w_a \mathbf{curl}(\mathbf{T_s}) d\Omega \quad \text{(Eq. 1)}$$

$$\int_{Dc} (jw[\sigma] \mathbf{grad} w_n (A + \mathbf{grad}\varphi))\, d\Omega = 0 \quad \text{(Eq. 2)}$$

Where $\mathbf{A}$ and φ are respectively the magnetic vector potential and the primitive in time of electric scalar potential. w_a and w_n are respectively denoted edge shape function and nodal shape function [6]. $[\sigma]$ represent the electrical conductivity of the pli i having a fiber orientation θ^i. The matrix form of the discrete $\mathbf{A}-\varphi$ formulation reads :

$$\begin{pmatrix} \mathbf{R}^t M_{ff}^{1/[\mu]} \mathbf{R} + jwM_{aa}^{[\sigma]} & jwM_{aa}^{[\sigma]} \mathbf{G} \\ jw\mathbf{G}^t M_{aa}^{[\sigma]} & \mathbf{G}^t M_{aa}^{[\sigma]} \mathbf{G} \end{pmatrix} \begin{pmatrix} \mathbf{A_a} \\ \Phi_n \end{pmatrix} = \begin{pmatrix} \mathbf{R} M_{ada} T_a^s \\ 0 \end{pmatrix} \quad \text{(Eq. 3)}$$

Where **R**, **G** are respectively the discrete counterparts of rot and grad operators $M_{ff}^{1/[\mu]}$, $M_{aa}^{[\sigma]}$ and M_{aa} are denoted mass matrix. They are given by:

$$M_{ff}^{1/[\mu]}(i,j) = \sum_{i,j=1...F} \int_{Dc} \frac{1}{[\mu]} w_f^i w_f^j d\Omega \quad \text{(Eq. 4)}$$

$$M_{aa}^{[\sigma]}(i,j) = \sum_{i,j=1...A} \int_{Dc} [\sigma] w_a^i w_a^j d\Omega \quad \text{(Eq. 5)}$$

$$M_{aa}(i,j) = \sum_{i,j=1...A} \int_{Dc} w_a^i w_a^j d\Omega \quad \text{(Eq. 6)}$$

2.2 Thermal model

The eddy current losses in the specimen whose volumic power in each element i of the volume mesh can be determined by:

$$P = \mathbf{J}_{ind}{}^{t} \cdot [\sigma]^{-1} \cdot \mathbf{J}_{ind} \quad \text{(Eq. 7)}$$

Eq. 4 represent the heat source of the thermal problem. The latter is defined by:

$$\rho C_P \frac{\delta \mathbf{T}}{\delta t} + div(-[\lambda]\mathbf{gradT}) = P \quad \text{(Eq. 8)}$$

Where T is the temperature, C_P is the specific heat, ρ is the specific mass and $[\lambda]$ is the tensor of thermal conductivities. $[\lambda]$ takes the same form than the electric conductivities $[\sigma]$. In thermal problem only the composite domain is considered.
We consider only the load domain in the thermal problem. So, weak formulation of thermal problem is given by:

$$\int_{Dc} \left(\rho C_P \frac{\delta \mathbf{T}}{\delta t} - div([\lambda]\mathbf{gradT}) \right) w_n d\Omega = \int_{Dc} P w_n d\Omega \quad \text{(Eq. 9)}$$

The matrix form of thermal weak formulation reads:

$$(\mathrm{M}_{nn}^{[\rho c]} + \mathbf{G}^t M_{aa}^{[\lambda]} \mathbf{G} + M_{nn}^{[h]}) \Delta T_{n,i+1} = M_n^{[P]} + M_{nn}^{[\rho c]} \Delta T_{n,i} \quad \text{(Eq. 10)}$$

Where $\Delta T_{n,i}$ is the temperature rise on the n at the instant i with respect to the initial temperature. The matrix $\mathrm{M}_{nn}^{[\rho c]}$, $M_{ee}^{[\lambda]}$, $M_{nn}^{[h]}$ and $M_n^{[P]}$are given by [7]:

$$\mathrm{M}_{nn}^{[\rho c]}(i,j) = \sum_{i,j=1...N} \int_{Dc} \frac{\rho C_p}{\Delta t} w^{ni} w^{nj} dD \quad \text{(Eq. 11)}$$

$$M_{aa}^{[\lambda]}(i,j) = \sum_{i,j=1...A} \int_{Dc} [\lambda] w^{ni} w^{nj} dD \quad \text{(Eq. 12)}$$

$$M_{nn}^{[h]}(i,j) = \sum_{i,j=1...N} \int_{Dc} h w^{ni} w^{nj} dD \quad \text{(Eq. 13)}$$

$$M_n^{[P]}(i,j) = \sum_{i=1...N} \int_{Dc} P w^{ni} dD \quad \text{(Eq. 14)}$$

Δt is the time step and is fixe. The parameters h takes into account the natural convection and the radiation on the surface of the plate.

3. Simulation investigations

Nondestructive testing by induction thermography through adhesive bonding have been simulated using our laboratory Finite Element code. Simulation tests were made with a circular coil where the informations are recorded in the Tab. 1.

In table 2, 3 and 4, we present the numerical parameters of the thermal problem. We have numerical parameters of: composite plates, Epoxy and defects.

Table 2: Thermal parameters of composites

Parameter	Value	Unit
ρ	1599	kg/m^3
C_p	930	$J/(kgK)$
$\lambda_{\parallel}$	4.6	$W/(mK)$
$\lambda_{\perp}$	4.6	$W/(mK)$
λ_z	0.5	$W/(mK)$
$h=h_{conv}+h_{radi}$	10	$Wm^2K)$

Table 3: Thermal parameters of Epoxy

Parameter	Value	Unit
ρ	1215.8	kg/m^3
C_p	1000	$J/(kgK)$
λ	0.1661	$W/(mK)$

Table 4: Thermal parameters of defects

Parameter	Value	Unit
ρ	8500	kg/m^3
C_p	375	$J/(kgK)$
λ	109	$W/(mK)$

The table 5 shows the parameters of the electromagnetic problem.

Table 5: Electromagnetic parameters

Parameter	Value	Unit
$\sigma_{\parallel}$	4×10^4	S/m
$\sigma_{\perp}$	4×10^4	S/m
σ_z	8	S/m
σ_{def}	16.69×10^6	S/m

The parameters in the table 6 have been chosen in the simulation.

Table 6: Simulation parameters

Parameter	Value	Unit
Frequency	$[2\times 10^3 ; 2\times 10^6]$	Hz
Current	200	A
Lift off	2	mm

The mesh used in the simulation is shown in fig. 4. The simulation results are shown in the figure 5. The curves were plotted on a logarithmic scale.

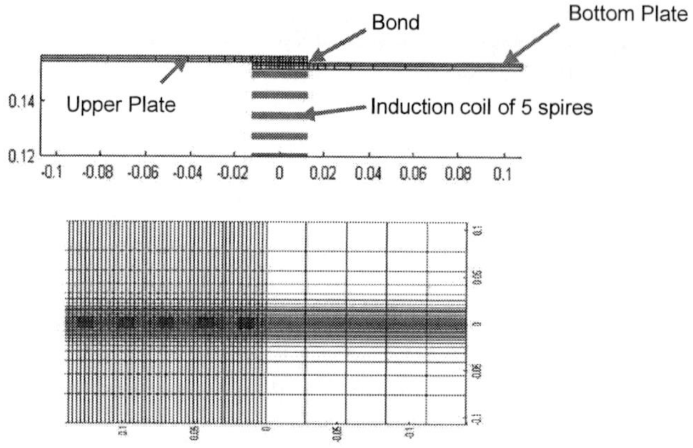

Figure 4: Mesh in the simulation

The circular inductor is placed below the second defect from the middle of the samples. We have five (5) defects in the sample. But only one defect will be inspected. The rectangular red color in the fig. 4 are the defects and the circular one is the coil.

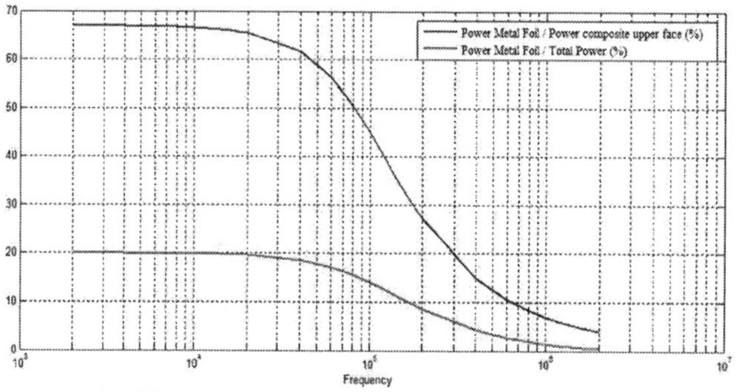

Figure 5: Ration of powers in the metal foil and composites

Fig. 5 shown that from 2 kHz to 40 kHz we have a good ratio for the two cases. For the first one, the ratio is about 70 percent and 20 percent for the second one. For all frequencies between 2 kHz and 40 kHz, we get a good ratio. We choose the frequency of **20 kHz** in order to get a good ratio of powers, in the following. At frequency 20 kHz with an excitation current 200 A, we observe that the powers are too small in each layers.

Since the powers are too small, we increase the current. The latter is chosen ten times larger i.e. 2000A. After simulation, the power in the metal foil goes from 0.08 W (200 A) to 8 W (2000 A) and the total power from 0.3995 W to 39.95 W. In this case, the power of the metal foil is greater than the powers of the upper layers, on the observation side of the infrared camera. The fig. 6 shown the powers in each layers.

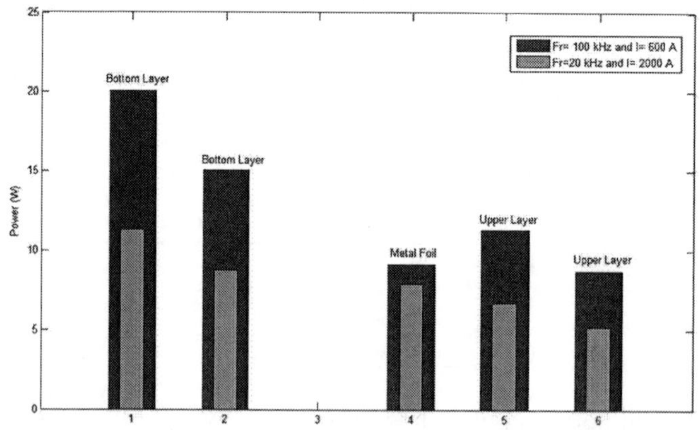

Figure 6: Comparison of powers in each layers (fr=20 kHz; I=2000A) and (fr=100 kHz; I=600A)

As we can see in the fig. 5, the ratio of power is about 45 percent when we choose 100 kHz as frequency. When we work with **100 kHz**, the power increase from 7 W to 64 W respectively with currents 200 A and 600 A. The total power (64 W) in the case 100 kHz and 600A is more important than the case 20 kHz and 2000A which is 39.95 W.

The simulated temperature at the surface (upper face) at frequency 100 kHz and current 600 A is shown in fig. 7. The composite plate is heated in 2s. The evolution of the temperature at the surface (upper face) is calculated for 60s. As we can see in the fig. 7, the temperature in the case with defect reached a maximum at t=6s and the temperature is the most important over almost the measurement period. The figure 8 shown the temporal evolution of the temperature at the middle of the metal foil.

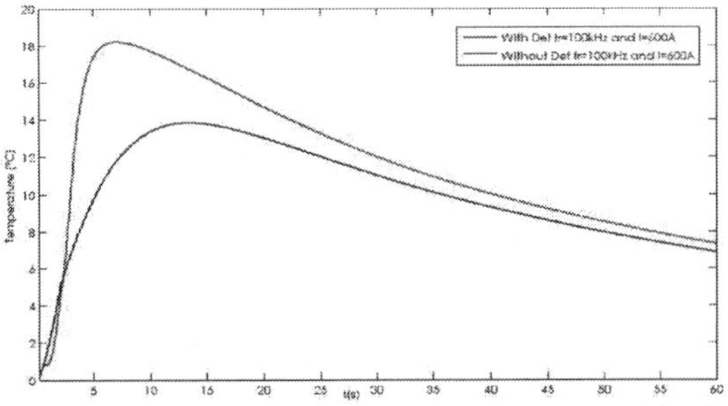

Figure 7: Temporal evolution of the temperature at the center of the defect upper face

Fig. 8 shows that the metal has heated well during the period when the generator is running. The temperature reached maximum at t=2s (heating time). The maximum of temperature is 35 °C. After shutting off the generator, the temperature at the middle of the metal foil begins to decrease. We can say that the power is well focused in the metal foil. This allows us to do well the inspection of the specimen.

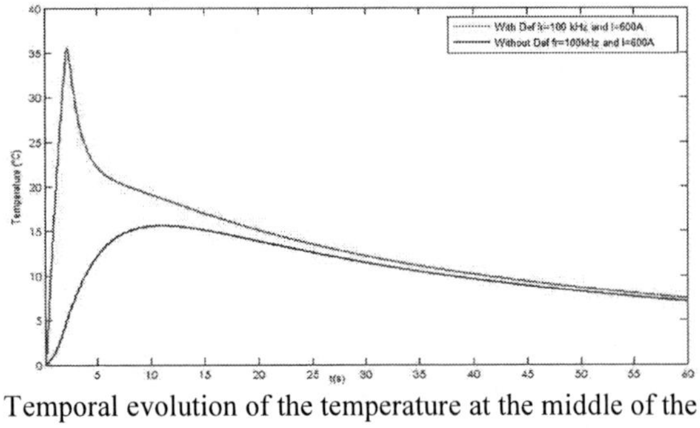

Figure 8: Temporal evolution of the temperature at the middle of the Metal Foil

3.1 Defect detection by absolute contrast image

Amplitude contrast technique consists of substraction between thermal images simulated on defective piece and healthy one, using Fast Fourier Transform (FFT). The non-null contrast areas on the resulting images allow for detection of defects. The Fourier transform is performed by the following well know formula:

$$\boldsymbol{F}(\mathrm{f}) = \frac{1}{\mathrm{Nt}} \sum_{\mathrm{i}=0}^{\mathrm{N}-1} \mathrm{T}(\mathrm{i}) \exp\left(-\frac{\mathrm{j}2\pi\mathrm{fi}}{\mathrm{Nt}}\right) = \mathrm{Re}(\mathrm{f}) + \mathrm{Im}(\mathrm{f}) \qquad \text{(Eq. 15)}$$

$$\mathrm{A}(\mathrm{f}) = \sqrt{\mathrm{Re}^2(\mathrm{f}) + \mathrm{Im}^2(\mathrm{f})} \qquad \text{(Eq. 16)}$$

$$\mathrm{C}^{\mathrm{Amp}} = \mathrm{A}_{\mathrm{with}_{\mathrm{Def}}}(\mathrm{f}) - \mathrm{A}_{\mathrm{without}_{\mathrm{Def}}}(\mathrm{f})$$

Some images of amplitude contrast are shown in the fig. 9.

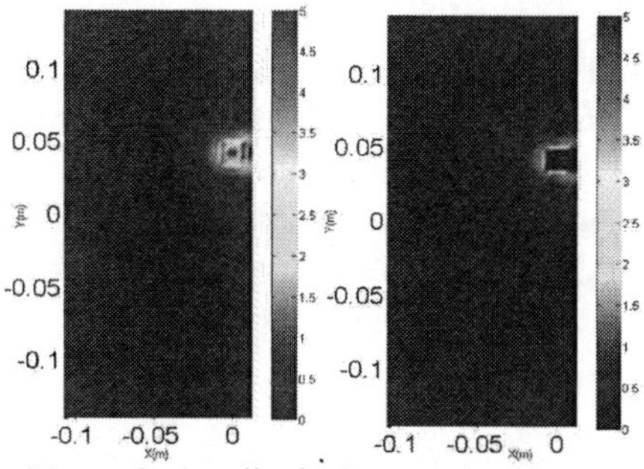

Figure 9: Amplitude Contrast Image FFT

4. Conclusion

Numerical investigations are used in this paper to evaluate the detectability of artificial defect metal foil on the interface of composite-adhesive bond. It was shown that the electromagnetic frequency and the current of the heating source are both significant parameters that have to be optimized in order to detect defects more clearly. Also, the thermal parameters (heating time, observation time, …) should be also optimized for should be also optimized for better resolution of the defects. In the future study, experimental investigations will be done by taking into account the parameters that are obtained from the simulation results.

Acknowledgments

This research was funded by NDTonAIR project from the European Union's Horizon 2020 research and innovation program under the Marie Sklodowska-Curie grant agreement No 722134-NDTonAIR.

References

[1] Bui H K, Wasselynck G, Trichet D and Berthiau G 2016 Study on flaw detectability of NDT induction thermography technique for laminated CFRP composites *Eur. Phys. J. Appl. Phys.* **73** 10902

[2] Yilmaz B, Jasiūnienė E and Mažeika L 2018 Evaluation of bonding quality by using ultrasonic waves *12th Eur. Conf. Non-Destructive Test. (ECNDT 2018), Gothenbg.* 2–4

[3] Bui H K, Berthiau G, Trichet D and Wasselynck G 2015 Inductive thermography nondestructive testing applied to carbon composite materials: multiphysics and multiscale modeling *8th Int. Conf. Electromagn. Process. Mater.* 403–6

[4] Ramdane B, Trichet D, Belkadi M, Saidi T and Fouladgar J 2010 3D numerical modeling of a new thermo-inductive NDT using pulse mode and pulsed phase methods *Eur. Phys. J. Appl. Phys.* **52** 23301

[5] Bui H K, Wasselynck G, Trichet D, Ramdane B, Berthiau G and Fouladgar J 2013 3-D modeling of thermo inductive non destructive testing method applied to multilayer composite *IEEE Trans. Magn.* **49** 1949–52

[6] Meunier G 2008 *The Finite Element Method for Electromagnetic Modeling* (ISTE Ltd, John Wiley & Sons Inc.)

[7] Bui H K, Senghor F D, Wasselynck G, Trichet D, Fouladgar J, Lee K and Berthiau G 2018 Characterization of Electrical Conductivity of Anisotropic CFRP Materials by Means of Induction Thermography Technique *IEEE Trans. Magn.* **54**

Simulation-supported Analysis of Complex Radiographs

Nick BRIERLEY [1]
[1] The Manufacturing Technology Centre (MTC), Ansty Business Park, Coventry, UK
Contact e-mail: nick.brierley@the-mtc.org

Abstract. The analysis of radiographic images acquired for samples of complex geometry (such as those enabled by additive manufacturing techniques), where the imprint of the sample's geometric complexity in the radiograph is likely to undermine the ability to identify defect indications, is very challenging. An algorithm has been developed to assist in such circumstances, exploiting knowledge of the component geometry and approximate inspection configuration to generate a simulated baseline image with which a quantitative comparison can be conducted – in principle, allowing the geometric complexity to be subtracted out. This technique however requires the simulation to be accurately calibrated to the experimental configuration, necessitating the use of a numerical optimisation to fit the simulation parameters. As a by-product, the parameters of the imperfectly specified experimental set-up are recovered. The algorithm design is described, followed by results from increasingly complex test problems that illustrate the value of the approach.

Introduction

Additive Manufacturing (AM) enables extremely complex geometries to be manufactured [1]. The inspection of these geometries is a major challenge, and given the lack of a precursor stage at which the component might be more readily inspectable, radiographic imaging techniques are frequently the only option for a detailed volumetric analysis [2]. X-ray Computed Tomography (XCT) is widely applied to AM components, but the technique is relatively time-consuming and costly [3]. 2D radiography physically offers the same advantages as XCT, given that XCT relies on the acquisition of an extended sequence of radiographs, but has the potential to be cheaper, faster and more flexible. One of the challenges however for the deployment of 2D radiography on complex-shaped samples is that the images obtained will contain an imprint of the sample's geometric complexity, which can easily obfuscate defect indications. The work presented here aims to overcome the challenge of geometric complexity in 2D radiographs by exploiting prior knowledge of the inspection configuration.

Complexity in inspection images arising from background structures can be handled in several ways. The simplest approach is to apply global image-processing operations, such histogram thresholding, but this can only work satisfactorily for negligible background variability relative to defect signals [4]. If several experimental baseline measurements are available, these in principle can be used as a statistical reference point, an approach taken in structural health monitoring [5]. However, the scope for taking this approach is limited in the context of very short production runs and part customisation. If the sample and inspection exhibits significant symmetries, these can be exploited to suppress background signals without the need for supplementary data [6,7], but this approach does not apply to the majority of use cases. The approach considered most promising for the present challenge and investigated further is based on the exploitation of prior knowledge about the inspection: generally, the sample geometry (as a CAD file) and approximate inspection configuration are known, allowing baseline images to be simulated, in principle enabling a quantitative comparison to the experimental images. The aspiration is to be able to subtract out the geometric complexity, leaving an image that is easier to interpret manually or automatically.

To be able to use a simulated image as a meaningful baseline however relies on the reference modelling providing an accurate representation of the experimental data collection. This hinges on the modelling capability and the calibration of the simulation to the experimental configuration. To achieve the required match, in this work, a numerical optimisation is used to fit the key parameters of the simulation to the experimental image(s) to be analysed. This process also returns the parameters of an imperfectly specified experimental set-up, which is itself a useful output, for instance enabling the use of visualisation tools such as that shown in Fig.1. Note the test geometry / sample shown here and used throughout this paper is courtesy of GRM Consulting Ltd., K-Tech & MTC.

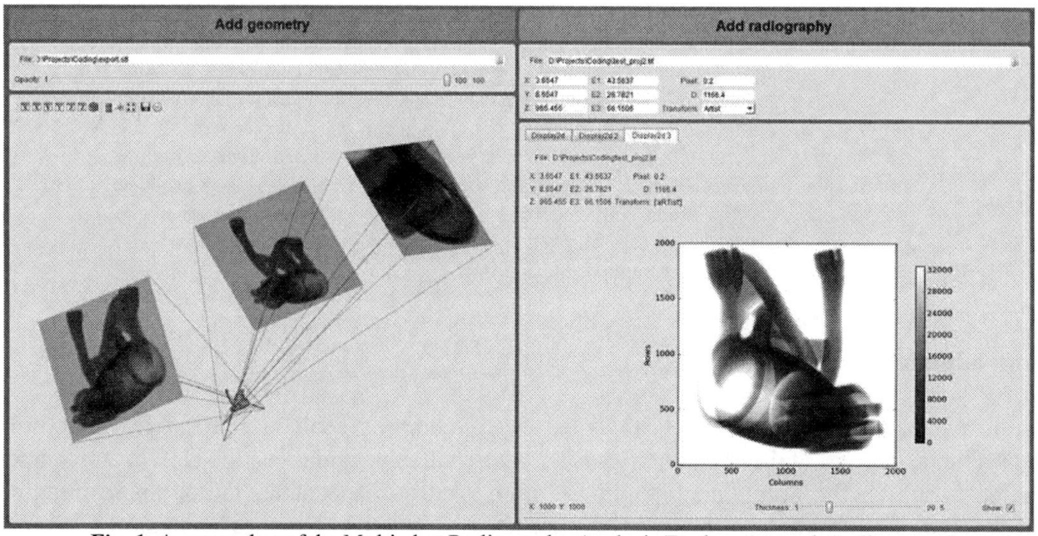

Fig. 1. A screenshot of the Multi-shot Radiography Analysis Tool prototyped. It allows an inspector to interactively explore the spatial relationships between pixels of different radiographic images acquired for a complex-shaped sample. The scene on the left can be moved about freely, whilst selecting any point on one of the radiographs on the right adds the corresponding ray through the sample rendering on the left.

2. Method

2.1. Overview

The algorithm developed takes as an input one or more experimental radiographs, a CAD description of the geometry being inspected, and a specification of the (approximate) configuration used to obtain each of the images provided. The calibration is then performed as illustrated in Fig 2. Once a good fit has been reached, the simulated image(s) can be used as a quantitative baseline to support the analysis of the experimental image(s), enabling unexpected deviations from the idealised configuration (for example caused by defects in the physical sample) to be highlighted efficiently.

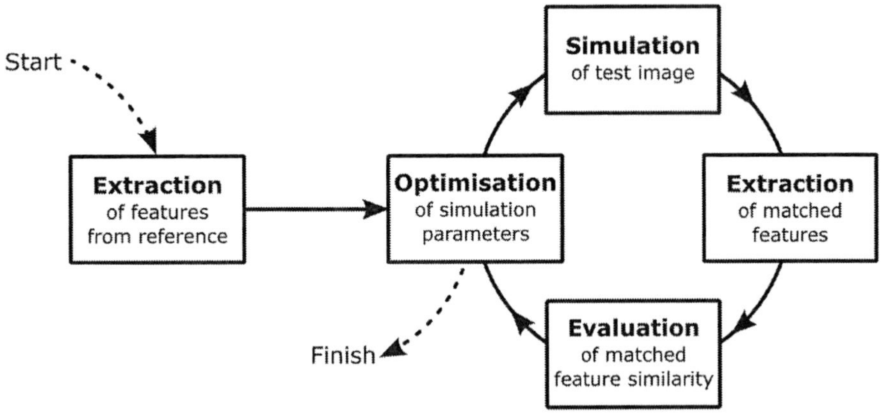

Fig. 2. A flow diagram of the algorithm's operation.

This work takes a similar approach to that in [8], but while that work relied on a custom simulation capability and its output was only demonstrated for a thin plastic sample, here the focus is on usage of a commercially available simulation engine to develop an otherwise imaging modality-independent framework geared towards (and demonstrated for) the inspection of complex-shaped metal AM components.

2.2. *Details of Algorithm*

Here some of the key features of the algorithm are described, based on the key operations shown in Fig. 2. A more comprehensive description is provided by [9].

2.2.1. *Simulation*

This component is able to return the expected radiographic image for an arbitrary inspection configuration of a provided CAD geometry. The aRTist [10] radiographic simulator engine developed by BAM [11] is used, controlled via a scripting interface enabled by BAM within a preceding project [12]. This is the only element of the program that is strictly imaging modality dependent. In the present set-up, an individual radiographic projection is specified by a set of 16 parameters, spanning both geometric (pose) and radiographic (such as the X-ray beam current and source-sample-detector spectral behaviour) parameters.

2.2.2. *Feature Extraction*

Features are extracted from the reference and simulation images to enable an efficient assessment of image similarity. Importantly, the features selected should be robust to noise and must not be compromised by potential defect indications, as this would mean the fitting process could undermine the subsequent defect detection effort that the calibration is meant to enable.

Here spatially well-distributed image patch features containing high gradients (i.e. edges) were selected, and the local image entropy computed. In the simulated images a match of features to those from the reference image is assured by identifying the corresponding patch locations.

2.2.3. *Evaluation of Similarity Metric*

To assess the similarity of paired features across the two images being compared in a quantitative manner and hence allow a numerical optimisation to operate with this as the metric to be minimised, the Structural Similarity Index is used. This is a more robust way of comparing images than for instance the straightforward Mean Squared Error (MSE) [13]. This is scaled so that a value of 0 represents a perfect fit, and a positive value of 1 or more corresponds to a complete mismatch.

2.2.4. Optimisation

The numerical optimisation based on the well-known genetic global optimisation algorithm architecture [14] was implemented to minimise the described image similarity metric. The implementation has several features specific to the application here. This includes a capability to break known symmetries in the objective function space [15] to reduce the parameter space to be explored. Additionally, the structure allows the computation to proceed in two phases, initially focusing the optimisation on the geometric parameters that will dominate the image similarity until a good fit is achieved, before expanding the parameter space to span the full set of imaging settings.

3. Results and Discussion

Here a preview of the results obtained is provided, starting with a synthetic test case, where the reference image was itself simulated, thus providing a performance validation, followed by a complex experimental example to demonstrate the benefits of the approach in practice.

3.1. Simple Synthetic Test Case

The simulated reference image in this instance was distorted across the full 16 parameters describing the inspection configuration for an individual radiograph, but no noise or similar distortion was introduced, such that the true global optimum of the optimisation was known precisely.

The results are illustrated in Fig 3: difference images (subtracting the reference image from the image simulated) are shown for the initial state and the best metric reached by the optimisation. Clearly, the optimisation has succeeded in almost eliminating the substantial mismatch seen initially, demonstrating that the algorithm is operating appropriately. However, the faint imprint of the geometry still visible after the computation shows that the computation has not quite reached the global optimum.

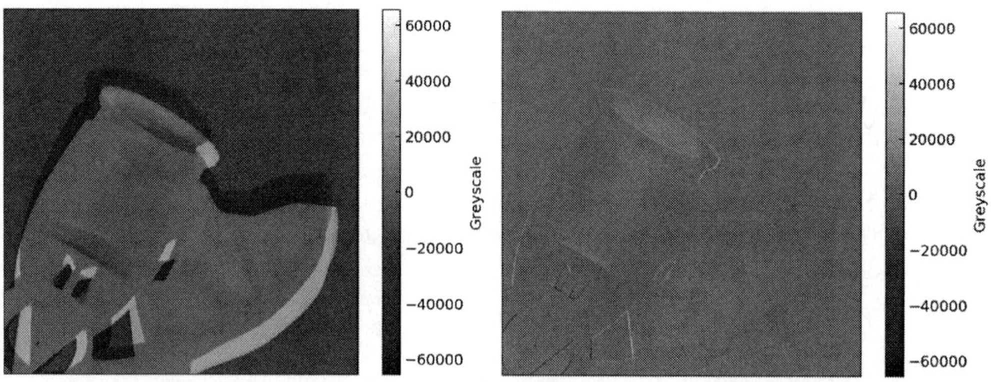

Fig. 3. Difference images showing the performance of the optimisation algorithm for a synthetic example noise-free reference image distorted in all parameters (pose & radiographic) compared with the nominal parameter combination. On the left the pre-optimisation state is shown, whilst on the right the post-optimisation state is illustrated.

3.2. *Complex Experimental Test Case*

In this specific scenario three experimental radiographs of the physical component corresponding to the previously seen geometry were used to calibrate the simulation. The component was made by laser powder bed fusion in stainless steel – its bounding box measures 7.9 x 6.9 x 2.9 cm. The radiographs were obtained on a Nikon XTH 225 ST XCT machine [16]. All images were obtained at the same radiographic settings and using the same fixture, meaning that the reference images include linkages amongst the 16 parameters per projection in the simulation. As a consequence the full parameter space for the optimisation was 24-dimensional.

There are two key differences here compared with the previous synthetic test case. Firstly, the correct parameter combination (and hence, location of the global optimum) is not known. Secondly, the performance will be constrained by the limitations of the modelling applied, not only in terms of the limitations of the simulation engine and the computational assumptions made, but also in terms of the physical geometry deviating from the nominal CAD used.

To assess the potential scale of the last issue, a structured light scan of the physical sample was completed (using a GOM ATOS Triple Scan [17]) and the surface thus obtained compared to the nominal geometry, as shown in Fig. 4. The identified maximum deviation of 0.8 mm is expected to be large enough relative to overall dimensions to impact on the calibration process.

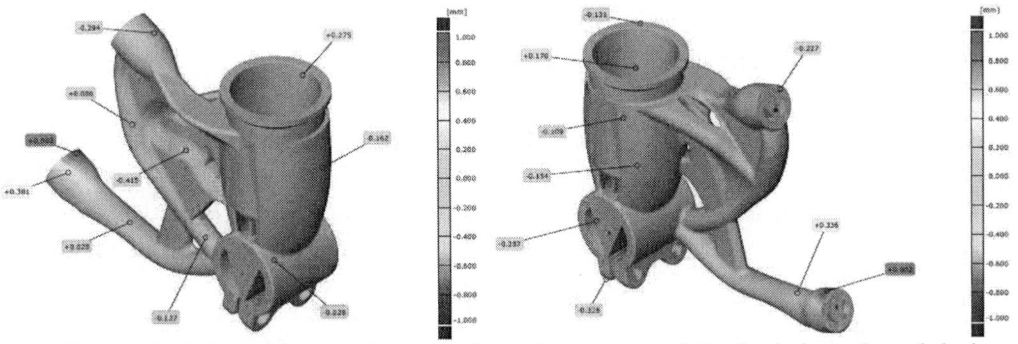

Fig. 4. The deviations between the nominal sample geometry, and the physical sample made by laser powder bed fusion additive manufacturing, for both sides of the sample, based on a best-fit alignment. The largest deviation recorded is 0.8 mm, at the end of the long horizontal elongation.

The computation is illustrated focusing on a single one of the three reference images for brevity. Fig. 5 presents the initial raw experimental image alongside the pre-optimisation difference image, based on nominal parameters of the inspection configuration. The post-optimisation difference image on the same greyscale as before is shown below, highlighting the extent to which the optimisation has suppressed gross image mismatches seen initially. Finally, the same post-optimisation state image is presented with an adjusted greyscale, as a means of showing an example of the enhanced analysis facilitated by the calibrated simulation baseline image. Whilst the imprint of the complex geometry has not been eliminated, compared with the original images, significant additional details (such as the surface texture of the component on some surfaces) are apparent without needing to dynamically explore the greyscales.

It is also noteworthy that the largest discrepancies seen in the post-optimisation images relate precisely to the element of the geometry for which the greatest nominal-actual deviations were recorded, showing that this simulation-based analysis provides a means of checking the (coarse) geometric conformance of components, in addition to their structural integrity.

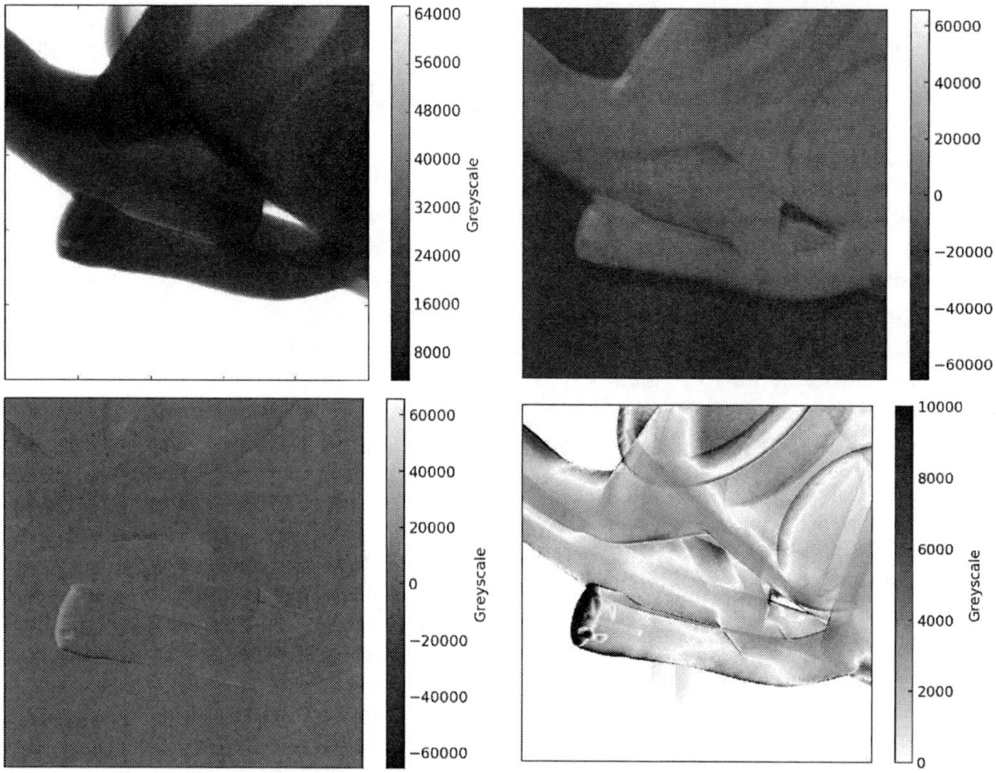

Fig. 5. Illustration of the action of the algorithm on an experimental radiograph (one of three calibrated against simultaneously). Top left is the original image acquired on the Nikon XTH 225 ST hardware; top right is the difference image for the pre-optimisation state based on the nominal parameters of the acquisition; bottom left is the difference image for the post-optimisation state (on the same greyscale as before, to emphasise the improved match); bottom right then is the same data, but presented as an absolute difference, with a compressed greyscale.

4. Conclusions

This paper has described an algorithm developed for the calibration of a radiographic simulation to one or more experimental images using a numerical optimisation, as a means of obtaining a quantitative baseline with which to compare the experimental images to support their analysis. The approach taken is readily adaptable to alternative imaging modalities, subject to the provision of a relevant simulation engine.

The results are qualitatively very promising, with the approach revealing both fine structural details and gross geometric distortions of the test component in the experimental image processed using the algorithm. The hope is that this enhanced analysis capability will allow the more widespread and effective use of 2D radiographic inspection on highly complex geometries, as an alternative to X-ray Computed Tomography.

Further work includes a refinement of the simulation model to include complex effects presently ignored such as the scattering contributions to the image, as well as improvements to the robustness and speed of the optimisation.

Acknowledgements

The author thanks colleagues Borja Lazaro Toralles, Silvia Estelles-Martinez and Matteo Villa for taking on some supporting tasks. This work was completed as part of a Core Research Programme project at the MTC, and funded, including the article processing charge, by industrial membership fees and the UK High Value Manufacturing Catapult. The author is grateful to Imperial College London and Trevor Tippetts, formerly of Imperial College London (now Los Alamos National Laboratory), for permitting the code base developed during the author's doctorate in collaboration with Trevor Tippetts to be reused and built on for the purposes of this project. The project is also indebted to all contributors to the open source code libraries that the program has been built on.

References

[1] N. Hopkinson, R. Hague, and P. Dickens, Rapid manufacturing: an industrial revolution for the digital age. John Wiley & Sons, 2006.
[2] E. Todoro et al., America makes: National additive manufacturing innovation institute (NAMII) project 1: Nondestructive evaluation (NDE) of complex metallic additive manufactured (AM) structures, http://www.dtic.mil/dtic/tr/fulltext/u2/a612775.pdf, 2014.
[3] L. D. Chiffre et al., Industrial applications of computed tomography, CIRP Annals Manufacturing Technology, vol. 63, no. 2, pp. 655-677, 2014.
[4] A. Staude et al., Quantification of the capability of micro-CT to detect defects in castings using a new test piece and a voxel-based comparison method," NDT and E International, vol. 44, no. 6, pp. 531-536, 2011.
[5] T. Clarke, Guided wave health monitoring of complex structures, Ph.D. thesis, Imperial College London, Department of Mechanical Engineering, 2009.
[6] M. I. Haith, P. Huthwaite, and M. J. Lowe, Defect characterisation from limited view pipeline radiography, NDT and E International, vol. 86, pp. 186-198, 2017.
[7] N. Brierley, T. Tippetts, and P. Cawley, Data fusion for automated non-destructive inspection, Proceedings of the Royal Society of London A: Mathematical, Physical and Engineering Sciences, vol. 470, no. 2167, 2014.
[8] A. Marinovszki, J. D. Beenhouwer, and J. Sijbers, An efficient CAD projector for X-ray projection based 3D inspection with the ASTRA toolbox, 8th Conference on Industrial Computed Tomography, 2018.
[9] N. Brierley, Simulation-based analysis of complex radiographic images, Journal of Nondestructive Evaluation, 2018, in review.
[10] aRTist - Analytical RT Inspection Simulation Tool, http://artist.bam.de, 2015. [Accessed 14/08/18]
[11] Federal Institute for Materials Research and Testing, Germany. {https://www.bam.de/Navigation/EN/About-us/Organisation/Organisation-Chart/President/Department-8/Division-83/division83.html, 2018. [Accessed 14/08/18].
[12] N. Brierley, C. Bellon, and B. Lazaro Toralles, Optimised multi-shot imaging inspection design, Proceedings of the Royal Society of London A: Mathematical, Physical and Engineering Sciences, vol. 474, no. 2216, 2018.
[13] Z. Wang et al., Image quality assessment: from error visibility to structural similarity, IEEE Transactions on Image Processing, vol. 13, pp. 600-612, 2004.
[14] K. Jones, Comparison of genetic algorithm and particle swarm optimisation, International Conference on Computer Systems and Technologies, Varna, Bulgaria, 2005.
[15] T. Walsh, General Symmetry Breaking Constraints, pp. 650-664. Springer, 2006.
[16] Nikon XTH 225 ST, https://www.nikonmetrology.com/en-gb/product/xt-h-225-st, 2017. [Accessed 13/09/18].
[17] GOM, \GOM ATOS Triple Scan." http://www.gom.com/metrology-systems/atos/atos-triple-scan.html, 2017. [Accessed 17/08/2017].

Semi-analytical modeling of eddy current inspection of stratified and homogenized composite materials

Houssem.CHEBBI[1], Denis.PRÉMEL[1]
[1] Commissariat à l'Energie Atomique et aux énergies alternatives (CEA) LIST, Gif-sur-Yvette cedex, France
Contact e-mail: houssem.chebbi@cea.fr

1 abstract

In the present paper, we deal with the development of a semi-analytical model for the fast computation of a quasi-static field induced by a 3D eddy current probe in a flawless composite material. After a prior phase of homogenization, the workpiece is assumed to be homogenized and stratified. Therefore, each layer of the material is characterized by a conductivity tensor with axial or biaxial form according to the main orientation of the fibers. The approach is based on a modal decomposition of the Electromagnetic (EM) field in each layer. To take into account the wave propagation through the multilayered structure and boundary conditions at each interface, a recursive scattering matrix (S-matrix) algorithm is implemented. Numerical results provided by the model are compared to other simulated data obtained by a Finite Element commercial software.
Keywords:
computational electromagnetics, semi-analytic model, composite material, scattering matrix, homogenization.

2 introduction

Composite materials, such Carbon Fiber Reinforced Polymer (CFRP), are increasingly used in the aeronautic industry to reduce the weight of aeronautical structures. Therefore, the inspection of these complex materials for characterization or the detection of potential defects is a major challenge due to their complex physical behavior, to ensure their good performance in different operations.

In one hand, Eddy Current testing technique (ECT) is sensitive to the variations of the conductivity of the material, it can be applied for inspection (crack detection, thickness measurements...) and characterization (bulk conductivity, fiber orientation...). In the other hand, the interactions between electromagnetic waves and composite materials are highly depending on the microstructure of the material itself. This gives rise to challenging modeling issues. Many homogenization approaches were proposed [1, 2] to overcome high anisotropy, random positionning of the fibers. Therefore, each layer of the stratified media is characterized by an equivalent conductivity tensor and equivalent permittivity tensor. In the same framework, many research works [3, 4, 5] present a variety of EC techniques providing interesting results in characterizing homogenized composite material.

In general, analytical models are restricted to cases with simple geometries, although providing exact solutions. In contrast, purely numerical methods are so general and used for complex problems, however, considering the computing time and memory storage, they are quite expensive. In our modeling approach, we consider a compromise that we call semi-analytical (SA) model, to insure accuracy, efficiency and rapidity. Our semi-analytic modeling framework consists in writing Maxwell's equations in each layer, in the Fourier domain in order to obtain an algebraic form. Moreover, the EM field is expanded as sum of Eigen functions. Applying boundary conditions (BC) leads us to estimate the coefficients of the expansion and to reconstruct the EM field in any region. A stable recursive algorithm like the S-matrix one, is implemented in order to take into account the wave propagation through the entire structure and BC at each interface.

the paper is organized as follows, first we describe the formalism used for the SA model, then we illustrate the keystone of the approach which is the modal decomposition of the EM field in both isotropic and anisotropic media. Next we introduce the S-matrix algorithm used for stratified media. Finally, the numerical results of the model are compared to those of a FE commercial software and results from literature.

3 semi-analytical model framework

3.1 summary

Our semi-analytical modeling framework is based on previous research works [6, 7, 8]. This work constitutes the preliminary task for the development of a complete numerical model, which has the capability to address stratified and anisotropic layered media with geometrical complexity. As a start point, we adopt the case of homogenized and anisotropic semi-infinite media excited by a 3D eddy current probe placed in air. Then we move to the case of one layered media and multilayered media with different orientation of the fibers. We attribute to each layer an equivalent conductivity tensor taking into account the preferred anisotropy orientation.

3.2 material characteristics

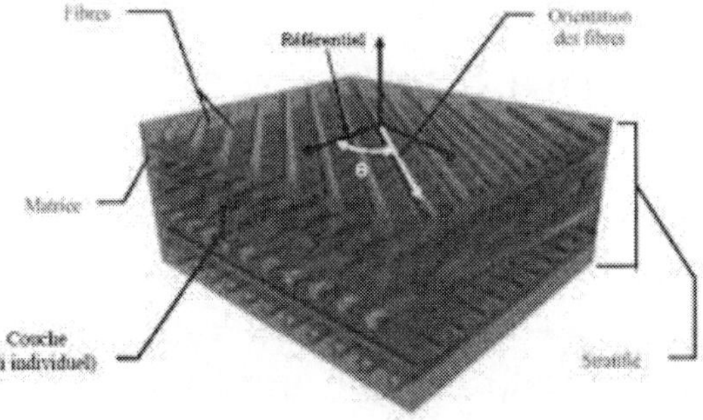

Figure 1: Structure of the considered CFRP

The structure of CFRP is usually described as in Fig1. These materials are made of several unidirectional plies of about 125 μm thickness, stacked with different fiber orientations. Each unidirectional ply is constituted of carbon fibers embedded in an electrically non conductive matrix. The conductivity in the direction transverse to the fibers is not null since there are contacts between fibers not perfectly aligned. The conductivity tensor of each ply depends on the type, the volume fraction, and the orientation of the fibers in the ply; it varies between $5.10^3$and 5.10^4 S/m in the direction of the fibers, and between 10 and 5.10^2 S/m in the transverse direction [9]. When the plies are assembled, there are also contacts between the fibers of adjacent plies, giving rise to a cross ply conductivity which is of lower value being typically half the transverse value [10]. Conductivity effects are therefore introduced in the conductivity tensor by involving the longitudinal and transversal conductivities (σ_l, σ_t) of the plies, the orientation θ of each ply, and the cross ply conductivity σ_n [11]:

$$\bar{\bar{\sigma}} = \begin{bmatrix} \sigma_l cos^2(\theta) + \sigma_t sin^2(\theta) & \frac{\sigma_l - \sigma_t}{2} sin(2\theta) & 0 \\ \frac{\sigma_l - \sigma_t}{2} sin(2\theta) & \sigma_t cos^2(\theta) + \sigma_l sin^2(\theta) & 0 \\ 0 & 0 & \sigma_n \end{bmatrix} \quad (1)$$

Thus, we define the permittivity tensor as :

$$\bar{\bar{\varepsilon}} = \begin{bmatrix} \varepsilon_{xx} & \varepsilon_{xy} & 0 \\ \varepsilon_{xx} & \varepsilon_{xy} & 0 \\ 0 & 0 & \varepsilon_{zz} \end{bmatrix} = \varepsilon_r \bar{\bar{\mathbf{I}}} - \frac{i\bar{\bar{\sigma}}}{\omega\varepsilon_0} \quad (2)$$

where ε_0 is the permittivity in air. Capacity effects may be introduced by different values of the the relative permittivity ε_r. In our case we choose $\varepsilon_r = 1$ but it is not required in general.

3.3 formulation

Assuming the absence of external electric source, working in the harmonic regime and with a non magnetic material, we focus on Maxwell-Faraday law for induction and Ampère's circulation law equations. we introduce the permeability in the air μ_0, the impedance coil in the air $Z_0 = \sqrt{\frac{\mu_0}{\varepsilon_0}}$ and the wave number $k = Z_0 \omega \varepsilon_0$, thus:

$$\nabla \times \mathbf{E} = -ik\mu_r Z_0 \mathbf{H} \quad (2)$$

$$Z_0(\nabla \times \mathbf{H}) = ik\bar{\bar{\varepsilon}}\mathbf{E} \quad (3)$$

4 model decomposition (M-D)

In this section, we illustrate the way we decompose the EM field into sum of eigenmodes in both sides of a separating interface. we start with the isotropic media which is the air in our case then the anisotropic media which is the slab.

4.1 M-D for homogenized isotropic material

In isotropic material, we consider a Cartesian coordinate system and planar interfaces, thus we adopt Rayleigh's decomposition. We divide the field into transversal (t index) and longitudinal (l index) component $(\mathbf{E_t}, \mathbf{H_t})$ and (E_z, H_z), so that we obtain a new expression of Ψ_z (Ψ could be E or H)in function of transversal components. At this stage, we use the 2D discrete Fourier transform ($TF(f) = \tilde{f}$) and we define the truncate numbers M_u and M_v as the numbers of modes along the 2 dimensions X and Y. In what follows and to avoid ambiguity, $f = \tilde{f}$.

Conventionally, $\partial_i \equiv \frac{\partial}{\partial i}$ and in Fourier domain the partial derivatives become $\partial_x \equiv -i\alpha$ and $\partial_y \equiv -i\beta$. We take also the assumption of variable separation: $f(x,y,z) = f(x,y)e^{-i\gamma}$ so $\partial_z \equiv -i\gamma$. Therefore we notice that the z components verify Helmholtz's equation:

$$[\partial_z^2 + k_c^2 - \alpha^2 - \beta^2]\Psi_z = 0 \tag{4}$$

Where $k_c^2 = k^2\mu_r\varepsilon_r$. therefore, we have two scalar potentials:

$$E_z = a^{TM-}(\alpha,\beta)e^{-i\gamma z} + a^{TM+}(\alpha,\beta)e^{i\gamma z} \tag{5}$$
$$H_z = a^{TE-}(\alpha,\beta)e^{-i\gamma z} + a^{TE+}(\alpha,\beta)e^{i\gamma z} \tag{6}$$

Coming back to Maxwell equations, we use a $\mathbf{TE}_z(H_z = 0), \mathbf{TM}_z(E_z = 0)$ decomposition to rewrite the transversal components of the EM field in function of modes.

$$\begin{bmatrix} E_y \\ Z_0 H_x \\ Z_0 H_y \\ E_x \end{bmatrix} = a^{TE\pm}\Phi^{TE\pm} + a^{TM\pm}\Phi^{TM\pm} \tag{7}$$

Such:

$$\Phi^{\pm} = \frac{1}{k_c^2 - \gamma^2}\begin{bmatrix} \pm\gamma\alpha E_z + \mu_r \alpha Z_0 H_z \\ \varepsilon\beta E_z \pm \gamma\alpha Z_0 H_z \\ -\varepsilon\alpha E_z \pm \gamma\beta Z_0 H_z \\ \pm\gamma\beta E_z - \mu_r Z_0 H_z \end{bmatrix} \tag{8}$$

4.2 M-D for homogenized anisotropic material

In the general case of biaxial anisotropy form of the permitivity tensor, Maxwell's equations can be written as two sets of three equations each:

$$E_{sys} = \begin{cases} \partial_y E_z - \partial_z E_y = -ik\mu_r Z_0 H_x \\ \partial_z E_x - \partial_x E_z = -ik\mu_r Z_0 H_y \\ \partial_x E_y - \partial_y E_x = -ik\mu_r Z_0 H_z \end{cases} \tag{9}$$

$$H_{sys} = \begin{cases} \partial_y H_z - \partial_z H_y = \frac{ik}{Z_0}(\varepsilon_r^{xx} E_x + \varepsilon_r^{xy} E_y) \\ \partial_z H_x - \partial_x H_z = \frac{ik}{Z_0}(\varepsilon_r^{yy} E_y + \varepsilon_r^{yx} E_x) \\ \partial_x H_y - \partial_y H_x = \frac{ik}{Z_0}\varepsilon_r^{zz} E_z \end{cases} \tag{10}$$

Likewise, we write $E_z = fct(H_y, H_x)$ and $H_z = fct(E_y, E_x)$. Thus we determine two decoupled matrix systems of dimension (2x2) to be solved, verified by the transversal components of the electric field in one hand and the magnetic field in the other hand presented in (11) and (13). In the same framework, Sabbagh and M. Robert in [12, 13, 14, 15] present a similar approach but do not take into account the decoupled variables, they resolve a (4x4) matrix system to describe the EM field in the layers.

$$\partial_z \begin{bmatrix} E_x \\ E_y \end{bmatrix} = \mathscr{L}_{EG} \begin{bmatrix} H_x \\ H_y \end{bmatrix} \tag{11}$$

where

$$\mathscr{L}_{EG} = \begin{bmatrix} -\frac{Z_0 \partial_x \partial_y}{ik\varepsilon_r^{zz}} & \frac{Z_0 \partial_x^2}{ik\varepsilon_r^{zz}} - ik\mu_r \\ -\frac{Z_0 \partial_y^2}{ik\varepsilon_r^{zz}} + ik\mu_r & \frac{Z_0 \partial_y \partial_x}{ik\varepsilon_r^{zz}} \end{bmatrix} \tag{12}$$

$$\partial_z \begin{bmatrix} H_x \\ H_y \end{bmatrix} = \mathscr{L}_{GE} \begin{bmatrix} E_x \\ E_y \end{bmatrix} \tag{13}$$

where

$$\mathscr{L}_{GE} = \begin{bmatrix} \frac{\partial_x \partial_y}{iZ_0 k\mu_r} + ik\varepsilon_r^{yx} & -\frac{\partial_x^2}{iZ_0 k\mu_r} + ik\varepsilon_r^{yy} \\ \frac{\partial_y^2}{iZ_0 k\mu_r} - ik\varepsilon_r^{xx} & -\frac{\partial_x \partial_y}{iZ_0 k\mu_r} - ik\varepsilon_r^{xy} \end{bmatrix} \tag{14}$$

By applying ∂_z to the first matrix equation verified by the transversal components of the Electric field and the TF, we obtain the eigenvalues system to be resolved in the Fourier domain.

$$-\gamma^2 \begin{bmatrix} E_x \\ E_y \end{bmatrix} = \mathscr{L}_{EG} \mathscr{L}_{GE} \begin{bmatrix} E_x \\ E_y \end{bmatrix} \tag{15}$$

This system provides two eigenvalues $\pm\sqrt{\lambda_1}$ et $\pm\sqrt{\lambda_2}$ corresponding to the direction of the propagation of the wave, and two associated eigenvectors $\vec{v_1}$ et $\vec{v_2}$. So E_x and E_y are a linear combination of eigenvectors.

$$\begin{bmatrix} E_x \\ E_y \end{bmatrix} = \begin{bmatrix} v_x^- \\ v_y^- \end{bmatrix} \begin{bmatrix} a_x^- \\ a_y^- \end{bmatrix} + \begin{bmatrix} v_x^+ \\ v_y^+ \end{bmatrix} \begin{bmatrix} a_x^+ \\ a_y^+ \end{bmatrix} \tag{16}$$

The eigenvalues are used to calculate the wave attenuation factors $\psi_i^{\pm}(z)$ for several depths. From (13) we reconstruct the $\mathbf{H_t}$ components:

$$\begin{bmatrix} H_x \\ H_y \end{bmatrix} = \frac{-i}{\gamma} \mathscr{L}_{GE}^{-1} \begin{bmatrix} E_x \\ E_y \end{bmatrix} = \begin{bmatrix} w_x^- \\ w_y^- \end{bmatrix} \begin{bmatrix} a_x^- \\ a_y^- \end{bmatrix} + \begin{bmatrix} w_x^+ \\ w_y^+ \end{bmatrix} \begin{bmatrix} a_x^+ \\ a_y^+ \end{bmatrix} \tag{17}$$

Finally the modal decomposition in the anisotropic layer is defined by:

$$\begin{bmatrix} \mathbf{E}_t \\ Z_0 \mathbf{H}_t \end{bmatrix} = a_x^{\pm} \Phi_x^{\pm} + a_y^{\pm} \Phi_y^{\pm} \tag{18}$$

Such

$$\Phi_x^{\pm} = \begin{bmatrix} v_x^{\pm} \\ v_y^{\pm} \\ w_x^{\pm} \\ w_y^{\pm} \end{bmatrix} \tag{19}$$

4.3 applying Boundary Conditions (air-plate interface)

In the absence of any surface current ($J_s = 0$), the BC of an EM field at a separating interface are defined by the continuity of the tangential components of the EM field:

$$\mathbf{E}_1 \times \mathbf{n} = \mathbf{E}_2 \times \mathbf{n} \tag{20}$$

$$\mathbf{H}_1 \times \mathbf{n} = \mathbf{H}_2 \times \mathbf{n} \tag{21}$$

At this stage, we introduce a reference state field Φ_{ref} which a field du a source term in air:

$$\Phi_{ref} = \begin{bmatrix} E_{0y} \\ Z_0 H_{0x} \\ Z_0 H_{0y} \\ E_{0x} \end{bmatrix} \tag{22}$$

And we aim to evaluate a perturbation field so that :

$$Phi = \Phi_p + \Phi_{ref} \tag{23}$$

Fig. 2 illustrates the incidence, the reflected and the transmitted field with the unknown coefficients to be calculated by applying BC.

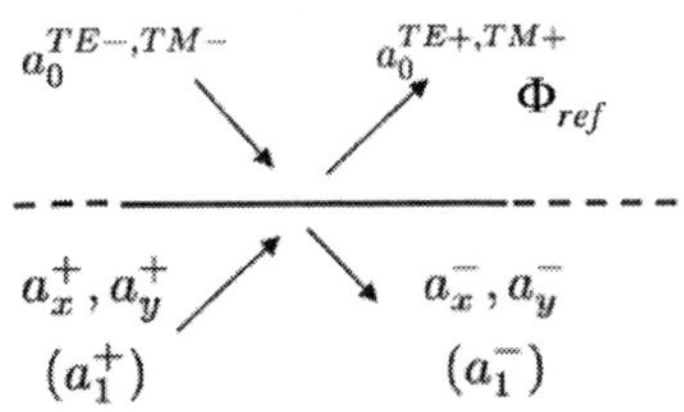

Figure 2: BC illustration

Then we write:

4.3.1 in the air

$$\begin{bmatrix} \mathbf{E}_t \\ \mathbf{H}_t \end{bmatrix} = \begin{bmatrix} \Phi_0^{TE+} & \Phi_0^{TM+} \end{bmatrix} \begin{bmatrix} a_0^{TE+} \\ a_0^{TM+} \end{bmatrix} + \Phi_{ref} \tag{24}$$

4.3.2 in the plate

$$\begin{bmatrix} \mathbf{E}_t \\ \mathbf{H}_t \end{bmatrix} = \begin{bmatrix} \Phi_x^- & \Phi_y^- \end{bmatrix} \begin{bmatrix} a_x^- \\ a_y^- \end{bmatrix} \tag{25}$$

Numerically, the reference field is calculated by some external modules already implemented into CIVA. the definition of Ψ_{ref} leads to assume that $a_x^- = 0$ and $a_y^- = 0$. From (20) and (21) applied at the first separating surface, we have:

$$\begin{bmatrix} \Phi_0^{TE+} & \Phi_0^{TM+} \end{bmatrix} \begin{bmatrix} a_0^{TE+} \\ a_0^{TM+} \end{bmatrix} + \Phi_{ref} = \begin{bmatrix} \Phi_x^- & \Phi_y^- \end{bmatrix} \begin{bmatrix} a_x^- \\ a_y^- \end{bmatrix} \tag{26}$$

The constants are determined by resolving the system :

$$\begin{bmatrix} a_0^{TE+} \\ a_0^{TM+} \\ a_x^- \\ a_y^- \end{bmatrix} = -\begin{bmatrix} \Phi_0^{TE+} & \Phi_0^{TM+} & \Phi_x^- & \Phi_y^- \end{bmatrix}^{-1} \Phi_{ref} \tag{27}$$

finally, we have the total field in the whole region but in the Fourier domain, therefore, we apply the inverse Fourier transform to reconstruct the perturbation field in the spatial domain and from (23) we obtain the total field.

4.4 scattering matrix approach

the main idea of the S-matrix algorithm is to represent the different interfaces as four-port black boxes with two inputs and two outputs (respectively (a_p^-, a_{p+1}^+) and (a_{p+1}^-, a_p^+) for the $(p+1)^{th}$ interface). the inputs-outputs relationship of such a box is then entirely described by a (4x4) matrix, called intermediate S-matrix and denoted by S(p+1,p) for the $(p+1)^{th}$ interface). The so-called scattering matrices algorithm is a recursive routine that aims at linking all the intermediate S-matrices in order to compute the global S-matrix, which relates directly (a_0^-, a_N^+) and (a_N^-, a_0^+), thus leading to a black box representation of the global mock-up, including all the internal reflections.

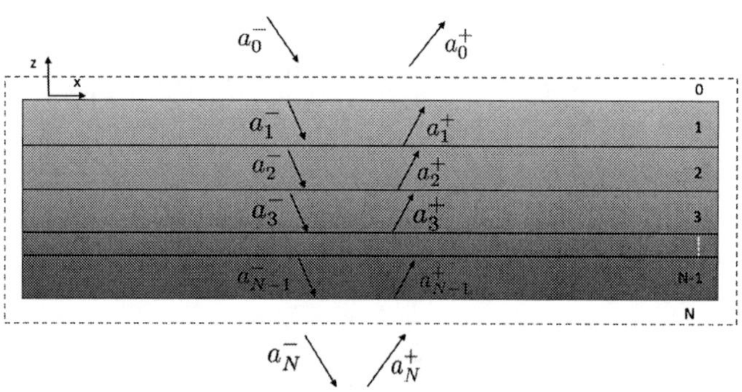

Figure 3: Stratified media

We refer the reader to [16] for more detailed set up of a general S-matrix algorithm as numerical scheme and implementation.

5 results and discussion

In this section, we illustrate the modal validation through comparison with FEM results. Two test-cases was considered:

- one layered media with $\theta = 0°$ and $45°$
- 4 layered slab with different orientations:
  - $[0°, 0°, 90°, 90°]$
  - $[0°, 90°, 0°, 90°]$
  - $[0°, -45°, 90°, 45°]$

5.1 semi-analytic (SA) model features:

Figure 4: CIVA configuration for reference field calculation

The model is developed on Matlab software. The test case configurations and parameters we adopted here are:

- Circular coil: its characteristics are described in Tab. 1.
- Anisotropic plate: its characteristics are described in Tab. 2.
- reference field is calculated on a surface delimited by the red outline.
- M_u=25, M_v=25.
- Spatial domain: [-23, 32]x[-32,32] mm^2.

R_{int}	R_{ext}	Height	Frequency	Intensity	Lift-off
4 [mm]	6 [mm]	1 [mm]	10^4 [Hz]	0.4 [A]	0.6 [mm]

Table 1: Coil characteristics

Thickness	$(\sigma_{xx}, \sigma_{yy}, \sigma_{zz})$	depths	θ
1[mm]	$(10^4, 10^3, 10^3)$ [S/m]	7	0°

Table 2: specimen characteristics

5.2 FEM-(SA) Model validation

In the representation of the model results, we are limited to the H_z components which is the most complicated one to reconstruct since it depends on the two transversal components of the electric field E_x and E_y, so if there is a good agreement between data, it will confirm that the other components also agree with FEM results. in figure (5) and (6) we illustrate the 3D view of the real part $\Re(.)$ and imaginary part $\Im(.)$ of H_z. It is obvious the effect of the anisotropy with the orientation. This allows us to be able to detect the fiber orientation is each layer by ploting the current density. In figure(7) we show a slice view of H_z at 7 different depths in the slab.

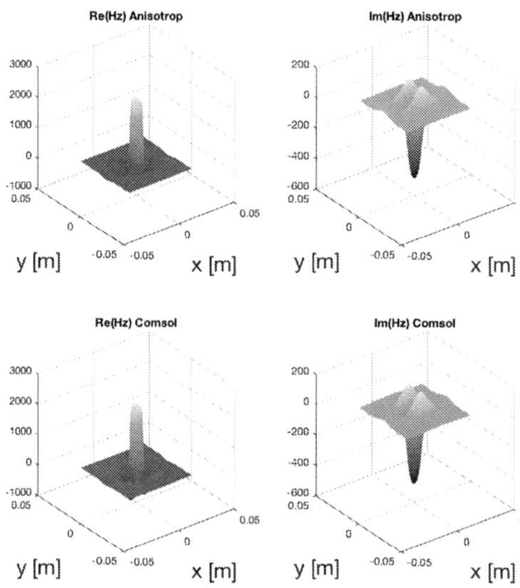

Figure 5: 3D view of H_z for one layer case with $\theta = 0°$

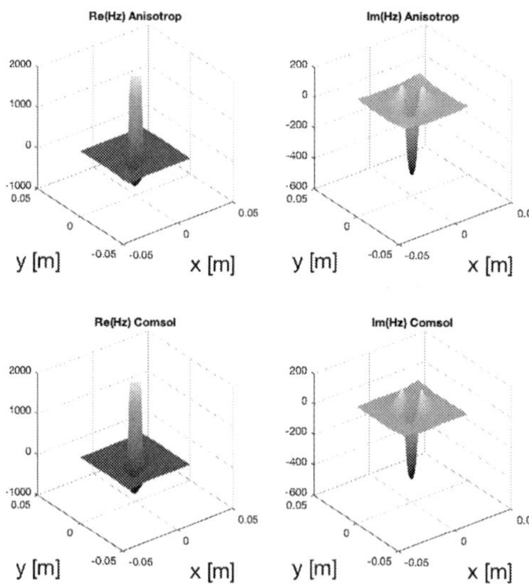

Figure 6: 3D view of H_z for one layer case with $\theta = 45°$

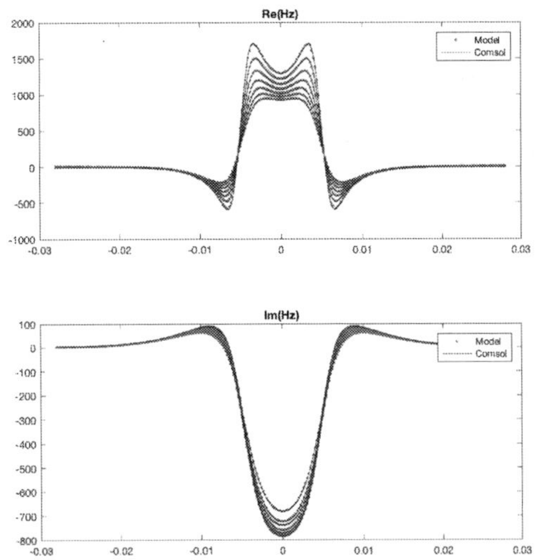

Figure 7: 3D view of H_z for one layer case with the sequence $[0°, -45°, 90°, 45°]$

In Tab. 3 we illustrate the efficiency of the semi-analytic model developed through the calculation time. it's obvious that the model present a exceptional rapidity.

FEM	SA Model
30 min to 2h depending on the mesh	55 [s]

Table 3: Comparison of calculation time

6 conclusion

The work presented in this paper constitutes a contribution to a three-dimensional numerical modeling for the simulation of the interaction of an Eddy Current 3-D probe with a anisotropic and stratified planar media, as an homogenized composite material, for simulating eddy current non destructive testing applications. We have developed a semi-analytic model that solves steady-state Maxwell's equations for materials with biaxial conductivity based on modal decomposition of the field. we started with the semi infinite case and then we use the recursive S-matrix algorithm to resolve for the multi-layered case. Finally, the model is numerically validated by using FE Data. The EM field generated in a plate with a 3D air-core cylindrical coil is clearly reproduced at different depths. By using all previous developments already implemented into the CIVA platform, we will be able soon to tackle any kind of EC probe (rectangular, racetrack, D-coil, spiral circular . . .).

acknowledgment

This paper is supported by European Union's Horizon 2020 research and European Training Network NDTonAIR in NDT and SHM of aircraft structures H2020-MSCA-ITN-2016 under grant agreement No 722134.

References

[1] O. Ouchetto, S. Zouhdi, A. Bossavit, G. Griso, B. Miara and A. Razek, "Homogenization of structured electromagnetic materials and metamaterials". Journal of Materials Processing Technology 181 (2007) 225-229.

[2] G. Wasselynck, Didier Trichet and Javad Fouladgar, "Determination of the Electrical Conductivity Tensor of a CFRP composite Using a 3-D Percolation Model". IEEE Transaction on Magnetics, Vol. 49, No. 5, May 2013.

[3] H. Menana and Mouloud Féliachi, "3-D Eddy Current Computation in Carbon Fiber Reinforced Composites". IEEE TRANSACTIONS ON MAGNETICS, VOL. 45, NO. 3, MARCH 2009.

[4] W. Yin, Philip J. Withers, Umesh Sharma and Anthony J. Peyton, "Noncontact Characterization of Carbon Fiber Reinforced Plastics Using Multifrequency Eddy Current Sensors". IEEE Transactions on instrumentation and measurement, Vol. 58, No. 3, MARCH 2009.

[5] T. M. Roberts, Harold A. Sabbagh and L. David Sabbagh, "Electromagnetic scattering for a class of anisotropic layered media". Journal of mathematical physics, 29, 2675 (1988); doi: 10.1063/1.528060.

[6] L. David Sabbath and Harold A. Sabbagh, "Eddy-Current Modeling and Flaw Reconstruction". Journal of nondestructive evaluation, Vol. 7, No. 1/2, 1988.

[7] H.A. Sabbagh, L. David Sabbagh and Thomas M. Roberts, "An Eddy Current Model For Three-Dimensional Nondestructive Evaluation Of advanced Composites". Naval Surface Weapons Center Research and Technology department, 28 July 1985.

[8] T.M. Roberts, Harold A. Sabbagh and L. David Sabbagh, "Electromagnetic Interactions with an anisotropic Slab". IEEE TRANSACTIONS ON MAGNETICS, VOL. 24, NO. 6, NOVEMBER 1988.

[9] T.M. Roberts, "Explicit Eigenmodes for Anisotropic Media". IEEE Transactions on Magnetics, Vol. 26, No. 6, Nov. 1990.

[10] F. Caire, D. Prémel and G. Granet, "Fast computation of the field diffracted by a multilayered conductor with non-parallel rough interfaces. Application to Eddy Current non Destructive Testing simulation". IEEE Transactions on Magnetics, Vol.51, No.3, March 2015.

[11] F. Caire, D. Prémel and G. Granet, "Semi-Analytic computation of quasi-static field induced by a 3-D eddy current probe scanning a 2D layered conductor with parallel rough interfaces". International Journal of Numerical Modeling: Electronic Networks, Devices and fields int. J. Numer. Model. 2013; 00:1-13.

[12] D. Prémel and G. Granet, "Fast computation of the response of any 3D Eddy Current probe scaling a 3D stratified conductor characterized by a set of arbitrary interfaces of complex shape". International Journal of Applied Electromagnetics and Mechanics 41 (2014) 1-0.

[13] D. Prémel and G. Granet, "Development of the curvilinear coordinate method for the computation of quasi-static fields induced by an eddy current probe scanning a 3D coductor of complex shape characterized by an arbitrary 2D surface". International J. Numer Model. 2017; e2219.

[14] R. C. Rumpf, "Improved Formulation of Scattering Matrices for Semi-analytical Methods that is consistent with convention". Progress In Electromagnetics Research B, Vol. 35, 241-261, 2011.

[15] E.l. Tan, "Note on formulation of the enhanced scattering-(transmittance-) matrix approach". J. Opt. Soc. Am. A, Vol. 19, No. 6, JUNE 2002.

[16] L. Li, "Formulation and comparison of two recursive matrix algorithms for modeling layered diffraction grattings". IJ. Opt. Soc. Am. A, Vol. 13, No. 5, JUNE 1996.

[17] L. Li, "Note on the S-matrix propagation algorithm". IJ. Opt. Soc. Am. A, VOL. 20, NO. 5, April 2003.

[18] Shaoni Jiao, Jian Li, Fei Du, Lei Sun and Zhiwei Zeng, "Characteristics of Eddy Current Distribution in carbon fiber Reinforced Polymer". Journal of Sensors, Volume 2016, Article ID 4292134, 8 pages.

[19] Wuliang Yin, Philip J. Withers, Umesh Sharma and Anthony J. Peyton, "Noncontact characterization of carbon fiber reinforced plastics using multifrequency eddy current sensors". IEEE Transactions on instrumentation and measurement, Vol. 58, No. 3, MARCH 2009.

[20] J. B. Park, T. K. Hwang, H. G. Kim, and Y. D. Doh, "Experimental and numerical study of the electrical anisotropy in unidirectional carbonfiber-reinforced polymer composites," Smart Mater. Struct., vol. 16, pp. 57–66, 200.

[21] S. B. Pratap and W. F. Weldon, "Eddy current in anisotropic composites applied to pulsed machinery," IEEE Trans. Magn., vol. 32, no. 2, pp. 437–444, Mar. 1996.

[22] G. Granet "Analysis of diffraction by surface-relief crossed gratings with use of the Chandezon method: application to multilayer crossed gratings". Journal of the Optical Society America A 1998; 15:1121–1131.

[23] N P K. Cotter et al. "Scatteringmatrix approach to multilayer diffraction". Journal of the Optical Society America 1995; 12: 1097–1103.

[24] L. Li "Formulation and comparison of two recursive matrix algorithms for modeling layered diffraction gratings". Journal of the Optical Society America A 1996; 13: 1024–1035.

Guided Waves Simulation Capabilities in Isotropic and Anisotropic Structures in View of Structural Health Monitoring Applications

Ramanan SRIDARAN VENKAT, Christian BOLLER, Alexander ERNST, Mohamed EL BAKOURI
Lehrstuhl für Zerstörungsfreie Prüfung und Qualitätssicherung, Universität des Saarlandes, Saarbrücken, Germany
Ramanan.sridaran@uni-saarland.de

Abstract. Guided waves are a means very much sought in the context of Structural Health Monitoring (SHM). In view of a structure's damage tolerance and hence light weight potential a reliable SHM system has to be configured. Compared to classical NDT where the options of data sampling are virtually endless, SHM is fixed to a network of sensors and possibly also actuators, which hence has to be optimum. To determine this optimum, numerical simulation is possibly the best option. Specifically, when a structure is designed damage tolerant the allowable size and location of a damage must be known. As a consequence guided wave propagation in a damage tolerant structure can be simulated and since the difference in the signals between a damaged and an undamaged condition is the key information to be 'harvested' by a sensing system it is this difference which has to be identified in view of configuring the optimum sensor network for monitoring. There are various bases on how such differential signals can be determined, being not just limited to the guided waves' amplitude and phase but being also extended to so called damage indices, which have been continuously developed over the years to specifically retrieve features from guided wave signals. This paper initially provides examples on how to identify such optimum sensor networks through simulation with regard to isotropic materials based on what is called 'differential imaging' before moving to anisotropic materials such as CFRP, where numerical simulation capabilities are quickly reaching limitations in case no specific measures can be taken, which will be briefly discussed.

1. Introduction

If SHM is due to become successful in application then the expectations towards simulation will be high. It is therefore worth to explore and demonstrate the current capabilities of simulation with respect to Guided Waves (GW) and SHM and to possibly determine what might be required such that numerical simulation can play its expected role in the longer term. Ultrasonic waves are a principle widely considered in SHM and here specifically in the form of GW being long-range ultrasonic waves useful for monitoring large, thin prismatic structures as found in different aerostructures' applications. GW may even be considered for monitoring complicated structures where in SHM fixed PWAS (Piezoelectric Wafer Active Sensors) are either attached to or even integrated into the structures. GW ultrasonic inspection using PWAS has a fixed sensor network which limits the probability of arbitrarily detecting damage. However, if a structure is designed damage tolerant, size and location of the damage are defined and a GW PWAS based SHM system can be designed that reliably detects the tolerable damage. Various numerical simulation tools have been widely applied in the field of NDT/SHM in recent years [3, 5, 7, 8, 9, 13, 15, 18]. Among those FEM-based tools are the commercially ones available for performing GW simulations.

There are numerous ways along which one can achieve an optimum actuator-sensor pattern for given tolerable damages to be present in the structure where some are reported in [4, 5, 6, 11, 15, 17, 19, 20]. Theoretically, an increase in the density of the sensor network is considered to detect damages more reliably. However, such an approach becomes highly impractical to implement since the volume of data to be processed will be big for a large as well as complex structure. For a given component, the accuracy of the FEM solution for a GW simulation additionally depends on the complexity of the material (i.e. isotropic/anisotropic) and the complexity on how to model the damage to be detected by the SHM system. Unlike isotropic materials, anisotropic materials such as Carbon Fibre Reinforced Polymers (CFRP) composites face a considerable challenge in apropriately defining material properties resulting from the stacking sequence of the laminate layers. Modelling each layer with a size of 100-200 μm each can result in a remarkable computation challenge. In addition damages in CFRP composites need to be modelled accurately to get an optimum result with respect to GW simulation. Even with commercial FEM tools it is challenging to model damages such as interply-delaminations, matrix cracks and fibre breakages. Overviews of modeling the damages in composites have been reported in [10, 16]. In the highlight of seeking an SHM solution through GW simulation, this paper describes possibilities of performing numerical simulation and data processing through *differential imaging* and *Damage Indices- maps (D.I)* for an isotropic aluminium notched plate and how far similar approaches could be tackled with an anisotropic composite plate.

2. Test samples and material properties

For the two types of plates analysed the mechanical properties for aluminium 2024 are given in Table 1 and for a CFRP composite in Table 2 respectively. The geometry of the aluminium plate with three holes and a crack of 14 mm on the right side is shown in Fig. 1 and of the composite plate in Fig. 2(a) including the 24-layer stacking sequence shown in Fig. 2(b). An individual layer and a homogenized layer condition have been considered for the simulation for which the stiffness matrices are shown in Table 2. The piezoelectric properties of PWAS are given in Table 3.

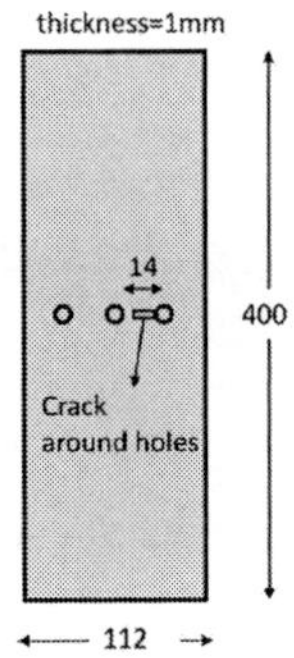

Figure 1. Aluminium plate with artificial crack on the right side

Density [kg/m³]	Young's Modulus [GPa]	Poisson's Ratio [-]
2700	70	0.33

Table 1. Material properties of aluminium 2024

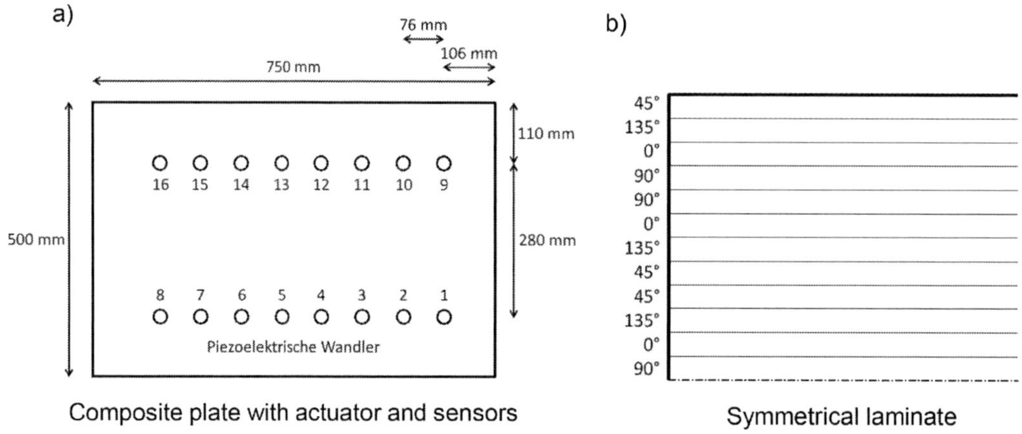

Figure 2. a) Model of the composite plate b) symmetrical layers

Orientation	Elastic matrix [GPa]					
0°	165,4	6,8	6,8	0,0	0,0	0,0
		14,5	6,5	0,0	0,0	0,0
			14,5	0,0	0,0	0,0
				3,9	0,0	0,0
					5,2	0,0
						5,2
Homogenized	66,6	30,2	6,6	0,0	0,0	0,0
	30,2	66,6	6,6	0,0	0,0	0,0
	6,6	6,6	14,5	0,0	0,0	0,0
				4,5	0,0	0,0
					4,5	0,0
						23,4

Table 2. Stiffness matrices of the 0° layer and homogenised layer

Elasticity matrix [GPa]						Coupling matrix [C/m2]			Relative permittivity [-]		
E_{11}, E_{22}	E_{33}	E_{44}, E_{55}	E_{66}	E_{12}	E_{13}	C_{15}, C_{24}	C_{33}	C_{31}, C_{32}	ε_{11}	ε_{22}	ε_{33}
120.4	110.9	21.1	22.6	75.2	75.1	12.3	15.8	-5.4	919.1	919.1	826.6

Table 3. Properties of the piezoelectric material

3. Numerical simulation of guided waves

The *piezoelectric devices* and the *structural mechanics* modules in the COMSOL software are used to send and receive GW into a given structure. The monitoring principle of the sensing approach is shown in Fig. 3. The respective constitutive equations can be found in various textbooks such as [18].

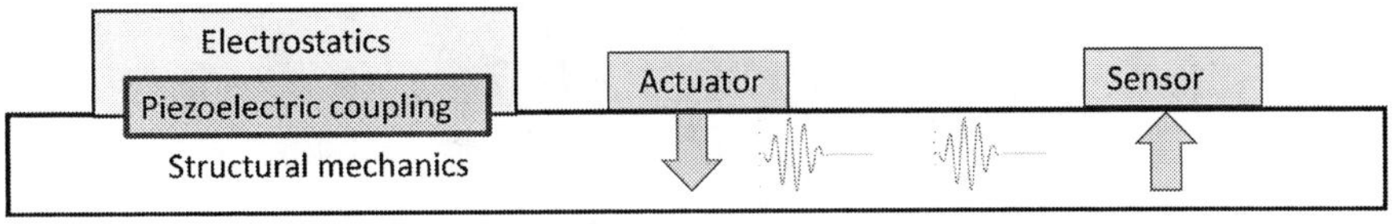

Figure 3. Monitoring approach applied

A *Hanning window* function of the form described in Eq. (1) has been used to minimize the frequency band [18] which is written as

$$V(t) = \frac{1}{2}\left[1 - cos\left(\frac{\omega_0 t}{N_C}\right)\right] sin(\omega_0 t), t \in \left[0, \frac{N_C}{f}\right] \quad (1)$$

and where the following parameters represent: $V(t)$- excitation voltage, ω_0 - angular frequency and N_C - number of counts that matches the length of the Hanning window.
The resulting excitation signal is shown in Fig. 4. The criterion for the selection of the mesh and time step can be referred to in [4, 13].

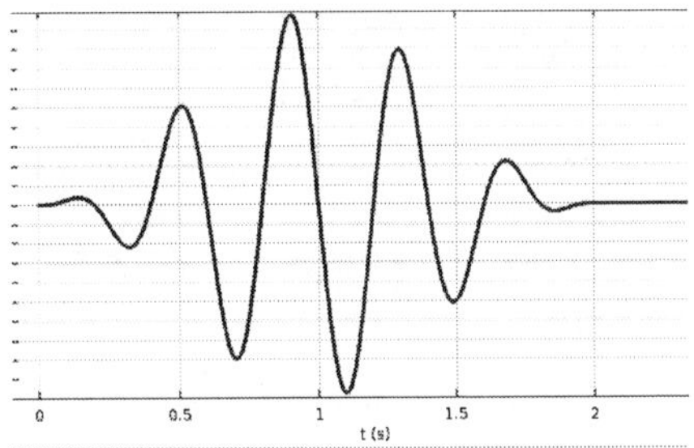

Figure 4. Five cycles Hanning window excitation function

4. GW simulation of aluminium plate

GW propagation in the aluminium plate with a crack described above is shown for different time frames in Fig. 5. If one needs to know in detail how a GW signal changes in the presence of damage, the differential imaging method is opted. The principle of differential imaging is that for a time incident 'x' the wave propagation image for a given damage configuration is subtracted from the wave propagation of a pristine condition. Thus, the resultant image shows the areas of maximum and minimum displacements caused due to the scattering of the GW in the presence of damage. The application of differential imaging in search of optimal sensor locations for isotropic and anisotropic materials has been reported in [12]. The logic and results of the differential imaging screenshots for the given aluminium plate is shown in Fig. 6. What can be seen is that at which incident of time and at which location a difference due to a change such as damage can be observed in the guided wave signal.

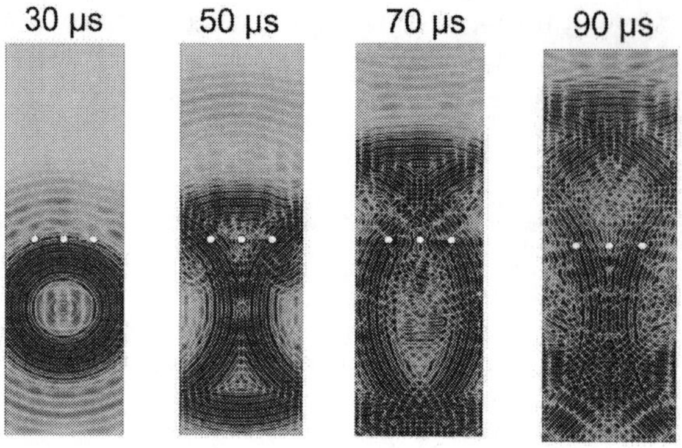

Figure 5. GW propagation in a damaged condition after a) 30 µs, b) 50 µs, c) 70 µs & d) 90 µs

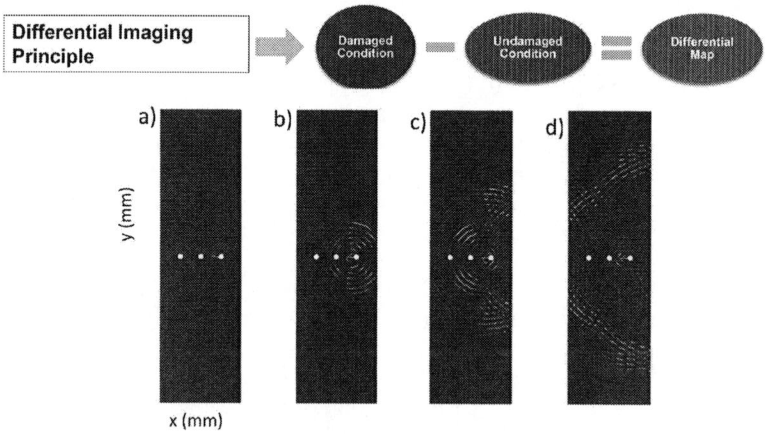

Figure 6. Principle and differential images of model with a crack at the right side at a) 30 µs, b) 50 µs, c) 70 µs & d) 90 µs

Damage indices or damage metric values are used to measure the amount of damage present in a structure through the mathematical processing of the resultant signal with the signals of a pristine state. If any damage is present in the structure, the spectral content of the signal will be affected. The resultant scalar value is called a damage index and may be another indicator on where to retrieve a maximum of information being related to a damage to be monitored. Differential Energy (DE), Correlation Coefficient Deviation (CCD), Root Mean Square Deviation (RMSD) and Mean Absolute Error (MAE) are examples of such damage indices that have been considered here. The procedure to obtain a *D.I. map* is to virtually assume a sensor to be present on each point of the plate where the displacements are recorded as a result of the GW propagation. For the given time range, all the sensor points in the plate record displacement signals and subsequently D.I.s are evaluated for all points to obtain a *D.I. map* as shown in Fig. 7. The detailed description of the damage indices can be found in [14].

Similar to differential images, the *D.I. map* shows points of high intensity (red colours) representing locations where the influence of damage can be significantly recorded. By correlating the *differential images* and the *D.I. maps*, optimum sensor positions can now be identified which in the present case is on the upper right side of the plate. An experimental proof of this result has also been given in [14]. The flavour of the D.I.s is that the information results from an integration of the signal being recorded and that no specific time window has to be set.

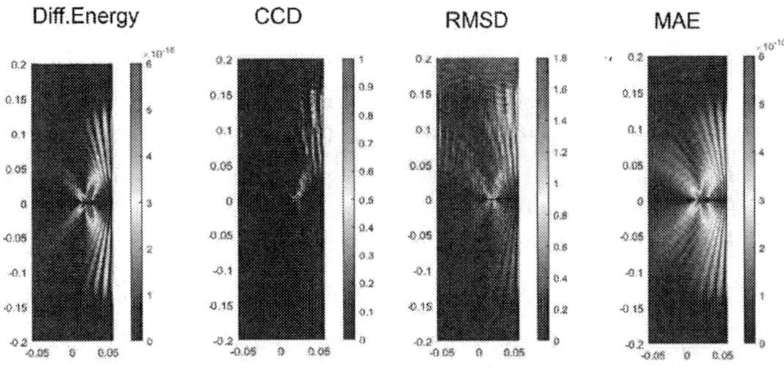

Figure 7. D.I. maps plotted for the corresponding damage indices

However, with some of the D.I.s such as DE and MAE the relevant information mainly stays close to the damage itself, which in the sense of SHM may not always be the desired solution. The further away from the damage the damage information can be retrieved the more flexible the SHM solution could be handled. This looks to be true for CCD and RMSD as well as on the basis of the pure time domain signal with regard to the differential imaging at the price that a specific time window has to be set in the latter case.

5. GW simulation of composite plates

Modeling each of the layers of a composite results in an increased number of degrees of freedom (DOF) and an exponential increase in computation effort and time. Details of this problematic have been elaborated and reported in [2]. An alternative is to homogenise the stiffness matrices of all the layers such that a single layer of the homogenised stiffness matrix is obtained. Fig. 2(a) shows the PWAS with the numbers 1 to 8 to be actuators and 9 to 16 to be sensors respectively. All the sensors record the excited signal of each actuator and the signals were stored for further analysis. The FE model uses tetragonal elements and when considering 20 elements per wavelength for a 200 kHz signal, the element size is approximately 1.4 mm, which easily exceeds a RAM size of 32 GB and limited the mesh size to 3 mm at present.

Interesting contributions concerning modelling of delaminations can be obtained from [15] where the following three different mechanisms have been used; 1. Stiffness Reduction Method (SRM), 2. Duplicate Node Method (DNM) and 3. Kinematic Based Method (KBM). SRM is based on the idea that the presence of delamination causes a reduction of stiffness in a structure. Thus, the stiffness loss at the delamination area can be incorporated by modifying the material properties such as Young's modulus, shear modulus, density and others. DNM uses either 1D beam elements or 2D plane stress/strain finite elements to create duplicate nodes, which leads to high bending-stretching coupling in the mid-plane of the element. This results in a mode conversion at the delamination region. KBM is more useful in modelling multiple delaminations and fibre breakages. The approach involves enforcing kinematics to the nodes surrounding the flaw. This is achieved using rigid links between the FE nodes that share the flaw. The procedure for modelling delaminations and fibre breakage are described in [15]. In this paper, the Thin Elastic Layer (TEL) constraint is considered for modelling delaminations and this method can be related to KBM as mentioned before. TEL is commonly used to model thin elastic layers and damping properties are included with this constraint. Thus, the elastic constraint ensures the connection between two components (two intermediate layers) and both edges of thin layers are connected to elastic and viscous forces that act in opposite directions. The forces are proportional to the displacement or velocity. The TEL layer is infinitely thin and can be modelled to any geometrical shapes. Besides this, various types of spring constants can be selected to implement the damping characteristics. The main criterion to define a TEL is to have at least two layers, so the homogenised sample is divided into two segments with the same material properties and the "contact pair" boundary condition is set on the free surfaces to prevent the penetration of opposing surfaces.

The GW propagation in the above mentioned composite plate with a 20 mm circular delamination is shown in Fig.9 a) and the position of the delamination in Fig. 9 b) respectively. The fundamental Lamb wave modes (S0 and A0) and possible reflections of S0 can be clearly seen. A comparison of signal from simulation and experiment for the actuator-1 and sensor -12 for an undamaged condition is shown in Fig.10 a) where the differences in phase and amplitude at the beginning of the wave packets are less while this difference increases with increasing time. This becomes obvious after the third wave packet for the reflected A0 mode, being a consequence of the mesh size for the A0 mode to be at least half of the S0 mode, which the 3 mm mesh size is not able to completed. Furthermore the S0 wave velocity plays a significant role, which has been 6240 m/s in the simulation versus 5730 m/s in the experiment. Fig. 10 b) compares the signals for the pristine and damaged condition where a minor change in amplitude and phase is observed after 200 µs. Lack of experimental results has not allowed further comparison so far.

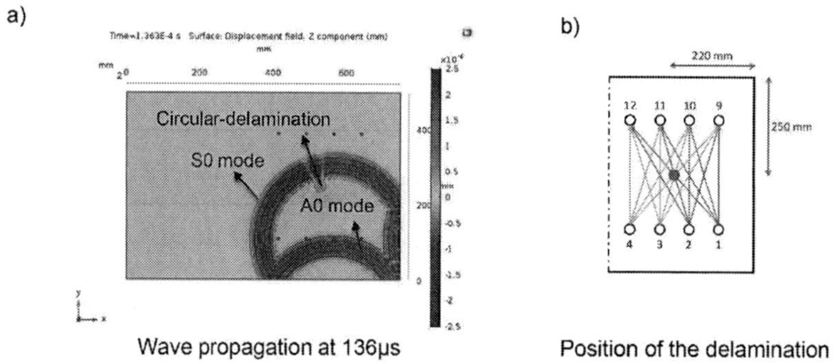

Figure 9 a) GW propagation in composite plate with delamination b) Position of the delamination related to right half of the plate

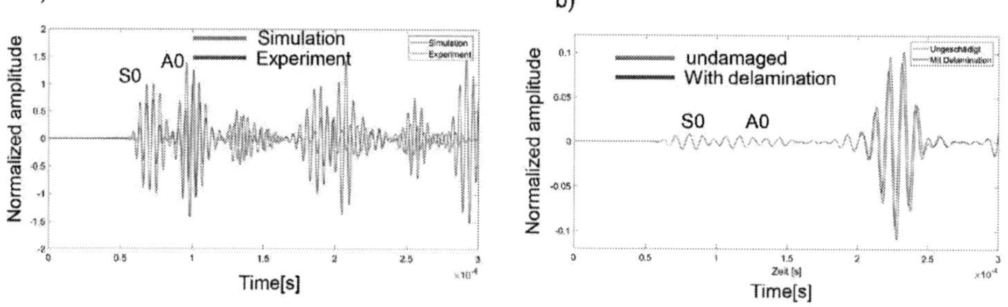

Figure 10 Signal comparison between a) simulation and experiment b) with and without delamination for simulation only

6. Conclusions

Numerical simulation is crucial with respect to configuring a GW based SHM system. *Differential imaging* and *D.I. mapping* are ideal means to visualize information of a tolerable damage on a structure having to be retrieved in view of an SHM system. However, limitations in simulations can be easily seen in terms of speed and computation power. To get this overcome the following points need to be considered:

1. Irrespective of homogenization there should be scope in FE analysis to model individual layers. Appropriate boundary conditions between the layers need to be identified to ensure continuity of displacement fields.
2. Mesh size is critical for the accuracy of the solution, where $\geq$ 20 elements per wave length should be a must.
3. Damage mechanisms in composites need to be accurately defined in the FEM model using connecting interfaces. Interply-delamination, fibre and matrix cracks require attention. The accuracy depends analogous to the first point stated above.
4. Modelling large CFRP structures needs to take advantage of symmetrical aspects and to use 'hybrids' such as the SAFE (Semi Analytical Finite Element) approach.
5. Attenuation of the GW signal has to be considered where information can be retrieved from experiments.

Since simulation can never be perfect, experimental validation is required from where differences can be retrieved which may then lead again to an improvement in the numeric model resulting possibly in some sort of iteration loop. Aspects to be considered do include anisotropy of the material or ray tracing simulation just to name two examples where precision of the simulation is improved on the one side and speed of the simulation on the other, being two drivers where numerical simulation regarding SHM has a specific need.

References

[1] A. Bulavinov, R. Pinchuk, T. Gurieva"Robot-based In-Process Examination of ITER Dome and First-Wall Panels based on Novel Ultrasonic Tomography Approach" Proceedings in 19th World Conference on Non-Destructive Testing (2016)

[2] A. Ernst, " Simulationsgestützte Ermittlung von Strukturzuständen in Faserverbunden zur beschleunigten Detektion von Schädigungszuständen" , Master Thesis Saarland University (2018). (in German)

[3]A. Ernst, R. Sridaran Venkat, C. Boller,: "Grenzen und Herausforderungen bei der Simulation geführter Wellen in Faserverbundstrukturen" Proc. DGZFP Jahrestagung, (2018). (in German)

[4] A. Taltavull, R. Sridaran Venkat, C. Boller, C. Dürager, "Structural health monitoring of repaired metallic aircraft panel modeling-based approach", *8th Internat. Symp. on NDT in Aerospace*, Bangalore/India, (2016).

[5] B. C. Lee, W. J. Staszweski, "Sensor location studies for damage detection with Lamb waves ", *Journal on Smart materials and structures*, Volume 16, pp 399-408 (2007).

[6] C. Fendzi, J. Morel, M. Rebillat, M. Guskov, N. Mechbal, G. Coffignal, "Optimal sensor placement to enhance damage detection in composite plates", *7th European Workshop on Structural Health Monitoring (EWSHM 2014),* 8-11 July, Nantes/France (2014).

[7] J.H. Nieuwenhuis, J. Neumann, D.W. Greve, and I.J. Oppenheim, "Generation and detection of guided waves using PZT wafer transducers", *IEEE Trans Ultrasonic Ferroelectr Freq Control.* Vol.52 Issue.11.(2005).

[8] L. Qiu, R. Sridaran Venkat, C. Boller, S. Yuan, "Multiphysics Simulation of Guided Wave Propagation under Load Condition", *8th Internat. Symp. on NDT in Aerospace* Bangalore/India (2016)

[9] M. Zennaro *et al.* "Experimental and numerical analysis of a transducer for the generation of guided waves". In: IXth NDT in Progress (2017).

[10] Nag A et al,"A spectral finite element with embedded delamination for modelling of wave scattering in composite beams", *Composites Science and Technology*, Volume 63, pp 2187-2200 (2003).

[11] P. Malinowski, T. Wandowski, W. Ostachowicz, "Investigation of sensor placement in Lamb wave based SHM method", *Key engineering materials* .Vol.518, pp 174-183 (2012).

[12] R. Sridaran Venkat, C. Boller, "NDT approaches to optimize acoustics based SHM systems for anisotropic composite structures" *Internat.l Workshop on Structural Health Monitoring*, Stanford CA, USA (2017).

[13] R. Sridaran Venkat, C. Boller, Lei Qiu, D.R. Mahapatra, N.B. Ravi, N. Chakraborty, "Integrated approach to demonstrate optimum sensor positions in a guided wave based SHM system using numerical simulation", *8th Internat. Symp. on NDT in Aerospace*, Bangalore/India. (2016)
[14] R. Sridaran Venkat, M. El Bakouri and C. Boller, "Damage Indices as a Measure for Optimal Sensor Placement in a Guided Wave Monitored Cracked Notched Plate Based on Numerical Simulation versus Scanning Lased Doppler Vibrometry", *9th European Workshop on Structural Health Monitoring (EWSHM 2018),* July, Manchester,UK (2018).
[15] S. Gopalakrishnan, M. Ruzzene, S. Hanagud," Computational Techniques for Structural Health Monitoring", Springer publications (2011).
[16] S. Kumar et al, "A spectral finite element for wave propagation and structural diagnostic analysis in a composite beam with transverse cracks" *Finite Elements in Analysis & Design*, Vol. 40, pp 1729-1751 (2004).
[17] V. Ewald, R. M. Groves and R. Benedictus," Transducer Placement Option of Lamb Wave SHM System for Hotspot Damage Monitoring", *Aerospace -Open access Journal for Aeronautics and Astronautics* (MDPI), Vol.5(39) (2018).
[18] V. Giurgiutiu, "Structural Health Monitoring with Piezoelectric Wafer Active Sensors" 2nd ed. (2014).
[19] V. Mallardo, M. H. Aliabadi, Z. Sharif Khodaei, "Optimal sensor positioning for impact localization in smart composite panels", *J. of Intelligen Material Systems & Structures*, Vol. 24 Issue 5, pp 559-573 (2012).
[20] Z. Sharif Khodaei, "An optimization strategy for best sensor placement for damage detection and localization in complex composite structures", *8th European Workshop On Structural Health Monitoring (EWSHM 2016)*, 5-8 July, Spain (2016).
[21] Z. Y. Shi, S. S. Law, L. M. Zhang,"Optimum sensor placement for damage detection", *Journal of Engineering Mechanics,* Volume.126, Issue 11 (2000).

From data-driven NDT of systems to BIG DATA-based modeling

Dirk SÖFFKER [1]
[1] University of Duisburg-Essen, Duisburg, Germany

Contact e-mail: soeffker@uni-due.de

Abstract

The world is changing. New approaches from the IT-field allow engineers to replace well experienced first-principle-based modeling by data-driven approaches to conclude to final statements like: system healthy yes/no? or fault type no. n exists yes/no? More or less the same approaches allow connections to be built between variables from different fields and therefore complex modeling like:
build relations between systems usage and Remaining useful lifetime, and economical parameters from this field (industry 4.0). Using unstructured data everything seems to be connectable (BIG DATA).

Using practical examples the keynote will introduce into NDT-related tasks like fault detection and isolation of composites and remaining lifetime estimation of technical systems being in use. The related filtering of measured AE-data as well as related supervised classification approaches are introduced as an example for a NDT technique as well as briefly compared to classical NDT approaches. Using non-supervised approaches within this BIG DATA world new insights and relations can be found defining not only a new quality of modeling but also extending the view to system limitations and relations. Advantages and disadvantages will be discussed.

The presentation is mainly based on the following publications and

References:

Söffker, D.; Wei, C.; Wolff, S., Saadawia, M.S.: Detection of Rotor Cracks: Comparison of an old Model- with a new Signal-based approach. Nonlinear Dynamics, Vol. 83, 2016, pp. 1153-1170.

Baccar, D.; Söffker, D.: Wear Detection by Means of Wavelet-based Acoustic Emission Analysis. Mechanical Systems and Signal Processing, Vol. 60-61, August, 2015, pp. 198-207.

Söffker, D.; Rothe, S.: New Approaches for Supervision of Systems with Sliding Wear: Fundamental Problems and Experimental Results Using Different Approaches. Applied Sciences, Vol. 7, 2017, pp. 843.

Baccar, D.; Söffker, D.: Identification and Classification of Failure Modes in Laminated Composites by using a Multivariate Statistical Analysis of Wavelet Coefficients. Mechanical System and Signal Processing, Vol. 96, 2017, pp. 77-87.

Wirtz, S.F.; Beganovic, N.; Söffker, D.: Investigation of damage detectability in composites using frequency-based classification of Acoustic Emission measurements. Structural Health Monitoring, 2018, pp. 1-12.

Rothe, S.; Wirtz, S.F.; Kampmann, G.;Nelles, O.; Söffker, D.: Ensure the reliability of damage detection in composites by fusion of differently classified Acoustic Emission measurements. In: Chang, F.K.; Kopsaftopoulos (Ed.): Structural Health Monitoring 2017, Stanford, USA, September 12-14, 2017, pp. 1380-1387.

Beganovic, N.; Söffker, D.: Remaining lifetime modeling using State-of-Health estimation. Mechanical Systems and Signal Processing, Vol. 92, 2017, pp. 107-123.

Jihin, R.; Söffker, D.; Beganovic, N.: Integrated Prognostic Model For RUL Estimation Using Threshold Optimization. In: Chang, F.K.; Kopsaftopoulos (Ed.): Structural Health Monitoring 2017, Stanford, USA, September 12-14, 2017, pp. 640-647.

Incorporating Inductive Bias into Deep Learning: A Perspective from Automated Visual Inspection in Aircraft Maintenance

Vincentius EWALD [1], Xavier GOBY [1], Hidde JANSEN [1], Roger M. GROVES [1], Rinze BENEDICTUS [2]

[1] Aerospace Non-Destructive Testing Laboratory, Faculty of Aerospace Engineering, Delft University of Technology, Delft, Netherlands

[2] Structural Integrity and Composites Group, Delft University of Technology, Faculty of Aerospace Engineering, Delft, Netherlands

Contact e-mail: V.Ewald@tudelft.nl

Abstract. Narrow artificial intelligence, commonly referred as 'weak AI' in the last couple years, has developed due to advances in machine learning (ML), particularly deep learning, which has currently the best in-class performance among other machine learning algorithms. In the deep learning framework, many natural tasks such as object, image, and speech recognition that were impossible in the previous decades using classical ML algorithms can now be done by a typical home personal computer.

Deep learning requires a rapid collection of a large amount of data (also known as 'big data') to create robust model parameters that are able to predict future occurrences of certain event. In some domains, large datasets such as the CIFAR-10 image database and the MNIST handwriting database already exist. However, in many other domains such as aircraft visual inspection, such a large dataset of damage events is not available, and this is a challenge in training deep learning algorithms to perform well to recognize material damage in aircraft structures.

As many computer science researchers believe, we also think that in order to achieve a performance similar to human-level intelligence, AI should not start from scratch. Introducing an inductive bias into deep learning is one way to achieve this human-level intelligence in the aircraft inspection for damage. In this paper, we give an example of how to incorporate domain knowledge from aerospace engineering into the development of deep learning algorithms. We demonstrate the suitability of our approach using data from fatigue testing of an aerospace grade aluminum specimen to build a deep convolutional neural network that classifies crack length according to the crack propagation curve obtained from fatigue test. The results of this network were then compared to the same network that was not trained with domain knowledge and the biased learning achieved a validation accuracy of 97.55% on determining crack length, while unbiased network selected the unwanted parameter of sunlight intensity, however with 99.45% accuracy.

Keyword: *deep learning, convolutional neural network, image recognition, inductive bias, crack propagation, visual inspection, non-destructive testing*

1. Introduction

1.1. State-of-the art

As of 2018, the inspection of aircraft is primarily done manually with 80% of inspection being carried out visually [1]. This suggests that damage detection via image recognition systems has potential to assist in aircraft inspection. Ever since the thawing of the last, so called "AI-Winter", which took place during the early 90's, the field of artificial intelligence (AI) has experienced a dramatic reinvigoration. Since then, research and development in this field have been exponentially growing, particularly in the field of deep-learning, a broad family of machine learning algorithms within the field of AI. One of the reasons why deep learning is become increasingly more popular is that it eliminates hand-picked features, a time-consuming process in classical machine learning.

Among the many deep learning architectures, convolutional neural network (CNN) is one of the most popular architecture for image recognition [2]. In previous work, we have applied 1-hidden layer CNN for phase discontinuity predictions in NDT [3], however this was not with a deep architecture. To our knowledge, the very first deep CNN application for visual inspection was performed by Zhang et al. for road crack detection on concrete structure by using 6 layers ConvNet [4].

Since then, similar works using CNN for crack detection have been performed in civil infrastructure such as building, [5], pavement [6], and concrete surface [7 – 9]. While these results are promising and show the first step in advancing image recognition for crack detection by CNN, a generalization of the incorporation of domain knowledge is required to have a reasonable justification of the interpretable outcome of deep learning.

1.2. Objective

Taking into account the state-of-the art provided in [3 – 9], the objective of our paper is to show how to incorporate inductive bias in deep learning by injecting bias from aircraft maintenance. Precisely, we built and tested a deep CNN that classifies crack length based on crack propagation curves obtained from fatigue test. Thus, unlike image segmentation and thresholding, the network is built on the physical basis of crack propagation behavior.

In order to test the performance of injecting domain knowledge, the result of crack image classification of the CNN architecture with domain knowledge was compared to the result of image classification of the same network that was not injected by domain bias, but rather pre-clustered by unsupervised machine-learning.

Our paper is structured as follow: Section 2 describes the theoretical background of deep learning, particularly the deep CNN and the formalization of inductive bias. In section 3, we briefly describe the fatigue testing procedure and the machine learning algorithm we used. The results and discussion are presented in section 4, while our summary and conclusion are presented in section 5.

2. Machine and Deep Learning

2.1. Deep Convolutional Neural Network

A convolutional neural network is a class of feedforward artificial neural networks that has been proven to be an extremely effective tool in the field of computer vision. LeCun et al. (1998) [10] developed the first deep CNN back in the late 90's which was called the LeNet-5 architecture and consisted of 7 layers including the output layer as depicted in Fig. 1.

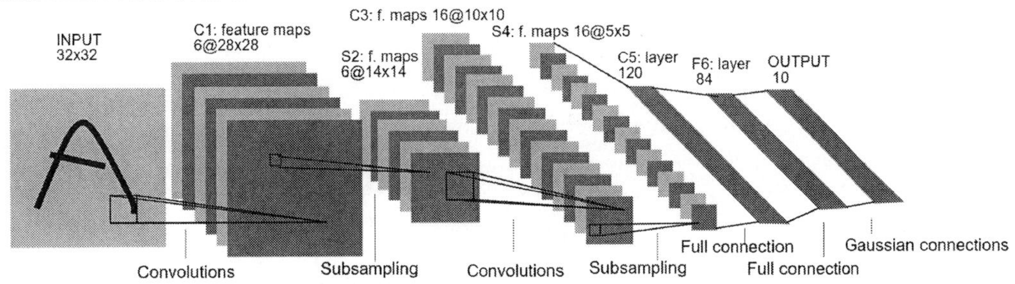

Fig. 1: LeNet-5 [10]

A CNN consists of multiple successively increasingly refined data filters – the layers of a CNN. Within the context of this analogy, the input data to a CNN, for instance a single black and white image can be represented as 3D tensor of size: height ×width × channels of pixel values ranging from 0 to 255. This tensor is multiplied by activation functions, and as it passes through each layer of the network, is transformed into more abstract representations for the networks to produce a prediction output. The transformations performed by each layer are parametrized by its neural weights. Then, the difference between the predicted output and the true output is computed. This difference is called the loss function and the central task of machine learning is to minimize the loss by adjusting the weights through an optimizer during the backpropagation.

There are two types of layers that are normally used in CNN core architecture: the convolutional layer and the pooling layer. The core CNN is typically attached to the fully-connected layer. The convolutional layer is made of a filter (or kernel) which performs a convolutional operation to the input tensor that it receives from previous layer for feature extraction. The filter performs the process of feature extraction by sliding across the input data by several pixels, also called the stride, as depicted in Fig. 2.

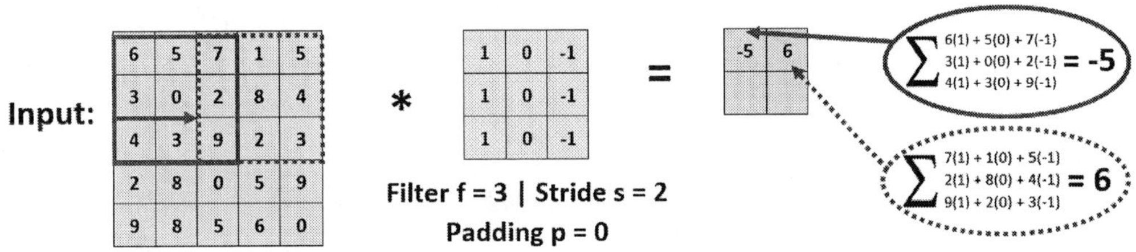

Fig. 2: Convolution operation in CNN using stride s = 2 and zero-padding

2.2. Inductive Bias

The inductive bias of a learning algorithm is the set of assumptions that the learning algorithm $\boldsymbol{A}$ uses to predict outputs for a given input that it has not yet encountered. Formally defined, the inductive bias of learning algorithm $\boldsymbol{A}$ is "*any minimal set of assertions* $\boldsymbol{B}$ *such that for any target concept* $\boldsymbol{c}$ *and corresponding training data* $\boldsymbol{D}$" [11] that following formula holds:

$$\forall \boldsymbol{x}_i \in \boldsymbol{X} : (\boldsymbol{B} \wedge \boldsymbol{D}_c \wedge \boldsymbol{x}_i) \vdash \boldsymbol{A}(\boldsymbol{x}_i, \boldsymbol{D}_c) \quad (1)$$

Where c the target concept, $\boldsymbol{D_c}$ set of training examples, $\boldsymbol{x_i}$ is $\boldsymbol{i}$-th instance of input set $\boldsymbol{X}$, and $\boldsymbol{A}(\boldsymbol{x_i},\boldsymbol{D_c})$ the classification assigned to $\boldsymbol{x_i}$ by $\boldsymbol{A}$ after training on set $\boldsymbol{D_c}$ is. To understand Eq. (1) in more precise way, we consider:

y_i:	*i*-th instance of output space Y
P:	Probability distribution on $X \times Y$
Q:	Distribution on P
L:	Loss function that maps $Y \times Y \rightarrow \mathbb{R}$
$\mathcal{H}$:	Hypothesis space which is a set of function $h: X \rightarrow Y$
$\mathbb{H}$:	Family of hypothesis space where $\mathcal{H} \in \mathbb{H}$
$\Delta_Q(\mathcal{H})$:	Loss of hypothesis space $\mathcal{H}$ on Q
$\Delta_P(h)$:	Loss of function h on distribution P

The hypothesis space $\mathcal{H}$ is the hypotheses set among which the approximated output space Y is searched, while the loss function, also sometimes called cost the function, is the error between the actual and predicted output. For a regression problem, there are several functions that can be chosen such as mean squared error, average absolute deviation, or mean difference. For a classification problem, it is more common to choose loss functions that map an output probability between 0 and 1, such as cross-entropy error or hinge loss [12]. The goal of bias learning is to find the hypothesis space $\mathcal{H} \in \mathbb{H}$ that minimizes $\Delta_Q(\mathcal{H})$:

$$\begin{aligned} \Delta_Q(\mathcal{H}) &\equiv \int_P \inf_{h \in \mathcal{H}} \Delta_P(h)\, dQ(P) \\ &= \int_P \inf_{h \in \mathcal{H}} \int_{X \times Y} L(h(x), y)\, dP(x, y)\, dQ(P) \end{aligned} \tag{2}$$

Typically, minimizing $\Delta_Q(\mathcal{H})$ cannot be done directly, therefore sampling n times from P according to Q is needed to yield: $P_1, \ldots, P_n$. Furthermore, we sample m times from $X \times Y$ according to each P_i to yield the pairs: $\{(x_i^1, y_i^1), \ldots, (x_i^m, y_i^m)\}$. In the sequel, an (n,m)-sample is denoted by z and written as a matrix in Eq. (3). In order to choose a hypothesis space $\mathcal{H} \in \mathbb{H}$, one needs to minimize the empirical loss on z $\Delta_z(\mathcal{H})$ as in Eq. (4).

$$z \equiv \begin{matrix} (x_{11}, y_{11}) & \cdots & (x_{1m}, y_{1m}) \\ \vdots & \ddots & \vdots \\ (x_{n1}, y_{n1}) & \cdots & (x_{nm}, y_{nm}) \end{matrix} = \begin{pmatrix} z_1 \\ \vdots \\ z_n \end{pmatrix} \tag{3}$$

$$\Delta_z(\mathcal{H}) \equiv \frac{1}{n} \sum_{i=1}^{n} \inf_{h \in \mathcal{H}} \Delta_{z_i}(h) \tag{4}$$

The bias induced learning algorithm $\hat{A}$ is then defined as [13]:

$$\hat{A}: \bigcup_{n>0, m>0} (X \times Y)^{(n,m)} \rightarrow \mathbb{H} \tag{5}$$

3. Experimental Procedure and Methodology

3.1. Fatigue Testing

To simulate crack growth in an aircraft fuselage, we selected aluminium 7075-T6 for the fatigue testing as it is commonly used for aircraft structures and because of its well known crack propagation behavior. The material parameters for aluminium 7075-T6 are shown in Table 1 (left column) and the test was performed according to the ASTM-E647 standard [14]. The fatigue test parameter is shown in Table 1 (right column). The test setup is shown in Fig. 3, while the dimension of the test specimen is shown in Fig. 4 and the experimental and empirical crack propagation curve are depicted in Fig. 5, respectively. The empirical crack propagation curve was calculated from AFGROW analysis software from Lextech, Inc. which is available at our department at TU Delft.

Table 1. Left: Material Data [15]. Right: Fatigue Test Parameters According to [14].

Density	2.81 [g/cm³]	Specimen type	M(T)-Spec.
Ultimate Tensile Strength	572 [MPa]	Specimen thickness	6.4 [mm]
Young's Modulus	71.7 [GPa]	Specimen width	100 [mm]
Poisson's Ratio	0.33	Hole diameter ø	12 [mm]
Fatigue Strength	159 [MPa]	Pre-crack length ($2a_0$)	16 [mm]
Fracture Toughness T-L	25 [MPa√m]	Stress ratio R	0.1
Shear Modulus	26.9 [GPa]	Tensile force	32 [kN]
Shear Strength	331 [MPa]	Frequency	2 [Hz]

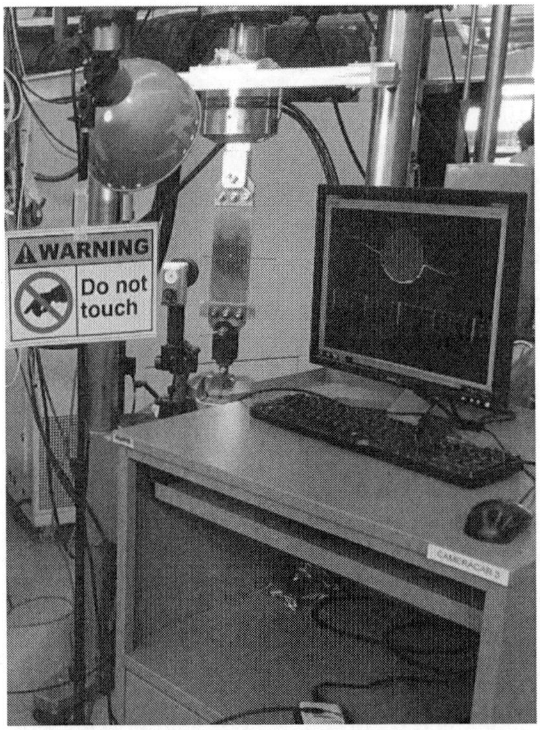

Fig. 3: Fatigue Test Setup

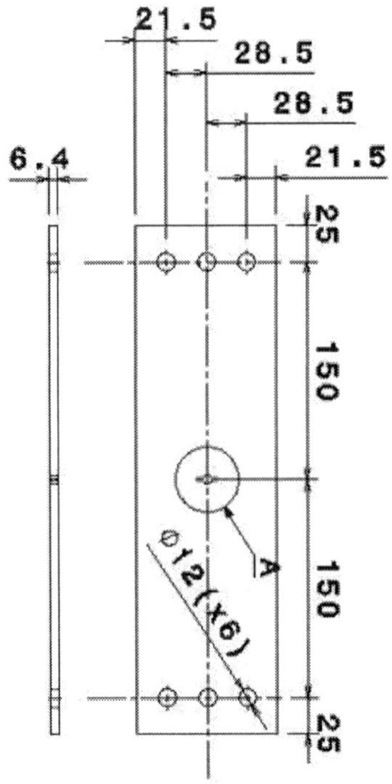

Fig. 4: Specimen Dimension

The test coupon was first modelled in AFGROW and the estimated critical crack length for the coupon was determined to be 80 mm and it would reach 122000 load cycles before failure. The design of experiment was based on this model, however since the hard-drive of the recording computer did not have enough space to store all raw images, it was decided to capture the image at every 10th cycle. During the experiment, the coupon failed at 118030 load cycles and consequently 11803 images which summed up to 47 GB of raw bitmap RGB images were captured by the camera.

We decided to have 4 crack length categories according to the data provided in Fig. 4: no crack, small crack, medium crack, and long crack. The criteria are given in Table 2. Note that fewer or more categorizations can be simplified into binary classification (no crack vs crack) or modified into more categorization (e.g. more than 8 or 12 crack lengths, etc.) depend on the classification need.

Table 2: Crack length categorization criteria, initial notch length ($2a_0$) = 16 mm		
No crack	0 to 44000 cycles	Pre-crack length (16 mm)
Short crack	44001 to 74000 cycles	16.1 mm to 24.0 mm
Medium crack	74001 to 111000 cycles	24.1 mm to 55 mm
Long crack	111001 cycles to failure	more than 55.1 mm

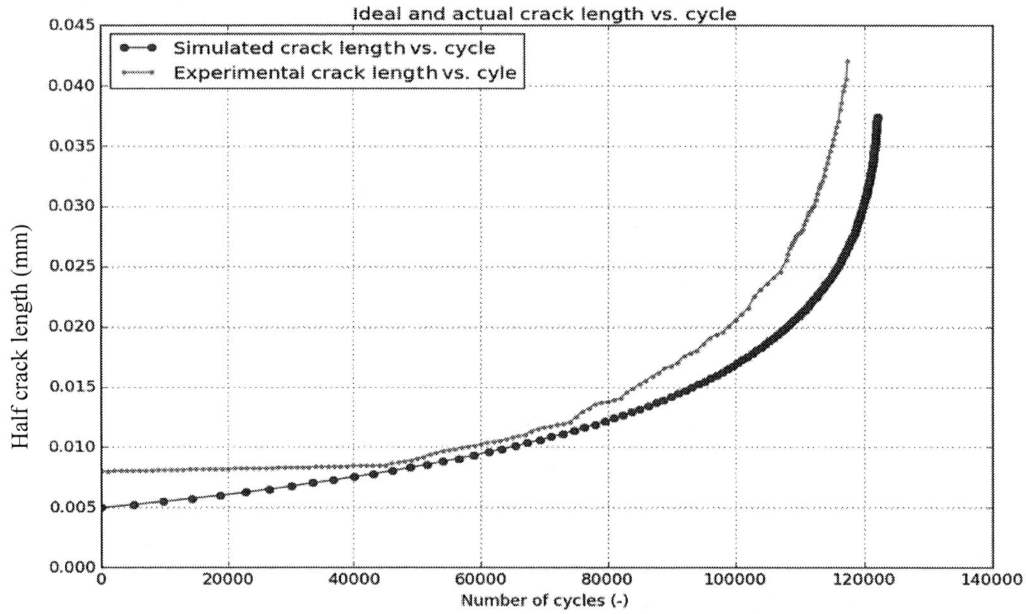

Fig. 5: Crack Propagation Curve

3.2. Pattern Discovery by Clustering Method

In the beginning, we talked about comparing the performance of the biased and non-biased CNN. Instead of manually dividing the fatigue test data in 4 different crack categorization, semi-supervised learning can be introduced to categorize the data according to patterns detected by the computer by pre-training using unsupervised learning.

A widely known method where unsupervised learning is often combined with supervised information to obtain patterns in a particular dataset is clustering, and this combination is called semi-supervised learning since it combines both learning algorithms in a single pipeline. Including a small amount of labelled data to unsupervised learning can lead to a considerable improvement in learning accuracy, but it requires assumptions to be made. The clustering method is based on the assumption that if datapoints share the same cluster, there is a high probability that they are in the same class [17].

Categorizing images with clustering can be performed with the uncomplicated K-means algorithm [18], which is a nearest centroid clustering technique. Concretely, this has athe objective to minimize the sum of the Euclidian distance δ between *n*-datapoints of the *i*-th cluster x_i and the cluster centroid μ_j for *m*-clusters according to:

$$\arg\min \sum_{i=1}^{m}\sum_{j=1}^{n} \delta = \arg\min \sum_{i=1}^{m}\sum_{j=1}^{n} \left(\left\| x_i - \mu_j \right\| \right)^2 \tag{6}$$

Afterwards the mean of the datapoints in each cluster becomes the cluster's new centroid and this process is then repeated for multiple iterations. The human influence in this clustering method is the choice of the number and the locations of the initial centroids. For the categorization of the fatigue test data the middle image of each group of crack lengths from Table 2 were taken as initial centroids.

3.3. Building a biased-CNN for Aircraft Visual Inspection

From the 11803 RGB images, 9204 were dedicated for training (78%), 1947 for validation (16%) and 652 for testing the trained model (6%). Prior to feeding the CNN with training data, the RGB images were first pre-processed by converting them into grayscale images (i.e. pixel-averaged value) to reduce the computational work.

Furthermore, assuming that the images are rotation and translation invariant, every image was rotated over a uniformly distributed range of angles between -0.2° and 0.2° and zoomed by a factor of 1.0 to 0.2 of the original size. Lastly, half of all the images were flipped horizontally. The pre-processing methods were done by Keras 2.1.6, a high-level neural network API wrapper that simplifies library function commands which is built on top of Tensorflow 1.9 by Google [16], the current most popular back-end open source machine learning library with the most up-to-date contributions. The computational work was performed on an Intel® Core™ i7-6700 HQ CPU.

The architecture of our CNN is depicted in Fig. 5 and consists of four layers of 2D convolutional layers with identical filter size of 3 × 3 pixel, stride of 1, 32 and 64 filters for the first two and last two layers, respectively. A 2D max pooling layer with a pooling size of 2 x 2 pixel was attached to each convolutional layer. The intermediate result was flattened and fed into four dense layers of 16 hidden neurons. The final classification layer consists of only 4 units for the 4 class labels that represent initial, small, medium and large crack and these are the inductive bias that map the input images into 4 crack classifications as per Eq. (5). Thus, any unseen images such as those in the test dataset will be classified according to the crack propagation curve depicted in Fig. 5.

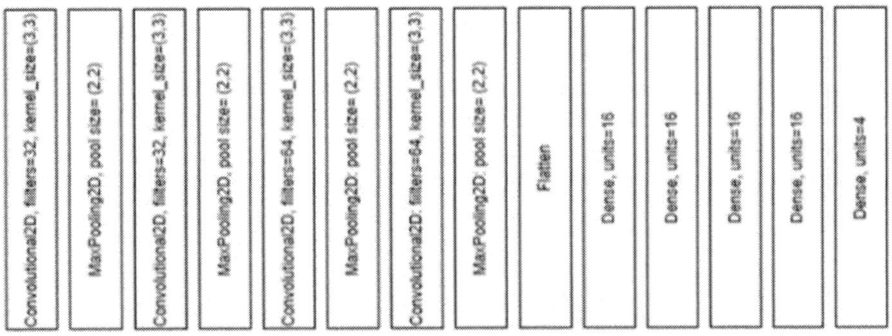

Fig. 5: The architecture of our CNN

4. Results and Discussion

4.1. Semi-Supervised Learning

The clustering method described in section 3.3 was applied to the fatigue test data but the desired result was not achieved. Instead of classifying the images according to their crack lengths, the result after five iterations of the K-means algorithm was that the images were categorized according to the light intensities in the images as depicted in Fig. 6 a – d. Due to large variations in the light intensities during the 23 hours of testing in the TU Delft aircraft hall, the computer did not cluster based on the the crack length.

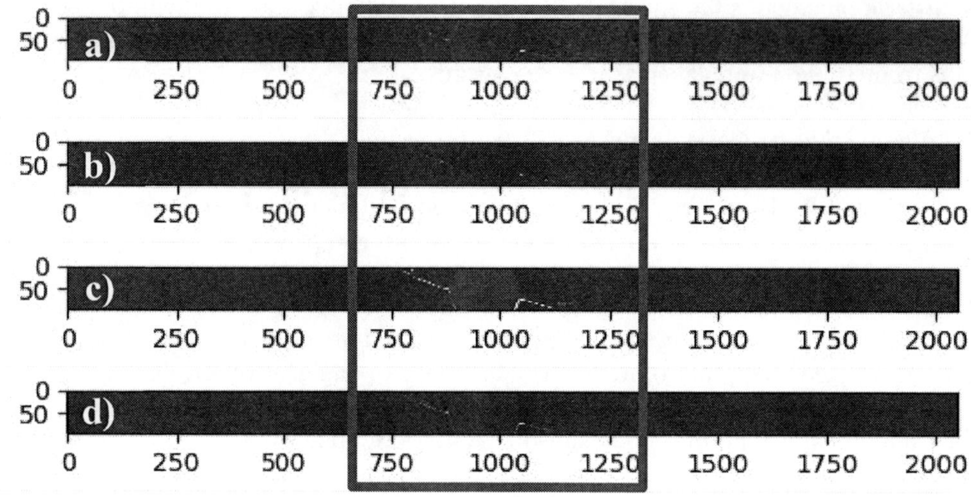

Fig. 6 a – d: Sample images taken from each cluster. X-Axis: Length in pixel, Y-Axis: Width in pixel Difference light intensities through the holes are shown in red rectangle.

Although the four different light intensity classes did not have any meaning from a structural point of view, the images were still fed into our CNN as depicted in Fig. 5 resulting in a 99.45% accuracy in the sixth epoch. This result confirms that semi-supervised learning leads to a considerably high learning accuracy, however it may not make any logical sense from a structural point of view.

4.2. Biased CNN from Supervised Learning

Training the biased CNN described in section 3.2 on the data set consisting of 9204 images for training and 1947 images for validation resulted in a final trained model possessing a training accuracy of 89.15% and validation accuracy of 97.55%, which is slightly lower than the unbiased CNN in Section 4.1.

In addition to this, the confusion matrix, shown in Fig. 7, was also generated to better visualize the correct and incorrect predictions. As it can be seen in Fig. 7, the final trained model experienced the greatest difficulty in making accurate predictions for small/no crack images.

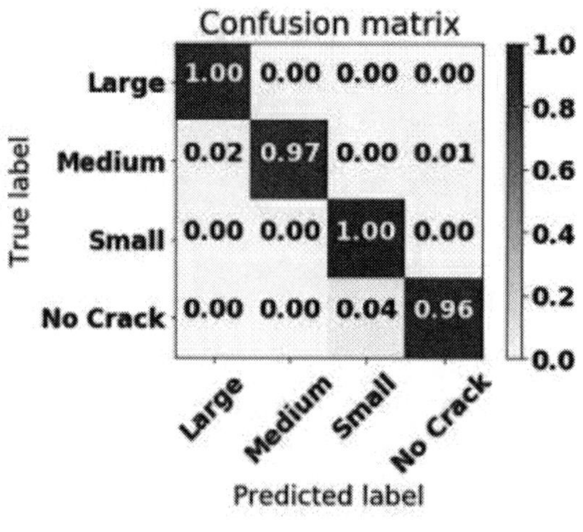

Fig. 7: Normalized confusion matrix of predicted and actual labels

5. Summary and Conclusion

In this paper, we have demonstrated the importance of domain knowledge in automated visual inspection for building a justifiable prediction from deep learning.

The clustering algorithm managed to cluster the 11803 images into 4 different clusters, however not according to 4 different crack lengths as per Table 2, but according to light intensity changes during the image recording. The network trained based on light intensity changes reached 99.45% accuracy. On the other side, the biased CNN that is trained according to different crack length achieved 97.55% accuracy.

These results are consistent with our prior assumption that AI-assisted NDT or Structural Health Monitoring (SHM) should involve human influence rather than just relying on pure unsupervised learning. Nevertheless, we would like also to point out that human decision can also be wrong – therefore we believe that the future NDT and SHM should be based on reciprocal human – computer interaction rather than computer only or human only.

References

[1]. Rice M, Li L, Ying G, Wan M, Lim ET, Feng G, Ng J, Jin Li MT, Bab VS. *Automating the Visual Inspection of Aircraft*. Singapore Aerospace Technology and Engineering Conference, Singapore, 2018.

[2]. Albawi S, Mohammed TA, Al Zawi S. *Understanding of a Convolutional Neural Network*. International Conference on Engineering and Technology (ICET), Antalya, 2017.

[3]. Sawaf F, Groves RM. *Phase Discontinuity Predictions Using Machine-Learning Trained Kernel*. Applied Optics, 2014. Vol. 53: pp 5439 – 5447.

[4]. Zhang L, Yang F, Zhang YD, Zhu YJ. *Road Crack Detection Using Deep Convolutional Neural Network*. IEEE International Conference on Image Processing (ICIP), Phoenix, 2016.

[5]. Cha YJ, Choi W, Büyüköztürk O. *Deep Learning Based Crack Damage Detection Using Convolutional Neural Networks*. Computer Aided Civil and Infrastructure Engineering, 2017. Vol. 32: pp 361 – 378.

[6]. Gopalakrishnan K, Khaitan SK, Choudhary A, Agrawal A. *Deep Convolutional Neural Networks with Transfer Learning for Computer Vision-Based Data-Driven Pavement Distress Detection*. Construction and Building Materials, 2017. Vol. 157: pp 322 – 330.

[7]. Da Silva WRL, De Lucena DS. *Concrete Cracks Detection Based on Deep Learning Image Classification*. 18th International Conference on Experimental Mechanics (ICEM18), Brussels, 2018.

[8]. Li S, Zhao X. *Convolutional Neural Networks-Based Crack Detection for Real Concrete Surface*. SPIE Smart Structures and Materials + Nondestructive Evaluation and Health Monitoring, Denver, 2018.

[9]. Yokoyama S, Matsumoto S. *Development of an Automatic Detector of Cracks in Concrete Using Machine Learning*. Procedia Engineering, 2017. Vol. 171: pp 1250 – 1255.

[10]. LeCun Y, Bottou L, Bengio Y, Haffner P. *Gradient-Based Learning Applied to Document Recognition*. Proc. of IEEE, 1998. Vol. 86: pp 2278 – 2324.

[11]. Mitchell TM. Machine Learning. McGraw Hill, 1997.

[12]. Rosasco L, De Vito E, Caponnetto A, Piana M, Verri A. *Are Loss Functions All The Same?* Journal of Neural Computation, 2003. Vol. 16: pp 1063 – 1076.

[13]. Baxter J. *A Model of Inductive Bias Learning*. Journal of Artificial Intelligence Research, 2000. Vol. 12: pp 149 – 198.

[14]. Standard Test Method for Measurement of Fatigue Crack Growth Rates: ASTM E647 – 15E1. American Society for Testing and Materials.

[15]. Aluminum 7075-T6; 7075-T651. ASM Aerospace Specifications Metals Inc. Site available online: http://asm.matweb.com/search/SpecificMaterial.asp?bassnum=ma7075t6 Last accessed: 30-AUG-2018

[16]. Tensorflow. Site available online: https://www.tensorflow.org/ Last accessed: 30-AUG-2018..

[17]. Chapelle O, Schölkopf B, Zien A. *Semi-Supervised Learning*. The MIT Press, 2006. pp 5 – 6.

[18]. Ju W. *Advances in K-means Clustering: A Data Mining Thinking*. Springer Publishing Company, 2012.

PCRT Resonance Solutions for Additive Manufacturing

Thomas KÖHLER [1]
[1] Vibrant GmbH, Elz, Germany

Contact e-mail: tkoehler@vibrantndt.de

Abstract. Additive Manufacturing (AM) processes are being used increasingly in aerospace to produce even critical components to the detriment of legacy manufacturing systems like casting, forging or machining. Yet quality assurance tools have to catch up to the challenges posed by these new manufacturing methods: process variation, potential manufacturing defects, long-term stability and capability… all are essential concepts still insufficiently understood that need reliable solutions.

Process Compensated Resonance Testing (PCRT) offers Resonance Solutions to a variety of AM challenges. PCRT can:

• Monitor component consistency to provide manufacturing process control data, correlating final component attributes to process parameters, material (powder) batches, and machine-to-machine variation. PCRT's precise, repeatable part-level data feed big data analytics, and combine with manufacturing and operational data to provide insight not available with other inspection methods.

• Monitor the consistency of critical processing operations such as hot isostatic pressing (HIP) and heat treatment (HT). These operations are critical to the confidence in AM components and manufacturers need certainty that they affect all components similarly.

• Validate models and quantitatively compare legacy and reverse-engineered components. Resonance is an integrity fingerprint to quantitatively verify that what was originally designed has been actually printed.

Emerging PCRT studies of AM parts show detection of porosity-related defects, powder supply variation, build process variation, retained powder and correlation to performance testing. PCRT assures part quality at every state, from verifying that the part printed is the part designed, to tracking the consistency of each and every part produced.

In addition, PCRT is production-line ready, capable of testing parts in seconds, integrated with parts-handling automation, and providing Pass/Fail results without highly trained inspectors.

All these turn PCRT into a powerful tool for the AM community to increase confidence in the manufacturing and inspection environment to allow them to take advantage of its tremendous potential.

Vibrant

PCRT Resonance Solutions for Additive Manufacturing

10th Symposium on NDT in Aerospace
Dresden 24-26 Oct 2018

VibrantNDT.de

Who is **Vibrant**?

Founded in 2006, Vibrant's **PCRT services** support the worldwide aerospace, automotive and power generation industries, as well as materials laboratories with unparalleled non-destructive testing (NDT).

CERTIFIED:
AS9100 Rev D and ISO 9001:2015.

Resonant Ultrasound Spectroscopy (RUS) and PCRT application are described by ASTM Standard Guide E2001-13 and ASTM Standard Practice E2534-10 and E3081-16.

PCRT is FAA-approved for gas turbine engine blade inspection.

Vibrant

How Does PCRT Work?

Ultrasound technology, not focused but a full body inspection using Natural Resonances via swept sine method

- Based on ASTM E2001-13, E2534-15 and E3081-16

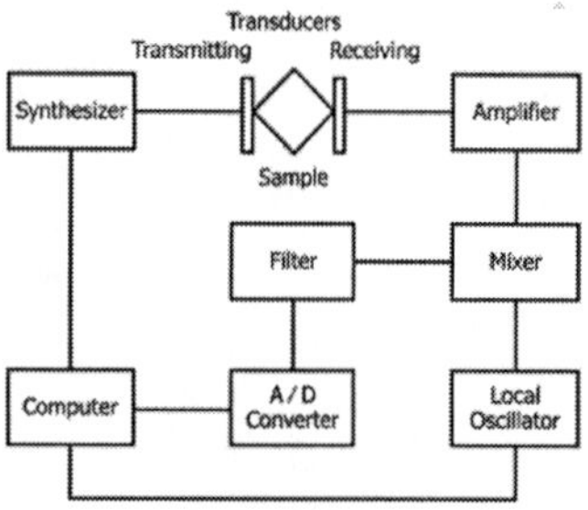

A – Swept Sine method

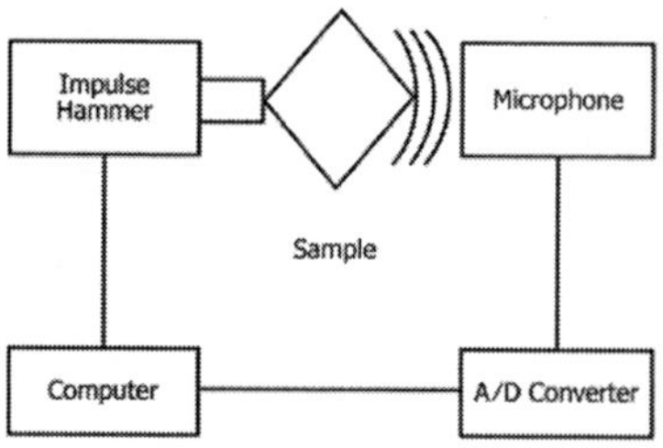

B – Impulse method

Vibrant

What Does PCRT Measure?

A **multi-frequency pattern** is the **fingerprint** or certificate of a component.

Fingerprint data is used to classify parts at different stages (during production process, during service life)

Advanced mathematics for evaluation

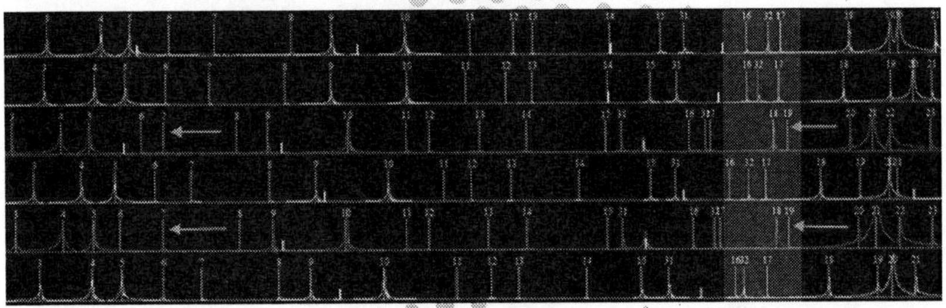

Vibrant

Resonance features

$$f_r \approx \sqrt{\frac{k}{m}}$$

- **Depend on stiffness and mass**
  - Material Properties
  - Production history (powder, laser speed, HIP)
  - Thermal history (heat treatment)
  - Fatigue history
  - Service life
- Integrating: whole body measurement
- Relative: reference set required
- Indirect: no defect localization
- Automatable: independent of user qualification

Vibrant

Additive Manufacturing examples (1)

Use PCRT to ...

- Monitor component consistency
  - print position
  - batch to batch
  - powder to powder

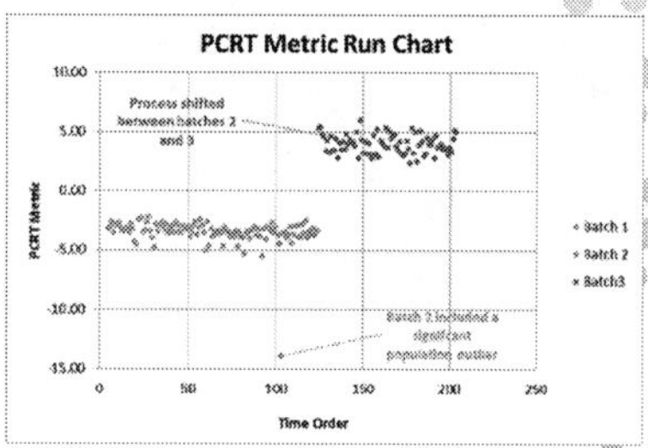

- Monitor critical operations
  - Hot isostatic pressure (HIP)
  - Heat Treatment

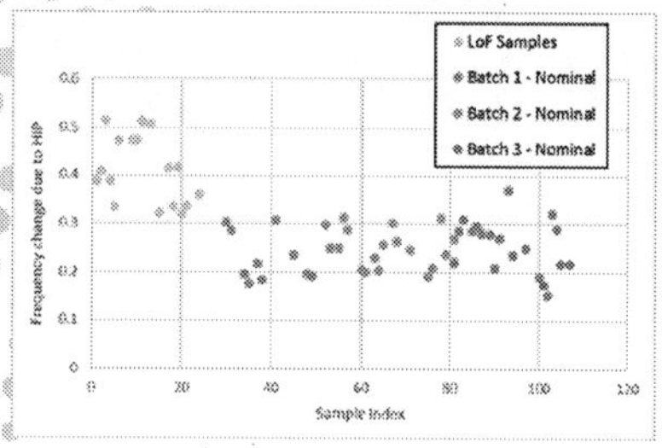

Vibrant

Additive Manufacturing examples (2)

Use PCRT to ...

- Detect outliers
  - Parts with extensive residual stress

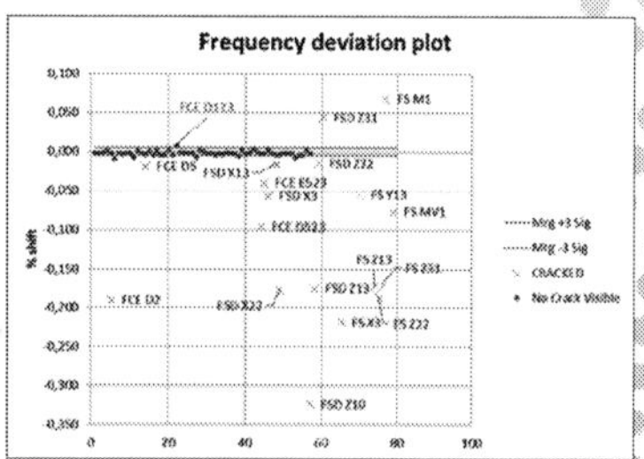

- Validate (Reverse Engineered) Models and Components

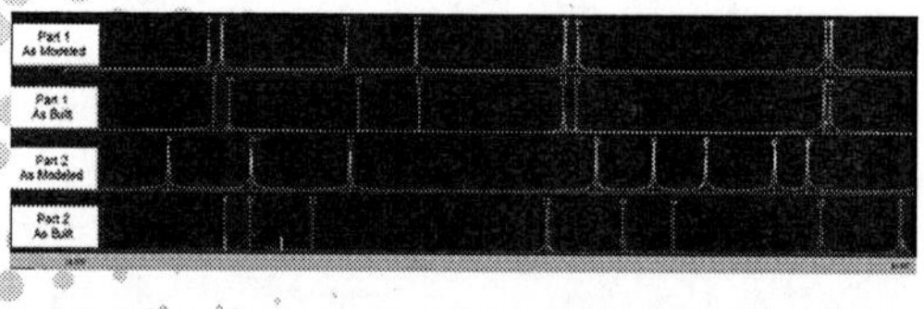

Comparison of resonance spectra 'as modeled' and 'as built' for 2 samples. Part 1 is much more representative of the modeled properties and dimensions than Part 2.

Vibrant

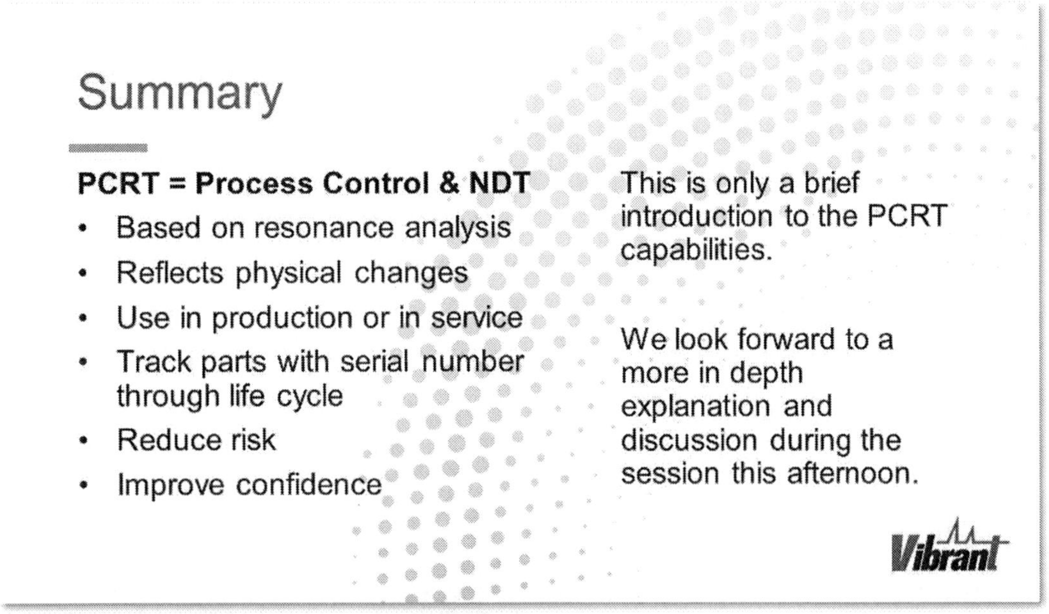

References

[1] ***"America Makes: National AM Innovation Institute (NAMII) – Project 1: Non- destructive Evaluation (NDE) of Complex Metallic AM (AM) Structures,"*** June 2014 Interim Report. AFRL-RX-WP-TR-2014-0162. Todorov, E., Spencer, R., Gleeson, S., Jamshidinia, M., Kelly, S.

[2] ***"Process Compensated Resonant Testing in Manufacturing Process Control,"*** Material Evaluation, 63 736-739, (July 2005). J. Schwarz, J. Saxton, and L. Jauriqui.

[3] ***"Process Compensated Resonance Testing JT8D-219 1st Stage Blades,"*** ATA NDT Forum 2008, (September 24, 2008). D. Piotrowski, L. Hunter, and T. Sloan.

[4] ***"Delta TechOps Reduces Engine Inspection Costs by Nearly $2m Annually,"*** Vibrant Corporation and Delta TechOps. (April 2014).

[5] ***"Enhancing Reliability with Process Compensated Resonance Testing at Delta TechOps,"*** ATA NDT Forum 2016, (September 28, 2016). D. Piotrowski, G. Weaver.
[6] ***ASTM E2001-13 Standard Guide for Resonant Ultrasound Spectroscopy***—Outlines capabilities and applications of several resonant inspection methods
[7] ***ASTM Standard Practice E2534-15***—Describes auditable method for application of PCRT Targeted Defect Detection inspection
[8] ***ASTM Standard Practice E3081-16***—Describes auditable method for application of PCRT Outlier Screening inspection
[9] ***Federal Aviation Administration Approved***—Since July of 2010 for the detection of micro-structural changes indicating over-temp of turbine blades (JT8D-219 HPT)

Non-Destructive Testing for Detection, Localization and Quantification of Damage on Composite Structures for Composite Repair Applications

Pratik SHRESTHA [1], Roger M. GROVES [1]
[1] Aerospace Non-Destructive Testing Laboratory, Faculty of Aerospace Engineering, Delft University of Technology, Delft, Netherlands
Contact e-mail: P.Shrestha@tudelft.nl

Abstract. Composite materials are being widely used for manufacturing aircraft components due to their superior material properties such as high strength, light weight, corrosion resistance, etc. However, compared to isotropic materials, composite materials exhibit complex damage characteristics. Moreover, when the composite material is impacted by a foreign object they are prone to barely visible impact damages such as delamination, matrix cracking, etc. Since composite materials are being increasingly used in aircraft component production the likelihood of composite damage occurrence during aircraft operation increases as well. Therefore, it is crucial to address the challenges associated with detecting composite damage and performing composite repairs. The focus of this research is the development of automated depot repair technology for composite structures, which combines; non-destructive testing (NDT) for damage size determination, damage removal by milling, repair by adhesive bonding of a repair patch and NDT for post repair assessment. In this study, a damaged curved CFRP panel with dimensions of 1.3 × 1.3 m was used for the development of algorithms for automated composite repair process. NDT using a laser line scanner was performed to acquire the composite panel's surface data, to assess features of the panel such as its shape, visible damage, etc., and the thermographic inspection was done to assess the extent and location of internal damage. Algorithms were developed to perform data fusion of the sensor data; a) to detect, localize, quantify and visualize the damage on the composite panel, through analysis of gradient changes between defined local sections of the panel, b) to generate a 3D model of the repair region based on the surface geometry and with design considerations that ensures the optimal structural integrity of the repaired panel, and c) to output suitable computer-aided design (CAD) files which can be imported to the milling tool, to perform the damage removal, and the CAD tool, to fabricate the repair patch. Finally, after the composite panel undergoes the milling and repair process, NDT inspections will be performed to ensure its safety and integrity.

Keywords: automated composite repair, NDT, laser line scanner, thermography.

1. Introduction

Aircraft structure damage can occur during flight or on the ground. Common causes are human error, impact from runway debris and bird strike. As the number of aircraft in operation has been increasing significantly [1], the frequency of aircraft structural damage event can also be expected to increase. Operational costs due to grounded aircraft can be significantly high; Ball et al reported that in the year 2007 fight delay cost airlines an estimated $32.9 billion [2]. As a result, an automated system capable of inspecting damage and for repairing damage to enable the earliest return of the aircraft to service can lead to large cost savings. The realization of an autonomous system capable of damage assessment for on-site repair involves considerable development of tools and algorithms for performing the desired on-site inspection and providing accurate data as input to the repair process.

The composite repair process involves damage assessment, which can be performed using several different types of sensors and non-destructive techniques (NDT), such as PZT sensors [3], shearography [4], phased arrays ultrasonic testing [5], thermography [6] and laser scanning [7] for inspecting visual and barely visible impact damages on composite structures [8]. Furthermore, a multi-sensor environment consisting of various NDT techniques can be used for obtaining data for feature extraction and then the use of data fusion [9] to yield accurate results for damage monitoring and diagnosis on aircraft structures. When a damage is detected on the composite structure, it can be repaired using the scarfing repair approach which involves removal of the damage region by machining the composite layers [10] and, subsequently, the repair can be completed by bonding the patch to the parent structure [11].

When an aircraft structure is damaged, an automated system that can inspect the damage to determine the extent of the damage, output the area that needs to be removed to repair the damage, perform repair and subsequently assess the repair is highly desired. Therefore, the topics covered in this paper include structural inspection using multiple NDT techniques, data fusion and analysis, and damage removal. The structure of the paper is as follows: a) in Section 2, the methodology, experimental set-ups and the algorithms developed to analyze the data acquired using the line-laser scanner and thermography inspection to automate the damage detection, localization and cut-out profile for damage removal is presented, b) in Section 3, the damage localization results and the cut-out profile model generated by the algorithms are presented, and c) the conclusions and future works are presented in Section 4.

2. Methodology

The methodology of the automated damage assessment and repair system involves inspection of the structure using multiple NDTs. After the data acquisition process, data fusion is done in order to perform damage analysis as consistently and accurately as possible. The data fusion and analysis will determine if any damage exists and if damage is detected then it will localize and quantify the damage on the structure. Subsequently, the file containing the damage parameter will be output so that the damage repair process can be performed. The damage repair process involves the damage removal and patch fabrication using the damage parameters provided by the damage analysis algorithm. After the damage removal is performed the patch will be bonded to the parent panel. Finally, the repaired structure will be accessed using the NDT. The automated damage assessment and repair methodology is presented in Fig. 1.

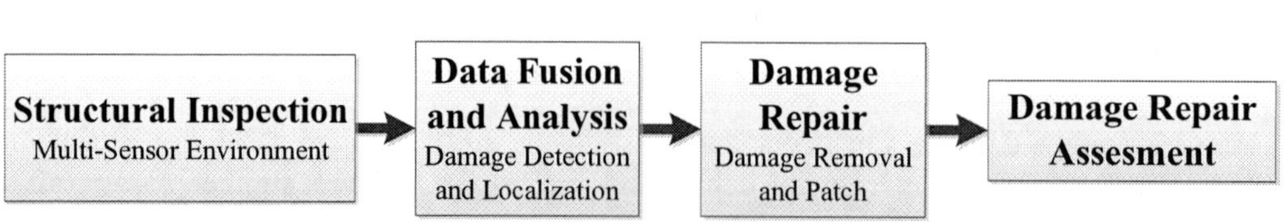

Fig. 1. Functions of automated damage assessment and repair system.

In this paper, the main research area focus is on the development of algorithms for data fusion and for the damage analysis process capable of assessing the data acquired by multiple NDTs used for the structural inspection. Moreover, the NDT techniques selection criteria for the data acquisition includes; 1) a suitable method for determining the structural geometry and damage information to create the cut-out profile for damage removal application and patch profile for patch fabrication process, and 2) an additional portable light-weight NDT technique for damage analysis in order to increase the consistency and accuracy of the damage analysis tool. The output provided by the algorithms will then be used for performing the damage removal. The experimental set-up and the damage detection and localization algorithms are presented in the following subsections.

2.1 Experimental Setup

The damaged curved CFRP panel, with dimensions of 1.3 × 1.3 × 0.0038 m and a radius of 2 m, was inspected using a laser line scanner and thermographic inspection. The laser line scanner was used to determine the composite panel's surface features such as its geometry and damage on the panel. A high-speed laser line scanner, Micro-Epsilon scanControl 2950-25, which has 2 μm z-axis resolution was fixed on the end of the arm of the Kuka robot, Kuka KR210, to scan the CFRP panel. The scan path was assigned by selecting reference points with respect to the panel's curvature to create the curved scan line path, to scan the entire CFRP panel. Subsequently, the KUKA robot arm's 3D position and angle on the scan path was determined to ensure that the laser line scanner was positioned normal to the surface of the CFRP panel, and the geometry of the CFRP panel was determined in the coordinate system of the manufacturing cell. The set-up for the line laser scan of the CFRP panel is shown in Fig. 2. The laser line scanner stores the acquired data in point cloud data (PCD) format, which consists of the x, y and z coordinates of each point scanned by the laser scanner. The PCD with coordinates of about 39.6 million points on the panel were acquired and stored for further processing.

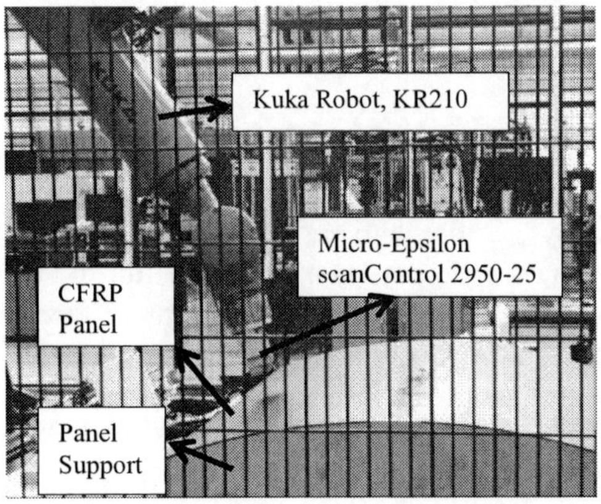

Fig. 2. CFRP panel scanning using the laser line scanner.

The PCD acquired using the laser line scanner provides the surface level information. Therefore, it is essential for determining the geometry of the structure and any defects on the surfaces. A portable thermography device, NDTherm NT (OPGAL), was used to determine the extent of the internal damage by TiaT Europe at TU Delft. As shown in Fig. 3, the infrared images were acquired using the NDTherm NT (OPGAL) placed on top of the region with visible damage of the CFRP panel.

The data acquired using the laser line scanner and the thermographic inspection were analyzed using the automated damage localization and quantification algorithms, developed at TU Delft (see Section 2.2), to determine the damaged area and to model the profile for damage removal. Finally, the output parameters from the algorithm were used as input parameters for the damage removal using the portable high precision 5 axis milling machine, MobileBlock device, provided by DMG MORI SAUER Ultrasonics GmbH. In order to perform the damage removal from the CFRP panel, the MobileBlock was placed on top of the CFRP panel and locked to the panel using the suction legs. The experimental set-up for the damage removal process is shown in Fig. 4.

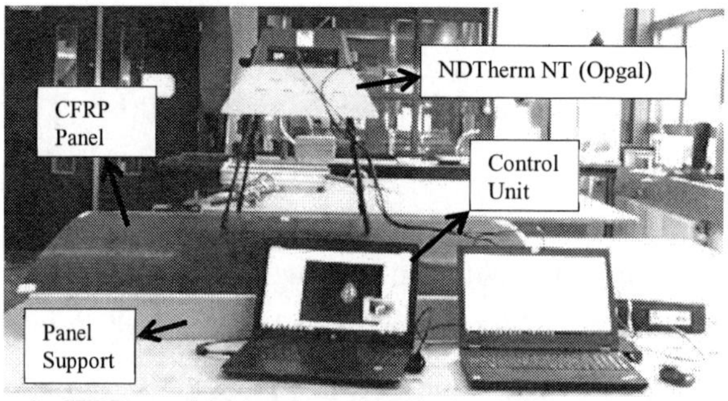

Fig. 3. CFRP panel inspection using the NDTherm NT (OPGAL).

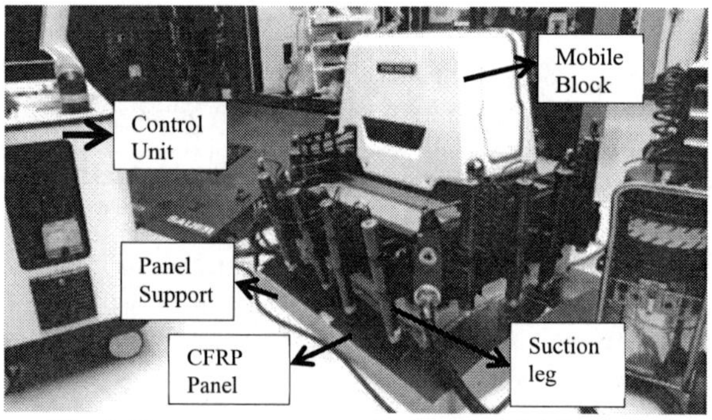

Fig. 4. Damage area removal using the MobileBlock.

2.2 Algorithm overview

The data acquired after the structural inspection was imported to the automated damage detection and localization algorithm for; 1) data pre-processing of the laser line scanner PCD data and infrared image, 2) damage detection and quantification and 3) providing the geometry and location of the scarf repair as input to the milling process. An overview of the algorithm is illustrated in Fig. 5. The first step of the algorithm involves pre-processing of the data acquired using the laser line scanner and thermographic inspection, in which; 1) the algorithm determines the geometry information of the curved CFRP panel based on the data acquired using the laser line scanner, 2) the panel's geometry data is used to reconstruct the 3D model of the curved CFRP panel, and 3) subsequently, the reconstructed model of the CFRP panel is then used to transform the pixel information of the infrared image to the coordinate system of the PCD of the CFRP panel obtained by the laser line scanner.

The laser line scanner and infrared image PCDs were analyzed independently to determine the damaged area. The damage length on the PCD obtained using the laser line scanner was determined by analyzing the gradient change in the z-axis by scanning the PCD in the x-direction and y-direction at x-interval and y-interval, respectively. The infrared image analysis for damage determination was also done by analyzing the pixel data by scanning the PCD in the x-direction and y-direction at specified intervals. After the damage localization procedure, the damaged areas are enclosed using an ellipse, which is created using the minimum and maximum x-values and y-values of the damage region. The ellipse covering the damaged area of the laser line scanner PCD and the infrared image PCD are compared to check if they are in agreement.

Finally, the damage region determined based on the results obtained from the laser line scanner and thermography image analysis is used for modeling the cut-out for milling the CFRP panel. The cut-out profile is modeled so that the size of the ellipse at the bottom of the panel is 15 mm apart from the detected damage. Additionally, a commonly used scarf ratio [12], 1:20, was selected for modeling the cut-out profile for bonded repair applications. Finally, the milling profile parameters are output in .stl format by the algorithm. In the last step, the output parameters of the scarf profile are used as an input for damage removal using the MobileBlock.

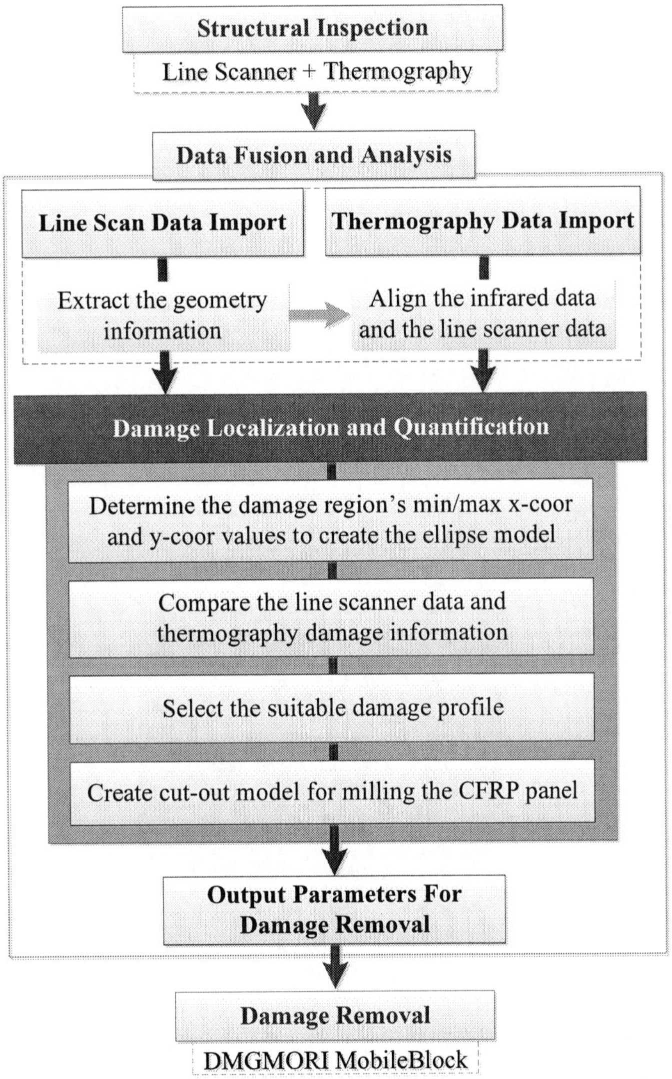

Fig. 5. Overview of the damage detection and localization algorithm.

3. Results

The PCD acquired after the entire panel was scanned using the laser line scanner is presented in Fig. 6 (a). The damage information on the PCD is represented by deviations in the z-value and missing point cloud elements, as shown in Fig. 6 (b). The infrared image acquired through the thermographic inspection is presented in Fig. 7. The white and the black regions, corresponding to hottest and coldest region, can be seen on the infrared image because of the uneven distribution of temperature due to the presence of the damage. These two data sets were imported to the automated damage localization and quantification algorithms for further analysis.

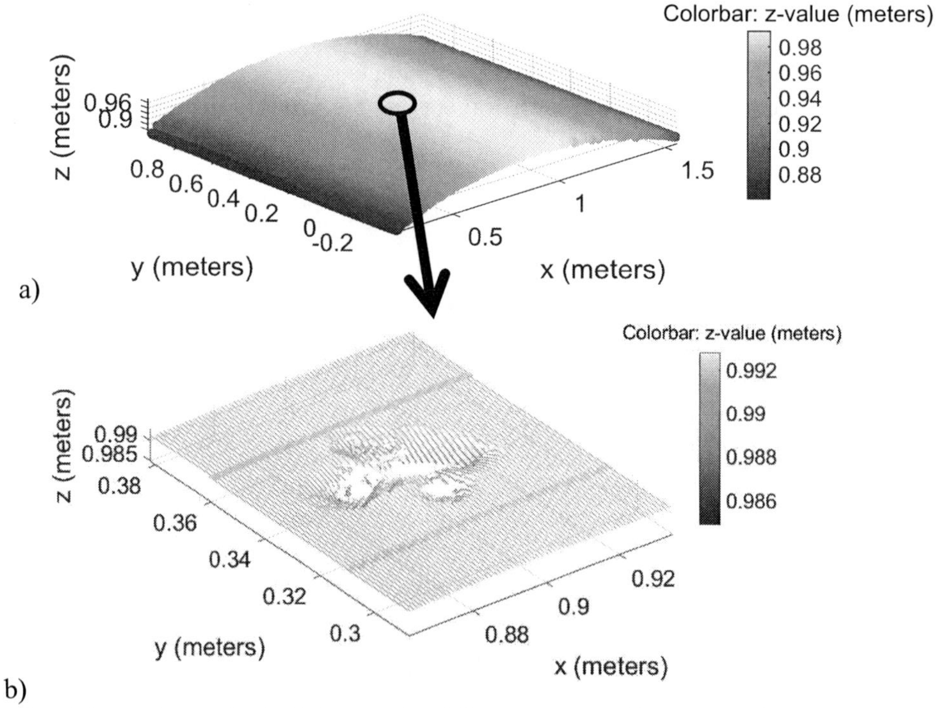

Fig. 6. a) Visualization of the PCD of the entire CFRP panel acquired by the laser line scanner and b) close-up of the damage region.

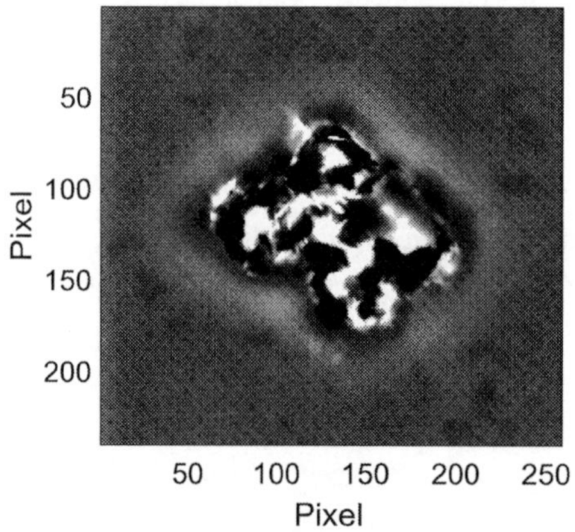

Fig. 7. Infrared image acquired using the NDTherm NT (OPGAL).

The damage localization algorithm was used to determine the damage location and size by analyzing the laser line scanner data and the infrared image. In case of the laser line scanner, the algorithm analyzed the PCD that lie between x-values of 0.7505 m and 1.049 m, and y-values of 0.24 m and 0.44 m. Whereas, the data range analysis for the infrared image was performed for: x-values between 0.8473 m and 0.9459 m, and y-values between 0.2977 m and 0.3794 m. Fig. 8 (a) shows the damage region, with a maximum x-length of 48.9 mm and a maximum y-length of 41.1 mm, detected by analyzing the PCD obtained from the laser line scanner. The algorithm then performed the conversion of the pixel information obtained using the thermographic inspection to PCD information with respect to the laser line scanner PCD information, as shown in Fig. 8 (b). In Fig. 8 (c), the damage region detected by analyzing the infrared image is compared with the results obtained by analyzing the laser line scanner data. Moreover, it can be seen that the thermographic inspection detects a larger damage area than the laser line scanner, as the damage detection algorithm determines the damage to be 22.4 mm and 12.3 mm longer in the x and y-directions, respectively.

Furthermore, since the damage location detected and quantified by these two different NDT techniques overlaps with each other, the certainty of the detected damage location is increased. As a result, based on the damage information determined by the algorithm the maximum damage length in the x-direction and y-direction is selected to model the cut-out profile for damage removal, as shown in Fig. 8 (d): with ellipses of maximum x-length of 253.2 mm at the top surface and 101.6 mm at the bottom surface of the panel, and maximum y-length of 235.4 mm and 83.8 mm at the top and bottom surface, respectively. The parameters output by the algorithm was then used for the damage area removal as shown in Fig. 9.

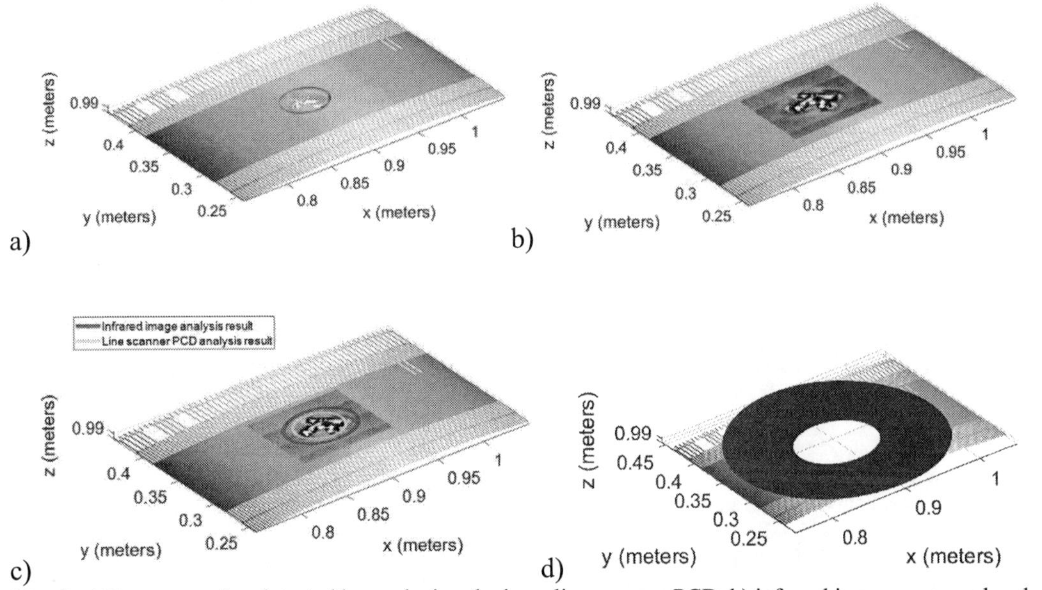

Fig. 8. a) Damage region detected by analyzing the laser line scanner PCD, b) infrared image processed and converted to the laser line scanner PCD coordinate system, c) infrared image and laser line scanner based damage region comparison and d) cut-out profile model generated by the algorithm.

Fig. 9. Damage removed using the MobileBlock.

4. Conclusions and Future Works

In this paper, the algorithms developed for processing PCD and infrared images for automated damage detection and localization was presented. The developed algorithm was used to detect, localize, quantify, visualize the damage, and to output the cut-out profile parameters to the milling machine. Moreover, the presented work covers the composite structure monitoring, data analysis, and damage removal tasks. The subsequent tasks in the project; inspection of the damage removal performed by the MobileBlock, composite repair by bonding the patch to the CFRP panel and repair assessments remains to be performed to determine the suitability of the performed repair work. Furthermore, the developed algorithm needs to be further tested for performing the automated repair works in order to realize a fully automated repair system.

Acknowledgements

This research was supported by the Defense Technology Project 'Development of Technology for Depot Repairs of Composite Structures', SID:11691, supported by the Royal Netherlands Air Force. The authors would like to thank Mr. Rik Tonnaer, TU Delft, for his assistance with KUKA robot, Mr. Pieter Troost, TiaT Europe BV, for his assistance with thermographic inspection, and DMG MORI SAUER Ultrasonics GmbH for their assistance with the MobileBlock machine.

References

[1] Airbus, "Global Market Forecast 2018-2037," 2018.

[2] M. Ball *et al.*, "Total delay impact study : a comprehensive assessment of the costs and impacts of flight delay in the United States," Washington, D.C, 2010.

[3] P. Ochôa, V. Infante, J. M. Silva, and R. M. Groves, "Detection of multiple low-energy impact damage in composite plates using Lamb wave techniques," *Compos. Part B Eng.*, vol. 80, pp. 291–298, 2015.

[4] Y. Y. Hung, "Shearography for non-destructive evaluation of composite structures," *Opt. Lasers Eng.*, vol. 24, no. 2, pp. 161–182, 1996.

[5] A. McNab and M. J. Campbell, "Ultrasonic phased arrays for nondestructive testing," *NDT Int.*, vol. 20, no. 6, pp. 333–337, 1987.

[6] A. P. Chrysafi, N. Athanasopoulos, and N. J. Siakavellas, "Damage detection on composite materials with active thermography and digital image processing," *Int. J. Therm. Sci.*, vol. 116, pp. 242–253, 2017.

[7] L. Laflamme, S, Turkan, Y and Tan, "Bridge structural condition assessment using 3d imaging," in *Proc. 2015 Conf. on Autonomous and Robotic Construction of Infrastructure*, 2015.

[8] T.-W. Shyr and Y.-H. Pan, "Impact resistance and damage characteristics of composite laminates," *Compos. Struct.*, vol. 62, no. 2, pp. 193–203, 2003.

[9] S.-F. Jiang, C.-M. Zhang, and S. Zhang, "Two-stage structural damage detection using fuzzy neural networks and data fusion techniques," *Expert Syst. Appl.*, vol. 38, no. 1, pp. 511–519, 2011.

[10] E. Paquet, S. Garnier, M. Ritou, B. Furet, and V. Desfontaines, "Implementation of a new method for robotic repair operations on composite structures BT - Advances on Mechanics, Design Engineering and Manufacturing : Proceedings of the International Joint Conference on Mechanics, Design Engineering & Advanced Manufact," B. Eynard, V. Nigrelli, S. M. Oliveri, G. Peris-Fajarnes, and S. Rizzuti, Eds. Cham: Springer International Publishing, 2017, pp. 321–328.

[11] A. Baker, "Bonded composite repair of fatigue-cracked primary aircraft structure," *Compos. Struct.*, vol. 47, no. 1, pp. 431–443, 1999.

[12] D. Holzhüter, A. Pototzky, C. Hühne, and M. Sinapius, "Automated Scarfing Process for Bonded Composite Repairs BT - Adaptive, tolerant and efficient composite structures," M. Wiedemann and M. Sinapius, Eds. Berlin, Heidelberg: Springer Berlin Heidelberg, 2013, pp. 297–307.

Simultaneous multi-frequency dielectric measurement technique for fast curing reactions

Steffen GEINITZ [1], Maximilian EBERHARDT [1], Dominik WALSER [1], Dominik GRUND [1], Bernhard HOLLAUS [2], André WEDEL [1]
[1] Fraunhofer-Einrichtung für Gießerei-, Composite- und Verarbeitungstechnik IGCV, Augsburg, Germany
[2] MCI MANAGEMENT CENTER INNSBRUCK, Innsbruck, Austria
Contact e-mail: maximilian.eberhardt@igcv.fraunhofer.de

Abstract. Dielectric Analysis has proven to be an important tool in various industry and research sectors. But new developments, in particular, highly reactive thermosetting resins and adhesive systems with UV radiation activation reveal limitations of commercially available measurement systems. These systems achieve dynamics which reach a conversion rate of up to 90% in less than one second.

In dielectric analysis the material properties are determined by measuring the reaction current and the corresponding phase angle as a result of an input voltage signal. By determining the complex impedance and phase angle, different material properties, such as relative permittivity or ionic conductivity, can be calculated to evaluate the material conditions. In order to be able to adequately evaluate manufacturing processes and curing processes, a broadband measurement of the electric variables mentioned before is necessary. Unfortunately, commercially available systems achieve spectroscopic results only by the use of a frequency sweep, which is not applicable to monitor properly the behaviour for highly dynamic UV-curable resins. Likewise, this is similar for rheological measurement methods, which are equally limited in their measurement point rate and dependent on the applied excitation frequency.

To attain sufficient resolution in the time domain even for highly reactive systems, a system which offers the possibility of spectroscopic examinations combined with an extremely high measuring point density was developed. The actual simultaneous recording of all applied frequencies (up to 20 different ones) reduces significantly the measurement time per measurement point. Thus creating the potential to analyze extremely dynamic reactive systems in detail even at low frequencies. Limits in speed are only given by the lowest frequency monitored. In addition, the measurement points are recorded time-synchronously for all frequencies, allowing immediate spectroscopic analysis. Overall, this opens up the potential, especially in the area of UV-curing highly reactive resin systems, both in the laboratory and in the process environment, to realize material characterization and condition monitoring.

Introduction

Dielectric analysis (DEA) has become a popular tool not only in science but in industry as well. Nevertheless, it has long been used to investigate dielectric dispersion phenomena for colloidal solutions [1, 2] and still is a standard method for this purpose [3]. In biology, the

use of dielectric analysis for the evaluation of dielectric behavior of the human or any other biological tissue and the interactions on a cellular level can be found in literature [4-6].
For some time, the use of dielectric analysis for industrial applications has become more common. In the aerospace industry, the method was used to monitor the curing process of fiber reinforced plastics within the resin transfer molding (RTM). In [7] a process control device based upon the dielectric analysis for an epoxy based resin is presented. Shepard et al. [8] demonstrate the feasibility for the cure monitoring and process control via DEA for polyester based sheet molding compound (SMC) respectively for bulk molding compound. More recently, the DEA was used to examine the curing behavior of a b-stage resin [9] and the phase change in neat as well as fiber reinforced PA6 [10].
An interesting application is the use of DEA for ultraviolet (UV) curable resins. Only little literature can be found investigating the conversion behavior at low frequencies [11-14]. The main reason are the fast curing cycles which can be less than one second. Therefore spectroscopic data is difficult to obtain. Hence, the standard procedure monitoring fast conversions for UV-curable resins was demonstrated by a solution based upon IR spectroscopy [15].
In this paper the concept for a device capable of quasi-spectroscopic monitoring for low frequencies using DEA is presented. With this approach, the use of DEA for highly reactive UV-curing systems shall be demonstrated obtaining synchronous data for up to 20 discrete frequencies within a bandwidth of 4 decades.

1. Methods

In the 1880's, Oliver Heaviside first introduced impedance spectroscopy (IS), thoroughly discussed in [16]. The basic idea behind IS is to apply an electrical (voltage or current) stimulus (periodically or non-periodically) to a specimen and observe the resulting signal. Usually the relations between the specimen properties and the response signal caused by periodic excitation is difficult to formulate within the time domain, often involving a complex solution of a system of differential equations. In this respect the Fourier transformation allows significant simplification by transferring the problem to the frequency domain. This allows a simplified calculation of the impedance magnitude by measuring the current and voltage signal, and subsequently calculating the phase difference between them for arbitrarily selected frequencies.
This more general concept is the basis for the DEA. Even though it is counted to the thermal analysis methods it has a special status among these. Classical thermal measurement techniques like differential scanning calorimetry (DSC) or rheological investigation, focus on the chemical conversion rate, respectively the mechanical stress within a specimen. Contrarily DEA examines the interactions between ions and molecules due to a changing electric field between the electrodes of a sensor. An excitation voltage is applied to the test specimen. By measuring the phase shift and amplitude of the current response signal and the known geometrical parameters of the used sensor, the conductivity, permittivity and further derived quantities can be obtained. These properties may be represented as a function of frequency, temperature and/or time, depending on the conducted experiment.

1.1 Basic schematic circuit layout

In the literature different approaches for the electrical circuit layout for the measurement and processing of the created voltage and current signals can be found, each with its own benefits and purpose [17, 18].
The chosen approach features a transimpedance amplifier (TIA) for the current-to-voltage conversion. The corresponding circuit diagram is shown in Fig. 1. The excitation of the

capacitive sensor drives an input current into the inverting input of the op amp. The output of this circuit is an amplified and inverted current-proportional voltage signal according to the formula $V_{out} = I_{in} \cdot Z$. Similar circuit designs for different applications have been introduced by [18], [19] and [20].

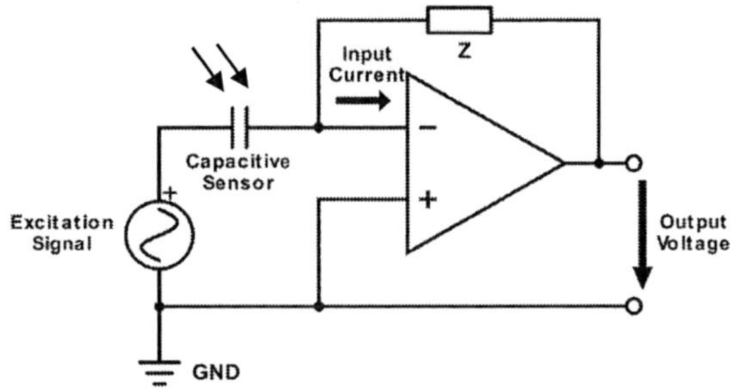

Fig. 1: Schematic representation of the basic circuit layout featuring a transimpedance amplifier.

1.2 Signal

Most commercial systems use pure sine waves as excitation signals as they provide the best signal quality [21]. For a spectroscopic analysis a frequency sweep needs to be conducted. For slow curing resins this method has proven great success as the changes in material properties proceed at an even slower rate than the sweep time. However, for highly reactive UV-curable acrylate based materials the curing cycles might be shorter than a full frequency sweep over the chosen bandwidth. Therefore another excitation signal is needed.

Recently, multi-frequency simultaneous (MFS) measurement techniques have become more popular. Using broadband excitation signals comprised of multiple discrete frequencies and evaluating the signal response by means of spectral analysis applying the fast Fourier transformation (FFT), the time to obtain the complete IS result is dramatically reduced. Different signal types used for MFS measurements have been reported in literature, e.g. binary sequences as maximum length binary sequences (MLBS), discrete interval binary sequences (DIBS) [22, 23], sine based signals as chirp [24] or multi-sine signals [25, 26].

Each signal comes with certain advantages as well as disadvantages and has to be carefully chosen for the intended application. The signal used in the presented work is a multi-sine, combining the possibilities of a simultaneous capturing of several discrete frequency components with a satisfying signal quality for a real-time monitoring application [21]. The signal is generated by adding up multiple sine waves, while each wave can be chosen with its particular frequency f_n, amplitude a_n and phase φ_n according to the following equation:

$$u(t) = \sum_{n=1}^{N} a_n \cos(2\pi f_n t + \varphi_n) \qquad (1)$$

Metrics are needed to evaluate the quality of the generated signals and the corresponding impedance spectra. To determine the signal quality in the time domain the crest-factor (CF) is used to determine how much signal amplitude is needed to introduce a certain amount of energy into a system. The CF is the ratio between the peak value of a signal and the effective root-mean-square value of the signal [21]. For a signal *s(t)* measured over a time interval [0; T] the CF is calculated by:

$$CF(s) = \frac{\max_{t\epsilon[0,T]} |s(t)|}{\sqrt{\frac{1}{T}\int_0^T |s(t)|^2 dt}} \qquad (2)$$

Complimentary to the crest-factor, the root mean square (RMS) voltage of the multi-sine for a limited voltage level is used as a metric for the energy density as proposed by [27]. For

further insight, the signal-to-noise ratio (SNR) is applied. The SNR is the ratio between the power of a signal and the power of present background noise.
The more frequencies are incorporated in the multi-sine signal, the better the spectral resolution of the impedance spectrum. However, the amount of energy being introduced to the system for a single frequency component is inversely proportional to the number of excited frequencies N. It was found during the research of this project that for a satisfying frequency resolution at least 10 different frequencies are needed. For N > 25, the single amplitudes, especially for the low frequencies, drop below a level where they cause a measurable effect in the specimen as the maximum value of the applied voltage signal was limited to 10 V_{PP}. A number of 20 frequencies has proven to be a good compromise between spectral resolution and sufficient energy input. The signal was additionally optimized in terms of amplitude, phase angle and frequency distribution.
The parameters indicate that the combination of a second order power function and phase offsets calculated according to Newman give the best result with the lowest CF (3.32) and simultaneously the highest RMS voltage (1.5091 V). Fig 2 shows the corresponding FFT spectrum of the created signal. Due to the repeating nature of the signal higher frequency components appear more often than low frequency components. To keep the energy input for each frequency component at a similar level, the amplitude decreases with rising excitation frequency.

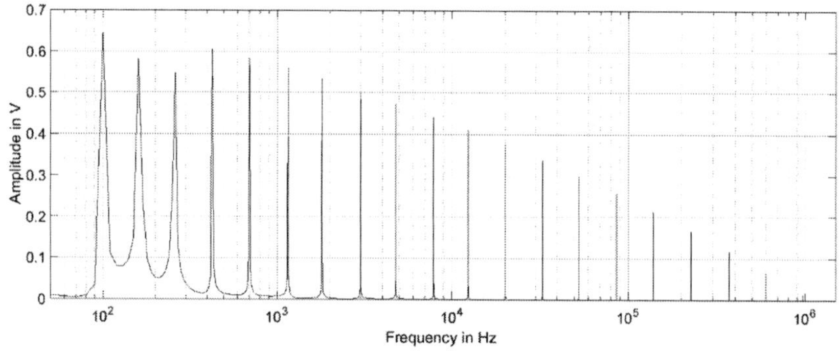

Fig. 2: FFT spectrum of the created multi-sine signal with decaying amplitudes.

1.3 Analysis

For the interpretation of the results, sensor parameters as the A/D ratio and the substrate capacitance C_{sub} are needed. Both parameters are unique to the used sensor. The A/D ratio is obtained by the data sheet while the substrate capacitance is measured and calculated according to [30]. The system specific gain and phase information for the TIA as well as the following analog filtered outputs is determined by a circuit simulation using SPICE software. Additionally, information about the used excitation frequencies is necessary for the calculation of the characteristic values later on and for the interpolation of the simulation data to fit the values of interest in the data analysis.
In terms of measurements, the excitation voltage signal and the converted current proportional voltage signals are necessary for the analysis. These signals, so far only time dependent, are Fourier transformed to gain access to the frequency related information. Further the impedance Z, respectively its inverse, the admittance Y, as well as the phase angle θ between current and voltage are calculated. From these two fundamental values, Z and θ, material specific parameters, among others conductance and capacitance are derived. This procedure is repeated for every single time sample. To assess the quality and information density of the acquired data the SNR is determined from the FFT data. As soon as the last data set is analyzed, the data is sent to plots for graphical interpretation of the results.

2. Experimental

2.1 Materials and sample preparation

During this investigation, a commercially available UV-radiation curable system provided by DELO Industrie Klebstoffe GmbH & Co. KGaA (DELO®) was examined. The type used within this research was the standard DELO® Photobond® PB437 (PB437). The PB437 is a one-component and solvent-free UV- and light curing adhesive based on a modified urethane acrylate. Typical fields of application are the bonding of plastics (PMMA, PC) but also glass or metal. Detailed information can be found on the DELO® website.
To avoid interference, the sensing electrodes were cleaned upfront using acetone and isopropanol. Any remaining solvent was removed by heating up the sensing electrodes to fully vaporize it. The use of a planar interdigitated sensor for the measurement requires the thickness of the sample to be greater than the spacing between the electrodes in order to avoid measurements of air or material on top of the sample. Further, only shielded wires were used as long and unshielded wires pick up electrical interference and noise from nearby devices.

2.2 Test setup

The experimental investigations were conducted using three different methods, differential scanning calorimetry (DSC), shear rheology and the proposed simultaneous multi-frequency dielectric measurement technique. The first two methods are used to provide comparability of the measurement results to commonly used material characterization techniques and demonstrate the capabilities of the developed measurement system.

2.2.1 Differential Scanning Calorimetry (DSC)

The DSC measurements were performed with a Netzsch DSC 204 Phoenix photo-DSC setup. This setup consists of a light source (OmniCure S2000SC) and fiber optics which are connected to the DSC furnace. The optics are placed into the top cover of the furnace to ensure reproducible distances between sample and reference. Samples with masses between 9 – 13 mg of the adhesive were placed into aluminum pans and both, sample and reference, were irradiated simultaneously. During this procedure, the sample temperature and the heat flow difference were recorded.

2.2.2 Rheology

Oscillating rheology measurements were conducted using an MCR302 rheometer from Anton Paar GmbH. The MCR302 is equipped with a parallel plate setup. The bottom plate consisted of a highly UV penetrable glass plate. As a light source, the same OmniCure S2000SC and fiber optics were used as in the DSC and DEA measurements. The measurements were conducted within the linear region of the storage modulus, which allows only for a maximum measurement frequency of 1 Hz.

2.2.3 Multi-frequency dielectric measurements

For the dielectric measurements the interdigitated sensors were placed in an aluminum pan and fully covered with the PB437 adhesive. As radiation source the OmniCure S2000SC was used with the fiber optics placed above the resin in the same distance as for the DSC and rheology measurements.

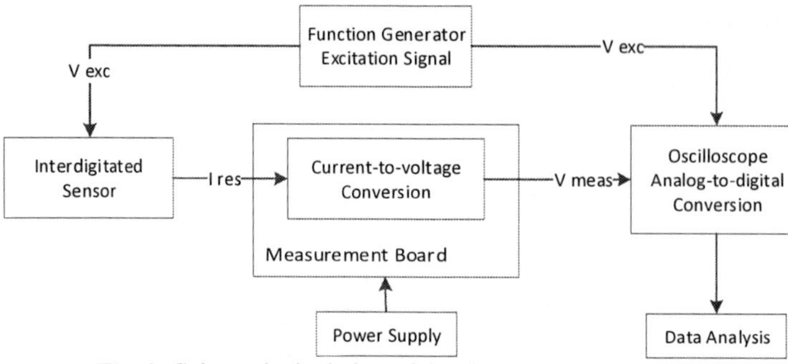

Fig. 3: Schematic depiction of the signal processing chain

Fig. 3 shows the schematic signal processing chain. The signal chain starts at the function generator (Agilent Trueform Series Waveform Generator 80 MHz), which provides the excitation signal V_{exc}. The excitation voltage causes a response current I_{res} in the capacitive interdigitated sensor. This current is subsequently converted into a voltage by the circuitry on the measurement board. The voltage V_{meas} is then forwarded to an oscilloscope (Keysight Infiniium S Series 500 MHz) where the analog current-proportional signal is converted to digital information for further data analysis. The same analog-to-digital conversion is performed for the excitation voltage V_{exc} that is measured at the function generator.

3. Results

Fig. 4 shows the calculated results of the electrical impedance generated with the developed system as a function of frequency *f* and time *t*. From the impedance data as a function of (f, t) and the phase angle as a function of (f, t) the relative permittivity ε' and the loss factor ε'' can be derived according to [30]. In addition, the dissipation factor *tan* δ, the ionic conductivity ν and its inverse, the ionic viscosity σ, can be calculated as well. This data is more appropriate for further material analytics than only the purely electrical parameters impedance and phase angle.

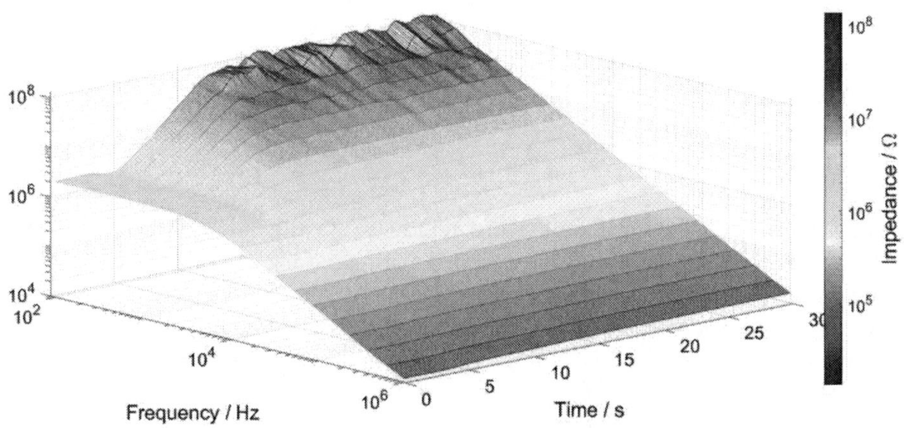

Fig. 4: Impedance change during the UV radiation activation of the PB437 test sample. The activation starts after 4 seconds as can be derived from the abrupt impedance change.

Due to this representation, the impedance changes at higher frequencies are hard to determine. Nevertheless, the signal to noise ratio is much better at higher frequencies and the relative change for the derived quantities is in the same order of magnitude.

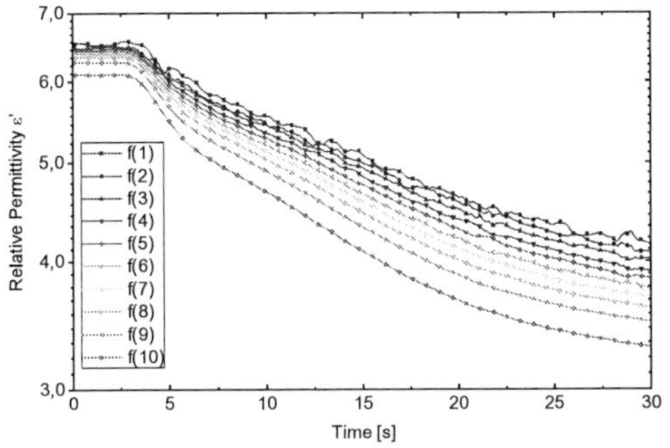

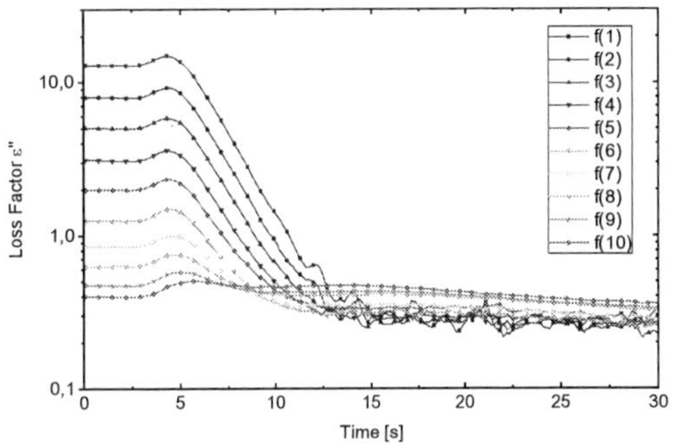

Fig. 5: Relative permittivity *ε'* for several frequencies. All frequencies were recorded simultaneously.

Fig. 6: Loss factor *ε''* for several frequencies. All frequencies were recorded simultaneously.

Fig. 5 and Fig. 6 show the relative permittivity *ε'* and the loss factor *ε''* for 10 different frequencies. With the proposed system, 20 frequencies were obtained at once ranging from 90 Hz to 1 Mhz. For better representation, however, only 10 frequencies logarithmically spaced between 5 kHz (*f(1)*) and 380 kHz (*f(10)*) were selected.

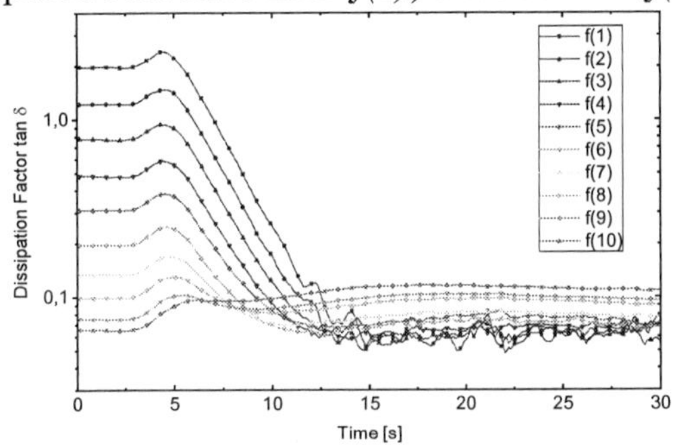

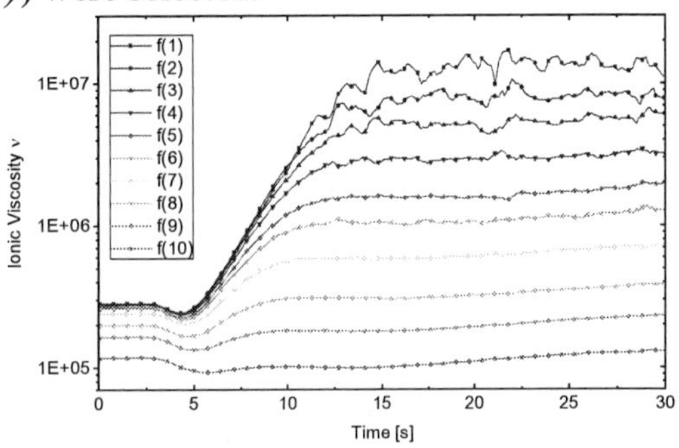

Fig. 7: Dissipation factor *tan δ* for several frequencies. All frequencies were recorded simultaneously.

Fig. 8: Ionic viscosity ν for several frequencies. All frequencies were recorded simultaneously.

As can be seen from Fig. 5 depending on the frequency different behavior within the reaction can be observed. While the behavior of the relative permittivity is constant for all frequencies monitored, except for an offset between the different frequencies. Nevertheless, the progression of the offset can be an indicator, as well being now accessible for further analysis. Fig. 7 and Fig 8 show the dissipation factor *tan δ* and the ionic viscosity *ν* for the same 10 frequencies used in the figures above. As the dissipation factor is *ε''* divided by *ε'* the curve characteristics show similarities to Fig. 5 and Fig. 6. In Fig. 8 the frequency dependency of the ionic viscosity of the material becomes quite obvious.
This data was then used to compare the proposed monitoring system with the rheology and DSC measurement. In Fig. 9 the heat flow measured with the DSC, the storage modulus from the rheology measurements and the loss factor *ε''* for 10 different frequencies are displayed. With *f(1)* at a low frequency of 5 kHz and *f(10)* with 380 kHz, the other frequencies are equally distributed. Each mark represents a distinct data point. The amount of data captured by the rheology setup is limited to the linear region of the complex viscosity, which is up to 1 Hz for the PB437. The heat flow from the DSC setup represents an integrated signal of the energy converted within the reaction. This allows a good distinction of different reactions if they take place after one another or differ strongly in time.
The system presented in this paper offers additional information as the short signal rise directly after the start of UV radiation, which could be reproduced for all other measurements of the PB437. Further, the data obtained from the DEA yield a more distinct representation of the reactions during the curing process, compared to rheology and DSC.

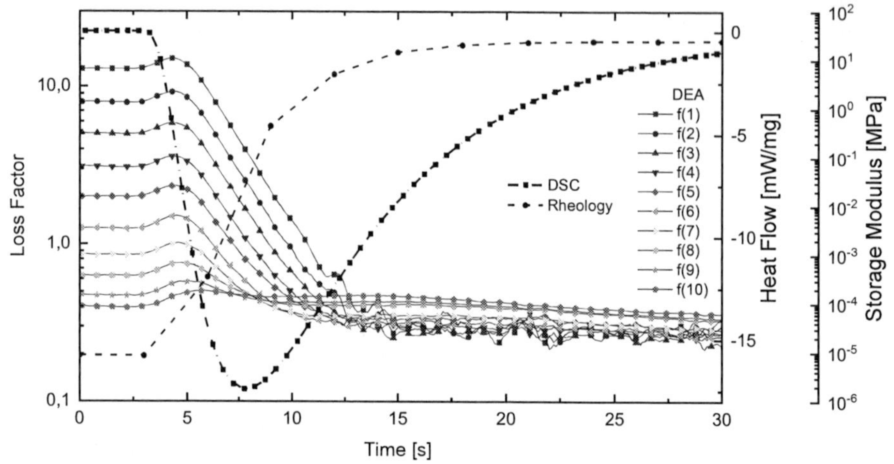

Fig. 9: Comparison dielectric measurements for 10 different frequencies with the storage modulus from the rheology and the heat flow from the DSC.

4. Conclusion

The presented device has proven its feasibility for the spectroscopic monitoring of reactive UV curing resins. Now the frequency dependent behavior becomes accessible in real-time. Because 20 data points over the entire bandwidth of the system are captured time synchronously a curve for each time step can be created. The lowest frequency used to monitor the reaction limits the time resolution. Therefore, the proposed system overcomes the limitations typically accompanied by the standard DEA systems commercially available. The same is true for commercial vector network analyzers or LCR-meters, which both use frequency sweeps for spectroscopic analysis.

The time synchronous data opens up new possibilities of reaction monitoring, e.g. to determine the conversion of reaction partners by the use of the frequency dependent permittivity.

Further development of the system will focus on the extension of the measuring range and a miniaturization of the system to avoid additional equipment for signal generation and acquisition. Additional signal processing optimizations are ongoing and other fields of application are currently considered.

5. Acknowledgments

This research was supported by the administration of Swabia and the Bavarian State Ministry for the Economy, Media, Energy and Technology. Further thanks goes to the Company DELO Industrie Klebstoffe GmbH & Co. KGaA for their support with material samples.

References

[1] H. P. Schwan, G. Schwarz, J. Maczuk u. H. Pauly: On the low-frequency dielectric dispersion of dolloidal particles in electrolyte solution. J. Phys. Chem. 66 (1962) 12, S. 2626–2635

[2] G. Schwarz: A theory of the low-frequncy dielectric dispersion of colloidal particles in electrolyte solution. J. Phys. Chem. 66 (1962) 12, S. 2636–2642

[3] P. J. Beltramo, R. Roa, F. Carrique u. E. M. Furst: Dielectric spectroscopy of concentrated colloidal suspensions. Journal of Colloid and Interface Science 408 (2013), S. 54–58

[4] C. Gabriel, S. Gabriel u. E. Corthout: The dielectric properties of biological tissues: I. Literature survey. Physics in Medicine & Biology 41 (1996) 11, S. 2231

[5] S. Gabriel, R. W. Lau u. C. Gabriel: The dielectric properties of biological tissues: II. Measurements in the frequency range 10 Hz to 20 GHz. Physics in Medicine & Biology 41 (1996) 11, S. 2251

[6] S. Gabriel, R. W. Lau u. C. Gabriel: The dielectric properties of biological tissues: III. Parametric models for the dielectric spectrum of tissues. Physics in Medicine & Biology 41 (1996) 11, S. 2271

[7] D. E. Kranbuehl, P. Kingsley, S. Hart, G. Hasko, B. Dexter u. A. C. Loos: In situ sensor monitoring and intelligent control of the resin transfer molding process. Polymer Composites 15 (1994) 4, S. 299–305

[8] D. D. Shepard, D. R. Day u. K. J. Craven: Applications of Dielectric Analysis for Cure Monitoring and Control in the Polyester SMC/BMC Molding Industry. Journal of Reinforced Plastics and Composites 14 (1995) 3, S. 297–308

[9] A. Chaloupka, R. Bezerra, V. Madaksira, I. Taha u. N. Rudolph: Detection and modelling of thermal and rheological transitions of a 2-step-curing thermoset using dielectric and standard measuring techniques. In: 20th International Conference on Composite Materials. Copenhagen 2015

[10] A. Chaloupka, A. Wedel, I. Taha, N. Rudolph u. K. Drechsler: Phase Change Detection in Neat and Fiber Reinforced Polyamide 6 Using Dielectric Analysis. Materials Science Forum 825-826 (2015), S. 944–951

[11] M. Rosentritt, M. Behr, S. Knappe u. G. Handel: Dielectric analysis of light-curing dental restorative materials—a pilot study. Journal of Materials Science 41 (2006) 10, S. 2805–2810

[12] UV & EB Technical Conference Proceedings. RadTech International 2004

[13] K. Resch, G. M. Wallner u. R. W. Lang: Cross-linking behavior of UV-curable acrylate resins. In: Journal of Plastics Technology. Carl Hanser Verlag 2008

[14] K. Zahouily, C. Decker, E. Kaisersberger u. M. Guener: Cure Monitoring of UV-Curable Free Radical And Cationic Systems by Using In-situ Dielectric Analysis and Real-Time FT-Infrared Spectroscopy. In: UV & EB Technical Conference Proceedings. RadTech International 2004

[15] C. Decker u. K. Moussa: A new method for monitoring ultra-fast photopolymerizations by real-time infra-red (RTIR) spectroscopy. Die Makromolekulare Chemie 189 (1988) 10, S. 2381–2394

[16] D. D. Macdonald: Reflections on the history of electrochemical impedance spectroscopy. Electrochemical Impedance Spectroscopy 51 (2006) 8, S. 1376–1388

[17] J. D. Menczel u. R. B. Prime (Hrsg.): Thermal Analysis of Polymers: Fundamentals and Applications. Wiley 2009

[18] J. Schröder, S. Doerner, T. Schneider u. P. Hauptmann: Analogue and digital sensor interfaces for impedance spectroscopy. Measurement Science and Technology 15 (2004) 7, S. 1271

[19] S. Kaufmann, A. Malhotra, G. Ardelt u. M. Ryschka: A high accuracy broadband measurement system for time resolved complex bioimpedance measurements. Physiological Measurement 35 (2014) 6, S. 1163

[20] B. Sanchez, G. Vandersteen, R. Bragos u. J. Schoukens: Basics of broadband impedance spectroscopy measurements using periodic excitations. Measurement Science and Technology 23 (2012) 10, S. 105501

[21] B. Sanchez, G. Vandersteen, R. Bragos u. J. Schoukens: Optimal multisine excitation design for broadband electrical impedance spectroscopy. Measurement Science and Technology 22 (2011) 11, S. 115601

[22] S. Gawad, T. Sun, N. G. Green u. H. Morgan: Impedance spectroscopy using maximum length sequences: Application to single cell analysis. Review of Scientific Instruments 78 (2007) 5, S. 54301

[23] Y. Yang, W. Zhang, F. Du, X. Tang, H. Wen u. Z. Teng: Broadband Bioimpedance Spectroscopy Based on a Multifrequency Mixed Excitation and Nuttall Windowed FFT Algorithm. Mathematical Problems in Engineering 2014 (2014), S. 9

[24] M. Nahvi u. B. S. Hoyle: Electrical Impedance Spectroscopy Sensing for Industrial Processes. IEEE Sensors Journal 9 (2009) 12, S. 1808–1816

[25] J. Ojarand u. M. Min: Efficient Excitation Signals for the Fast Impedance Spectroscopy. Electronics and Electrical Engineering 20 (2014)

[26] B. Sanchez, X. Fernandez, S. Reig u. R. Bragos: An FPGA-based frequency response analyzer for multisine and stepped sine measurements on stationary and time-varying impedance. Measurement Science and Technology 25 (2014) 1, S. 15501

[27] J. Ojarand, P. Annus, M. Min, M. Gorev u. P. Ellervee (Hrsg.): Optimization of multisine excitation for a bioimpedance measurement device. 2014

[28] D. J. Newman: An L1 Extremal Problem for Polynomials. Proceedings of the American Mathematical Society 16 (1965) 6, S. 1287–1290

[29] M. Schroeder: Synthesis of low-peak-factor signals and binary sequences with low autocorrelation (Corresp.). IEEE Transactions on Information Theory 16 (1970) 1, S. 85–89

[30] ASTM Standard E 2039; 31.07.2013. *Standard Test method for Determining and Reporting Dynamic Dielectric PropertieS*

Utilise AI for NDT Inspection

Peter CHOW [1]
[1] Fujitsu, Lewes, East Sussex, UK
Contact e-mail: peter.chow@uk.fujitsu.com

Abstract

Everyone is talking about machine learning (ML), deep learning (DL), and artificial intelligence (AI). In the consumer space it is used in many applications for our daily activities and conveniences. This technology marks a step change in productivities in virtually all areas of our lives. In this article we leverage the technology for Non-Destructive Testing (NDT) inspection. Today, virtually all manufactured products are inspected manually from visual (surface) to ultrasonic (internal) in quality control processes. NDT is one challenge and opportunity that could leverage AI for efficiency and productivity, and the potential of innovation in the manufacturing processes. In this presentation we will present real-world inspections achieved, including a customer case in wind turbine blade manufacturing.

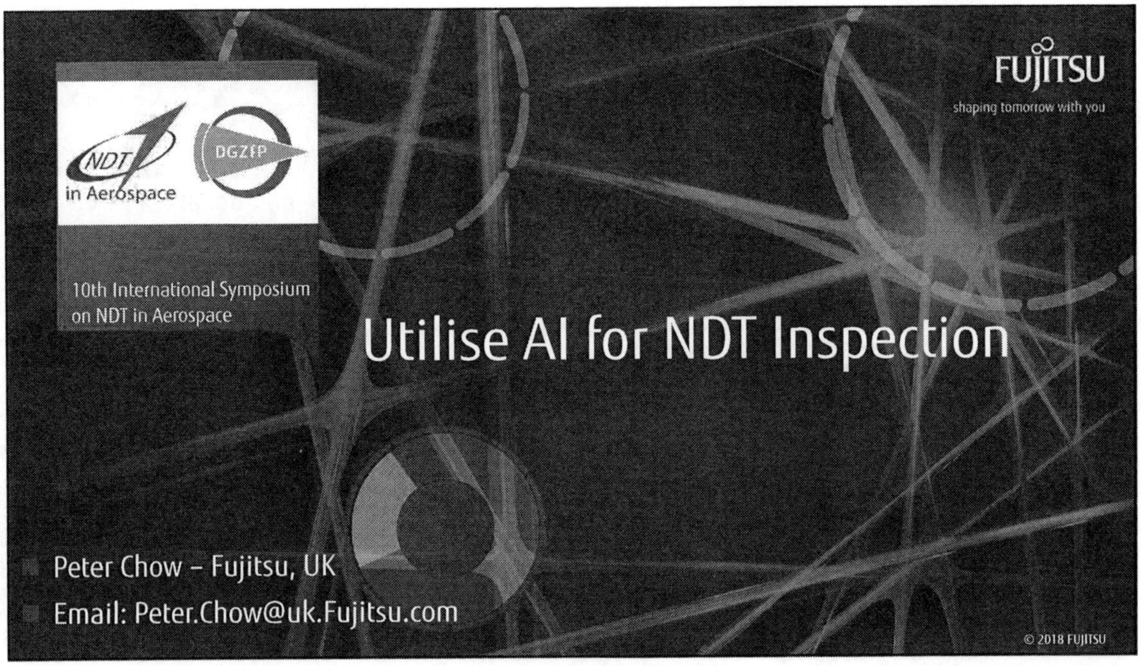
NDT in Aerospace
DGZfP
FUJITSU
shaping tomorrow with you
10th International Symposium on NDT in Aerospace
Utilise AI for NDT Inspection
Peter Chow – Fujitsu, UK
Email: Peter.Chow@uk.Fujitsu.com
© 2018 FUJITSU

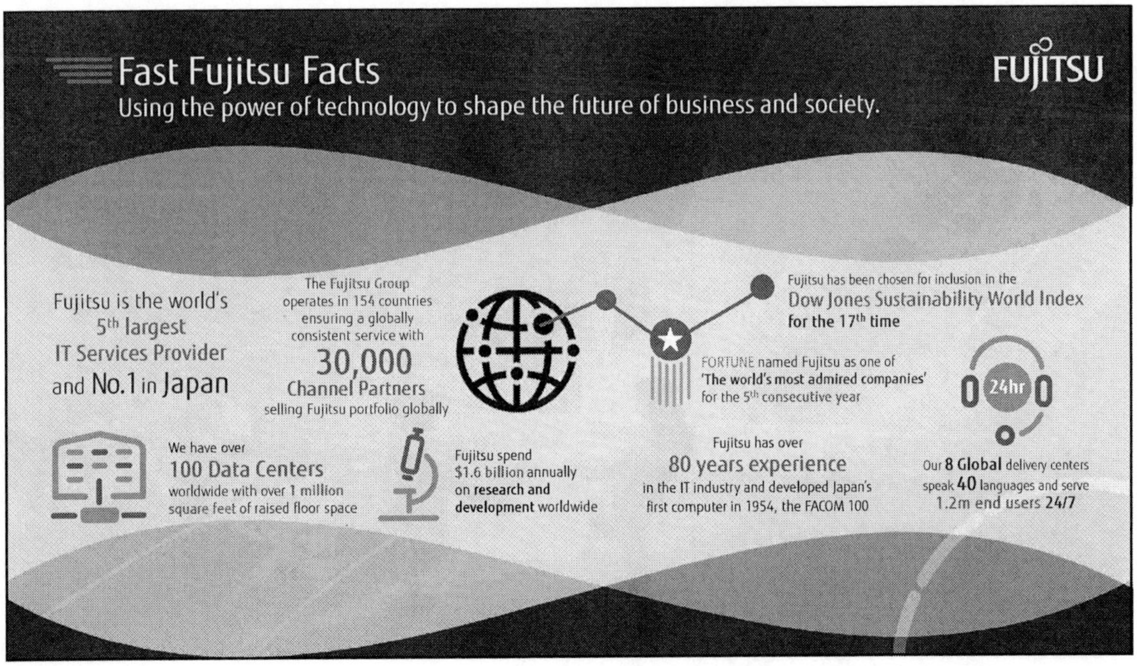
Fast Fujitsu Facts
Using the power of technology to shape the future of business and society.
FUJITSU
Fujitsu is the world's 5th largest IT Services Provider and No.1 in Japan
The Fujitsu Group operates in 154 countries ensuring a globally consistent service with 30,000 Channel Partners selling Fujitsu portfolio globally
Fujitsu has been chosen for inclusion in the Dow Jones Sustainability World Index for the 17th time
FORTUNE named Fujitsu as one of 'The world's most admired companies' for the 5th consecutive year
24hr
We have over 100 Data Centers worldwide with over 1 million square feet of raised floor space
Fujitsu spend $1.6 billion annually on research and development worldwide
Fujitsu has over 80 years experience in the IT industry and developed Japan's first computer in 1954, the FACOM 100
Our 8 Global delivery centers speak 40 languages and serve 1.2m end users 24/7

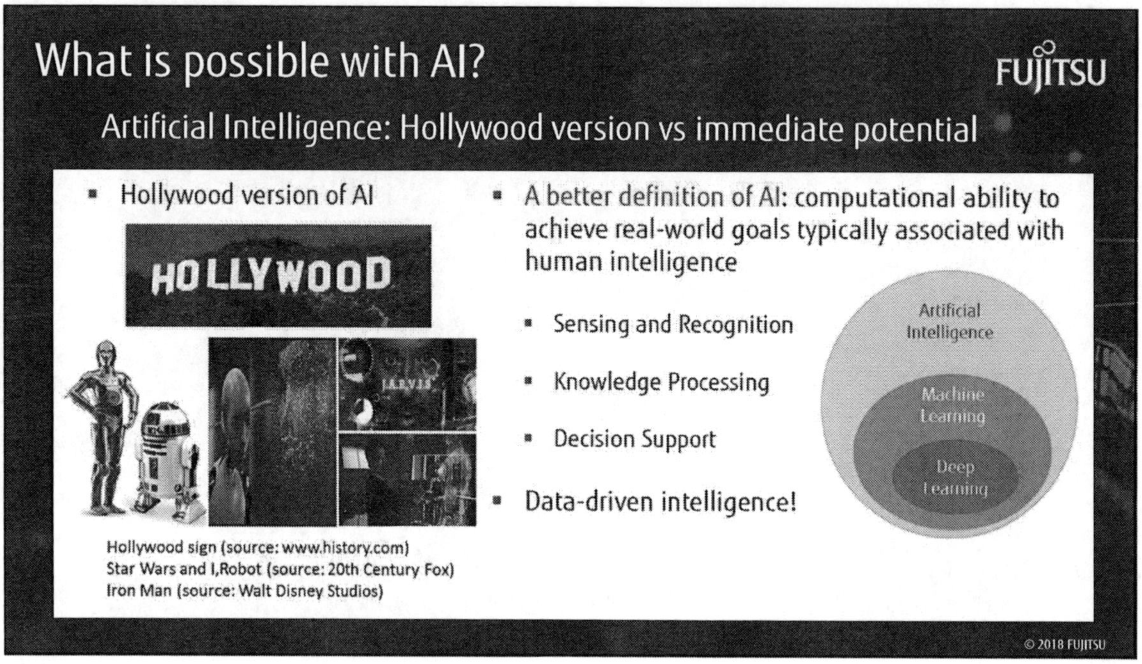
What is possible with AI?
FUJITSU
Artificial Intelligence: Hollywood version vs immediate potential
Hollywood version of AI
HOLLYWOOD
Hollywood sign (source: www.history.com)
Star Wars and I,Robot (source: 20th Century Fox)
Iron Man (source: Walt Disney Studios)
A better definition of AI: computational ability to achieve real-world goals typically associated with human intelligence
Sensing and Recognition
Knowledge Processing
Decision Support
Data-driven intelligence!
Artificial Intelligence
Machine Learning
Deep Learning
© 2018 FUJITSU

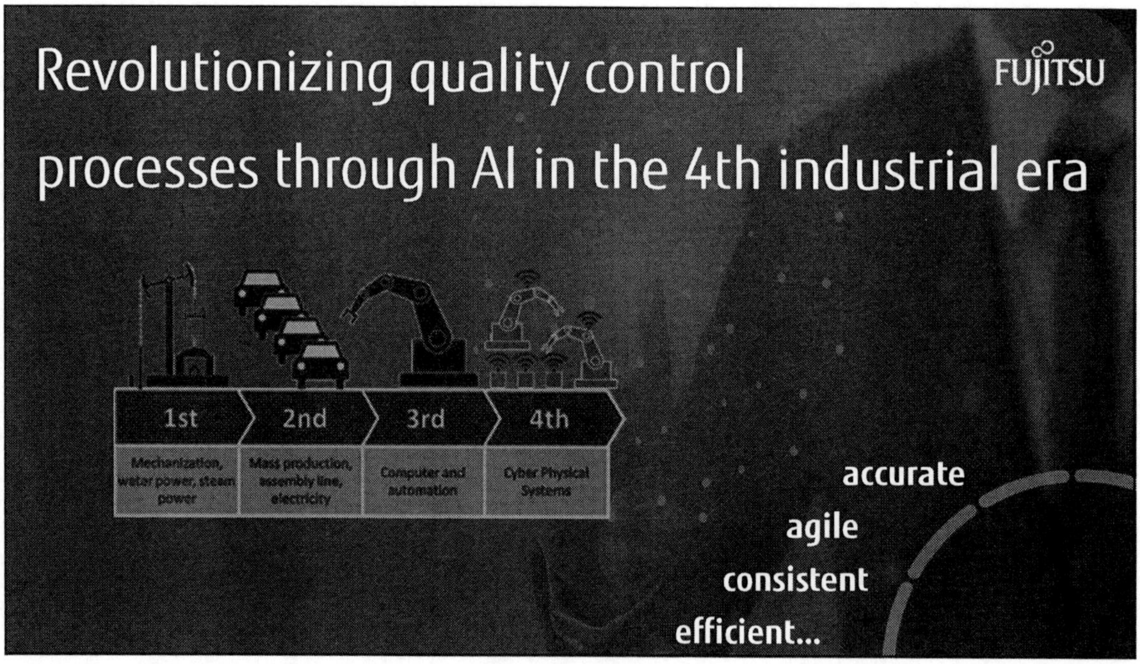
Revolutionizing quality control processes through AI in the 4th industrial era
FUJITSU
1st
2nd
3rd
4th
Mechanization, water power, steam power
Mass production, assembly line, electricity
Computer and automation
Cyber Physical Systems
accurate
agile
consistent
efficient...

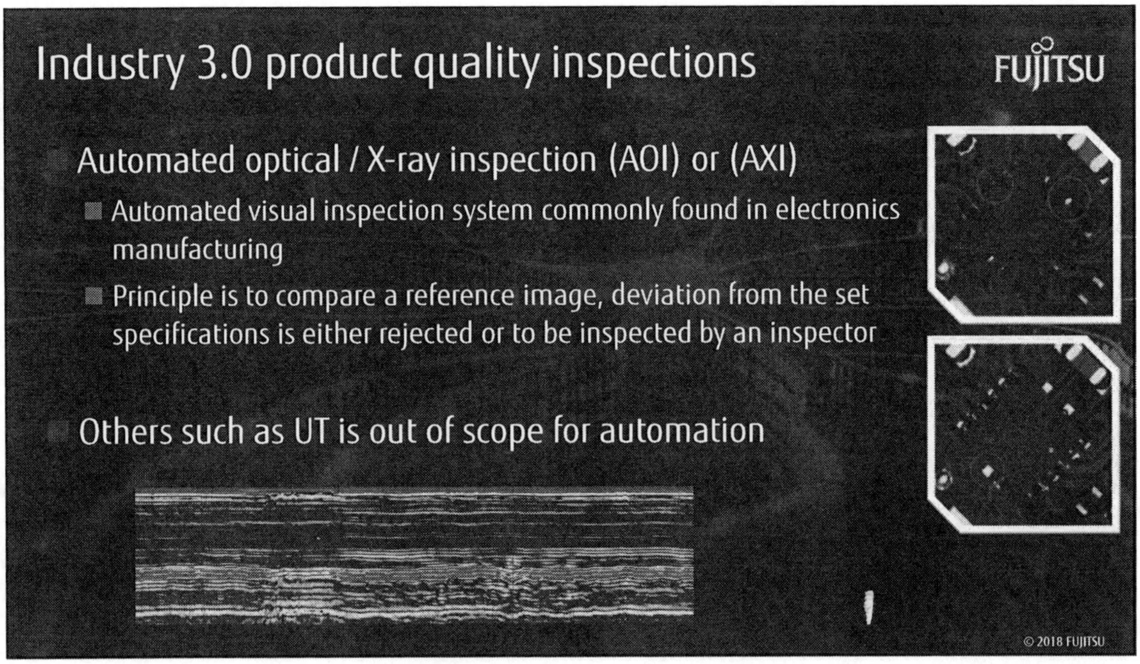
Industry 3.0 product quality inspections
FUJITSU
Automated optical / X-ray inspection (AOI) or (AXI)
Automated visual inspection system commonly found in electronics manufacturing
Principle is to compare a reference image, deviation from the set specifications is either rejected or to be inspected by an inspector
Others such as UT is out of scope for automation
© 2018 FUJITSU

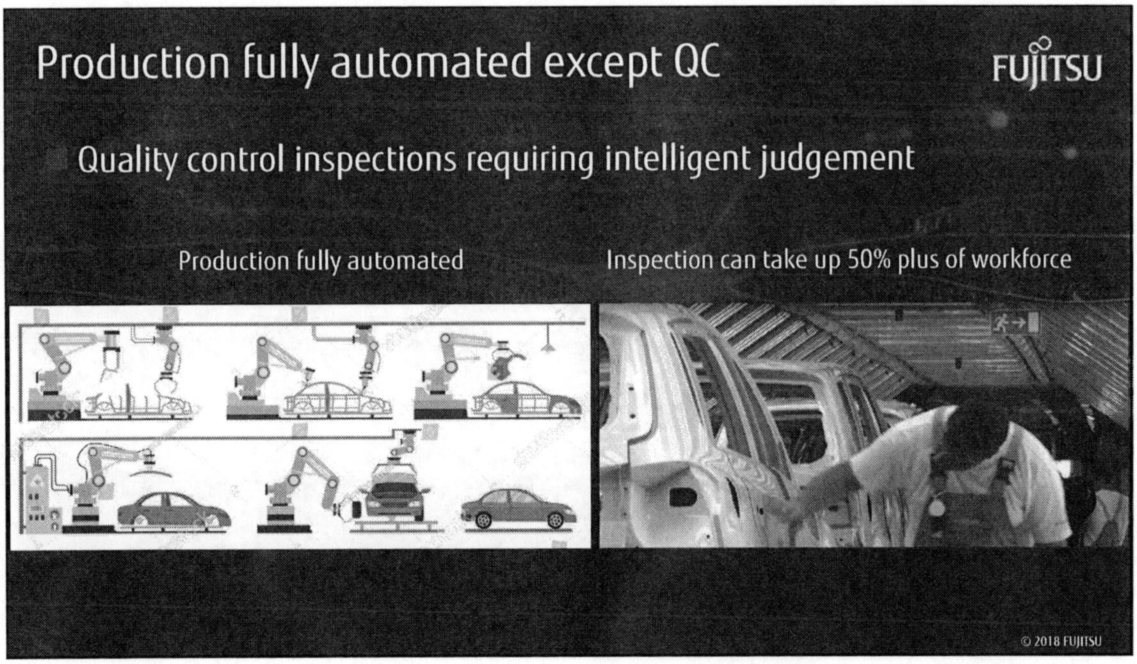
Production fully automated except QC
FUJITSU
Quality control inspections requiring intelligent judgement
Production fully automated
Inspection can take up 50% plus of workforce
© 2018 FUJITSU

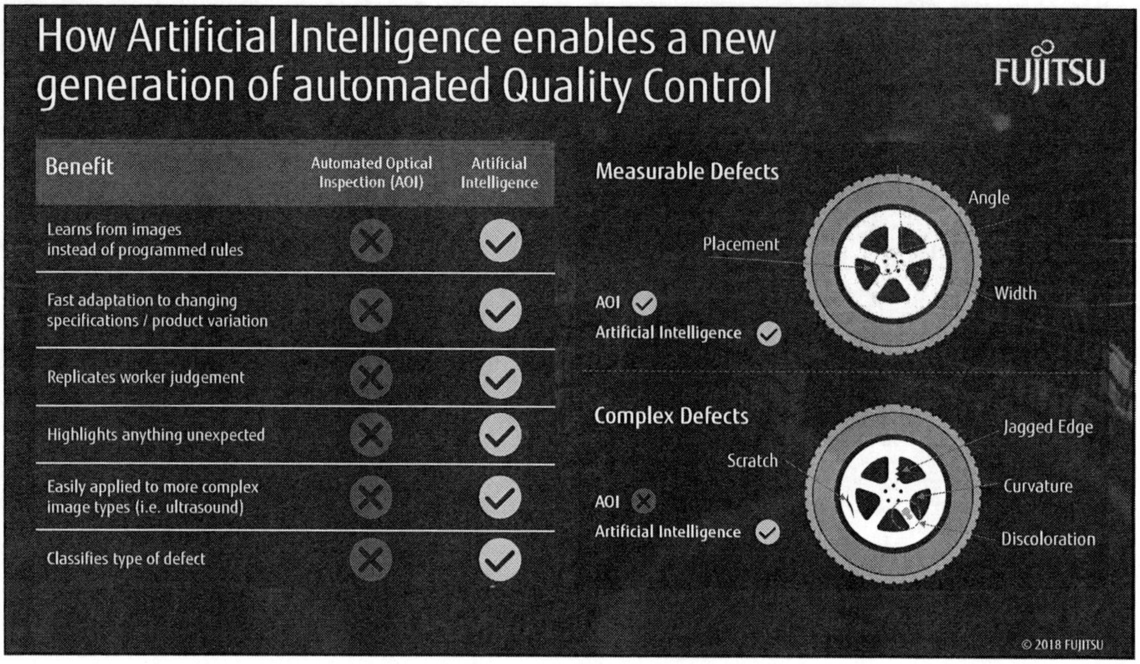
How Artificial Intelligence enables a new generation of automated Quality Control
FUJITSU
Benefit
Automated Optical Inspection (AOI)
Artificial Intelligence
Learns from images instead of programmed rules
Fast adaptation to changing specifications / product variation
Replicates worker judgement
Highlights anything unexpected
Easily applied to more complex image types (i.e. ultrasound)
Classifies type of defect
Measurable Defects
Angle
Placement
Width
AOI
Artificial Intelligence
Complex Defects
Jagged Edge
Scratch
Curvature
AOI
Artificial Intelligence
Discoloration
© 2018 FUJITSU

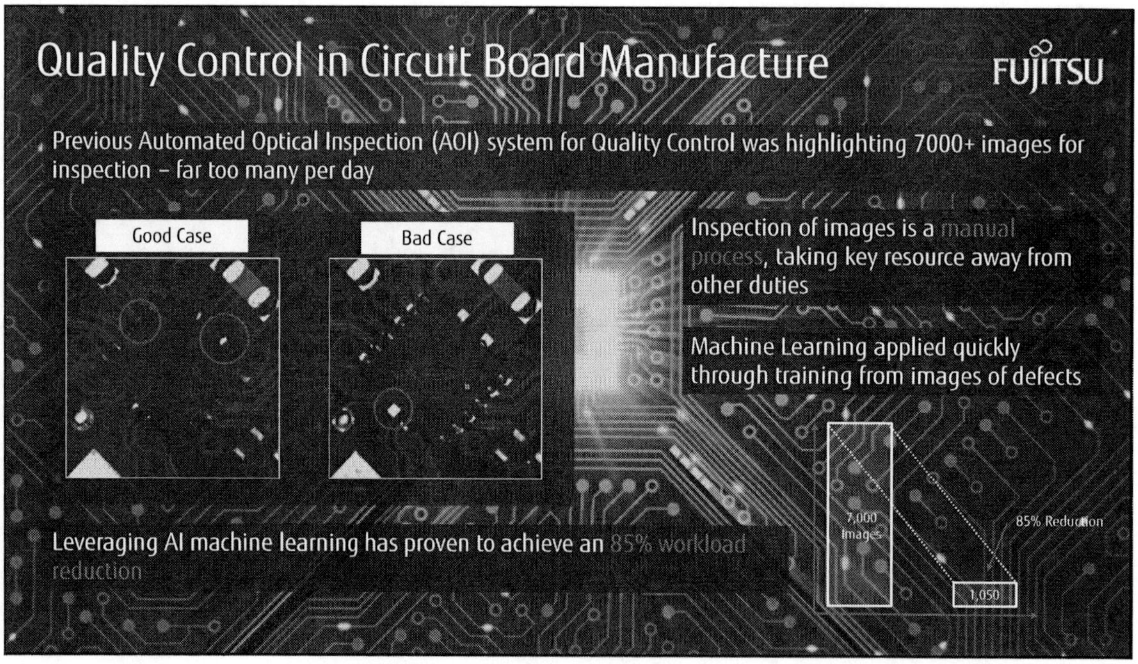
Quality Control in Circuit Board Manufacture
FUJITSU
Previous Automated Optical Inspection (AOI) system for Quality Control was highlighting 7000+ images for inspection – far too many per day
Good Case
Bad Case
Inspection of images is a manual process, taking key resource away from other duties
Machine Learning applied quickly through training from images of defects
7,000 Images
85% Reduction
1,050
Leveraging AI machine learning has proven to achieve an 85% workload reduction

Applying AI for Wind Turbine Manufacture

SIEMENS Gamesa RENEWABLE ENERGY FUJITSU

- Ultrasonic quality inspection generates massive amount of blade scan data for Quality Control
- Scan data evaluation process now completely automated
- High gains in time efficiency, enabling skilled operators to focus on the important part of the data
- Cuts inspection times for windmill turbine blades from six hours to just one and a half hours

'By adopting Fujitsu's ground-breaking AI technology it takes only a quarter of the time previously required to perform an inspection' Kenneth Lee Kaser, Head of SCM, Siemens Wind Power

Summary and Remarks

FUJITSU

- Appling AI machine learning to real-world product inspections
  - Achieved workload reduction of 75% to 85%
- AI for NDT is the next generation to support automated inspections
  - Taking quality control to the next level
- We are at the start of the journey
  - Many challenges ahead e.g. standardisation, digital transformation, R&D, etc.
  - New innovations and business opportunities

FUJITSU

shaping tomorrow with you

Development of a Prognostic Framework

Antonino Marco SIDDIOLO [1], Matthias BUDERATH [1]
[1] Airbus Defence and Space GmbH, Manching, Germany

Contact e-mail: antonino-marco.siddiolo@airbus.com

Abstract. This contribution reports the work carried out in "Airbus Defence and Space GmbH" with regard to prognostic. As a matter of fact, the prognostic field is industrially considered nowadays a key factor. For different reasons, however, real prognostic applications are still scarce in industry.
A successful implementation of prognostic might enable, among the other things: condition-based maintenance; maintenance schedule optimization; resource utilization optimization; reconfiguration strategy; extension of item useful life; mission planning optimization.
In the present paper, a few subjects will be reported like the development of mathematical approaches for performing prognostics, the obtained virtual and experimental results and the overarching prognostic framework. With reference to the mathematical approaches, two avenues are usually followed: the model-based approach (which will be not addressed here) and the data-based approach. However, in order to embrace all possible situations, our research has also focused on the development of a hybrid technique capable of coping - in our intention - with the lack of knowledge and the lack of data, usual condition in many situation and that is likely the reason of the mentioned lack of a widespread industrial implementation of prognostic-applications.
For the data-based approach a Gaussian Process formulation has been adopted, whereas for the hybrid approach, a Genetic Programming technique has been chosen. Since our results are in line with recent progress on the topic, as above anticipated, no model-based experience will be reported in this contribution.
The developed approaches – after being successfully tested against synthetic data – have also been applied to a real environment, related to the Remaining Useful Lifetime (RUL) estimation of ball bearings equipping an aeronautic fan. The results gained were satisfactory and allowed us to test and validate the effective usage of the conceived prognostic approaches in a representative environment. Results have showed a comparable and satisfactory application of the prognostic approaches developed for both the synthetic and experimental investigations

Introduction

The present contribution aims at summarizing the work that has been carried out along the previous years in Airbus Defence and Space GmbH; the work has as objective the setting of a prognostic framework, in order to verify and validate – also in a quantitative manner - the effective and efficient implementation of predictive capabilities [1 and 2].

The prognostic field is without any doubt considered nowadays a key win-factor in dealing with maintenance strategies and beyond. As a matter of fact, prognostic (see below Fig. 1.) should avoid inopportune maintenance costs, by allowing the optimization of the ware-house, by avoiding failures – and possible corresponding secondary failures – of the equipment during its working-phase, by reducing the time-to-repair (less trouble-shooting). Moreover - and depending on the type of business even more importantly – prognostic should result in increased system availability.

However, for different reasons, real prognostic applications are scarce in industry: that is mainly due to the difficulty of selecting the proper technology and to the several constraints that such a technology shall observe in an industrial environment. Modern systems are in fact characterized by high complexity to fulfil the high requirements on functionality and quality. A typical vehicle e.g. consists of "about 2,000 functional components, 30,000 parts, and 10 million lines of software code" [3].

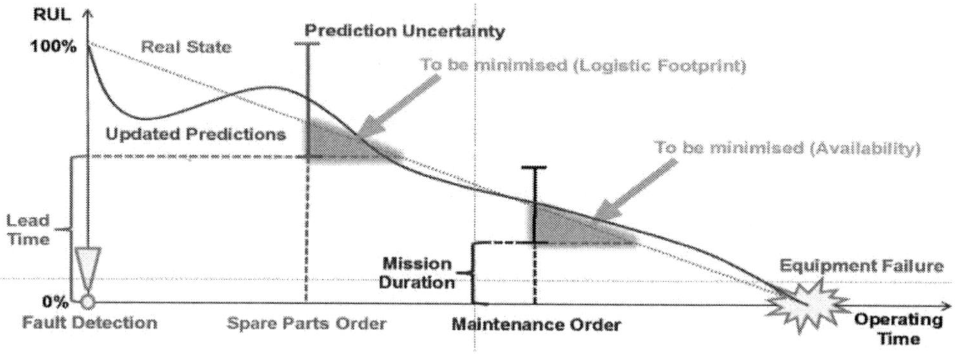

Fig.1. Challenges and benefits of prognostic capabilities

It could be useful to remind here more systematically – with the help of the above mentioned figure - the plenty of advantages that a proper implementation of prognostic will enable: 1) shift from a classical time-based preventive maintenance strategy to a condition-based maintenance strategy, in which maintenance events are dynamically allocated based on health indicators; 2) possibility of optimizing the maintenance scheduling; 3) possibility of optimizing the utilization of spare resources; 4) possibility of implementing a reconfiguration strategy to cope with the reduced operational & functional capabilities of components; 5) extend the useful life of a degraded component, by configuring the load on it by the environment; 6) possibility of optimizing the mission plan [4, 5, 6 and 7]

1. The prognostic module

Along with our tasks definition, it has been decided to consider the set of prognostic functionalities as a module which is going to be fed by the corresponding diagnostic module with the current value of a health feature: this parameter – which will be called Health-Grade (HG) and will range, if not somewhere else stated, between 100%, component healthy, and 0%, component failed - provides the actual health status of the component under observation.

The most characterizing feature of the prognostic module is the capability to make prediction, i.e. the capability to project into the future the (possible) trend of the equipment-condition (of the so-called HG). The prediction is based upon 1) historical recorded HGs; 2) historical recorded data (the "input-data", whose value has been recognized as being decisive for the HG estimation); when available, 3) an estimation of the future "input-data"; and lastly 4) a mathematical representation of the degrading process. In other words, it can be stated, that the prognostic module is characterized by a representation of the degrading process (4), which shall provide the means to forecast the health status of a component based on some inputs (1, 2 and 3).

Along with the approach followed by many researchers [8 and 9], a generic assumption is that the degrading process is going to predict the degradation increment that the component will withstand based on the actual values of the inputs (this is corresponding to the so-called "one-step ahead prediction"). The approach might also be reasonably understood considering the widespread leveraging of a Kalman-filtering environment when performing the prediction task. A Kalman-filtering mathematical frame does result *de-facto* as the preferred one also for the relatively simple integration of uncertainty considerations in the final prediction of the HG values [10 and 11]. The approaches that will be described in the following sections - as a matter of fact - are following the mentioned Kalman-based approach.

As will be clearer in the following sections, exploiting a common and fruitful mind-set in the prognostic-community [12 and 13], which assumes that no solution is the perfect one for all scenarios, but a reasoned compromise (ensemble) between different solutions might actually be the right choice, an approach named "Interactive Multiple Models - IMM" is going to be followed within the data-based approach. The hybrid approach is in principle following the same mind-set, but in actual fact implementing it slightly differently.

Finally, the final aim of the prognostic module is to make available to the downstream processes (e.g. to the on-ground Maintenance Planner/Optimizer) the Probability Density Function (or PDF) of the Remaining Useful Life (or RUL) parameter (The RUL is the predicted time at which a system or a component will no longer perform its intended function; as understandable, it is an important concept in decision making for contingency mitigation): from the PDF, the expected RUL can be evaluated, together with the associated confidence levels.

Therefore, coming back to the beginning of the present paragraph, following the above described framework, monitoring concepts and associated technologies, as well as possible algorithms conceived for post-processing raw data and deriving health-indicators are meant as functionalities belonging to the upstream diagnostic modules and will be not discussed herein.

Hence it is assumed being available a comprehensive "labelled" training data-set to be utilized within a well-known supervised learning approach, i.e. a data-set for which the known health grade of the item is given together with the corresponding values of the input data. As will be explained later on, this data set will be used to derive the mathematical representation of the degrading process.

2. The prognostic framework

2.1 Intro

Within the prognostic mind-set above described has to be placed and considered the developed prognostic framework (flow chart in Fig. 2.) that will be here discussed.

Two main subjects have been deeply investigated: the development of a tool for the synthetic generation of degradation trends ("Development of Degradation Models" in Fig. 2.) and the development of different and complementary mathematical approaches to handle the task of making a prediction ("Development of Prognostic Concepts" in Fig. 2.). Having at our disposal a tool for the generation of representative "run to failure" data has constituted an invaluable aid during the development stage of the prognostic concepts ("Model-based Verification" in Fig. 2.): the lack of "run to failure" data is well known, as well as the difficulties associated to such experimental activities aiming at gathering Run-to-Failure data (e.g. artificial aging of components, representativeness of laboratory condition, etc).

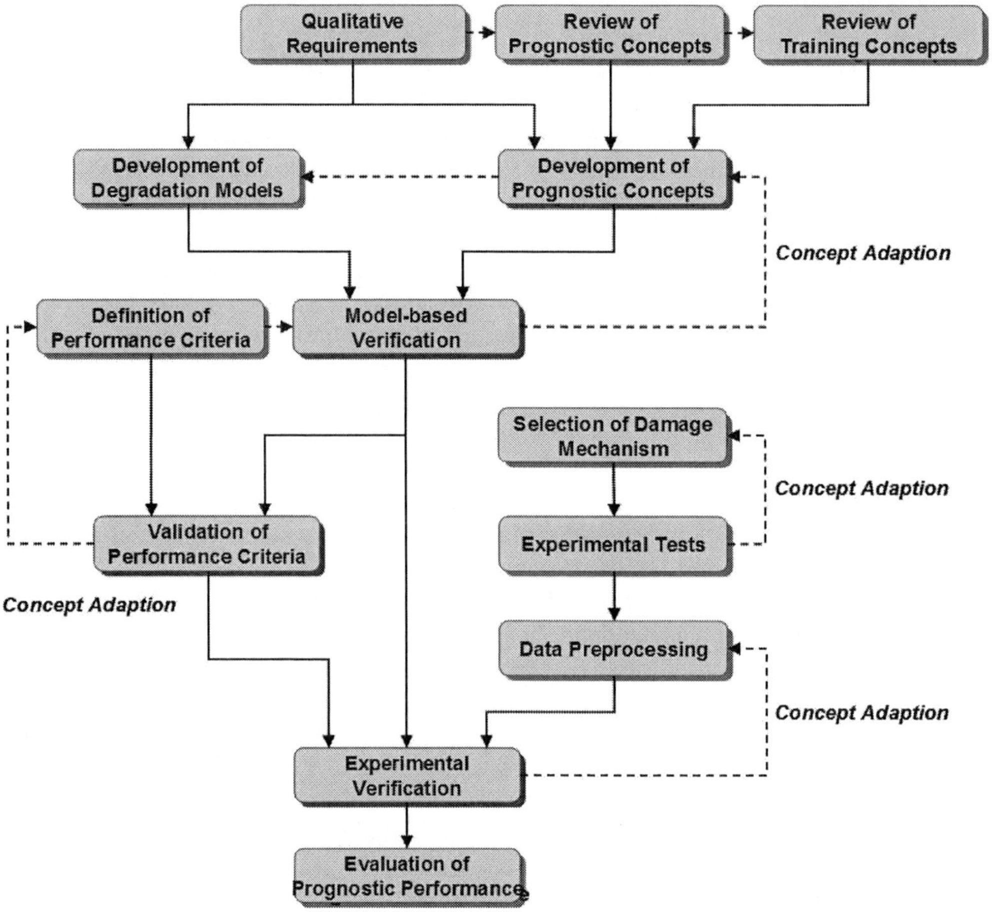

Fig.2. Prognostic Framework Flow-Chart

2.2 Prognostic approaches

In conceiving a prognostic concept ("Development of Prognostic Concepts" in Fig. 2.), two paths are usually followed: the model-based approach [14, 15, 16 and 17] and the data-based approach [18, 19 and 20] (see Fig. 3.).

In the model-based approach the physics underneath is well known and therefore physical modelling and advanced parametric identification techniques can be applied; however, the complexity of the observed systems is usually so high that first-principle approaches have limited application (A typical application is the prediction of the battery's discharge [17]).

On the other hand, data-based approaches rely on the usage of a comprehensive set of historical data, which is then used to properly train the algorithm.

Both approaches have pros and cons; however the above mentioned scarcity of real prognostic applications can be probably traced back to the general lack of statistical-meaningful historical data and to the poor knowledge of the physics behind complex systems. Therefore, in order to try to embrace all possible situations, our research has focused not only on the development of model-based and data-based approaches, but also of hybrid techniques capable – in our intention – of coping with the lack of knowledge and of data (shaded squared area close to the origin of the axes in Fig. 3.).

In particular, in the current contribution, the description of the gained achievements with respects to the model-based approach will be omitted. As a matter of fact, the topic has been addressed thoroughly by many researchers [14, 15, 16, and 17]. We are going on the other side to focus more on the data-based method, a typical approach which might be framed within the nowadays common big-data / data-analytics discussion, and on the hybrid approach, which represent our original contribution to the subject.

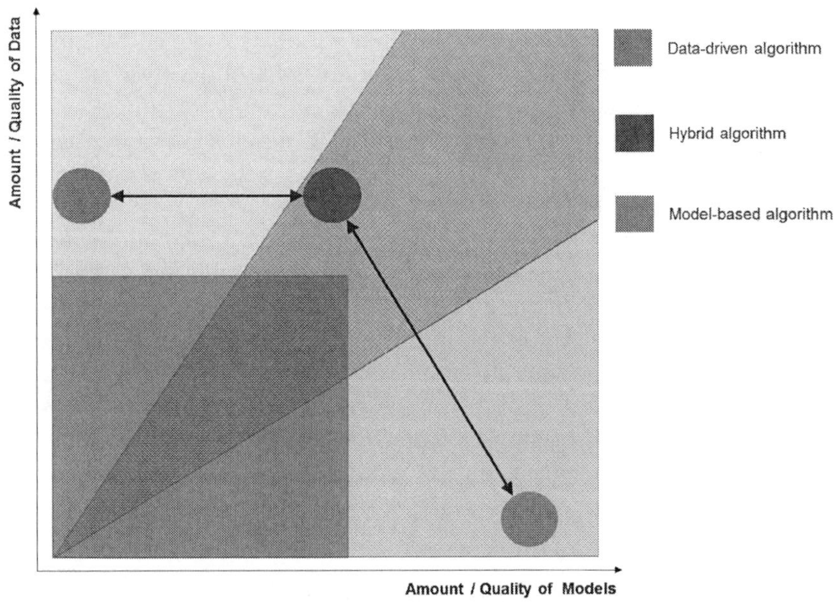

Fig.3. Prognostic approaches

Within the design loop a proper metric has been set up ("Definition of Performances Criteria" in Fig. 2.) and utilized in order 1) to quantitatively measure the performances of the approach under investigation, 2) to compare different concepts and 3) to value the effects of changes in the overall concept's design. The metric has been derived from common and well recognized means settled by different researchers belonging to the "Prognostic & Health Management" community [21].

2.2.1 Intro

In this context we will focus on the performances of the conceived data-based and hybrid approaches: the two developed approaches deeply rely on the adaptive filtering capabilities provided by the Kalman theory.

As said above, it is assumed having at our disposal, a statically representative training data set, to be utilized to design a prediction concept by means of a supervised learning approach. As it usually happens, just a few Run-to-Failure data are available; the benefit of having a tool for the synthetic generation of Run-to-Failure data has proven for this reason its unique advantages: as a matter of fact, based on the limited numbers of Run-to-Failure data available and its corresponding statistical characterization, a big set of "similar" Run-to-Failure data have been generated. By means of this bigger amount of failure-data a more reliable and robust concept could be accordingly designed.

The sketched process has been followed for the design of the prognostic concept for the real experimental tests, that we have performed and that we will describe later.

However, in what follows will also be described the results gained within a whole synthetic environment; different data-sets have been generated, each one corresponding to a typical degrading process (linear, convex, concave, etc…). The mathematical approaches on hand have then been tested against the different synthetic failure data.

2.2.2 Data-Based Approach - Gaussian Process

The approach is based upon the mathematical theory that goes under the name: the Gaussian Process. Apart from a little intro taken from Wikipedia, no further detail of the theory will be given here. "In probability theory and statistics, a Gaussian Process is a stochastic process (a collection of random variables indexed by time or space), such that every finite collection of those random variables has a multivariate normal distribution, i.e. every finite linear combination of them is normally distributed. The distribution of a Gaussian Process is the joint distribution of all those (infinitely many) random variables, and as such, it is a distribution over functions with a continuous domain, e.g. time or space." [22, 23 and 24].

Let's dive into the actual Gaussian Process based implementation that has been carried out. Several models, each one associated to an historical Run-to-Failure trend, have been built during the "training & testing" phase (yellow coloured rectangle in Fig. 4.): during this phase proper kernels are, in particular, selected and the corresponding hyper-parameters optimized. These different models constitute the process models from a Kalman perspective: as a matter of fact, each process model provides an estimation of the next degradation increment, giving as input the actual degradation value and other selected inputs considered of relevance for the application (e.g. the operative time, or the "load").

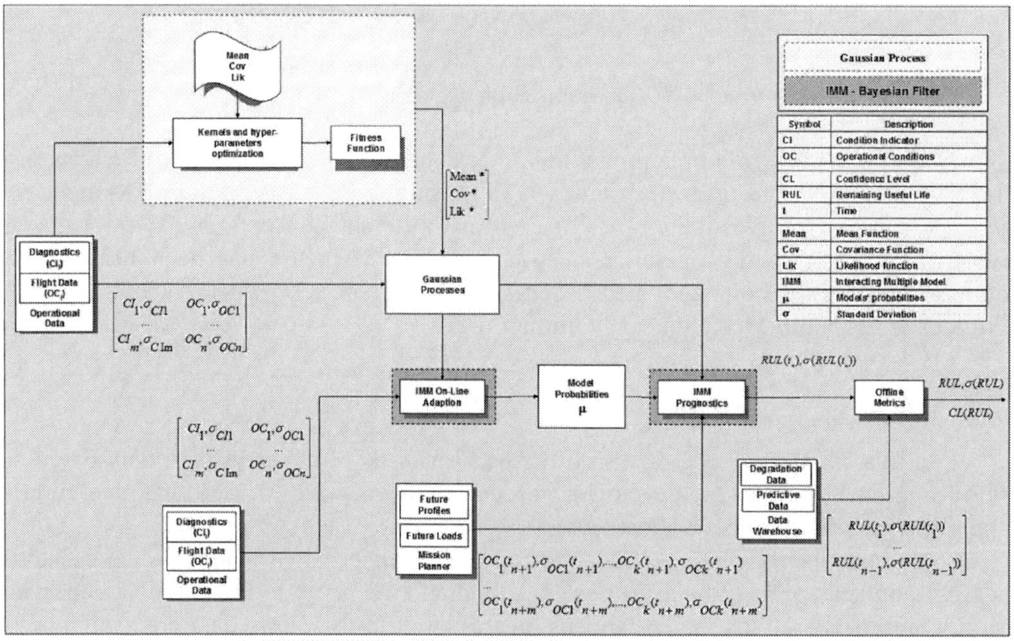

Fig.4. "Gaussian Process" Concept

Most importantly, the process models provide in addition an estimate of the uncertainty of the process itself: this is a quite remarkable capability that makes pretty attractive the utilization of Gaussian Processes within the prognostic framework by means of a Kalman paradigm. As a matter of fact, providing the uncertainty associated to the RUL is as important as the estimated value for the RUL itself.

As anticipated above, the built bank of Kalman filters - equipped with Gaussian Processes – is designed to work in parallel within an "Interactive Multiple Model - IMM" approach: in fact, during the observation stage ("IMM On-Line Adaptation" in Fig. 4.) an integral-in-nature information is estimated, which informs about the relevance of each model (i.e., a vector of weighting factors is estimated, which in the mentioned figure is indicate with the Greek letter μ) for the current observation.

This information (the last gained vector μ will be utilized), together with the bank of filters, which this time will work without the so-called correction step, since there is no measurement, is then utilized to project into the future the degradation value and its associated uncertainty. This provides then the basis for the final calculation of the PDF of the RUL (whose task is performed in the box: "IMM Prognostics") which of course takes advantage of the information coming from the "Mission Planner", like future profiles or loads.

2.2.3 Hybrid Approach - Genetic Programming

As said for the introduction to the "Gaussian Process", also the genetic approach is based upon a well-known mathematical subject: the Genetic Programming. Apart from a little intro taken from Wikipedia, no further detail of the theory will be given here. "In artificial intelligence, Genetic Programming (GP) is a technique whereby computer programs are encoded as a set of genes that are then modified (evolved) using an evolutionary algorithm – it is an application of (for example) genetic algorithms where the space of solutions consists of computer programs. The results are computer programs that are able to perform well in a predefined task. The methods used to encode a computer program in an artificial chromosome and to evaluate its fitness with respect to the predefined task are central in the GP technique and still the subject of active research." [25, 26 and 27].

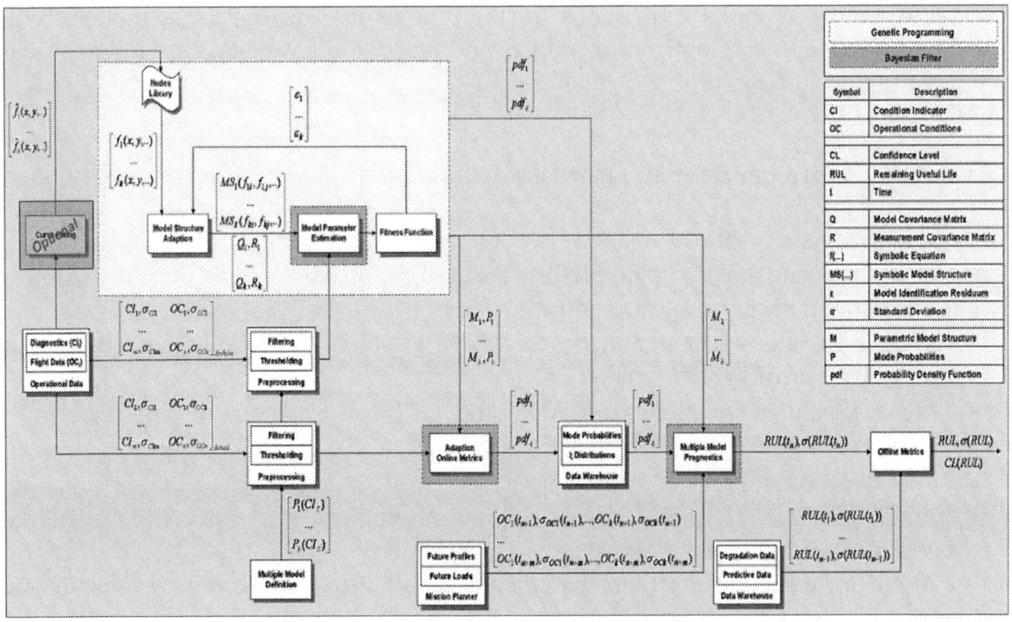

Fig.5. "Genetic Programming" Concept

One of the most limiting cons of the briefly discussed data-based approach in the previous section is however the need of a comprehensive data-base of Run-to-Failure data. On the other hand, the complexity of the handled systems does not usually allow – as remarked – a straightforward usage of first-principle laws. As before discussed and depicted in Fig. 3., the designed hybrid approach try to overcome these cons by mathematically incorporating and taking advantages of the positive characteristics of the model- and data-based approaches.

In fact, in the hybrid approach, we have tackled the task of deriving a set of mathematical expressions for the degradation process under observation by means of artificial intelligence techniques (the yellow green rectangle in Fig. 5. – as per the "Gaussian Process Concept" above - represents the training & testing phase of the Genetic Programming concept). In particular, Genetic Programming techniques have been leveraged to derive the process equations corresponding to different dynamic portions of the historical data at our disposal: this way, having a family of N degradation trends belonging to the very same failure mode (same family), instead of building N different process models (as performed by the application of the data-based approach), we seek for M (M<N) different process models, each one representative of a definite frequency window of the degradation family dynamic behaviour (As a matter of fact, we are assuming, by proceeding this way, common characteristics of the different degradation trends, that make them a family). This operation requires some pre-processing task ("Pre-processing" box in Fig. 5.), in which – automatically - the statistic characteristics of the above mentioned M dynamic ranges are collected and given as input to the "training & testing" box.

As said for the data-based approach, also the hybrid approach will be characterized by an adaptive stage (the observation phase, which has been called "Adaptation Online Metrics" in the Fig. 5.), in which the bank of filters is building up information corresponding to the current happening degradation, which is peculiar but still belonging to the same family.

Furthermore, at any point in time, the bank of filters can be used with no correction step, to project in time the degradation information, by using the last updated data related to the peculiarity of the observed degradation trend ("Multiple Model Prognostics" box in Fig. 5.). The uncertainty propagation is this time totally driven by the Kalman filtering itself: it is therefore based on initial estimations for the process and measurement models. When available, information coming from the "Mission Planner" might be utilized to increase the accuracy of the prediction.

3. Evaluation of prognostics and algorithm-training

As anticipated shortly in the last paragraph, a metric has been utilized to quantitatively measure and track the performances of the approach to be designed. Some of the most important key performances indicators are:

- Mean Absolute Percentage Error (or MAPE), which is the average of the absolute percentage error in RUL prediction;
- Mean Absolute Deviation (or MAD), which is the average spread of the error;
- False Positives (or FPs), which are unacceptable early predictions;
- False Negatives (or FNs), which are unacceptable later prediction and
- Prognostic Horizon (or PH), which is the normalized time after which the lower confidence limit remains stable.

A full understanding of these parameters would also require the introduction of a well-known performance plane, which is used to visualize the prediction capabilities of a concept. As in Fig. 1., the performance plane (See Fig. 6.) has got on the *x-axis* the prediction instants (usually normalized by the End of Live – EoL, and indicated with the Greek letter λ), whereas on the *y-axis* we have the estimated RUL values (typically in percentage). The line linking (0, 100%) to (1,0%) represents therefore the performance of an ideal algorithm predicting every time the proper RUL. The interested reader may refer to reference [21] to have a better understanding of the meaning of the key performances indicators as well as of the advantages and actual usage of the "performance planes".

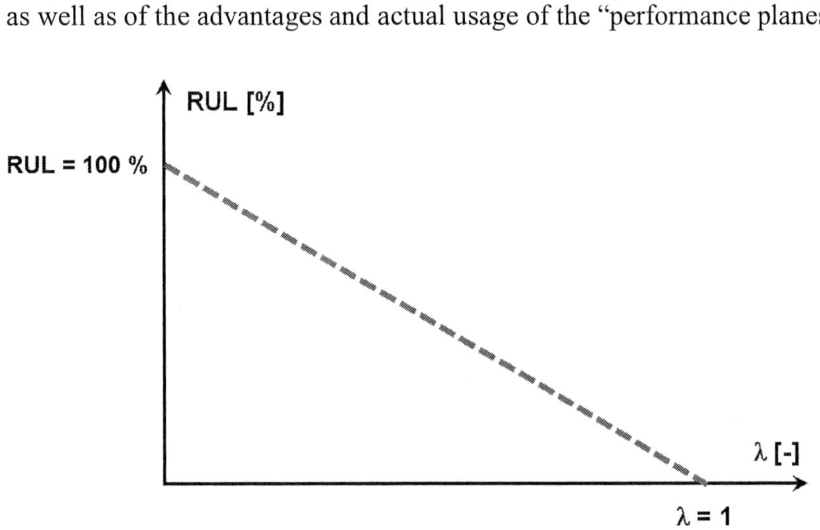

Fig.6. Performance-plane

In what follows the "performance-plane" will be also utilized to plot synthetically the result of our virtual and real experimentations.

The set of key-performances indicators has been extensively adopted within the training stage of the algorithm. In the following Fig. 7. a flow-chart of this stage is drawn. As usually done, the whole labelled data-set has been first decomposed in training and in testing data-set. Then the key-performances indicators are used first to guide the training steps and to finally quantitatively give a value to the solution conceived by using the testing data-set.

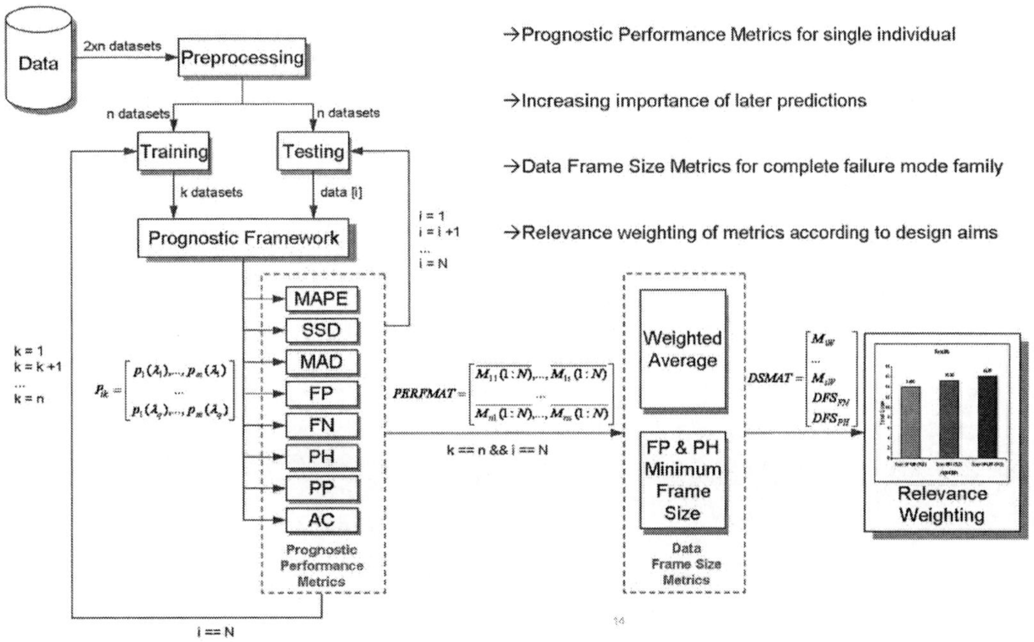

Fig.7. Algorithm training flow chart

4. Results

The whole above-described prognostic framework has been first successfully tested against synthetic data. Afterwards the same approach has been followed in a real environment: the RUL estimation of the ball-bearings equipping an Ametek fan for aeronautic application was the project's aim. These initial experimental investigations have proven once more the inherent difficulties linked with the execution of representative Run-to-Failure experiment.

4.1 Synthetic experiments

As anticipated before, one of our focuses has been the development of a tool for the generation of synthetic Run-to-Failure data-sets. The tool allows the setting of some initial parameters so to generate a set of Run-to-Failure trends that we can label as belonging to the same failure family. Therefore, we could generate trends of different shapes: linear and convex, for example. They may be characterized by an expected time of failures and moreover the degradation limit may be set to resemble some experimental results. These trends may depend on loading conditions and incorporate steps and plateaus; as a matter fact, one of the setting parameters is responsible to take into account the stochastic nature of the degradation.

All above said is synthetically summarized in the following Fig. 8. In it, one can see at the top-left typical loading conditions from which the degradation will depend, as well as the PDF of the degradation limit (on the top-right). Below, on the other hand, one can see the cumulative plot of two generated family of Run-to-Failure trends: a convex one (on the left) and a linear one (on the right). For both of them, the EoL was set to 100 hours (although the added stochastic nature is systematically decreasing this value), whereas the degradation limit has a pre-defined distribution.

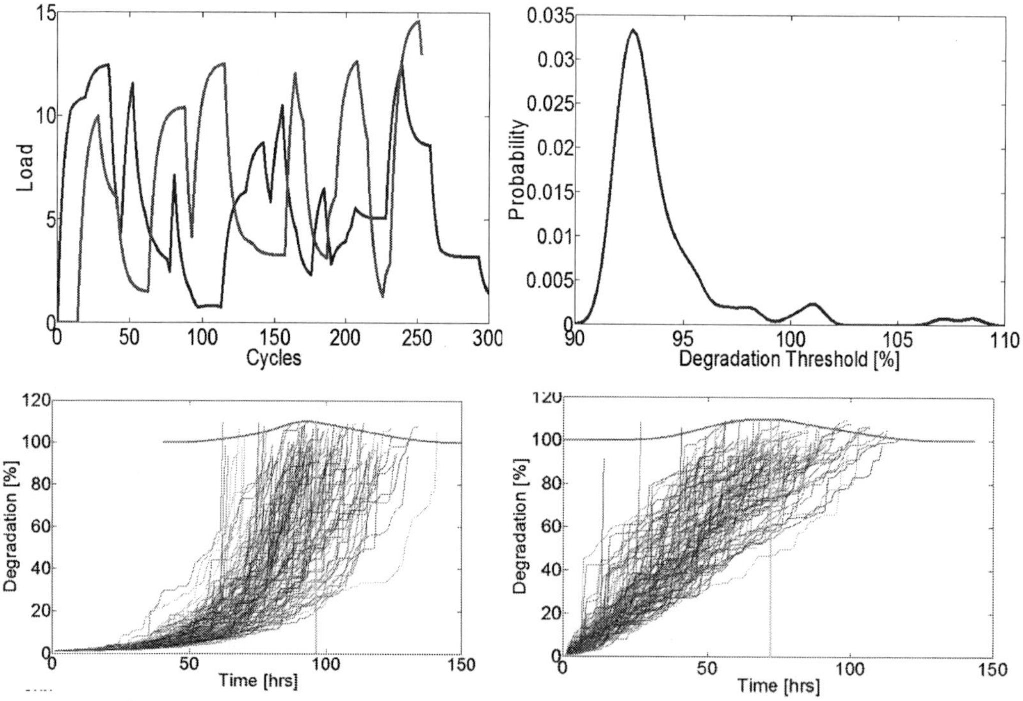

Fig.8. Synthetic tool for Run-to-Failure data generation

In Fig. 9., six different families are plotted. The top row is dedicated to the convex-type families; the bottom-one to the linear-type families. Within each typology we also distinguish between a family with high stochastic behaviour (the one on the left), a family with medium stochastic behaviour (the one on the right), and a last family (in the middle) with no variance at all.

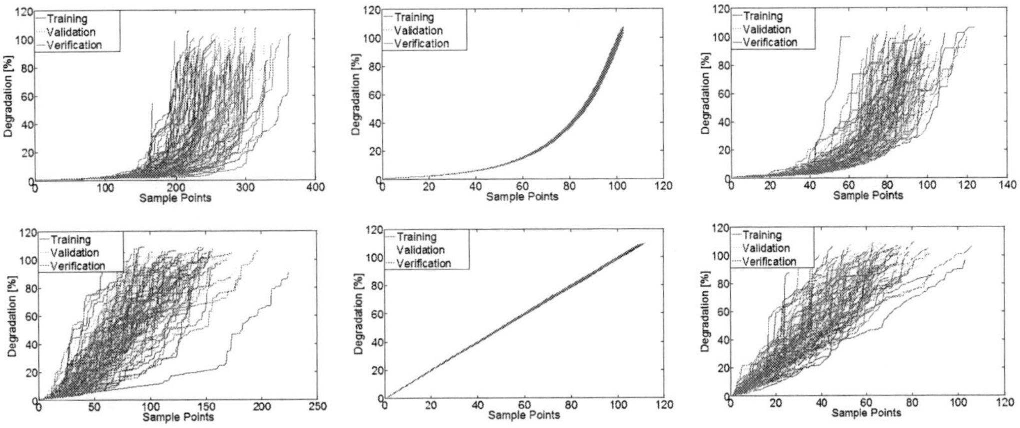

Fig.9. Synthetic data-sets with loading conditions

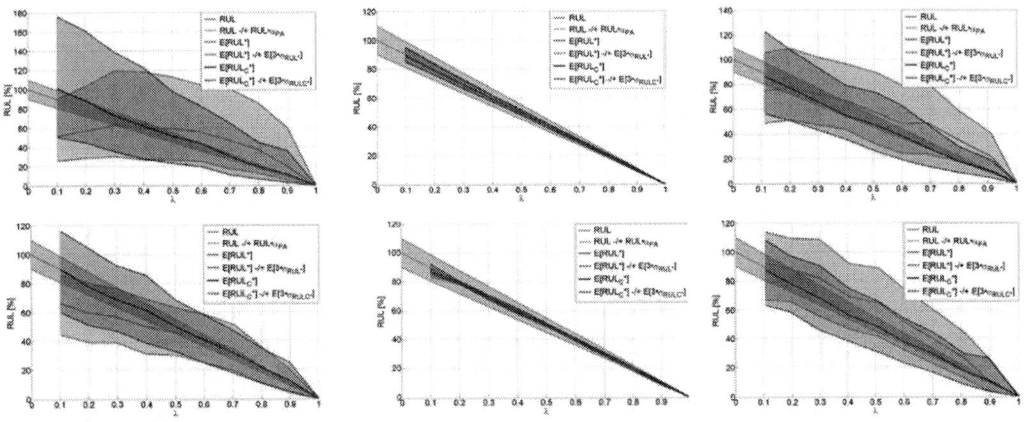

Fig.10. "Gaussian Process Concept" application on the synthetic data-sets

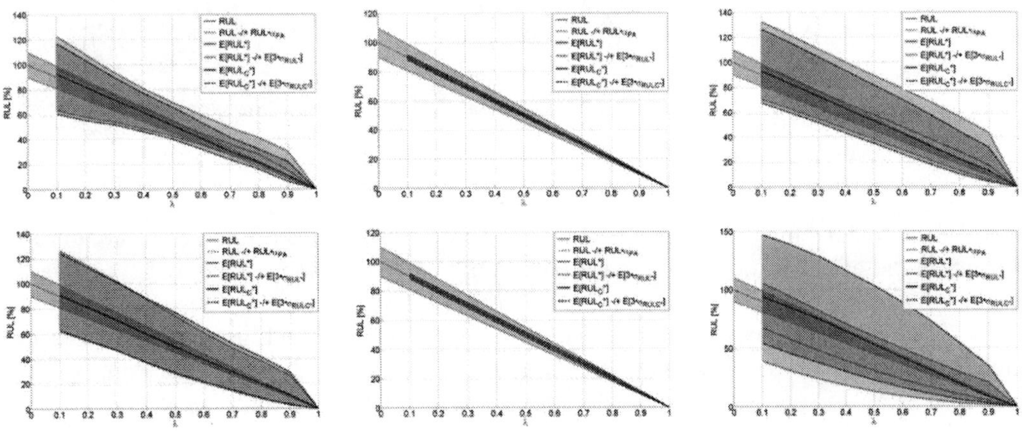

Fig.11. "Genetic Programming Concept" application on the synthetic data-sets

In the previous two figures, the results of the performances of the two concepts are plotted. For both concepts, an outer-correction loop has been implemented. In fact, a systematic error has been always found: this could be filtered out easily. Finally, in the below plots (Fig. 10.) you can see both results, the corrected (purple) and the not corrected (orange) one.

With regards to the convex families, the "Genetic" concept slightly outperforms the "Gaussian" one. However, when dealing with linear families the opposite holds.

The same conclusions are applicable to the experimentation on a real environment.

4.2 Real experiments

The following figures synthetically describe the set-up of the experiments: Fig. 12. shows the characteristics of the utilized aeronautic fan; Fig.13. describes the measurement set-up; whilst in Fig. 14. the procedure settled to artificially inject a failure in the ball-bearings is summarized (a precisely weighted amount of diamonds-powder was introduced inside the ball-bearing before starting the Run-to-Failure test).

- AMETEK Rotron MAX50021 2900X7
- 3 Phase asynchronous motor
- 115V line to line, 400Hz
- 11100-11300 rpm, 185-188 Hz
- Air-flow feedback controller: 235 l/s
- Fan used as cabin evaporator fan in Boeing 737

Fig.12. Fan characteristics

- Hall-Effect current sensors, SNR > 600
- Sampling rate 25 kHz
- DSpace PPC board
- 400 Hz SW notch-filter to suppress power supply current
- Three-axis accelerometer, ±50 g, 100 mV/g
- Sampling rate 10 kHz
- Integrated passive speed sensor

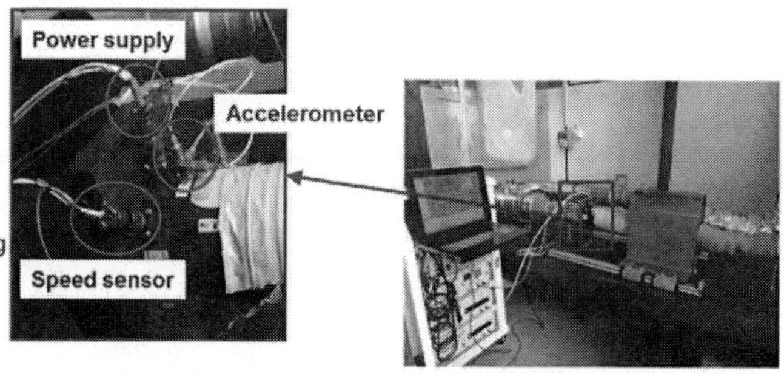

Fig.13. Fan measurement set-up

The tracked/measured quantity considered as an indicator of the health-condition of the fan is the acceleration. Furthermore, the component was considered to be on a failure condition as soon as the case temperature was reaching a predefined value.

The Fig. 15. shows the eleven Run-to-Failure trends, which have been successfully measured during several months of experimentations. Although maximum care has been taken to maintain and guarantee the same experimental conditions, as it can be easily recognized, a big spread in both EoL and degradation limit do exist: This to confirm once more the extreme difficulty in performing representative artificial aging tests in the lab.

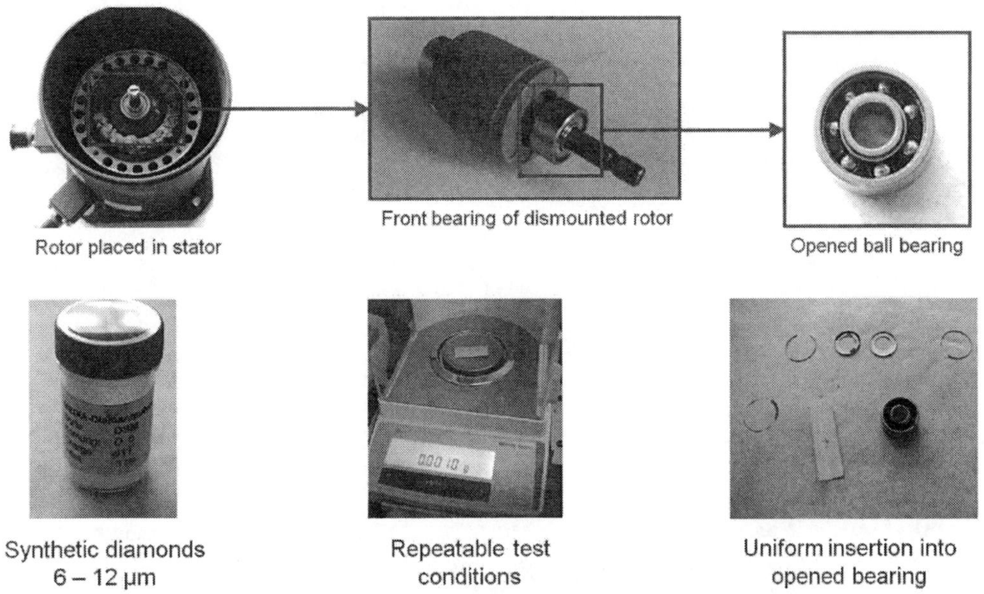

Fig.14. Failure injection procedure. Top: Rotor front bearing as "Use Case" for verification. Bottom: Fault insertion with synthetic diamonds

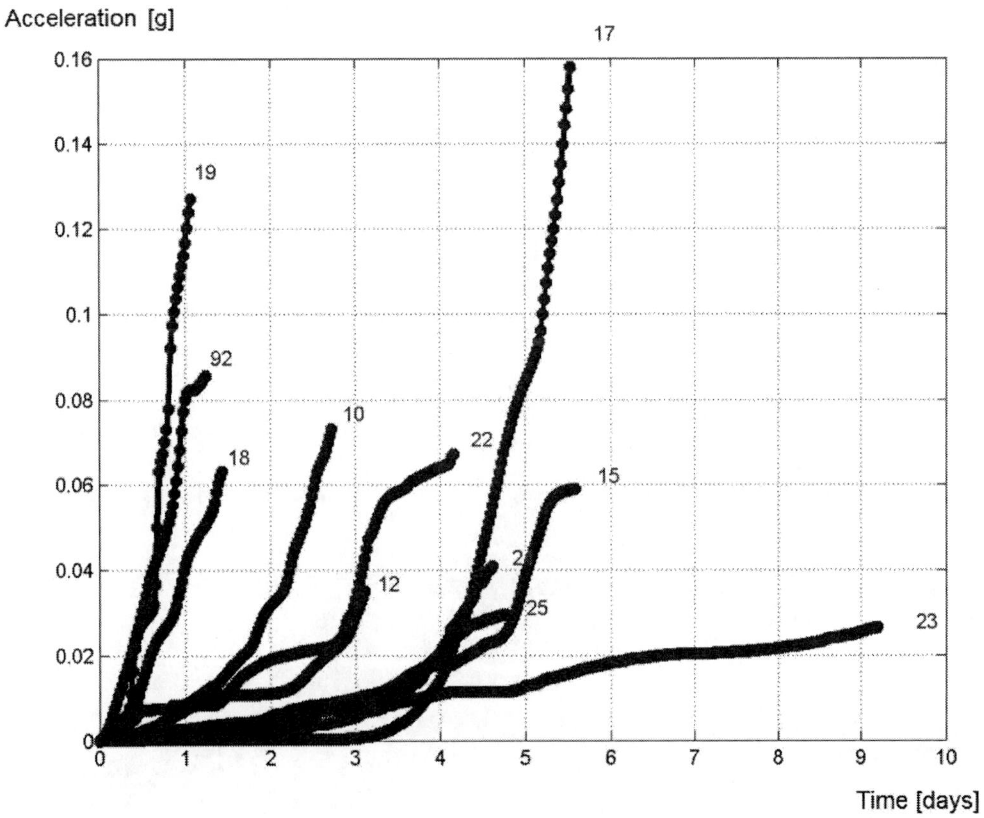

Fig.15. Family of recorded Run-to-Failure data

Nevertheless the results gained were satisfactory and allowed us to test and validate the effective usage of the conceived prognostic approaches in a representative environment.

Fig. 16. displays the resulting performance-planes for the application of the Gaussian Process Concept (top) and of the Genetic Programming Concept; whereas in Fig. 17. two "spider" plots are displayed. Spider plots have been proven to be very effective in synthetically showing the results of our investigation: each ray is related to a selected performance parameter of the overall developed and mentioned metric. The results have showed a comparable and satisfactory application of the prognostic approaches developed for both the synthetic and the real investigations.

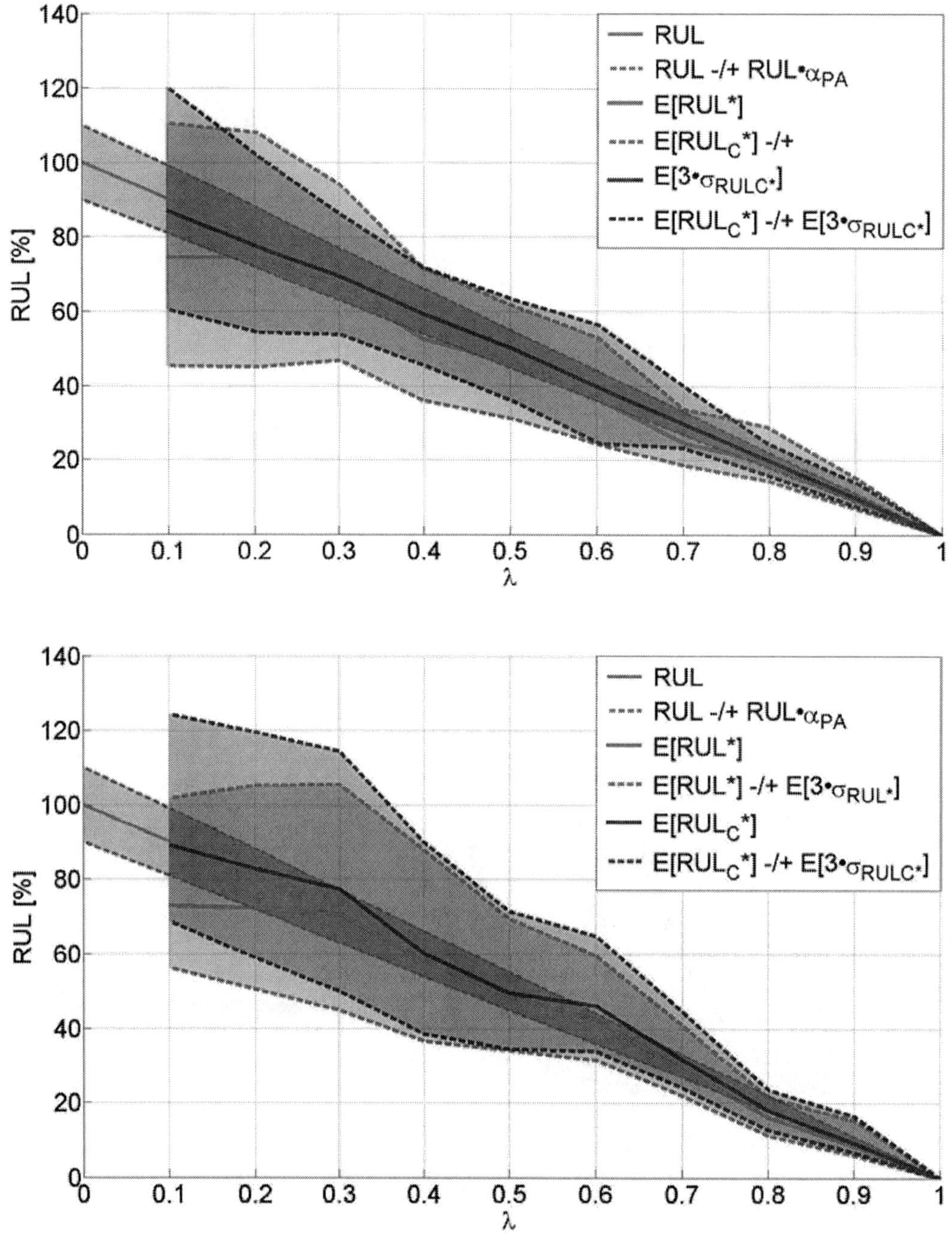

Fig.16. Performance planes for Gaussian Process Concept (top) and Genetic Programming Concept (bottom) application

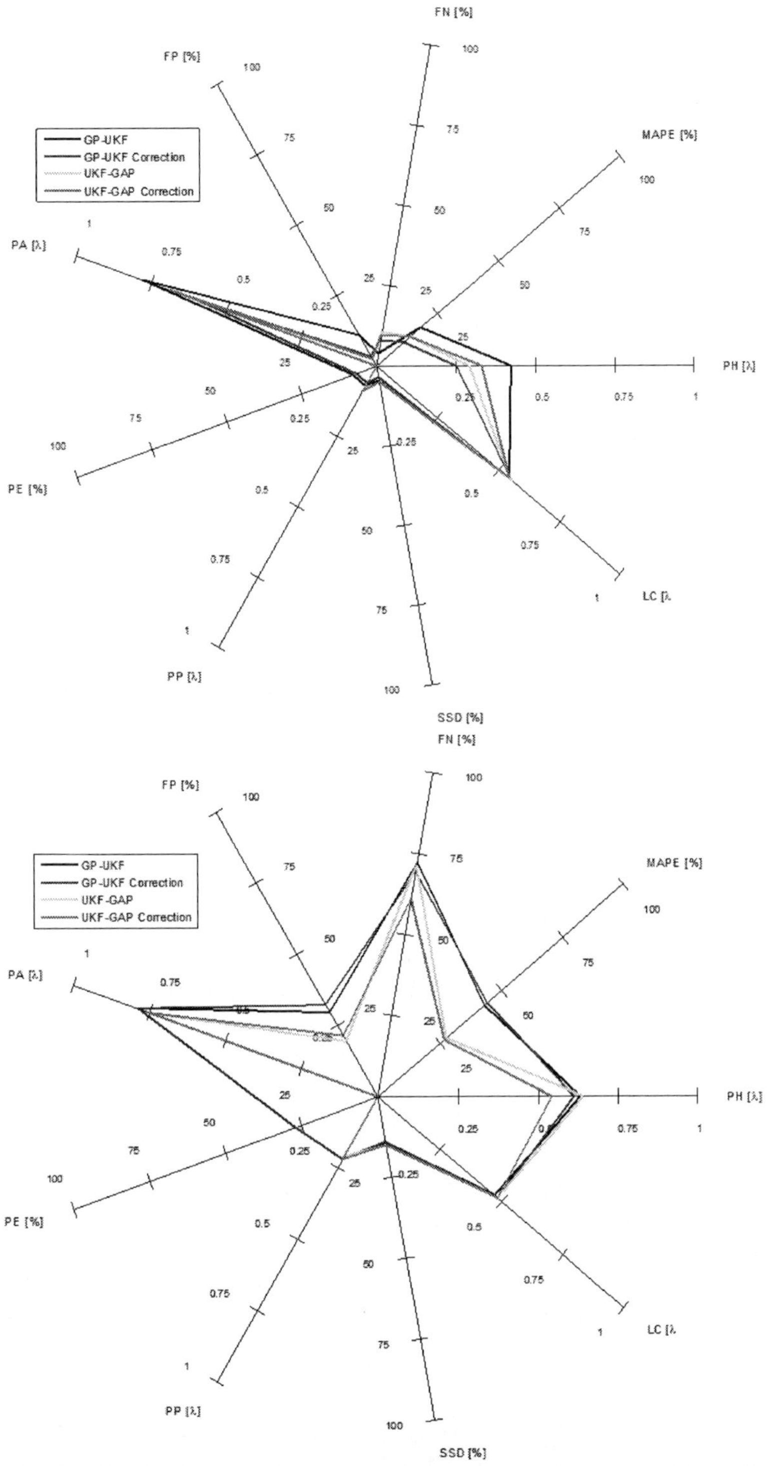

Fig.17. Spider plots for the synthetic campaign (top) and the real experimental tests (bottom)

5. Conclusion

This contribution has reported the work carried out in “Airbus Defence and Space GmbH" with regard to prognostic. In particular, a few subjects have been discussed: the tool for the synthetic generation of representative degradation trends, the development of mathematical approaches for performing prognostics, the obtained virtual and experimental results, and the overarching prognostic framework.

As well known, usually two approaches have been followed by the researchers to perform prognostic: a model-based and a data-based approach. In our discussion, we have omitted presenting outcomes of the model-based approach, and rather have focused on the data-based approach and on a hybrid approach. In fact, in order to embrace all possible situations, our research has also focused on the development of such a hybrid technique capable of coping in our intention with the lack of knowledge and the lack of data.

Therefore we have presented the implemented data-based approach, based on the leverage of the Gaussian Process mathematical framework: different models – each one corresponding to a typical way of degrading of the item under investigation - operate in parallel within an "Interactive Multiple Model" framework. The interaction is made possible by means of a bank of Kalman filters. In fact, the Bayesian approach - with its prediction and correction steps – allows the building up of a relevance-information, in order to pinpoint the most similar stored models to be utilized to forecast the observed degradation.

Despite the different advantages, like the elegant and effective handling of uncertainty, and the easy derivation of the degradation-process model, some cons (e.g.: the need to carry the whole set of historical data-base and/or the experienced model instabilities when the interested domain is not properly sampled – curse of dimensionality) have moved us towards the exploration of alternatives. The developed hybrid approach is based on the utilization of the Genetic Programming technique. This mathematical approach has been utilized to derive the process equations corresponding to different dynamic aspects of the degradation-data: in particular a stochastic component has been statistically estimated during the training stage for each model. Differently than as in the data-based approach, here the method aims at finding a similar behavior for selected frequency-windows and describe them in a compact way by means of equations plus a stochastic distribution which is taking over all the uncertain-behavior of the component degradation. When in operation, the adaptive stage is then choosing - from a Kalman perspective – both the most suitable models (i.e.: experienced frequency domain) to perform prognostic and proper values for the stochastic distribution.

Additionally, attention has been put on establishing effective means to measure the performances of the approaches.

Eventually, the approaches – after being successfully tested against synthetic data – have been applied to a real environment, related to the Remaining Useful Life (RUL) estimation of the ball bearings equipping an aeronautic fan. The results gained were satisfactory and allowed us to test and validate the effective usage of the conceived prognostic approaches in a representative environment. Results have showed a comparable and satisfactory application of the prognostic approaches developed for both the synthetic and experimental investigations

References

1. J. B. Bowles, "A survey of reliability-prediction procedures for microelectronic devices," IEEE Transactions on Reliability, vol. 41, no. 1, pp. 2–12, 1992.
2. K.-L. Tsui, W. Chiu, P. Gierlich, D. Goldsman, X. Liu, and T. Maschek, "A review of healthcare, public health, and syndromic surveillance," Quality Engineering, vol. 20, no. 4, pp. 435–450, 2008.
3. J. P. MacDuffie and T. Fujimoto, "Why dinosaurs will keep ruling the auto industry," Harvard Business Review, vol. 88, no. 6, pp. 23–25, 2010
4. G.W. Vogl, B.A. Weiss, M. Helu, "A review of diagnostic and prognostic capabilities and best practices for manufacturing", Journal of Intelligent Manufacturing, DOI 10.1007/s10845-016-1228-8

5. K. Goebel, A. Saxena, M. Daigle, J. Celaya, I. Roychoudhury, Introduction to Prognostics, First European Conference of the Prognostics and Health Management Society 2012
6. C.S. Byington, M. J. Roemer, T. Galie, "Prognostic Enhancements to Diagnostic Systems for Improved Condition-Based Maintenance", Proceedings, IEEE Aerospace Conference, Big Sky, MT, USA , 9-16 March 2002
7. R.M. Vandawaker, "Integrated Systems Health management as an enabler for condition based maintenance and autonomic logistic", Dissertation, Department of the air force air university, Air force institute of technology
8. J. Chiachío, M. Chiachío, D. Prescott, J. Andrews, "A Bayesian assessment for railway track geometry degradation", Vol 4 No 1 (2018): Proceedings of the European Conference of the PHM Society
9. Tran, Van Tung, Yang, BoSuk, Oh, MyungSuck and Tan, Andy Chit Chiow (2008) Machine condition prognosis based on regression trees and onestepahead prediction. Mechanical Systems and Signal Processing, 22 (5). pp. 11791193. ISSN 08883270
10. Lotfi Saidi, Jaouher Ben Ali, Mohamed Benbouzid, Eric Bechhoefer, "Wind Turbine Drivetrain Prognosis Approach Based on Kalman Smoother with Confidence Bounds", 2018 IEEE International Conference on Industrial Technology (ICIT)
11. Agusmian Partogi Ompusunggu, Jean-Michel Papy and Steve Vandenplas, "Kalman filtering based prognostics for automatic transmission clutches", IEEE/ASME Transactions on Mechatronics (Volume: 21, Issue: 1, Feb. 2016)
12. Sun, Jianzhong & Hongfu, Zuo & Haibin, Yang & Pecht, Michael. (2010). Study of ensemble learning-based fusion prognostics. 2010 Prognostics and System Health Management Conference, PHM '10. 1 - 7. 10.1109/PHM.2010.5414582.
13. James Kuria Kimotho1 and Walter Sextro2, "Comparison and ensemble of temperature-based and vibration-based methods for machinery prognostics", Annual Conference of the Prognostics and Health Management Society 2015
14. H.C. Vu, P. Do, M.S. Jha, D. Theilliol, and F. Peysson, "On the use of particle filters for prognostics in industrial applications", Vol 4 No 1 (2018): Proceedings of the European Conference of the PHM Society
15. H. Mol, A. Gabelli, "Model Based Detection and Diagnosis of Bearing Defects", Vol 4 No 1 (2018): Proceedings of the European Conference of the PHM Society
16. D. Gazizulin, R. Klein, J. Bortman, "Physics Based Methodology for the Estimation of Bearings' Remaining Useful Life: Physics-Based Models, Diagnostic Methods and Experiments", Vol 4 No 1 (2018): Proceedings of the European Conference of the PHM Society
17. Arachchige, Buddhi & Perinpanayagam, Suresh & Jaras, Raul. (2017). Enhanced Prognostic Model for Lithium Ion Batteries Based on Particle Filter State Transition Model Modification. Applied Sciences. 7. 1172. 10.3390/app7111172.
18. Mosallam, L. Laval, F.B. Youssef, J. Fulton, D. Viassolo, "Data-Driven Fault Detection for Neutron Generator Subsystem in Multifunction Logging-While-Drilling Service", Vol 4 No 1 (2018): Proceedings of the European Conference of the PHM Society
19. M. Baptista, C.L. Nascimento Jr., H. Prendinger, E. Henriques, "A Case for the Use of Data-driven Methods in Gas Turbine Prognostics" , Annual Conference of the Prognostics and Health Management Society, St. Petersburg, Florida, 2017
20. E. Jakobsson, E. Frisk, R. Pettersson, M. Krysander, "Data driven modeling and estimation of accumulated damage in mining vehicles using on-board sensors", , Annual Conference of the Prognostics and Health Management Society, St. Petersburg, Florida, 2017
21. Abhinav Saxena, Jose Celaya, Edward Balaban, Kai Goebel, Bhaskar Saha, Sankalita Saha, Mark Schwabacher, "Metrics for Evaluating Performance of Prognostic Techniques", 2008 International Conference on Prognostics and Health Management
22. https://en.wikipedia.org/wiki/Gaussian_process
23. S. Saha, B. Saha, A. Saxena and K. Goebel, "Distributed prognostic health management with gaussian process regression," 2010 IEEE Aerospace Conference, Big Sky, MT, 2010, pp. 1-8. doi: 10.1109/AERO.2010.5446841
24. Pavle Boškoski, Matej Gašperin, Dejan Petelin, "Bearing fault prognostics based on signal complexity and Gaussian process models", Prognostics and Health Management (PHM) 2012 IEEE Conference on, pp. 1-8, 2012.
25. https://en.wikipedia.org/wiki/Genetic_programming
26. Ludwig, Simone & Roos, Stefanie. (2010). Prognosis of Breast Cancer Using Genetic Programming. 6279. 536-545. 10.1007/978-3-642-15384-6_57. Metric
27. Liao, Linxia. (2014). Discovering Prognostic Features Using Genetic Programming in Remaining Useful Life Prediction. IEEE Transactions on Industrial Electronics. 61. 2464-2472. 10.1109/TIE.2013.2270212.

Application of a PHM-based reliability prediction for an UAV's control surface actuation system

Simon MEHRINGSKÖTTER [1], Henrik HEIER [1]
[1] Institute of Flight Systems and Automatic Control, TU Darmstadt, Otto-Berndt-Str. 2, 64287 Darmstadt, Germany

Contact e-mail: mehringskoetter@fsr.tu-darmstadt.de, heier@fsr.tu-darmstadt.de

Abstract. Unmanned Aerial Vehicles (UAVs) become more and more important for the civil sector as they foster new business models and promise cost-savings for many applications. However, the acceptance and successful integration of this new aircraft type into the existing airspace largely depends on the overall system safety. From an operator's point of view, it is therefore interesting to get a feedback on the current and future health status of flight critical systems.

Current methods used to obtain and maintain safety constraints are based on reliability and safety engineering methods, which largely rely on the statistical processing of historic failure events to estimate the mean time to failure. In addition to this, modern systems are often equipped with Fault Detection and Isolation mechanisms. While the first approach is rather static and does not consider any feedback from an individual system (i.e. the actual health state), the latter is unable to give any predictions about the expected time to failure.

With the discipline of Prognostics and Health Management (PHM) there exists an alternative approach to overcome these limitations. By assessing the actual degradation state of a system from sensor data (diagnosis) and predicting it (prognosis), the current and future health state can be estimated during flight. By further combining PHM results with conventional reliability methods, an improved prediction up to system level can be achieved.

In this paper, the application of a PHM-based system reliability prediction is described. The system under study is the control surface actuation system for a hybrid UAV. By aggregating the Remaining Useful Lifetime (RUL) estimations of all monitored components, the future health states and system capabilities become predictable. By comparing these to the mission objectives the UAV's operator gets a precise information on mission safety and is supported in the decision making process.

The applicability of this approach is shown for a representative UAV mission scenario with exemplary RUL data. This paper is considered as a contribution to the PHM research at the end of the OSA-CBM process to connect RUL predictions with the decision-making process.

Introduction

Current outlook studies from the FAA and EASA predict a rapid emerging sector of civil unmanned aerial vehicles (UAV) for the coming years. It is expected that the number of UAVs will grow up to 400 thousands in Europe during the next 20 years [1]. An acceptance for UAV's integration into the existing airspaces can only rise together with the satisfaction of the overall system safety. However, compared to the current commercial aviation the operators will have a diverse background because of a wide variety of services, which shall be fulfilled by these new aircraft type. Beside operators which conduct public safety and security missions, e.g. in catastrophic situations, there will be those whose missions follow commercial interests, e.g. parcel delivery or infrastructure inspection [1]. While operators for public affairs might be able to follow strict safety guidelines, this will not necessarily, be the case for each of the commercial operators. Therefore, it is of high interest to get a feedback on the current and future health status of the aircraft and its ability to fulfil a specified mission.

State of the art methods for a system's safety assessment are based on reliability and safety engineering methods, which rely on statistical data of historic failure events. They are mainly performed during the development phase [2]. Furthermore, methods, which maintain the system's health state when it is in operation, are based on corrective or scheduled maintenance actions. In case of the latter, a statistical estimation of a part's mean time to failure (MTTF) is assumed for maintenance planning [2]. Mitigation of faults, which are covered by corrective maintenance actions, is typically performed by fault detection and isolation mechanisms, which enable the system to degrade gracefully to a safe state [3].

The relatively new discipline of Prognostics and Health Management (PHM) is an alternative approach to enable a better scheduling of part's time to failure. Sensor data is acquired and used to determine the part's actual health state and use this information as an input for a prediction of its remaining useful life (RUL). Thereby, the operator can easily be informed that a part of the system will fail in a specific amount of time and corresponding maintenance actions can be planned accordingly. An advantage of PHM is that this approach can estimate this important information during flight [4].

However, depending on the operator's experience it might not be sufficient to know that a specific part will fail. It is also important to be aware of the impact on the system's reliability. In case the operator has an understanding of the system and its individual components, it could be assumed that the impact is known. Nevertheless, this cannot be supposed for everyone [4].

A general method, which combines PHM results with conventional reliability methods for an improved prediction up to the system level, was proposed by the authors in [5]. The application of the proposed method supports the operator in the decision-making process, which is aligned at the end of the widely known OSA-CBM [6]. Therefore, it directly addresses the problem of missing system understanding. In the following, the application of the method on the safety critical control surface actuation system of an UAV is demonstrated.

1. Review of Methods

Before the PHM-based reliability prediction method is applied to the UAV's control surface actuation system, a brief overview of the method and definitions is recapped.

1.1 Conventional Reliability Assessment

Reliability is defined in [7] as "the probability that an item will perform a required function under specified conditions, without failure, for a specified period of time". In order to apply the PHM-based reliability prediction it is necessary to distinguish different levels for which a reliability assessment is possible. Table 1 depicts the terminology for this paper, which complies with [7].

Traditional reliability methods for parts assess the reliability based on historical failure data which are acquired from experience or if no data are available from so-called failure rate handbooks such as the MIL-HDBK-217F [8]. From these data, a distribution for each part's reliability can be set up e.g. based on an exponential or Weibull distribution [9].

Table 1. Terminology used within this paper, based on [7]

Term	Definition
Part	Smallest entity for which a reliability can be stated.
Component	Any self-contained part or combination of parts, which performs a distinct function necessary to operate the system. A component has two states, either working or failed.
System	A combination of N inter-related components arranged to perform a specific function. A system can have up to 2^N failure states, depending on the combination of component failures.

The overall component's reliability is aggregated from the combination of the part's reliability using combinatorial models or state-based models. While combinatorial models rely on Boolean algebra and only consider the two states failed or non-failed, state-based models can consider multiple states. However, they can run into the so-called state-space-explosion problem [10].

A typical example for combinatorial models is the widely known fault tree model, which is often applied in aviation branch. In a fault tree, the top event – a failure – is recursively split into all possible combinations of logical AND or OR connections. Thereby, the fault tree models the part's fault combination, which leads to the failure of the component. As an input for a basic event of the fault tree, either a failure distribution or the MTTF can be used, which results from the previous analysing step of the part's reliability [7, 11].

1.2 Basics of PHM

PHM algorithms aim to predict the RUL of a given part. An exemplary prediction is shown in Fig. 1 and used to define the notation for this paper.

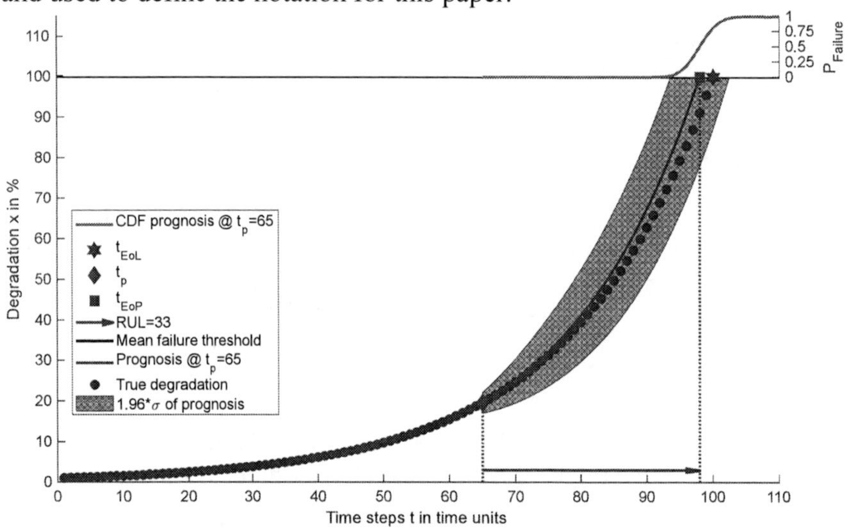

Fig. 1. Exemplary Prognosis

In Fig. 1, the prognosis starts at time $t_p = 65$. First, the current degradation level x needs to be determined. Second, the degradation path is predicted from the current degradation level until the predefined degradation threshold $x_T = 100$, based on trained prediction models. The time step when the threshold is crossed is called end of prediction t_{EoP} and the part is considered to have failed. Calculating the difference $t_{EoP} - t_p$ results in the RUL, i.e. the time that the component is assumed to stay alive. However, the true degradation path in Fig. 1 differs from the predicted and the true end of life t_{EoL} differs from t_{EoP}. Uncertainty of the prognosis mainly contributes to this difference and it is depicted as the shaded area around the prediction path in Fig. 1. Considering the prediction path and the uncertainty a cumulative distribution function (CDF) of the prognosis is determined, which estimates the probability of failure for each time step [6].

It is necessary to fulfil the four fundamental assumptions defined in [12] to apply a PHM method: The part has to degrade as a function of use, time and environmental conditions and the degradation is a monotonic process. Furthermore, signs of aging have to be visible and it has to be possible to fit a model for RUL estimation.

The described PHM process can be repeated as soon as a new estimation of the current degradation level is available. Ideally, the uncertainty bound becomes smaller for later predictions because the algorithm's precision increases (as shown in Fig. 7). The repetition leads to continuously updated CDFs, which are the major input of the dynamic reliability assessment method [5].

1.3 PHM-based Reliability Assessment

PHM-based reliability assessment as proposed in [5], aims for a prediction of the whole system's reliability considering different operating or failure states. It combines the conventional reliability assessment approach of a fault tree with predicted failure probability CDFs of a PHM approach. The resulting method is a hybrid reliability model consisting of multiple fault trees, which are merged into a system wide state-based model. Each fault tree describes an individual component. Fig. 2 depicts the proposed method's structure.

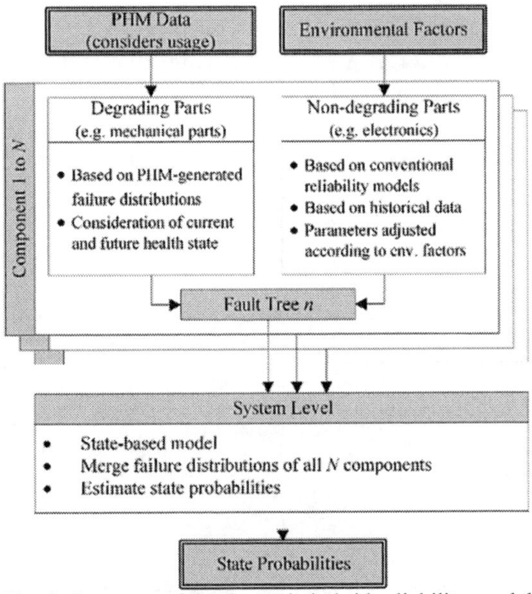

Fig. 2. Structure of the dynamic hybrid reliability model

As it is shown in Fig. 2, the method does not only include PHM results. It is able to consider the combination of parts, which are either monitored by a PHM approach or described through a conventional failure distribution function. The CDF, which results from the PHM algorithm, replaces the part's conventional static failure distribution. Non-degrading parts, e.g. most electronics, do not fulfil the PHM's fundamental assumptions as stated in [12]. Therefore, these parts are still considered using static probability distributions like exponential or Weibull distribution fitted on experience or failure rate handbook data.

The system can be operated in different states. A failure of an individual component must not necessarily result in a system's failure, but leads to a change in its operating state, e.g. a reduced operating performance. This is expressed by the state-based model on the system level, which estimates the state probabilities from the aggregation of the fault trees.

2. Application on an UAV's actuation system

The dynamic PHM-based reliability assessment method is applied to a hybrid UAV's control surface actuation system. A hybrid UAV is an aircraft, which can be operated either in a fixed-wing mode or in a multicopter mode. It has the advantages of a fixed-wing aircraft's long travel distance and a multicopter's vertical takeoff, landing and hover capabilities. Within this paper, the Quadcruiser of Airbus Group [13] – shown in Fig. 3 – is selected for the case study and especially its control surface actuation system.

Fig. 3. Quadcruiser [13]

2.1 Description of control surface actuation system

Airbus' Quadcruiser has six aerodynamic control surfaces. There are two control surfaces for each rotational axis, i.e. two aileron, two elevator and two rudder. All control surfaces are actuated by the same electromechanical actuator (EMA) in form of a servo. A servo has an output lever, which is actuated through a gearbox by a small motor that is controlled by an electronic board. The servo under examination is a redundant servo from Pegasus Actuators GmbH[1]. This servo has two brushless direct current (BLDC) motors, two electronic boards and two connectors for the servo's connection to the flight control computers. The servo is shown schematically and simplified in Fig. 4, with focus on the relevant parts.

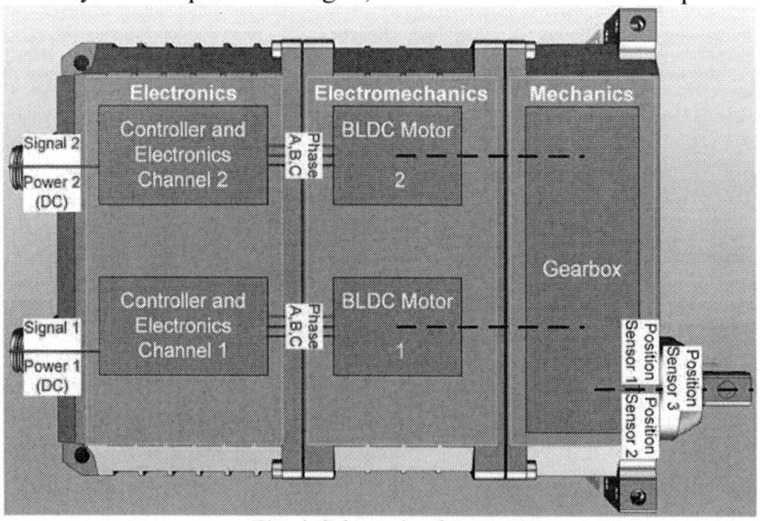

Fig. 4. Schematic of Servo

It can be seen that the two controller and both BLDC work independently, i.e. they are redundant. There is one gearbox connected to both BLDC motors. Therefore, this has to be handled as a possible single point of failure. At least two of the three position sensors need to deliver a valid signal in order to guarantee correct control of the output shaft's angular

[1] http://www.pegasus-actuators.com

position. Fig. 5 shows the Boolean structure as a fault tree. The servo's failure event occurs if both controller OR both BLDC OR the gearbox OR 2 out of 3 position sensors are broken.

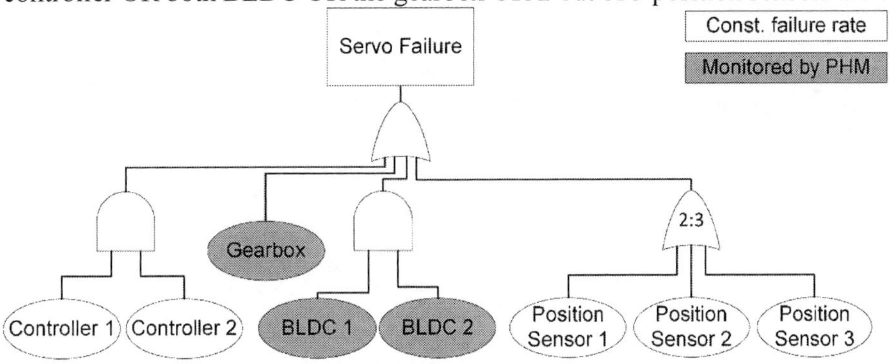

Fig. 5. Fault tree of servo

For the rest of the paper it is assumed that the gearbox and both BLDCs are monitored by a PHM system, as these are those components expected to degrade. The remaining parts are expected to fail on a more random basis and will be represented by constant failure rates.

2.2 States of the actuation system

Based on the described system the Quadcruiser's control surface actuation system has 2^6 states, which result from the failure combinations of the six EMAs. States can be grouped, depending on their criticality for the aircraft. For example, the failure of one aileron and one elevator will result in a degraded performance, while the failure of both ailerons will result in a loss of control (LoC) as there is no more control for this axis. With this approach, the initial 64 states are merged into the five remaining states presented in Table 2.

Table 2. Grouped States for the actuation system

State	Failure State	Effect	Criticality
S1	No failure	Normal Operation	None
S2	1-Axis degraded	Reduced Performance	Low
S3	2-Axes degraded	Reduced Performance	Medium
S4	3-Axes degraded	Reduced Performance	Medium (very unlikely)
S5	Axis-failure	Loss of Control	High

2.3 Test rig

To support the project in future with real data, there is currently a test rig under construction, which will deliver degradation data for the servo's degrading parts. The test rig's mechanical setup is shown in Fig. 6. At the left side, the servo of Pegasus Actuators GmbH is mounted to the test bed and coupled by a safety clutch to a torque sensor. A Kollmorgen synchronous torque motor (at the right side) is connected by another clutch to the other side of the torque sensor. The servo is commanded to a specified angular position and the torque motor applies a load. This setup enables commanding the servo to positions, which are typical for a flight mission, and apply the corresponding aerodynamical loads of the control surfaces at the same time.

Pegasus Actuators GmbH delivered a slightly modified version of the servo with additional sensors. It has temperature sensors at each BLDC, current and voltage sensors to measure the direct current (DC) at the connectors and the alternating current (AC) at each BLDC motor phase. A vibration sensor is applied on the housing of the servo in the area of the gearbox and the shaft's position can be determined by the servo's internal position sensors as well as the torque motor's encoder. The torque is captured by the torque sensor.

Fig. 6. Test rig mechanical setup

The data is acquired using a National Instruments CompactRIO (cRIO-9038) system with a high resolution and high sample rate analogue digital converter card (NI 9220). AC current and voltage of each phase and vibration data are acquired with a frequency of 25 kHz. Torque, DC current and voltage are acquired with a frequency of 2 kHz. Position data are measured with a 1 kHz frequency and temperature data with a sample time of 2 seconds. Accelerated life tests are planned for generation of degradation data, which can be used to train data-based PHM algorithms to determine the part's actual CDF.

As the test-rig is still in its final construction phase, simulated data is used for this paper. Its generation will be discussed in the following section.

2.4 Simulated input data

As stated in 2.1 the model will use input data from classical failure rates as well as from a PHM system. The initial data is based on a previously conducted Failure Modes and Effects Analysis (FMEA). The used failure rates[2] are shown in Table 3.

Table 3. Failures Rates λ in failure per million operating hours

Controller	Gearbox	BLDC	Position Sensor
$\lambda_{Con} = 5$	$\lambda_{Gear} = 2$	$\lambda_{BLDC} = 100$	$\lambda_{Sen} = 3$

For the electronic components controller and position sensor, which are not considered by PHM, constant failure rates are assumed and the probability of each failure is thus based on the exponential distribution defined as $F(t) = 1 - e^{-\lambda t}$. The gearbox and both servo's BLDCs on the other hand are assumed to be monitored by a PHM system. As there is no real PHM data available for the system under study at the point of writing, the PHM data is artificially generated. Therefore, for each of the degrading parts the true end of life is randomly sampled, based on the failure distribution characterized by the parameters in Table 3. Afterwards, Gaussian CDFs are generated, which shall represent the PHM CDFs at different prognosis time steps (t_p). Each PHM CDF tries to fit the true failure time (t_{EoL}) with increasing accuracy and precision for later predictions. In this way, the behaviour of an idealized PHM system is simulated.

In Fig. 7, the generated PHM CDFs for the left rudder servo are exemplary shown for all monitored parts (two BLDCs and one gearbox). Each plot shows the true failure CDF as a step function and the corresponding generated PHM CDFs based on predictions at different t_p with increasing accuracy. From these data, it can be concluded that the true failure of this EMA would be at $t_{EoL_{EMA}} = 2968$, when the second (remaining) BLDC has failed, after the first failed at $t_{EoL_{BLDC\,A}} = 1520$. This servo's gearbox survives all other parts due to its very low failure rate. In a similar fashion, the PHM CDFs for the remaining EMAs are generated.

---

[2] These values have been modified for this paper to maintain data nondisclosure

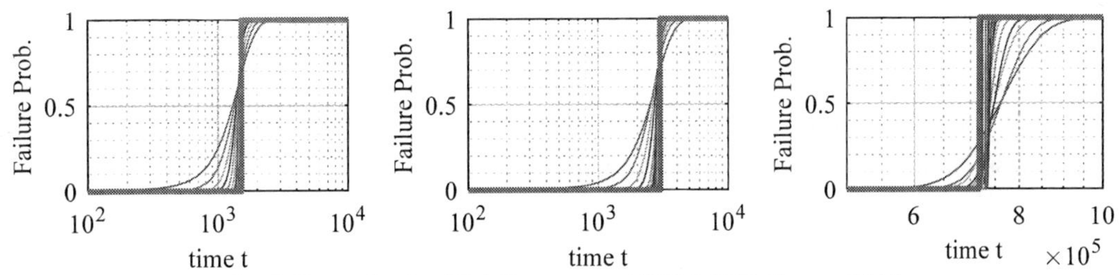

Fig. 7. Example of artificially generated PHM CDFs for one EMA (CDFs from left to right: BLDC A, BLDC B and the gearbox)

In Table 4 a comprehensive overview of the generated PHM input data is given. This will be used as a basis for the later discussion of the model's results. The data is sorted by the end of life (t_{EoL}) of each part. First, the BLDC A of the left elevator servo will fail at $t_{EoL} = 418$. Afterwards, the BLDC B of the left aileron will fail. However, both events will not cause a state change, as the motors are redundant. However, at $t_{EoL} = 1520$ resp. $t_{EoL} = 2986$ the BLDCs of the left rudder will fail, leading to a state change at $t = 2986$ as the left rudder is not controllable anymore. The controllability of the yaw-axis of the aircraft is in a degraded mode now.

Table 4. Part's sampled failure times

Part	Component	t_{EoL}	Part	Component	t_{EoL}
BLDC A	elevator left	418	Gearbox	elevator right	11610
BLDC B	aileron left	616	BLDC B	rudder right	16859
BLDC A	rudder left	1520	BLDC A	elevator right	17173
BLDC B	rudder left	2986	BLDC A	aileron right	38699
BLDC B	aileron right	4135	Gearbox	aileron left	113109
BLDC A	aileron left	5066	Gearbox	rudder right	283830
BLDC A	rudder right	5521	Gearbox	aileron right	540680

In the following, the hybrid reliability model is simulated with the input data from section 2.4, the fault tree(s) from 2.1 and the associated states defined in 2.2. The output of the model shows the estimated state probabilities. For very early estimations, where PHM data is still missing the conventional exponential failure distribution is used, as no more information is available at this point. As soon as PHM results are available for a certain part, these results will be considered by the model. In this way, the latest information about the actual health state and expected failure times of the used parts can be accounted for.

Compared to the static state estimations based on conventional failure distributions (which do not consider the true health state or any predictions), it is expected that state predictions based on PHM data will become more accurate as more PHM data becomes available.

3. Results

In the following section, the results of the simulated model are discussed. In Fig. 8, the results obtained from the model are shown. Each plot shows the estimated system state probabilities for different prognoses times ($t_P = 0, 400, 1500, 2000$). For the first prognosis at $t_P = 0$, no PHM data is available. Therefore, the model is calculated with the constant failures rates from Table 3. This result corresponds to the conventional reliability estimation method, based on historical failure data. In such a scenario the calculation would be done once. However, as PHM data is available for later epochs, this additional information shall be used. The second plot shows the results of a prognosis at $t_P = 400$. At this point PHM data is already available. Compared to the first plot, the curves of the state probabilities have slightly changed. Near $t = 400$ and $t = 600$ two jumps can be seen. Form the data given in Table 4 it can be concluded that two BLDCs (of different EMAs) will fail at $t_{EoL} = 418$ resp. $t_{EoL} = 616$. Both failures are considered by the model via the updated PHM CDFs, which results in the adjusted state probabilities, as the probability for a complete failure of an EMA and thus a state change is increased now. For the following state estimation at $t_p = 1500$ more accurate PHM data is available and it becomes apparent that one of the EMAs (left rudder) will fail soon. Thus the probability for a transition from state S1 (no failure) to S2 (1-axis degraded) increases. In parallel the probability for S2 (2-axis degraded) becomes more likely as well, while the probability to remain in S1 drops rapidly. The last plot shows the state estimation at $t_p = 2000$. This is shortly before the true failure of the left rudder EMA at $t_{EoL} = 2986$ (see Table 4). It is very likely now that the system will transit from state S1 to S2. As a reference the true state transition from S1 to S2 is indicated by the vertical dashed line in each plot.

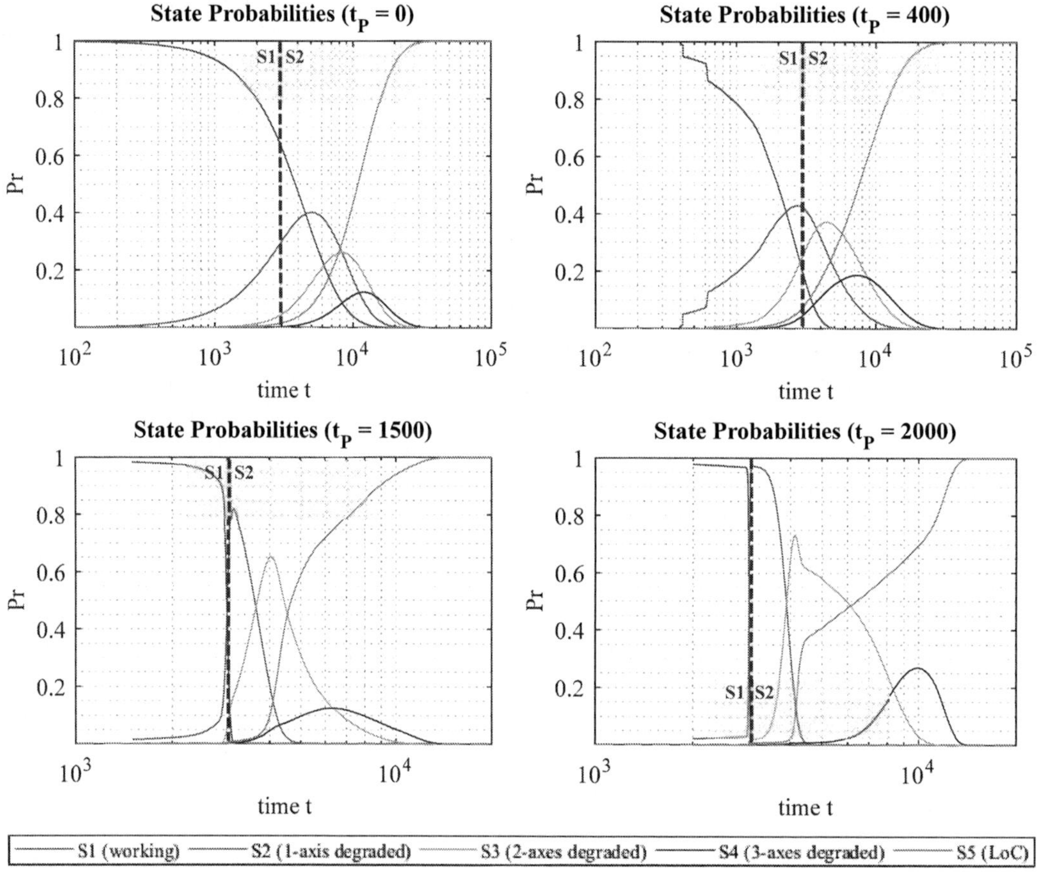

Fig. 8. Results obtained from the model

The results visualize how the model incorporates new PHM data as soon as new information are available. By recalculating the state probabilities, each time updated data is available, the model's state predictions become more and more accurate. In this way, the true health states of the system's monitored parts are considered. In contrast to conventional rather static reliability assessment methods, the presented approach provides to the operator of an UAV a dynamic reliability feedback, which includes recent and part-specific PHM data.

4. Summary and Conclusion

In this paper it was shown, how a PHM-based reliability prediction can be used to estimate the future system states of a safety critical aircraft system. As an exemplary application the control surface actuation system of the hybrid UAV Quadcruiser, consisting of six redundant EMA, was further investigated. The approach is based on the incorporation of PHM data from various subparts of the actuation system, which are aggregated and evaluated by means of a hybrid reliability model. The output of the model estimates the probabilities of the future system states. As more PHM data (with more accurate RUL data) becomes available the overall model becomes more certain about the system's state transitions. Since not only the point in time but also the most likely next state is predicted, the operator gets a meaningful feedback. As each state corresponds to a defined safety level the operator is continuously and online informed about the current and future safety of the monitored aircraft. Because of the consideration of the system's architecture, the operator does not need to have specific technical knowledge.

It has to be highlighted that the results in this paper were obtained with simulated data of an idealized PHM system. As the quality of the model's output largely depends on the performance of the used PHM algorithms and the structure of the system (e.g. number of monitored part vs. parts considered by a constant failure rate) this concept shall be further analyzed with real PHM data in the future. To support this, the authors currently work on the test rig described in section 2.3. Furthermore, it is planned to conduct realistic flight simulations on this test rig to simulate the usage and degradation of such EMAs. In this way the applicability of this method shall be further evaluated.

Acknowledgements

This paper contains results from the LuFo V-2 project "SiFliegeR". The authors want to thank the *German Federal Ministry for Economic Affairs and Energy (BMWi)* for funding this project. The test rig engineering was done by *Javier López*, whom the authors also like to thank. Furthermore, the authors want to thank *Airbus Defence and Space* and *Pegasus Actuators GmbH* for their cooperation.

References

[1] SESAR Joint Undertaking, Ed., "European Drones Outlook Study: Unlocking the value of Europe," 2016. [Online] Available: https://www.sesarju.eu/sites/default/files/documents/reports/European_Drones_Outlook_Study_2016.pdf.

[2] A. Birolini, *Reliability engineering: Theory and practice,* 6th ed. Berlin [u.a.]: Springer, 2010.

[3] I. Hwang, S. Kim, Y. Kim, and C. E. Seah, "A Survey of Fault Detection, Isolation, and Reconfiguration Methods," *IEEE Transactions on Control Systems Technology*, vol. 18, no. 3, pp. 636–653, 2010.

[4] L. R. Rodrigues *et al.,* "Use of PHM Information and System Architecture for Optimized Aircraft Maintenance Planning," *IEEE Systems Journal*, vol. 9, no. 4, pp. 1197–1207, 2015.

[5] H. Heier, S. Mehringskötter, and C. Preusche, "The Use of PHM for a Dynamic Reliability Assessment," in *IEEE Aerospace Conference*, 2018.

[6] N.-H. Kim, D. An, and J.-H. Choi, *Prognostics and Health Management of Engineering Systems: An Introduction,* 1st ed. Switzerland: Springer, 2017.

[7] *Guidelines and Methods for Conducting the Safety Assessment Process on Civil Airborne Systems and Equipment*, ARP4761, 1996.
[8] *Reliability Prediction of Electronic Equipment*, MIL-HDBK-217F, 1991.
[9] B. Schroeder and G. A. Gibson, "Understanding disk failure rates: What does an MTTF of 1,000,000 hours mean to you?," *Trans. Storage*, vol. 3, no. 3, 8, 2007.
[10] M. Malhotra and K. S. Trivedi, "Power-hierarchy of dependability-model types," *IEEE Transactions on Reliability*, vol. 43, no. 3, pp. 493–502, 1994.
[11] W. E. Vesely, F. F. Goldberg, N. H. Roberts, and D. F. Haasl, "Fault tree handbook," Nuclear Regulatory Commission Washington dc, 1981.
[12] S. Uckun, K. Goebel, and P. J. F. Lucas, "Standardizing research methods for prognostics," in *2008 International Conference on Prognostics and Health Management*, 2008, pp. 1–10.
[13] Airbus Group, *Qaudcruiser: An Innovative Hybrid Aircraft Concept.* [Online] Available: http://company.airbus.com/news-media/media~item=f2bf5171-ccd0-4863-80a6-bd1ee09731e5~.html. Accessed on: Aug. 20 2018.

Machine Learning based Data Driven Diagnostics & Prognostics Framework for Aircraft Predictive Maintenance

Partha ADHIKARI[1], Harsha GURURAJA RAO[1], Dipl.-Ing. Matthias BUDERATH[2]
[1]Airbus India, Bangalore, 560 048, India;
E-mail: partha.p.adhikari@airbus.com, harsha.rao@airbus.com
[2]Airbus Defence and Space, Willy-Messerschmitt-Strasse 81663 Munich,Germany
E-mail: matthias.buderath@airbus.com

Abstract. Recent trends of Digitalization across different industries have led to generation of massive amounts of data. As a natural consequence, there is a surge in Advanced Machine Learning techniques being applied to this Big Data. Simultaneously, there has been growing needs of increased Operational Reliability, driving down Maintenance costs and increase in Safety, due to which Predictive Maintenance is quickly becoming the most important strategy across many industries, especially in Aerospace. As newer Aircrafts are being equipped with more sensors, Machine Learning based Diagnostics and Prognostics techniques are becoming increasingly popular compared to conventional approaches in developing Predictive Maintenance solutions. Building a ML based Diagnostics & Prognostics Model requires massive amount of Run-to-Fail Sensor data, but the opportunity to capture this in-service failure related data is very limited in highly reliable & safety critical aircraft platforms in comparison to other domains. To address this challenge of lack of adequate & appropriate in-service failure data, Airbus DS has developed a Simulation Framework in the roadmap of Technology development of ISHM and Predictive Maintenance. To accelerate the development of Predictive Maintenance solutions for various aircraft systems, a Data Driven Diagnostics & Prognostics Framework has been developed. This paper outlines this unique framework and its validation using the data generated from the ISHM Simulation Framework.

Keywords: *Machine Learning, Data Analytics, Diagnostics, Prognostics, Predictive Maintenance, Simulation, Aircraft Systems, ISHM*

1 Introduction

Globally, aircraft operators as well as OEMs are facing major challenges in enhancing, even meeting specified Mission Success Rate (MSR) and Operational Support Cost targets due to unscheduled Maintenance leading to Operational Interruptions (OI). Lack of advanced warning of failure events, fault isolation / troubleshooting overhead and lack of integrated and optimized planning of fleet, maintenance and logistics are the key contributors leading to OI. So, Predictive maintenance / ISHM and related technologies are generally seen as potential strategies to address these problems.

Predictive maintenance provides an integrated solution of assessing present & future health (diagnostics and prognostics) of complete system and derives planning of operation, maintenance and logistics. With the rapid growth of technologies such as Internet of Things (IoT), Machine Learning (ML), Artificial Intelligence (AI) and Big Data technology which are inherent in Industry 4.0, Predictive maintenance is becoming smarter day-by-day. Thus, Data-driven and hybrid diagnostics and prognostics solutions are becoming more popular compared to purely model based diagnostics and prognostics. In order to keep up with the Digital transformation across the enterprise and to speed up the development of integrated Predictive maintenance solutions across various aircraft systems, a Machine Learning based Diagnostics and Prognostics (DnP) framework has been developed.

Given the highly reliable and safety critical nature of aircraft systems, it is very challenging to find sufficient amount of run to fail in-service data. Run-to-fail data is essential to train purely data-driven predictive model for calculating health grade and estimating Remaining Useful Life (RUL). Maturity and validation of ML based Diagnostics and Prognostics framework is supported by simulation strategy realized through digital twin of integrated predictive maintenance in Airbus DS.

Pundarikaksha Baruah, et. al, 2006 [1], has presented an innovative on-line approach for autonomous diagnostics and prognostics by developing a "generic" framework that is relatively independent of the type of physical equipment under consideration. The framework is based on unsupervised machine learning methods. Honglei Li, et. al, 2014 [2] presents a diagnostics and prognostics framework tested on data from a high fidelity equivalent circuit system simulation and empirical component degradation models. It adapts proven prognostics and health management techniques from machinery to electronic health management. Hamed Khorasgani, et. al, 2018 [3], claims that both data-driven and model-based diagnosis methods follow similar procedure and can by brought under one unifying framework. Their proposed framework for fault detection and isolation can integrate methods from both data-driven and model based techniques. Zhe Li, 2018 [4], suggests deep learning based framework for predictive maintenance.

The ML based Diagnostics and Prognostics (DnP) framework discussed within the scope of this paper (publication) is being developed keeping in mind several different applications and requirements of various stakeholders. One of them being an array of options available for conducting advanced ML based techniques for various benchmarking studies. Since the early stages of development, the framework has been consciously designed to be generic in nature, to minimize the efforts needed to customize it to different systems. The DnP framework is based on Python, a popular programming language for Data Science applications, providing easy deployment options to the Big Data ecosystem.

The subsequent sections describe the proposed ML based Diagnostics and Prognostics (DnP) framework in detail. Section 2, describes the simulation strategy for diagnostics and prognostics development. It is followed by section 3, which delves deeper in to the enablers for diagnostics and prognostics. Section 4 presents the proposed framework and subsections detail various algorithms incorporated in the diagnostics and prognostics framework. The framework is applied on NASA engine data and on the simulation of Electrical and Landing Gear systems within Airbus DS as use-cases and results are described in section 5.

2 Simulation Strategy for ML based Diagnostics & Prognostics Development

Airbus DS has developed a comprehensive integrated ISHM simulation framework which contributes in integrated demonstration of Proof of Enablers (PoE), training & maturity of diagnostics & prognostics algorithm, maturation of ISHM requirements (KPIs) and V&V of predictive maintenance (ISHM) functions. For end-to-end demonstration of ISHM, simulation framework hosts simulation of aircraft system with fault injection provision, on-board health assessment function, off-board analytics related to diagnostics & prognostics, fleet planning, maintenance/logistics planning, etc. in enterprise level. ISHM instead of IVHM is used because Health Monitoring and Management is considered as System of Systems approach.

There are three broad categories of condition based diagnostics & prognostics approaches: Data based, Model based and Hybrid. To develop data driven solution, ML based Diagnostics & Prognostics (DnP) Framework has been developed. DnP framework receives data from Big Data environment which is ingested with sensor and BITE data generated by ISHM simulation framework under different fault scenarios. DnP framework contains various steps of diagnostics & prognostics algorithms with having provision of selecting different options for benchmarking study. Initially, study in DnP framework provides feedback on model fidelity and data generation to ISHM simulation framework. Key objective of DnP framework is analysing & benchmarking various types of diagnostics & prognostics algorithms and providing selected trained algorithms for integration and demonstration of data driven diagnostics & prognostics algorithm in ISHM simulation framework and in-service deployment (Fig 1).

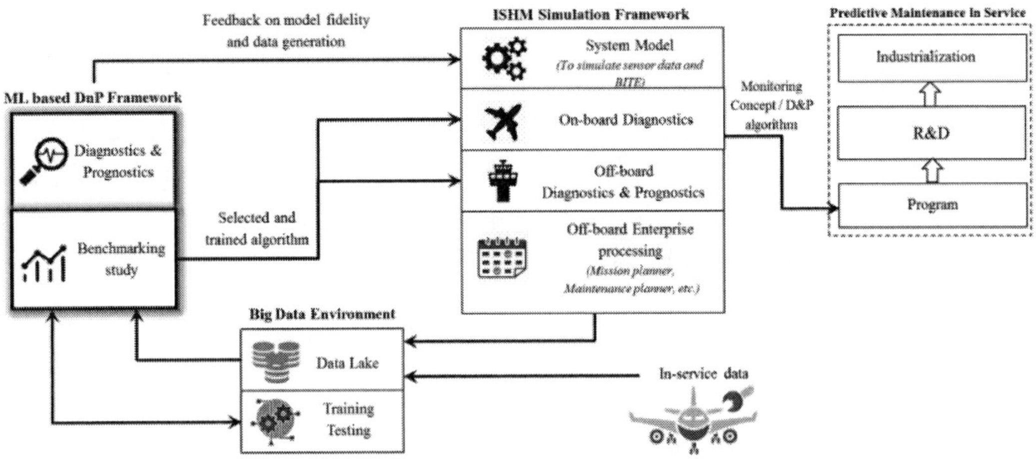

Fig. 1. ML Based DnP Framework as part of ISHM Simulation Framework

3 Enablers for Diagnostics & Prognostics

Diagnostics & Prognostics are key technology building blocks for predictive maintenance. Fig. 2 depicts different types of diagnostics & prognostics techniques. There exist the following three categories of diagnostics and prognostics.

Type I: Reliability Data-based which characterizes the expected lifetime of the average component operating in historically average conditions (e.g. Weibull Analysis). Type II: Stress-based which estimates the lifetime of the average component in a specific environment (e.g. Proportional Hazards Model). Type III: Condition-based which estimates the lifetime of the specific component in its specific operating environment.

Condition-based maintenance typically falls into three major categories viz. data-driven, model-based and hybrid. Data-driven Techniques often address only anticipated fault conditions, where a fault ''model'' is a construct or a collection of constructs such as neural networks, expert systems, etc. that must be trained first with known prototype fault patterns (data) and then employed online to detect and determine the faulty component's identity. Model-based Techniques, on the other hand, rely on an accurate dynamic model of the system and are capable of detecting even unanticipated faults. They take advantage of the actual system and model outputs to generate a ''discrepancy'' or residual between the two outputs that is indicative of a potential fault condition. Hybrid Techniques use knowledge about the physical process and information from observed data together to derive advantage of both model based and data driven methods.

Detailed enablers for data-driven diagnostics are shown in Fig. 2 and sections to follow, which constitute the proposed ML based diagnostics and prognostics framework.

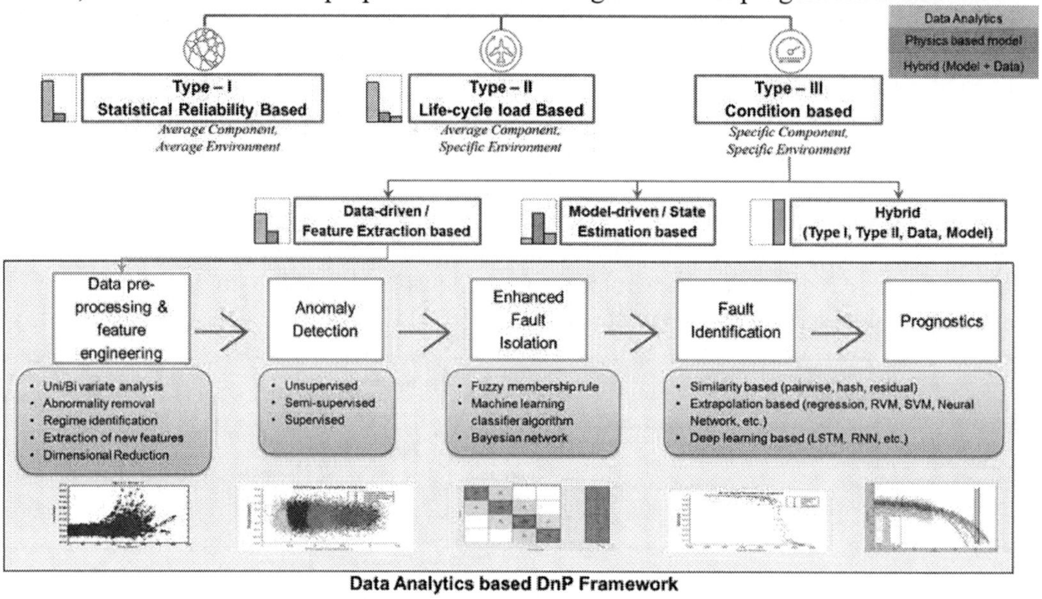

Fig. 2. Broad categories of Diagnostics & Prognostic techniques and different steps in Data Driven Diagnostics & Prognostics

4 ML Based Diagnostics and Prognostics (DnP) Framework

DnP Framework has mainly three functions: data pre-processing & feature engineering, diagnostics and prognostics, each with the provision of selecting techniques from a variety of different options. Following sub-sections summarize the landscape of various enablers for different steps of diagnostics & prognostics.

4.1 Data Pre-processing & Feature Engineering

The accuracy in ML based diagnostics and prognostics are primarily driven by the appropriate data pre-processing, data exploration and feature engineering which are to be systematically followed.

Data exploration starts with identification of predictor and target variables. In the context of predictive maintenance, predictor represents features whereas target represents the health grade or failure modes. Subsequent steps in data exploration are univariate analysis, bi-variate analysis, missing value treatment, outlier treatment and regime identification. Flight regime identification in data exploration stage for predictive maintenance may be necessary if failure propagation happens differently based on different flight regime. Flight regime are calculated by clustering operating conditions like take-off, cruise, landing, etc. Statistics of sensor data under different flight regimes are to be calculated and sensor data segregated in terms of flight regimes.

Next comes feature engineering (Fig. 3) which is a science (or art) of extracting more information from existing data. Original sensor data are primary features but in many situations generation of new derived features (FFT, Kurtosis, mean, etc.) may enhance failure observability of the model. Subsequent steps are feature benchmarking or dimension reduction.

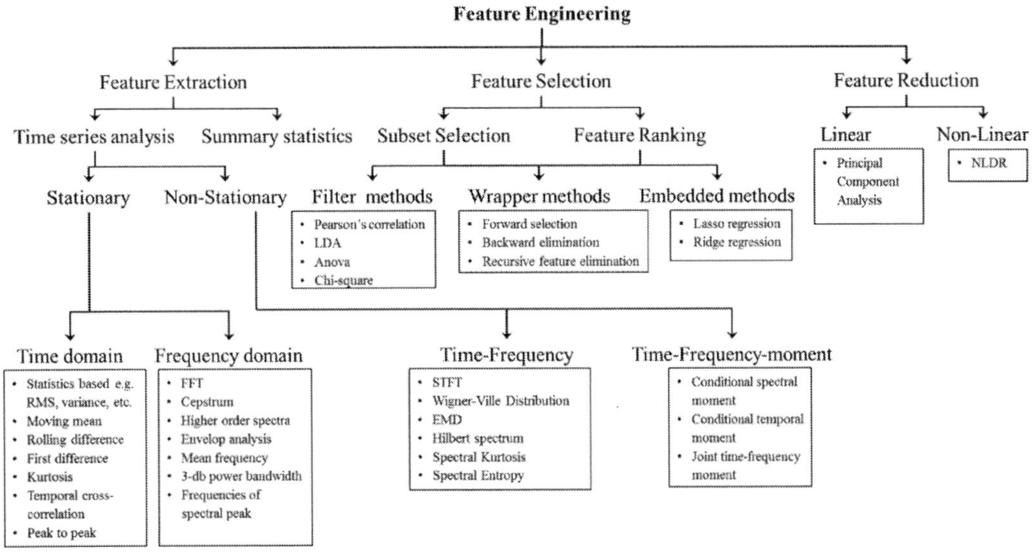

Fig. 3. Feature engineering overview

4.2 Anomaly Detection

In Diagnostics phase, the first important step is to detect the presence of an incipient failure or point of transition from normal state to degraded state within a sub-system component. Conventionally, this phase is referred as Anomaly Detection (AD). Along with Predictive Maintenance, there are various use cases for anomaly detection in areas of fraud prevention, fault detection, and monitoring across many industries such as finance, IT, security, medical, energy, e-commerce, and social media. Naturally, significant amount of research is ongoing, to develop advanced methods of Anomaly Detection.

4.2.1 Survey of state-or-art Anomaly Detection Techniques

Anomaly detection started out using statistical methods to detect outliers with "simple threshold" which was then improved to "dynamic thresholds" based on standard deviations, moving averages, comparison to a least squares model of the data, statistical comparisons of very recent data to older data, and histogram analysis. Lately the statistical approaches have been expanded by Machine Learning methods. Based on the way of training, machine learning based anomaly detection can be broadly classified in three categories: supervised, semi-supervised and unsupervised.

Supervised Anomaly Detection describes the setup where the data comprises of fully labelled training and test datasets. An ordinary classifier can be trained first and applied afterwards. Semi-supervised Anomaly Detection also uses training and test datasets, whereas training data only consists of normal data without any anomalies. The basic idea is that a model of the normal class is learned and anomalies can be detected afterwards by deviating from that model. Unsupervised Anomaly Detection is the most flexible setup which does not require any labels. Furthermore, there is also no distinction between training and test dataset

Unsupervised anomaly detection algorithms can be roughly categorized into the following main groups (Chandola V et al. 2009 [5]): Nearest-neighbour based techniques, Clustering based, Statistical algorithms, Subspace techniques, Unsupervised Neural Network, real time and time series analysis based algorithms. Fig. 4 summarizes the different categories of anomaly detection.

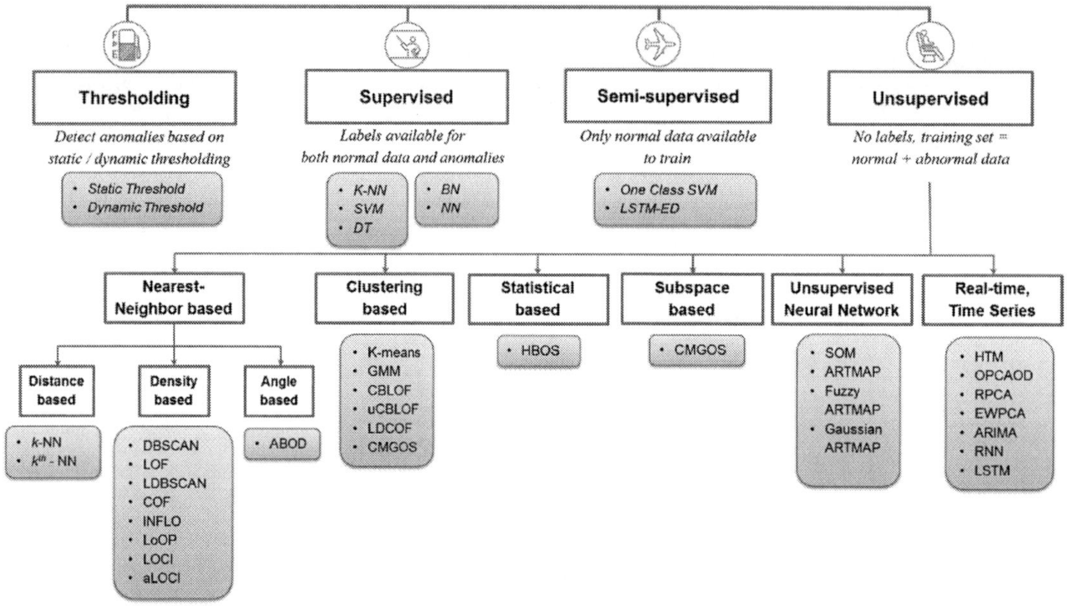

Fig. 4. Enablers for state of art anomaly detection

4.2.2 Anomaly Detection in the context of Predictive Maintenance

There are three types of anomalies: point anomalies, sequential anomalies and contextual anomalies. If a single point deviates from the considered normal pattern it is referred to as a point anomaly. If a sequence or collection of points is anomalous with respect to the rest of the data, but not the points themselves, it is referred to as a sequential or collective anomaly. If a point or a sequence of points is considered as an anomaly with respect to its local neighbourhood, but not otherwise, it is referred to as a contextual anomaly. In Predictive maintenance, sequential and contextual anomaly is applicable as it deals with sensor data under healthy stage of any asset to gradually degraded stages under different flight regimes.

Almost all types of algorithms as mentioned are applicable for predictive maintenance. Ramassol E., & Saxena A (2014) [6] surveyed extensively on different technique of anomaly detection related to predictive maintenance. Survey is summarized in the following table.

Table 1. Survey papers for Anomaly Detection

Category of Anomaly Detection	References
Supervised	Ramasso E. 2009 [7], Ramasso E. et. al. 2010 [8], Ramasso E. et. al. 2013 [9], Sarkar S. et.al 2011 [10], Xue Y. et al. 2011 [11], Zhao D. P. et al 2011 [12], Yu J 2013 [13], Lin Y. et al. 2013 [14], Tamilselvan P. et al. 2013 [15]
Semi-Supervised	Ramasso E. et al. 2013 [16], Ramasso E. et al 2013 [17], Alestra S. et al [18]
Unsupervised	Wang T. 2010 [19], Wang T. et al 2008 [20], Sarkar S. et al 2011 [10], Peng Y. et al 2012 [21], Serir L et al 2012 [22], Javed K. et al 2013 [23]

4.2.3 Time-series / Real-time Anomaly detection

Appropriate pre-processing, feature extraction and suitable customization of algorithms are needed to address time series data for predictive maintenance. Holt-Winters, Auto-Regressive Integrated Moving Average (ARIMA), and Support Vector Regression (SVR) are examples of statistical models that take into account seasonality. These models can be used as a time series dynamic threshold for anomaly detection. Alestra et al. used one class SVM for detection of degradation. For the modeling of multidimensional degradation behaviour i.e. degradation trending in time Autoregressive Integrated Moving Average (ARIMA) model is employed.

In recent years, for anomaly detection of time series data, deep learning Akash Sing, 2017 [24] has emerged as one of the most popular machine learning techniques, yielding state-of-the-art results for a range of supervised and unsupervised tasks due to its ability to learn high-level representations relevant for the task at hand. These representations are learned automatically from data with little or no need of manual feature engineering and domain expertise. For sequential and temporal data, LSTMs, RNNs have become the deep learning models of choice because of their ability to learn long-range patterns.

The ability to perform predictive analytics in real-time on streaming data is a game changer. Works Numenta, 2015 [25], Aggarwal C. et al, 2012 [26], Aggarwal C. C., 2005 [27], Aggarwal C. C. et al, 2011 [28] address application of real time anomaly detection. The most exciting technique is developed by Numenta is Hierarchical Temporal Memory (HTM) [25] which is a biologically inspired machine intelligence technology that mimics the architecture and processes of the neocortex. HTM is a promising candidate for real-time predictive maintenance.

4.3 Fault Identification & Prognostics

After an anomaly is detected, estimation of the degree of degradation is the next natural step. This is called Fault Identification which is followed by Prognostics. The main objective of prognostics is to provide an estimation of the Remaining Useful Life (RUL) of a degrading component or system. Anomaly detection and fault isolation are generally classification or clustering problem whereas Fault Identification & Prognostics are regression problems. Various techniques in Fault Identification and Prognostics are summarized in Fig. 5. Broadly there are three categories of prognostics techniques: similarity based, extrapolation based, and, model identification & estimation based. Table 2 lists some references for each broad category.

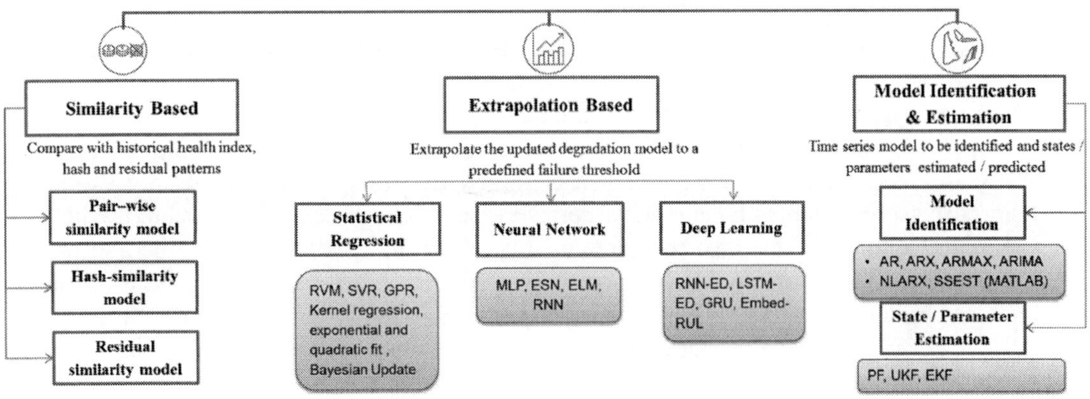

Fig. 5. Enablers for Fault Identification & Prognostics

Table 2. List of references for different prognostics methods

Category of Fault Identification & Prognostics	References	Remarks
Similarity based	Wang T. et al. 2008 [20], Wang T. 2010 [19], Peng et al. 2012 [21], Ramasso et al. [16], Ramasso E. (2014) [29]	[20] [19] →HI-based 3 similarity measures and kernel smoothing [21] → Similar to 36 and 37 using 1 similarity measure [16] → Feature-based similarity, 1 similarity measure, ensemble, degradation levels classification [29] → HI-based similarity, polygon coverage similarity, ensemble
Extrapolation based – Statistical Regression	Hu C. et al. 2012 [30], Coble J. 2010 [31], Coble & Hines, 2011 [32], Siegel 2009 [33], T. Wang 2010 [19], P. Wang et al. 2012 [34], Juan F. L et al. 2017 [35], Hu et al., 2012 [30], Liu K. et al. 2013 [36]	[31] [32] → Quadratic fit, Bayesian updating, [33] → Logistic Regression, [19] → Karnel regression, RVM [34] →RVM [35]→ GPR [30]→ RVM, SVM, RNN, Exponential and quadratic fit, Bayesian updating [36] → Exponential fit
Extrapolation based – Neural Network	Riad A. et al. 2010 [37], Jie Liu et al. [38], Marco R. et al. 2016 [39]	[37] → Application of Adaptive RNN [39] → ESN
Extrapolation based – Deep Learning	Gugulothu N. et al. 2017 [40], Jason Deutsch et al. 2017 [41]	[41] → Embed-RUL is used here.
Model Identification	Alestra S. et al. [18]	[18] → One class SVM, ARIMA
State Estimation	Bhaskar Saha et al. 2008 [42]	[42] → RVM, fitting exponential growth model, Particle Filter

In similarity based method, historical instances of trajectories of different features of all possible failures are used to create a library. For a given test instance similarity with instances in library is evaluated to generate a set of RUL estimates. Based on how the notion of similarity is defined and quantified there are three types of models [43]: Pairwise similarity model, Hash similarity model & Residual similarity model. Pairwise similarity model computes the distance between different time series, where distance is defined as correlation. Hash-similarity model transforms historical degradation data from each member of ensemble into fixed-size condensed information such as the mean, total power, maximum or minimum values, or other quantities. Residual-based estimation fits prior data to model such as an ARMA model or a model that is linear or exponential in usage time. It then computes the residuals between predicted model data and test data.

Unlike the similarity-based approach, the extrapolation-based approach employs the training data set not for the comparison with the testing data set but rather for obtaining prior distributions of the degradation model parameters. The testing data set is then used to update these prior distributions. An RUL estimate can be obtained by extrapolating the updated degradation model to a predefined failure threshold. Degradation models can be constructed using statistical regression, neural network and deep learning approach.

RUL can be estimated using dynamic model identification techniques like AR, ARX, ARMAX, ARIMA, etc. which are trained (i.e. model coefficients are estimated) using time series data [43]. Using time series model, features or parameters are predicted to compute RUL. There are also recursive estimators fitting models in real-time and process the data, such as recursive ARX and recursive AR. RUL estimation with state estimators such as Unscented Kalman Filter (UKF), Extended Kalman Filter (EKF), and Particle Filter (PF) works in a similar way. One performs state estimation on some time-varying data, and predict future state values to determine the time until some state value exceeds some specified limit or threshold indicating presence of failure.

4.4 Enhanced Fault Isolation

ML based component level fault isolation in early stage of degradation is carried out by supervised classification. In most cases residuals computed from sensor measurement and estimated measurement using physics based model are considered useful features along with other features e.g. sensor measurements and derived features from sensors. Health aggregation from subsystem component to subsystem, system and aircraft level is integral part of diagnostics. Lopez B. et al 2016 [44] implemented component level fault isolation and health aggregation using Bayesian network based HLR framework.

5 Use cases to validate D&P Framework

The proposed DnP framework shows a series of systematic steps to implement diagnostics and prognostics of different aircraft systems. Initially, this framework is used to implement diagnostics & prognostics of Landing Gear and Electrical systems. This in-turn helps to validate and mature the DnP framework. Turbofan engine simulated data from NASA [45] is used as third use case to benchmark the algorithms of the framework due to the popularity of this dataset which is mentioned in more than 100 publications related to diverse varieties of Diagnostics & Prognostics solutions. Details of fault and sensor measurements are mentioned for respective use-cases as shown in the Fig. 6.

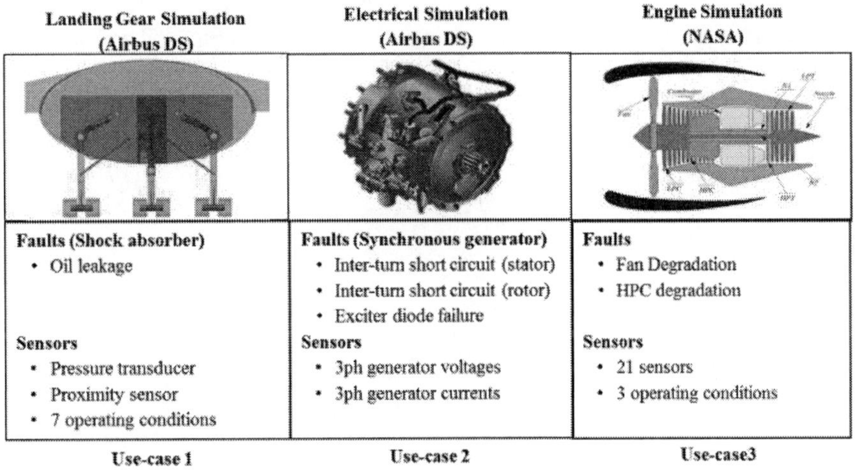

Fig. 6. Use-cases for DnP Framework

Various steps for use-case study driven by the DnP framework are shown in fig. 7. Feature engineering study for all systems are provided in Table 3. For Anomaly detection, One class SVM is chosen to apply for all use-cases, where model is trained using only selected features in normal operation. For fault isolation bench-marking study was carried out and K-NN classifier was selected for Engine whereas SVM was chosen for Electrical system. Different degradation levels (Healthy, Sub-healthy, Degraded, Fail) were obtained using K-mean clustering of PCA components of selected features. Time series health index for all training instances was computed using Euclidean distances from fail centroid. Then degradation model was generated using RVR. The degradation trajectory of testing data is compared with that of stored training data and the trajectory with closest match are selected for RUL calculation [46]. This Similarity based approach is used for Engine based on benchmarking study. For Landing Gear, ARIMA model was sufficient to predict RUL. Selected results are provided in Fig. 8 to 11 and Table 3, 4 and 5.

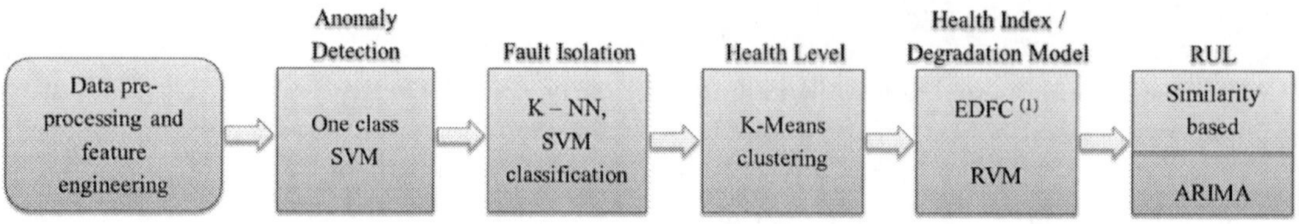

(1) Euclidian distance from fail centroid

Fig. 7. Various sequential steps for use-case study driven by DnP framework

Table 3. Feature Engineering results

	Landing Gear	Electrical	Engine
No. of original features (sensor)	9	6	24
New features	2 (Settling time, damping ratio)	66 (FFT – 36, Kurtosis – 6, Peak to Peak – 6, Lyapunov Exponent – 6, Entropy – 6, Correlation – 6)	24 (moving average)
Features after benchmarking	3 (Settling time, damping ratio, shock absorber height)	12 (Peak to Peak – 2, FFT – 10)	12, 8 PCA components

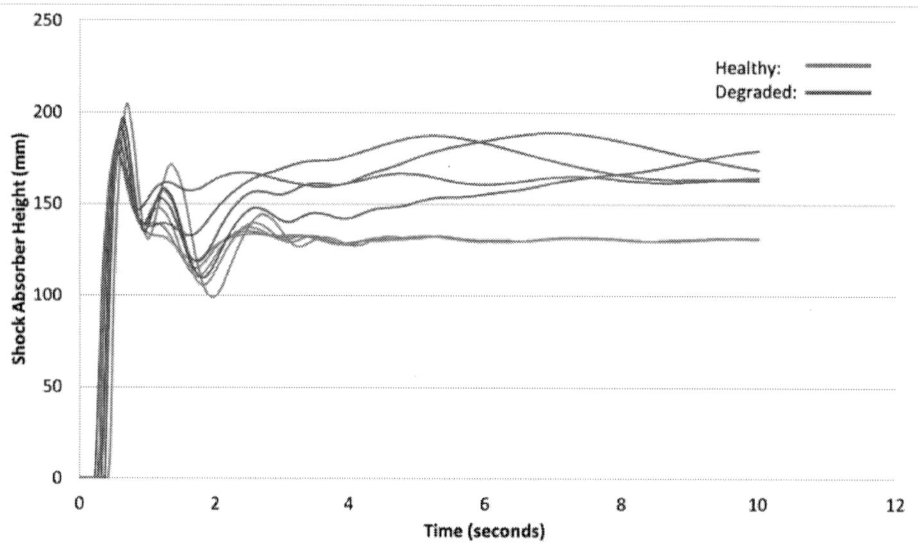

Fig. 8. Univariate analysis of Shock Absorber of Landing Gear

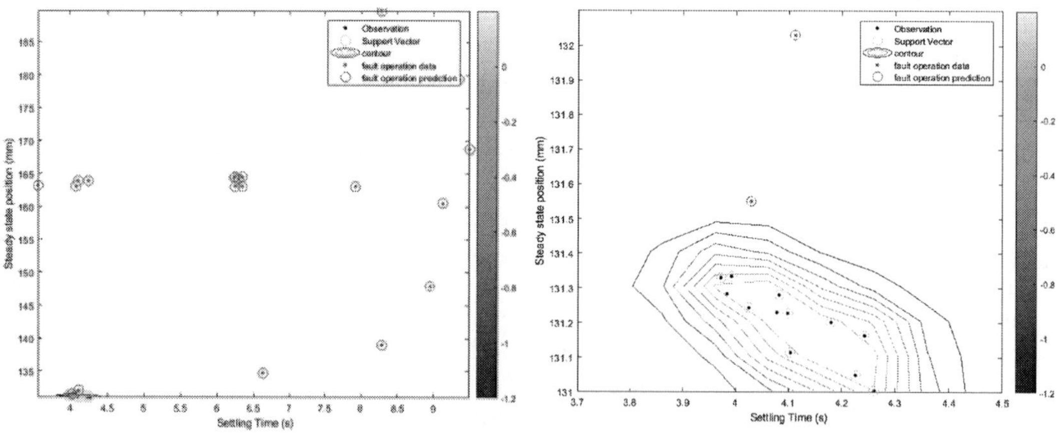

Fig. 9. Anomaly Detection of Landing Gear : degradation of shock absorber due to Oil leakage. Original - (left), Zoomed (right)

Table 4. Benchmarking for fault classification of Electrical System

Sr. No.	Algorithm Name	Accuracy	Precision	Recall	Training Time (S)
1	Decision Tree	0.99922	0.99922	0.99922	2.80
2	Naive Bayes	0.90429	0.90429	0.90429	0.41
3	SVM	1	1	1	8.54
4	Extra Tree	1	1	1	1.87
5	Random Forest	1	1	0.99929	7.89
6	ADABoost	0.99872	0.99872	0.99872	11.83
7	Logistic	1	1	1	6.54
8	Linear SVC	1	1	1	54.21

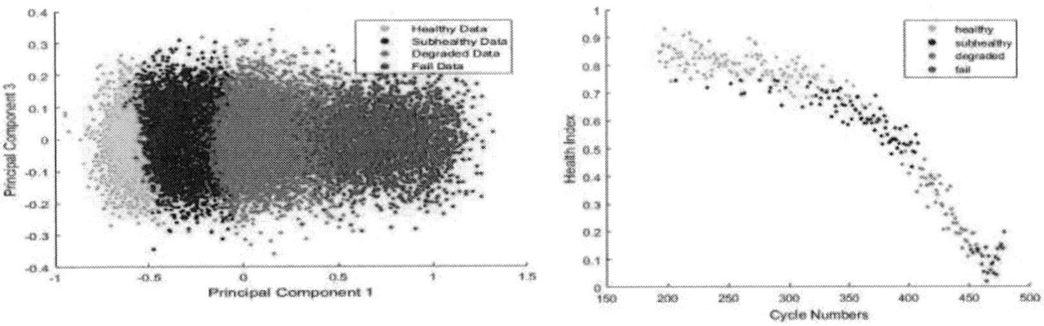

Fig. 10. Cluster features with health levels : Engine - (left)
Propagation of health index with cycle number : Engine - (right)

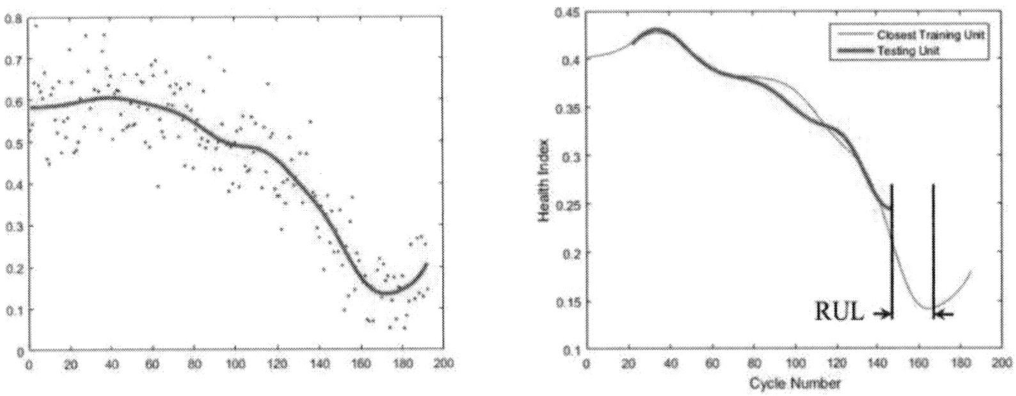

Fig. 11. Health degraded model with fitted RVR : Engine (left), Matched degraded pattern and RUL (right)

Table 5. Benchmarking of various algorithms for RUL calculation: Engine

Sr. No	Method	KPIs						
		PHM08	FP\ FN	MSE	MAPE	MAE	ME	MAD
1	Echo State Network (ESN) – LMS	2884	8 \ 68	1021	55.5	25.5	21.2	18.6
2	Echo State Network (ESN) – ABC	7635	11\69	1558	81.51	31.8	24.4	24.6
3	Regression Analysis	2884	8 \ 68	1021	55.5	25.5	21.2	18.6
4	MLP Neural Network (MLPNN)	1767	6\51	700	34.5	19.1	14.3	17.1
5	Recurrent Neural Network : RNN - NAR	3056	29\39	866	59.7	23.2	4.4	23.2
6	Recurrent Neural Network :RNN – NARX	1123	23\39	516	33	18.2	4.2	17.2
7	Similarity based	275	16\16	206	24.3	11.0	-2.5	11.0

6. Conclusion

Extensive survey has been done to find the state-of-the-art techniques of diagnostics and prognostics and selected algorithms are implemented in this ML based Diagnostics and Prognostics framework. Key highlights of this framework include its generic nature, options for benchmarking and ability to be customized for specific systems. This has been validated and matured with use-cases of diverse systems. However, this framework will undergo continuous maturity with experience gained in Predictive maintenance across different aircraft systems.

7 Acknowledgments

The authors are grateful to Gaurav Kumar, Elijah Toppo, Bibhudatta Patnaik, Manish Patwari, Pandiyaraj Subramani, Ajith Shenoy and Amey Muley for their support. We are also grateful to reviewers for many of the improvements to the document.

Nomenclature

ABOD	Angle Based Outlier Detection
AC	Advisory Circular
AD	Anomaly Detection
AI	Artificial Intelligence
aLOCI	Approximate Local Correlation Integral
AMC	Acceptable Means of Compliance
ARMAX	Auto-Regressive Integrated Moving Average
ARIMA	Auto-Regressive Integrated Moving Average
ARP	Aerospace Recommended Practice
ARTMAP	Adaptive Resonance Theory Mapping
AWR	Airworthiness Report
BIT	Built-in Test
BITE	Built-in Test Equipment
BN	Bayesian Network
uCBLOF	Cluster-Based Local Outlier Factor
CBM	Condition based maintenance
CC	Certification Coordinator
CMGOS	Clustering-based Multivariate Gaussian Outlier Score
COF	Connectivity-Based Outlier Factor
CS	Certification Specification
DBSCAN	Density-Based Spatial Clustering of Applications with Noise
DnP	Diagnostics and Prognostics
DT	Decision Tree
EAI	Enterprise Application Integration
EKF	Extended Kalman Filter
ELM	Extreme Learning Machine
EMD	Empirical Mode Decomposition
ESN	Echo state network
EWPCA	Exponentially Weighted Principal Component Analysis
FFT	Fast Fourier Transform
FHA	Functional Hazard Analysis
IVHM	Integrated Vehicle Heath Monitoring
K-NN	k – Nearest Neighbour
KPI	Key Performance Index
LDA	Linear Discriminant Analysis
LDBSCAN	Local-Density-Based Spatial Clustering of Applications with Noise
LDCOF	Local Density Cluster-based Outlier Factor
LOCI	Local Correlation Integral
LOF	Local Outlier Factor
LoOP	Local Outlier Probability
LRU	Line Replaceable Unit
LSTM	Long Short Term Memory
LSTM-ED	Long Short Term Memory Encoder Decoder
MAD	Mean Absolute Deviation
MAE	Mean Absolue Error
MAPE	Mean Absolute Percentage Error
ME	Mean Error
ML	Machine Learning
MLP	Multi-Layer Perceptron
MSE	Mean Square Error
NFLO	Influenced Outlierness
NLARX	Non-linear Auto Regressive Exogenous
NLDR	Nonlinear dimensionality reduction
NN	Neural Network
OCSVM	One Class SVM
OEM	Original Equipment Manufacturer
OI	Operational Interruptions
OPCAAD	Over-sampling PCA on-line anamoly detection
OR	Operational Reliability
ORA	Operational Risk Assessment
OSA	Open System Architecture
PCA	Principal Component Analysis
PF	Particle Filter
PHM	Prognostic Health Management

FMECA	Failure Modes, Effects, and Criticality Analysis	PoE	Proof of Enablers
		RCM	Reliability Centered Maintenance
FP/FN	False Positive / Negative	RMS	Root mean square
GMM	Gaussian Mixture Model	RNN	Recurrent Neural Network
GPR	Gaussian Process Regression	RNN-ED	Recurrent Neural Network Encoder Decoder
GRU	Gated Recurrent Units		
GUI	Graphical User Interface	RPCA	Real Time PCA
HBOS	Histogram-based Outlier Score	RUL	Remaining Useful Life
HILS	Hardware in Loop Simulation	RVM	Relevance Vector Machine
HLR	High Level Reasoning	RVR	Relevance Vector Regression
HTM	Hierarchical Temporal Memory	SHM	Structural Health Monitoring
HUMS	Heath Usage Monitoring System	SOM	Self-organizing Map
IA	Integrity Assessment	STFT	Short time fourier transform
ICA	Instruction for Continued Airworthiness	SVM	Support Vector Machine
		SVR	Support Vector Regression
INFLO	Influenced Outlierness	TRL	Technology Readiness Level
IoT	Internet of Things	UKF	Unscented Kalman Filter
ISHM	Integrated System Health Monitoring	V&V	Verification and Validation

References

[1] P. Baruah, R. B. Chinnam and D. Filev, "An Autonomous Diagnostics and Prognostics Framework for Condition-Based Maintenance" in *International Joint Conference on Neural Networks Sheraton Vancouver Wall Centre Hotel*, BC, 2006.

[2] Honglei Li, Xuerong Ye, Cen Chen and G. Vachtsevanos, A Framework for Model-Based Diagnostics and Prognostics of Switched-Mode Power Supplies, ANNUAL CONFERENCE OF THE PROGNOSTICS AND HEALTH MANAGEMENT SOCIETY, 2014.

[3] H. Khorasgani, Ahmed Farahat, Kosta Ristovski, C. Gupta and Gautam Biswas, Framework for Unifying Model-based and Data-driven Fault Diagnosis, ANNUAL CONFERENCE OF THE PROGNOSTICS AND HEALTH MANAGEMENT SOCIETY, 2018.

[4] Z. Li, "Deep Learning Driven Approaches for Predictive Maintenance: A Framework of Intelligent Fault Diagnosis and Prognosis in the Industry 4.0 Era," 2018.

[5] V. Chandala, A. Banerjee and V. Kumar, "Anomaly Detection: A Survey, ACM Computing Surveys" *University of Minnesota,* 2009.

[6] E. Ramasso and A. Saxena, "Performance Benchmarking and Analysis of Prognostic Methods for CMAPSS Datasets" *International Journal of Prognostics and Health Management,* vol. 5, pp. 1-15, 2014.

[7] E. Ramasso, "Contribution of belief functions to hidden Markov models with an application to fault diagnosis" *Machine Learning for Signal Processing,* vol. 2009, 2009.

[8] E. Ramasso and R. Gouriveau, "Prognostics in switching systems: Evidential Markovian classification of real-time neuro-fuzzy predictions" *Prognostics and Health Management Conference,* vol. 2010, 2010.

[9] E. Ramasso and R. Gouriveau, "RUL estimation by classification of predictions: an approach based on a neuro-fuzzy system and theory of belief functions" *IEEE Transactions on Reliability, Accepted,* 2013.

[10] S. Sarkar, X. Jin and A. Ray, "Data-driven fault detection in aircraft engines with noisy sensor measurements" *Journal of Engineering for Gas Turbines and Power,* vol. 133, p. 8, 2011.

[11] Y. Xue, D. P. Williams and H. Qiu, "Classification with imperfect labels for fault prediction" *Proceedings of the first international workshop on data mining for service and maintenance,* 2011.

[12] D. Zhao, R. Georgescu and P. Willett, "Comparison of data reduction techniques based on SVM classifier and SVR performance" in *Signal and Data Processing of Small Targets 2011*, 2011.

[13] J. Yu, "A nonlinear probabilistic method and contribution analysis for machine condition monitoring" *Mechanical Systems and Signal Processing,* vol. 37, pp. 293-314, 2013.

[14] Y. Lin, M. Chen and D. Zhou, "Online probabilistic operational safety assessment of multi-mode engineering systems using Bayesian methods" *Reliability Engineering & System Safety,* vol. 119, pp. 150-157, 2013.

[15] P. Tamilselvan and P. Wang, "Failure diagnosis using deep belief learning based health state

classification" *Reliability Engineering & System Safety,* vol. 115, pp. 124-135, 2013.

[16] E. Ramasso, M. Rombaut and N. Zerhouni, "Joint prediction of observations and states in time-series based on belief functions" *IEEE Transactions on Systems, Man and Cybernetics - Part B: Cybernetics,* vol. 43, pp. 37-50, 2013.

[17] E. Ramasso and T. Denoeux, "Making use of partial knowledge about hidden states in hidden Markov models: an approach based on belief functions" *IEEE Transactions on Fuzzy Systems,* vol. 10, p. 1109, 2013.

[18] S. Alestra, C. Bordry, C. Brand, E. Burnaev, P. Erofeev, A. Papanov and C. Silveira-Freixo, "Rare Event Aanicipation and Degradtion Trending for Aircraft Predictive Maintenance?" in *11th World Congress on Computational Mechanics (WCCM XI),* 6th European Conference on Computational Fluid Dynamics (ECFD VI.

[19] T. Wang, "Trajectory similarity based prediction for remaining useful life estimation" *Diss. University of Cincinnat,* 2010.

[20] T. Wang and others, "A similarity-based prognostics approach for remaining useful life estimation of engineered systems" *Prognostics and Health Management,* vol. 2008, 2008.

[21] Y. Peng, H. Wang, J. Wang, D. Liu and X. Peng, A modified echo state network based remaining useful life estimation approach, In Ieee phm conference, 2012.

[22] L. Serir, E. Ramasso and N. Zerhouni, An evidential evolving multimodeling approach for systems behavior prediction, In Annual conference of the phm society, 2012.

[23] K. Javed, R. Gouriveau and N. Zerhouni, Novel failure prognostics approach with dynamic thresholds for machine degradation, In Ieee industrial electronics conference, 2013.

[24] A. Sing, Anomaly Detection for Temporal Data using Long Short-Term Memory (LSTM), STOCKHOLM, SWEDEN: Information and Communication Technology, 2017.

[25] Numenta, "The science of Anomaly Detection - How HTM Enables Anomaly Detection in Streaming Data" *White Paper,* 2015.

[26] C. C. Aggarwal and K. Subbian, "Event detection in social streams" in *Proceedings of the 2012 SIAM international conference on data mining,* 2012.

[27] C. C. Aggarwal, "On abnormality detection in spuriously populated data streams" in *Proceedings of the 2005 siam international conference on data mining,* 2005.

[28] C. C. Aggarwal, Y. Zhao and S. Y. Philip, "Outlier detection in graph streams" in *27th International Conference on Data Engineering (ICDE),* 2011.

[29] E. Ramasso, "Investigating computational geometry for failure prognostics in presence of imprecise health indicator: Results and comparisons on c-mapss datasets" *PHME,* vol. 14, 2014.

[30] C. Hu, B. D. Youn, P. Wang and J. T. Yoon, "Ensemble of data-driven prognostic algorithms for robust prediction of remaining useful life," *Reliability Engineering & System Safety,* vol. 103, pp. 120-135, 2012.

[31] J. Coble, Merging data sources to predict remaining useful life - an automated method to identify prognostic parameters, University of Tennessee, Knoxville: Unpublished doctoral dissertation, 2010.

[32] J. Coble and W. Hines, "Applying the general path model to estimation of remaining useful life" *International Journal of Prognostics and Health Management,* vol. 2, pp. 1-13, 2011.

[33] D. Siegel, Evaluation of health assessment techniques for rotating machinery (Doctoral dissertation, University of Cincinnati, 2009.

[34] P. Wang, B. Youn and C. Hu, "Generic probabilistic framework for structural health prognostics and uncertainty management" *Mechanical Systems and Signal Processing,* vol. 28, p. 622, 2012.

[35] J. F. L_pez-L_pez, J. J. Cuesta-Ramirez and H. F. Quintero-Riaza, "Gaussian processes in ball bearing prognostics," *Dyna,* vol. 84, pp. 214-219, 2017.

[36] K. Liu, N. Z. Gebraeel and J. Shi, "A data-level fusion model for developing composite health indices for degradation modeling and prognostic analysis" *IEEE Transactions on Automation Science and Engineering,* vol. 10, pp. 652-664, 2013.

[37] A. Riad, H. Elminir and H. Elattar, "Evaluation of neural networks in the subject of prognostics as compared to linear regression model" *International Journal of Engineering & Technology,* vol. 10, pp. 52-58, 2010.

[38] J. Liu, A. Saxena, K. Goebel, B. Saha and W. Wang, An adaptive recurrent neural network for remaining useful life prediction of lithium-ion batteries, NATIONAL AERONAUTICS AND SPACE ADMINISTRATION MOFFETT FIELD CA AMES RESEARCH CENTER, 2010.

[39] M. Rigamonti, P. Baraldi and E. Zio, "Echo state network for the remaining useful life prediction of a turbofan engine" *annual conference of the prognostics and health management society,* vol. 2015, 2016.

[40] N. Gugulothu, T. V. Vishnu, P. Malhotra, L. Vig, P. Agarwal and G. Shroff, "Predicting Remaining Useful Life using Time Series Embeddings based on Recurrent Neural Networks" 2017.

[41] J. Deutsch and D. He., "Using deep learning-based approach to predict remaining useful life of rotating components" *IEEE Transactions on Systems, Man, and Cybernetics: Systems,* vol. 48, pp. 11-20, 2018.

[42] B. Saha, S. Poll, K. Goebel and J. (. Christophersen, "September). An integrated approach to battery health monitoring using Bayesian regression and state estimation" *In autotestconIEEE (p,* pp. 646-653, 2007.

[43] Mathworks, Predictive Maintenance Toolbox - User Guide, Natick, MA: Mathworks, 2018.

[44] B. S. Lopez, A. M. Siddiolo, P. P. Adhikari and M. A. Buderath, "Bayesian paradigm for aircraft operational capability assessment and improved fault diagnostics", EUROPEAN CONFERENCE OF THE PROGNOSTICS AND HEALTH MANAGEMENT SOCIETY, 2016.

[45] "Prognostics Center of Excellence Data Repository" NASA, [Online]. Available: https://ti.arc.nasa.gov/tech/dash/groups/pcoe/prognostic-data-repository/. [Accessed September 2018].

[46] Y. L. G. W. X. L. Jianming Shi, "Health index synthetization and remaining useful life estimation" *Maintenance and Reliability,* vol. 18 (4), p. 621–631, 2016.

Biographies

Partha Adhikari – Aerospace professional with more than 20 years of experience in Integrated Vehicle Health Management (IVHM) / Predictive Maintenance, Predictive Data Analytics, Navigation, Simulation & Modelling of Aircraft Systems and Avionics integration. He has served in different techno-management roles in various Aerospace organisations. Presently he is the IVHM domain lead in Airbus India, Bangalore and is working on design and development of Data driven Diagnostics and Prognostics, High level reasoning and tools related to decision support. He has published over 15 papers in the field of estimation, signal processing and IVHM in national as well as international conferences and journals.

Harsha Gururaja Rao – Software Engineer with more than 7 years of experience in designing & developing enterprise level Software applications and Predictive Maintenance solutions for Aircraft systems. Harsha has a Bachelor of Engineering Degree in Computer Science from the Visvesaraya Technological University. Harsha, in his current role as an Engineer at Airbus India, Bangalore is working on developing Data driven solutions to meet Aerospace & Defense business systems requirements in the areas of IVHM and Predictive Maintenance.

Matthias Buderath - Aeronautical Engineer with more than 30 years of experience in structural design, system engineering and product and service support. His main expertise and competency is in system integrity management, service solution architecture and Integrated System Health Monitoring and Management. Today he is head of Airbus Defence and Space R&T Strategy and Senior Expert Integrated System / Aircraft Health Monitoring and Management. He is member of international Working Groups covering Through Life Cycle Management, Integrated System Health Management and Structural Health Management. He has published more than 80 papers in the field of Structural Health Management, Integrated Health Monitoring and Management, Structural Integrity Programme Management and Maintenance and Fleet Information Management Systems.

Satellite battery degradation prognostic based on data analytics and big data infrastructure

Loïc BOUSSOUF [1], Clémentine BARREYRE [1], Anaïs CHARCOSSET [1]
[1] Airbus Defence & Space, Toulouse, France
Contact e-mail: loic.boussouf@airbus.com

Abstract. Communication satellites are complex systems remotely operated. As it is impossible to request any measure on the system that was not planned at design phase, it is assembled with hundreds of sensors. All those sensors capture many physical variables at a sampling rate of roughly one point per minute. Representing million points per sensor over years, this is the only information satellite operators have to diagnose the behavior of their satellite.

Satellite monitoring is today supported by expert definition of processing. It requires experienced understanding of physical or chemical behind a subsystem, and this knowledge is making use of available parameters. Being the most accurate method to monitor a subsystem, it is hardly applicable to a constellation of hundreds of satellites. Lessons learnt acquired from satellite operators and airlines lead as follows: while similar or even identical product at beginning of life, due to their operating conditions, each element of a fleet or a constellation rapidly behaves differently from the others. While still comparable, they are no longer identical. Expert approach not being scalable to a whole fleet or constellation at reasonable cost and in a reasonable planning, we investigate data analytics approach to support experts.

All battery data over all Airbus E3000 platforms represent about 300 years of observation. Benefiting from repetition of operating conditions due to the orbit and sun/shadow alternation, data was split in time windows. It is then demonstrated that Out Of Limits (OOL) monitoring can't capture all variability in data. Hence a functional decomposition is proposed to capture all signals dynamic and concentrate it into a smaller set of coefficients. Benefiting from this reduced dataset, a surrogate modeling approach is then applied to capture the correlation between battery data and battery age. This modeling technique accuracy is evaluated onto a test set and demonstrates good prognostic accuracy. It can be now an additional tool provided to experts to support decision on slowing or accelerating degradation of a given satellite battery.

This modeling does not rely on any strong assumption related to subsystem knowledge. It can be applied to other subsystems. But due to the data volume accessible and leveraged through this method, we investigate distributed storage and computation techniques on a Hadoop/Spark cluster. After a benchmark of storage techniques to store our time series, data analytics calculation is performed and measured. We present here final choice of techniques and time performance of overall process to support analytics on sensor data in a big data infrastructure.

Introduction

Satellite systems have specific characteristics that drive any monitoring approach. First they cannot be physically repaired because of their inaccessibility. Second, system observation is allowed only through existing sensors and implemented sensing sampling rate (with some flexibility about sampling rate). Last, there is legal frame imposing more and more deorbiting control (known as LOS or Law on Space Operations). It is not an option to use a space system until end of life and leave it in orbit. Such end must be anticipated to guarantee enough function is available to control de-orbit or graveyard orbit redirection maneuvers.
This made several actors in space domains to complete their current monitoring approach with recent advances in data analytics to support decision about satellite health status. This is performed by completing Out Of Limits detections with machine learning methods as in [3] by ESA, [1] by CNES, [2] by DLR-GSOC, [4] by Airbus Defence & Space. Those studies address the domain of event detection. We propose to address here the domain of degradation detection.
Both event and degradation detection are required to implement health monitoring approach. While event detection allows diagnosis, degradation detection allows prognosis. It relies partly on same algorithmic building blocks devoted at time series analytics, but the end use case is different from a space operations perspective, which requires a dedicated approach.
We describe here the approach adopted for degradation modeling, addressing first the algorithmic description and its performance on satellite data acquired in orbit. We then propose architecture to support such approach in an operational context, with preliminary lesson learnt, technologies benchmarks and results assessing its performance.

We acknowledge here the work performed on battery degradation modeling by our estimated colleagues: Jean François Gajewski from reliability team at Airbus, and André Cabarbaye from reliability team at CNES.

1. Battery degradation modeling: a data based approach

1.1 System and data description

For this study, we will consider the batteries that are on boarded geostationary telecom satellites, on the Eurostar E3000 platform developed and manufactured by Airbus Defence and Space. The geostationary orbit always benefit from sun light excepting two periods of about 45 days in the year known as eclipses phases, and centered on equinox dates as illustrated on **Fig. 1**. At this time, the satellite repeatedly enters earth shadow, every 24 hours for eclipse that can last up to 70 minutes. At those moments, batteries are highly solicited to perform service continuation on a platform designed for more than 15kW power consumption.

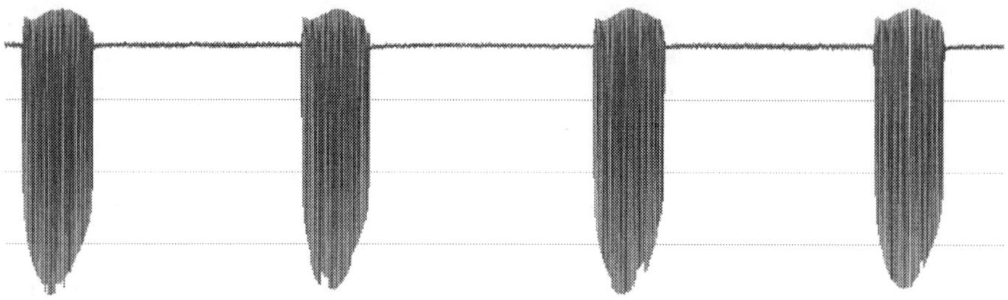

Fig. 1. Two years of geostationary battery voltage, 4 phases of eclipses (scales removed for confidentiality)

Battery on geostationary platform is well mastered by Airbus, and chosen specifically for this study due to no anomaly that would have corrupted the dataset. This system knows a nominal degradation due to its intensive usage all along the 15 years of satellite life. Hence during eclipse phases, end of discharge voltage level are lower and lower all along the years as illustrated in Fig. 2. This is mainly explained by loss of battery capacity, but those deeper discharge cycles can also be impacted by punctual higher consumption due to mission change.

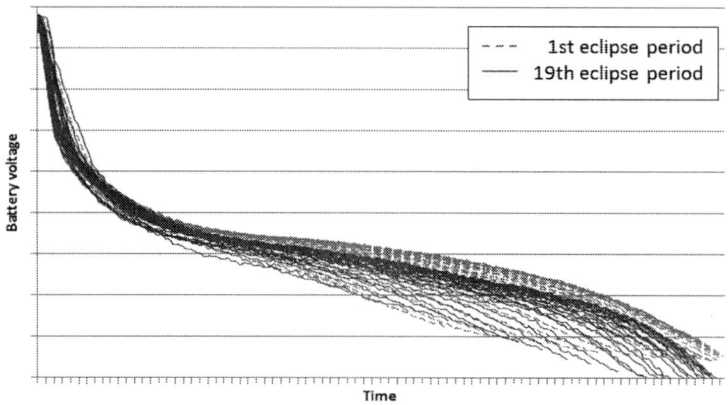

Fig. 2. Year 1 and year 10 compared through battery discharge voltage envelope

We observe that the data measured on the battery system is quite repetitive, but is not periodic. The battery discharge does not occur all year long, and when occurring, does not have the same length from one discharge to the following. We have a repetition, but no periodicity. This repetition is a characteristic of all space systems that know day/night alternation. A larger period can be observed at the scale of one year and is known as seasonality, due to sun/earth distance variation. This repetition assumption is the only assumption required to develop the following methodology. No taking any assumptions linked to the battery itself, it makes possible to apply this methodology to other systems. Once some repetition is observed on a given system, it may benefit from the proposed approach. The difficult part is then to extract those repetitions. Current work not presented here proposes to automate this repetitive pattern detection, recognition and extraction on any time serie. Once extracted, we propose to simplify the data before modeling the link between this pattern shape and system age.

1.2 Timeseries dimension reduction

As it can be observed, each discharge cycle is sampled with up to 170 samples. Given the signal is continuous, two successive points almost have the same value. Hence all samples may not be necessary to carry all the information from one discharge cycle.

To deal with such data, expert approach is to select relevant points depending on the analysis they want to perform. For the battery, it results in considering voltage before discharge, voltage at end of discharge and discharge slope between two specific inflexion points. But this is not transferable to another system, that requires another definition of points carrying the information. We propose a definition that is applicable to any system, ensuring it will carry a controllable level of information.

Given our signal is the sampled representation of a function, we decompose this along a functional basis. It exists many functional bases: Fourier, wavelets … All those bases are known prior to the data, those are called fixed basis decomposition. In order to better capture information in our signal, it is better to use a basis computed from the data. Principal Component Analysis (PCA) method is chosen. Further work presented in [4] compares different functional decomposition. It can support future work to combine the most appropriate bases depending on the anomaly category one wants to spot out.

Applying the PCA decomposition to the N discharge cycles f_i sampled through $y_n = [f_n(t_1), \dots, f_n(t_m), \dots, f_n(t_M)]$ results in M basis function ϕ_j so that:

$$f_n(t_m) = \sum_{j=1}^{M} \alpha_j . \phi_j(t_m)$$

The basis function ϕ_j sampled at the t_m are the eigen vectors of the covariance matrix of the observations y_n. The importance of variance explained by eigen vector j is measured through the ratio:

$$\frac{\lambda_j}{\sum_{i=1}^{M} \lambda_i}$$

Hence, if eigen values λ_j are sorted by decreasing order, it is considered that:

$$f_n(t_m) \approx \sum_{j=1}^{l<M} \alpha_j . \phi_j(t_m)$$

Is a good approximation of $f_n(t_m)$. Approximation quality is controlled by level l of eigen vectors kept in the decomposition. It is now possible to represent each discharge cycle through the set of coefficient α_j from its decomposition along the PCA basis. It is also possible to reconstruct the signal based on a truncated set of coefficients as demonstrated in Fig. 3. Such reconstruction reveals a maximum error of 0.2% when 20 coefficients are kept out of 170.

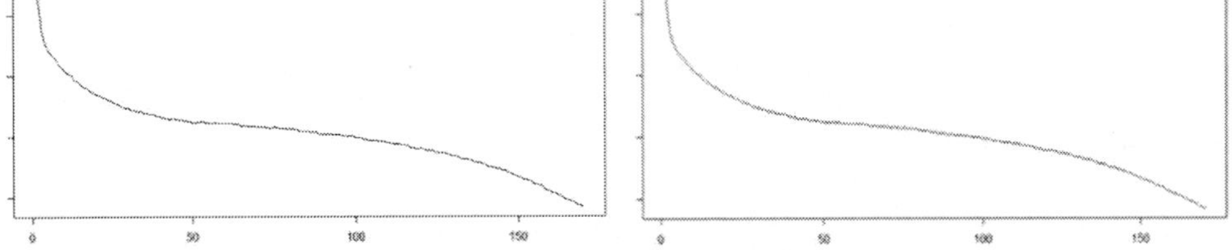

Fig. 3. One discharge cycle (left) and reconstruction based on 20 coefficients out of 170 (right). Max error about 0.2% of signal magnitude

1.3 Degradation modeling

Once the data properly reduced and keeping all valuable information in a smaller set of coefficient, we want to capture the relation between age and observable. We apply a surrogate modeling approach, that consist in fitting a parameterized model on top of observations and validating this fitting on an independent set of observations. We avoid simplistic models such as linear or polynomial models as no a priori on the data allows us to make the assumptions the data will follow a linear or polynomial model. We prefer a parameterized model able to catch nonlinear relationships. Neural network, radial basis functions and Gaussian process modeling, known as kriging, are possible surrogate modeling techniques. Gaussian process is preferred as requiring less observable than neural networks to support the training phase, and more flexible than radial basis function as allowing varying correlation between samples. Moreover, it was proven to perform very well in complex physic modeling such as mechanical, thermal, and fluid mechanic coupling within aircraft engines as in [5]. To sum up, we propose to build the function:

$$battery_{age} = f(discharge\ cycle\ PCA\ coefficients)$$

through the following methodology:

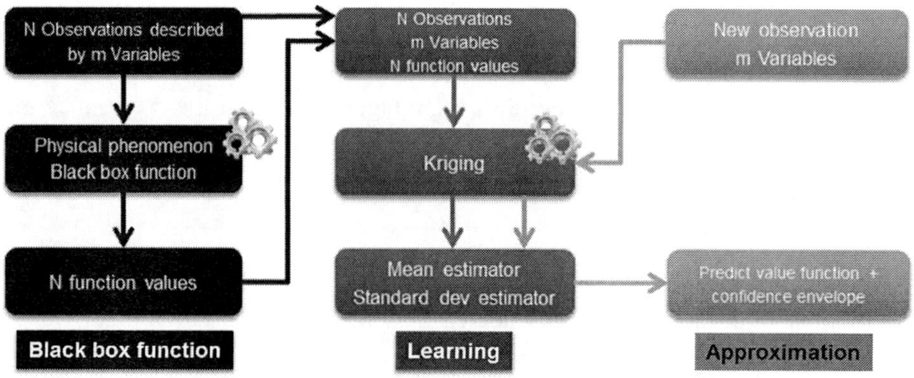

Fig. 4. Surrogate modeling approach to approximate a black box phenomenon

We will now apply this to real battery data, and estimate the quality of the model.

3. Battery degradation modeling: application on equivalent of 282 years of observations

To test this methodology, we extracted the 5 longest discharge cycles for each eclipse phase (twice a year) on 30 satellites supported by a Eurostar 3000 platform. This results in 2821 discharges cycles and half a million data rows. We then apply reduction dimension and degradation modeling.

3.1 Dimension reduction

Once PCA is applied, we first have a look at variance cumulative proportion to select the right number of component to keep.

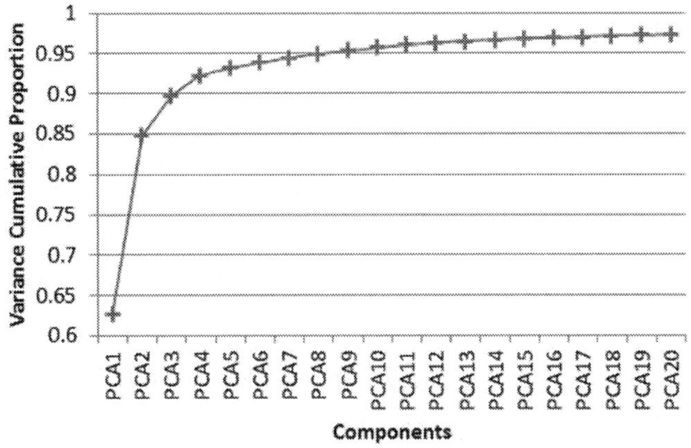

Fig. 5. Variance cumulative proportion

This leads to keep 20 first components that carry 97% of total variance. Benefiting from this reduced set of components, we look at some three dimensional projections of the discharge cycle to check no cluster effect will pollute further modeling and assess homogeneity of input data.

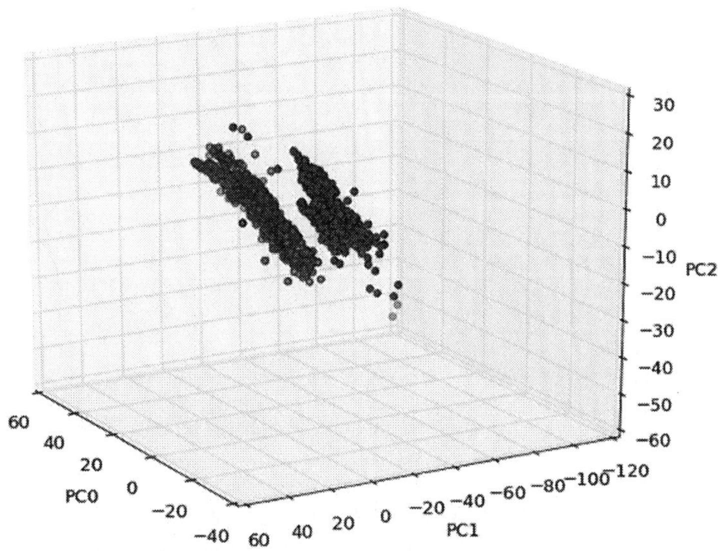

Fig. 6. Families discovered through PCA projections

A projection on the first 3 PCA components (carrying about 90% of total variance according to Fig. 5) makes appear distinctly at least 2 families through two cluster. Trying to capture relationship between degradation and PCA components on the two families through the same PCA basis will be less accurate than performing it separately on the two PCA basis computed on the two clusters. Other projections revealed a third cluster in red in Fig. 6. Hence, we develop the methodology two times on each cluster. The red cluster is considered too small to allow any modeling.

After exchanging with battery experts, it appears that the two clusters were matching two battery technologies: NiH2 batteries in blue (family 1), and Li-Ion batteries in green (family 3). Hence even with no design knowledge or battery assumptions, the mathematical modeling through the dimension reduction makes us able to separate significant different behavior.

After this dataset preparation and reduction, we apply degradation modeling.

3.2 Degradation modeling

When applying this step, one should pay attention to two details. First, because data is coming from an experimental process, a regression model should be preferred over an interpolation model. Second, because of the possible large number of observations, the training phase will result in an optimization problem with multiple local minima. A robust regression approach that can face high number of samples in high number of dimensions must be carefully chosen. The product pi-seven from the company Datadvance has been benchmarked and chosen to support this implementation.

After building one model for each of the two families identified, we score each of them by predicting age of discharge cycles kept apart from the learning step. We would expect that age of each discharge cycle (the year of its observation minus the date of launch) will match the prediction. We make this representation for the first family below.

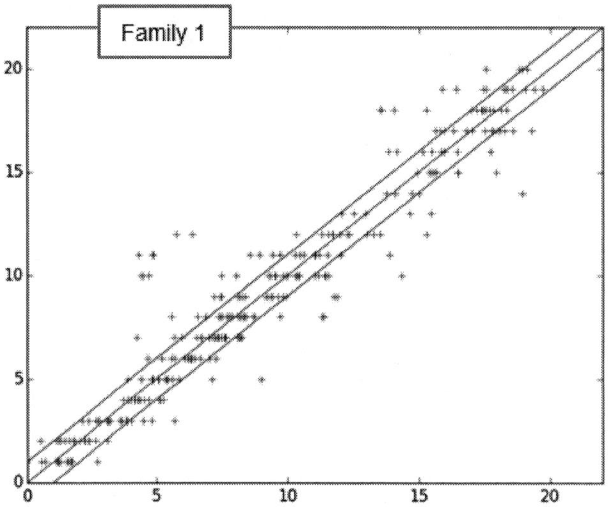

Fig. 7. Degradation modeling performance (scale in eclipse phases number: 2 eclipses phases = 1 year). Real age (x axis) vs prediected age (y axis)

The red line represents an envelope of ±1 eclipse phase error, meaning ±6 month error. Most of the discharge cycle used for validation lies within those boundaries. Below are the detailed results for both families:

Table 1. Degradation modeling accuracy, ratio of validated discharge cycle within each interval

	± 6 months	**± 12 months**
Family 1	67.5%	88.9%
Family 3	38.9%	67.4%

While modeling is quite accurate for Family 1, it degrades significantly for Family 3. Ratio of validation samples lying within an interval is a good measure of model accuracy, but is also a recommendation for application within operations as illustrated in Fig. 8. If method output is within confidence interval, it should not be considered as degradation. If result is beyond confidence interval, it should be raised to system expert for proper investigation.

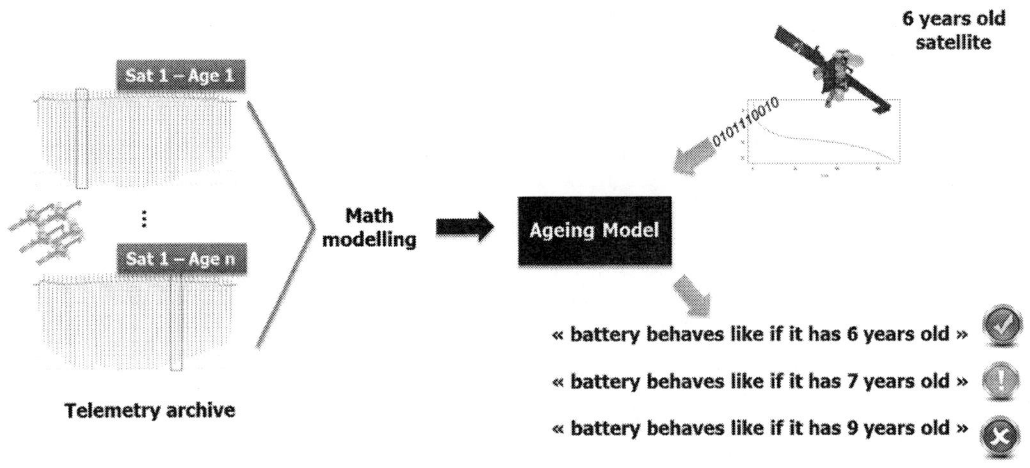

Fig. 8. Overview of method in operational context

Due to the results obtained for family 3 compared to family 1, several improvements were attempted. First trying to model both battery age and index within eclipse cycle, or considering additional observable such as bus current to take into account the mission variability supported by the battery. While no significant improvement was obtained, it is expected that considering a larger number of observables related to the system under observation can improve the model. Not limiting to bus current only, but extending to commands, and non-electrical measures such as temperatures is still to be explored. Such a multivariate approach can stay as far as system knowledge by relying on covariance analysis to infer the dependencies within the system as illustrated in [6].

4. Data based space system operations

Based on those results that can be further improved, and because this methodology is relying on few assumptions and can be applicable to other systems, it appears that classical computation means should be extended. Those cannot support 10000 telemetries sampled every 20 seconds for every satellite, multiplied by the size of a fleet or a constellation. We then started a migration to a distributed storage and computation environment. We present here preliminary results and technologies evaluated. This is far from being exhaustive due to the numerous technologies existing and arising in data analytics field. On the other hand, the set of technologies chosen to scale up this methodology is easy to implement in various context and does not rely on proprietary component. We have explored both distributed storage and distributed computation.

4.1 Distributed storage

We limit our choice to what was available in a standard Hortonworks distribution, this choice being the one recommended and deployed by Airbus Defence and Space Digital Transformation Office. Within this environment, we benchmarked for storage 3 technologies: Parquet file in HDFS, Hive database, and HBase database.
The use case measured here is the retrieval of data already stored. Storing data is a onetime action, while retrieving the data is supposed to occur very often. The performance of this step is considered critical. The data retrieved here is composed by 100 telemetries sampled every minute over 10 years, representing about half a billion data point, sampled by a long timestamp and a double value.

The results are consolidated below:

Table 2. Data retrieval performance

	Read data
Parquet on HDFS	7.5s
Hive	10.5s
HBase	75s

To get such performances, we took into account a particularity of our data which is its repeatability. Given the orbit, the whole data was split in 24 hours windows, the orbit period for geostationary platforms. This split was used as a partition rule in all those storage. Such partitioning improved greatly the data retrieval performances.

Based on those results, Parquet storage on HDFS is the privileged data format for storing satellite telemetry. Performant enough with flexibility, it will allow storing other data type than only time series and avoid to choose a niche technology dedicated to time series.

4.2 Distributed computation strategy

For distributed computation Spark is chosen as part of Hortonworks distribution. At that time, a major question arose: what do we want to distribute. Indeed, Spark API allows distributing calculations onto a cluster and proposes as well distribution of mathematical operators as well. Any linear algebra operation, like inverting a matrix can be distributed onto a cluster. In our case, we have linear algebra to perform. We then have two choices: distribute whole telemetry and process it fully on a core, or distribute each linear algebra operation over the cores. It appears that we were dealing with data about 100 to 200MB, observed size for one telemetry over 10 years. This is far below the memory capacity of each core of a standard Hadoop and Spark cluster (several GBs). That means we can limit distribution at telemetry level. If going beyond and distribute linear algebra operations as well, this results in major time overhead due to transfer of a lot of small information through nodes. This may not be questioned unless dealing with telemetries about several GBs. This illustrates that data is stored for a given processing, and data may be duplicated following different partitioning according to the processing to be applied.

4.3 Distributed computation performance

Once stored and partitioned, data is ready to be processed. For assessing computation performance, we rely on a simpler algorithm devoted at outlier detection. It relies on functional decomposition as the method presented here, but scores each observable with a different metric not catching degradation. Its full description is available in [4].

To give an order of magnitude of expected computation time, we make this algorithm run on a comparable amount of data. While different in its objectives – degradation or outlier detection, it relies on the same operators and comparable data sizes. It takes about 4.5 hours to process 10 years of telemetry archives in order to produce a score (predicted age or outlier score). Migration of algorithm presented here to pySpark language to allow its distribution is ongoing and is expected to require the same computation time.

4. Conclusion

We described here a non-destructive approach to support health monitoring diagnosis on a battery system. This method requires no assumption or knowledge on the system observed except repeatability through its observables. This makes a good candidate to apply health monitoring at system of system scale. Preliminary results are encouraging, while varying from one battery family to another. Those results and other algorithms developed for space system monitoring pushes us to anticipate a big data platform. We then investigated technologies to support distributed storage and computation.

While such data based approach is now demonstrated as feasible, the integration of method outputs in decision processes has to be worked out. Indeed, it is now quite easy to define an appropriate algorithm for a specific need and to measure its accuracy on massive amount of data. It is not required to be an expert to demonstrate this. But to transfer such capabilities from support decision to decision itself, expert should take attention at defining rigorously the datasets they want the algorithm to be assessed on. Such algorithm qualification process relevance is highly depending on expert involvement in it. And will be the key of success to benefit from Actionable Artificial Intelligence in industry.

References

[1] Improving Spacecraft Health Monitoring with Automatic Anomaly Detection Techniques, S Fuertes & al., SpaceOps 2016 Conference

[2] Applications of Deep Learning Neural Networks to Satellite Telemetry Monitoring, Corey O'Meara & al., SpaceOps 2018 Conference

[3] Enhanced Telemetry Monitoring with Novelty Detection, Jose Martinez-Heras and Alessandro Donati, 2014.

[4] Statistical Methods for Outlier Detection in Space Telemetries, C Barreyre & al., SpaceOps 2018 Conference

[5] Surrogate Based Optimization for Multidisciplinary Design, L Boussouf, 2011 SAE Technical Paper 2011-01-2507, 2011

[6] Detecting Outlier Detection in Multivariate telemetries thanks to Covariance Analysis, Clémentine Barreyre, 2017, Proceedings from Big Data from Space conference, Toulouse

Data Visualisation using Augmented Reality with a Focus set on Head Mounted Displays and Collaborative Tasks

Tobias DREY [1]
[1] Airbus Defence and Space
Wörthstr. 85, 89077 Ulm, Germany
Contact e-mail: tobias.drey@airbus.com

Abstract. For data analytics a proper visualisation of the data is important to provide the user a good overview and help him to find and understand relationships in the data. Furthermore, appropriate visualisations can help to improve the efficiency of tasks. There are a lot of different approaches for 2D as well as 3D visualisation, which can be applied to existing rich client applications. Many of these rules are applicable for augmented and virtual reality, too. But especially when head mounted displays are used additional factors have to be considered and new possibilities exist.

This paper will therefore describe related work regarding 2D and 3D visualisation. Factors like perspective, performed task, ambiguity, depth cues and the general user interface are discussed. Examples will be given for map centric applications and tasks, like the perception accuracy of objects on 3D maps. Furthermore, head mounted displays are categorised. They can provide stereoscopic vision, which is essential for human's binocular vision and adds benefits for some tasks.

With augmented and virtual reality new concepts for collaborative tasks could be created. This paper will provide an overview of related literature and discuss the advantages and disadvantages of augmented and virtual reality. Especially augmented reality can support collaboration which is linked to the fact that all users can see each other.

After the scientific basics, a concept for a big data visualisation is explained. For air traffic control it is important to use all available information to observe and secure the airspace. Automatic dependent surveillance – broadcast (ADS-B) is now equipped at most aircraft. It can be fused with the radar sensor data to generate an air picture. Historical data could be analysed with data analytics to recognize pattern and make predictions. This helps to improve live air surveillance. For collaborative decision making an augmented reality workspace was created. It offers an ADS-B flight data visualisation in 3D on a holographic geographic information system. Due to the holographic environment multiple persons can discuss together incidents in a way as the data would physically exist.

1. Introduction

Big data analytics uses algorithms to gain knowledge out of massive amounts of data. After the analytical process has finished the gained knowledge is visualised to the users. This is very useful for air surveillance where historical data could be analysed to recognize pattern, validate live data and predict incidents. This work will introduce such a system with the focus set on the visualisation of results. Furthermore, a concept will be shown to create an augmented reality (AR) workspace to ease collaborative discussion and decision making.

It will therefore provide an overview of visualisation rules from related work. These rules should be considered when using AR or virtual reality (VR) head mounted displays (HMDs). Besides visualisation benefits, AR can improve collaboration, which will be furthermore discussed and proven by related research findings. The related work section in this paper is based on parts of the related work section of the master's thesis of Drey [1]. After that, the big data air surveillance application for state-of-the-art workstations is introduced as well as a collaborative AR workspace. As AR HMD the Microsoft HoloLens [2] is used.

2. Related Work

2.1 Data Visualisation using Augmented and Virtual Reality

2.1.1 Definition of Head Mounted Displays, Augmented and Virtual Reality

HMDs can be classified regarding their capabilities. Typical types are monocular and binocular, opaque and see-through as well as video and optical. They were ordered in a matrix by Laramee et al. [3]. Applying this classification matrix to the Microsoft HoloLens it is a binocular optical see-through HMD.

To further categorise HMDs Miligram et al. [4] defined a reality-virtuality continuum (see Fig. 1). AR is linked to a real environment and VR to a virtual environment. The continuum merges in the middle. The widespread area between these two extremes has a lot of levels and is called mixed reality (MR). To distinguish between AR and VR this work will not use the term MR. As the Microsoft HoloLens is a see-through HMD it is classified as AR HMD. In this work the 3D stereoscopic visualisation of the Microsoft HoloLens is named as hologram.

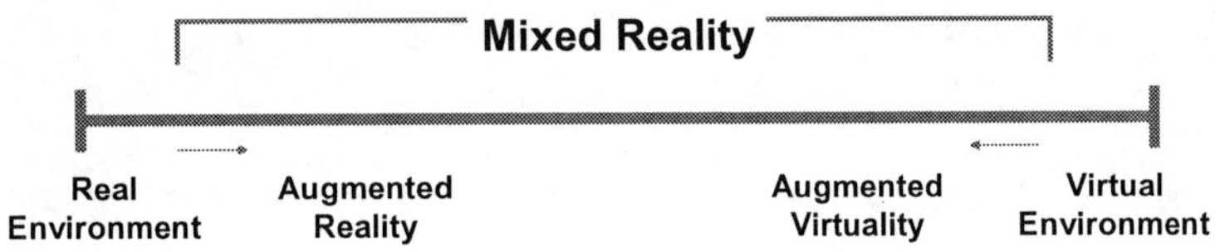

Fig. 1. Reality-virtuality continuum [1, 4]: Augmented reality (AR) is on the left side of the continuum and linked to the real environment. Virtual reality (VR) is on the right side and linked to a virtual environment. AR and VR merge in the middle. The definition of mixed reality (MR) is very widespread and addresses all levels in between.

2.1.2 2D vs 3D in Visualisation

AR and VR HMDs can offer 2D as well as 3D visualisation. Therefore, general visualisation rules apply as well. Such rules were evaluated in related work. An example is Tavanti et al. [5]. They used a spatial memory task with hierarchical data and compared a 2D with a 3D visualisation. The performance of the participants using 3D was better.

Evaluating the usefulness of 3D perspective St. John et al. [6] used simple block shapes as well as maps showing terrain data. Their hypothesis is that 2D should be used for relative positioning tasks and 3D when shape understanding is important. Similar results were evaluated by Todd et al. [7] in a study where shapes of objects had to be adjusted to match a reference object. Furthermore, they found out that especially the human perception of 3D metric structures is not accurate.

Nevertheless, studies exist with contrary results. St. John et al. [6] provide a literature review with studies that have different results compared to their hypothesis. As all these studies vary in tasks and interfaces used there seems to be a high use case dependency. In addition, other factors than the visualisation or the used dimensions could be more important for the task performance. Sebrechts et al. [8] compared 2D and 3D for displaying search results and evaluated that the available features and the user interface have a greater impact on the task performance of the participants than the used dimensions.

Another point that should be considered to choose between a 2D and a 3D visualisation are 3D ambiguity issues. An overview is provided by St. John et al. [6]. It is proven by Todd [9] that different shapes can look from a specific angle completely equal. In such a case the viewing position has to be changed to bypass such ambiguities [10]. Depth cues are important for 3D visualisation and a taxonomy which is split in physiological and pictorial depth cues is provided by MacEachren [10]. The physiological depth cues relate to the perception of a 3D environment by human stereoscopic vision. Pictorial depth cues describe principles to create a depth perception for non-stereoscopic vision such as perspective, motion, shading or shadows [10]. An issue of pictorial depth cues is that they can interfere with the visualised objects. For example a brightness variation caused by an illumination of a map can interfere with the colour gradient [11, 12].

The position ambiguity in 3D is another issue that has to be considered. As example, for a 3D geographic information system (GIS) operator it is important to know the position of an aircraft on the map. To clearly visualise the position on the ground plane drop shadows or drop lines can be used as means [13, 14, 15]. Fig. 2 shows a drop line to visualise the position of the sphere on the map. Hendrix et al. [16] compared different visualisation means such as shadows and drop lines in a study. The depth judgement was improved by 200% when drop lines were used, shadows improved it by 30%. Therefore, this work uses drop lines to visualise the position of floating objects on a map.

Fig. 2. Drop line (blue) used in a 3D geographic information system (GIS) [1]: Drop lines are more efficient to visualise the position of floating 3D objects on the map than other depth cues such as shadows [16]. The picture shows map data from © GeoBasis-DE / BKG 2017 (data changed) [34].

Furthermore, binocular HMDs can provide stereoscopic view. Rapp et al. [17] conducted a study where they compared the terrain insight of participants. They showed them typical map visualisations including depth cues such as shading with mono- and stereoscopic visualisation. The terrain awareness of the participants with the stereoscopic visualisation was better.

Summing up these findings it is assumed that 3D visualisation does not improve visualisation or user interfaces in general. It is highly task depended if a 2D or a 3D visualisation has a better task performance. A smart user interface should offer a combination of multiple factors where the visualisation is only one aspect. This assumption is similar to the conclusion of Smallman et al. [13]. As previously stated, related work results indicate that 3D should be used especially for tasks with a high need for shape understanding and 2D for tasks where relative positions are important. Furthermore, stereoscopic vision, as HMDs can provide it, can improve shape understanding as well.

2.2 Using Head Mounted Displays for Collaborative Tasks

Collaboration with HMDs is related to collaborative virtual environments (CVEs) which were defined by Churchill et al. [18] as a space that supports collaborative activities. Therefore, HMD environments that provide the same objects to multiple persons are CVEs with the benefit that each user has its own perspective of the shared space [19].

CVEs could be implemented with many technologies and not only in AR and VR, but this work will focus on AR HMDs. AR can provide benefits compared to VR in collaborative workspaces which will now be discussed. Billinghurst et al. [20] and Henrysson et al. [21] provide arguments to explain this. Natural communication with body language, gestures or facial expressions is possible due to the fact that all participants can see each other and are aware of what the others are doing. Resulting advantages are, that collaborative AR can decrease the task completion time compared to VR [22]. Furthermore, it enhances communication and face-to-face discussion [23].

The communication behaviour of collaborative AR users was evaluated in a study by Kiyokawa et al. [24]. Non-verbal communication and involvement of the other participant was observed most when the task space was placed between the two participants. Nevertheless, this placement was not linked to the working efficiency of the drawing task. Whenever non-verbal communication was difficult, speech was used as compensation. One characteristic of this study is that participants were sitting and not allowed to change their position. They had to perform the task with different fixed placements of the objects in between. The AR prototype introduced in this work will not limit the users in their position. They are free to walk around and change their position to the hologram and to each other. This is beneficial because Kiyokawa et al. [24] had the result that a task space in between facilitates communication and collaboration, but that the same perspective on the hologram can help to simplify communication.

When discussing about CVEs it is important to distinguish between symmetric and asymmetric collaboration. The arguments so far apply to symmetric AR collaboration. Asymmetric collaboration could be performed by remote operators that work on state-of-the-art workstations and operators that work in the holographic workspace, but this is not in the scope of this work.

Due to the previously mentioned benefits of AR collaboration this work focuses on AR CVEs. Nevertheless, VR CVEs exist as well. In many cases avatars are used to represent other collaborating users. Research exists that evaluates how real such avatars should look for good interaction quality [25]. Summing up, it depends on the task whether AR or VR HMDs should be used.

3. Air Traffic Big Data Visualisation for Augmented Reality

For air traffic control (ATC) situational awareness is very important. This is achieved by calculating a valid air picture as well as a proper visualisation of it which is displayed to the operators. This paragraph will first of all explain the process to create an air picture including the data sources and the data fusion. Afterwards, the visualisation of automatic dependent surveillance - broadcast (ADS-B) data is shown and a concept for an AR workspace for map based data visualisation is introduced.

3.1 Big Data Analytics for Air Traffic Control

The first step of the process to calculate an air picture for live ATC is the connection to multiple data sources, for example radar sensors or ADS-B receivers. ADS-B provides messages broadcasted by an aircraft with its own position, kinematics and further information to make tracking possible. These data is processed and a multi-source integration is performed. This means that a sensor data fusion is performed to generate improved track information. Based on the fused data tracking, identification and classification is performed. All these information is then visualised to the ATC operator as live air picture.

This is the common way ATC systems work. An improvement is the usage of historical data of ATC centres to perform big data analytics. The air picture could be enriched by machine and deep learning algorithms, rule based anomaly detection and enhanced prediction. Data analytics discovers hidden information in large sensor datasets or provides assistance functions for the operators. The goal is to learn from the past and predict the future. This prediction can be combined with live data to provide predictive surveillance.

Data analytics enhances the situational awareness thanks to behaviour analysis, pattern recognition, anomaly detection as well as prediction and anticipation. All information is displayed to the operator, but it has to be clustered to reduce overload and simplify decision making. Therefore, concepts to display information on workstations as well as AR workspaces are described in the following. As this paper focuses on visualisation and AR collaboration the used data analytics techniques are not described more in detail.

3.2 Visualisation of ADS-B Data Including Predictions

State-of-the-art ADS-B data visualization is very common since commercial platforms like flightradar24.com [26] are available for everyone in the internet. For the data analytics backend several frontends were created for workstations, mobile and web clients. Furthermore, a prototype for a collaborative AR workspace exist that will be explained in the next section. Fig. 3a shows the web client currently displaying live air traffic around Frankfurt airport in Germany. The ADS-B data transmitted from an aircraft can be viewed in a detail dialog.

When flight paths from historical data are visualised this provides a picture with a lot of information which is difficult to understand. An example of historical data is shown

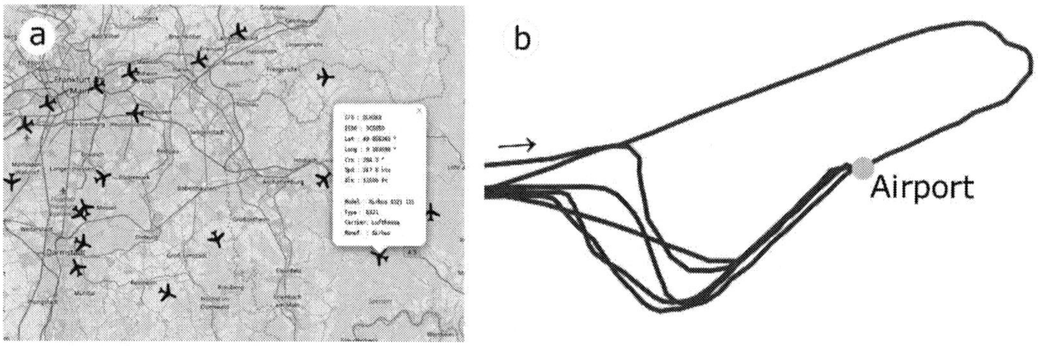

Fig. 3. Automatic dependent surveillance - broadcast (ADS-B) data visualisation: Live air traffic around Frankfurt airport in Germany based on ADS-B data is shown in (a). Detailed information transmitted from the aircraft can be seen in a detail dialog. In (b) an analysis was performed to recognize approach pattern of aircraft on an airport. They vary depending on weather conditions, daytime and air traffic. Picture (a) shows map data from © OpenStreetMap contributors [35]. Leaflet [36] is used as library.

in Fig. 3b. The picture is already filtered and shows only a few flight paths that were detected by pattern recognition. An analysis was performed to recognize approach pattern of aircraft on an airport. They vary depending on weather conditions, daytime and air traffic. Multiple airports could be analysed and the pattern could be compared. As described in the next section such information could be discussed in a 3D AR workspace.

Predictive surveillance was introduced in the previous section. Data analytics algorithms can be trained to predict based on historic data where an aircraft will fly. This can be seen in Fig. 4. It shows from left to right (a-c) how the algorithm predicts the flight path. The prediction adapts by the time. In the images pre-recorded data is used to compare the prediction with the real flight path (yellow crosses) during calculation. The blue dots visualise the input ADS-B data for the algorithm. The light blue boxes show the prediction area where the algorithm thinks the aircraft is going to fly. One box predicts one minute, which means that the next ten minutes are predicted. Such predictions can help to recognize anomalies during ATC and create alarms so that an operator turns his attention on this possible incident. As this work focuses on visualisation purposes the details of the algorithm will not be discussed.

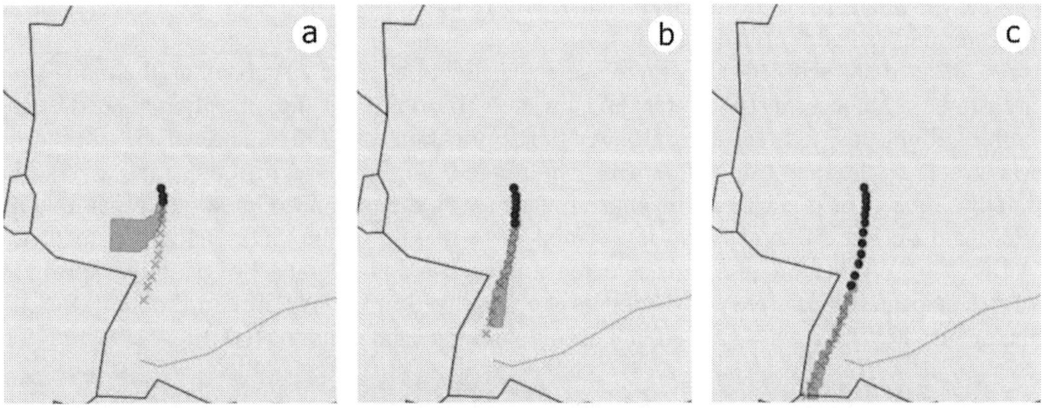

Fig. 4. Aircraft flight path prediction: Data analytics is used to predict the flight path of an aircraft. From left to right it is shown how the algorithm result adapts by time. Pre-recorded data is used to compare the prediction with the real flight (yellow crosses) during calculation. The blue dots are the input ADS-B data of the algorithm and the light blue boxes show the prediction. One box predicts one minute. Cartopy [37] is used as library.

3.3 Augmented Reality Workspace for Map Based Data Visualisation

The previous section shows how ADS-B data could be visualised on 2D user interfaces. As described in the related work section 3D can add benefits regarding shape understanding. Furthermore, an AR HMD can provide stereoscopic vision which improves depth perception and can, as related findings show, improve terrain awareness. AR is selected because the discussed related work indicates that AR can improve collaboration between users due to natural communication. The state-of-the-art 2D user interface is designed as single user workplace, whereas the AR workspace is designed to facilitate cooperation. ATC operators can discuss incidents or predictions in the collaborative workspace as well as conduct briefings.

Therefore, it is important to select an HMD that does not limit natural communication. As the Microsoft HoloLens occludes some parts of the face it limits the possibility of participants to see facial expressions. Furthermore, interaction with the hologram is limited to two gestures at the moment. Nevertheless, it was selected due to its visualisation and tracking capabilities. As the AR workspace is a prototype, potential drawbacks in the design or due to the selected hardware could be changed during development.

Several related applications exist which display holographic maps. Boing created an application for holographic unmanned aerial vehicle (UAV) control [27]. HoloTerrain [28] and HoloMaps [29] are applications quite similar to Google Earth [30]. Their user interfaces were used as reference during the design of the AR workspace. Furthermore, UC Davis [31] created an augmented sandbox to interactively teach topographic maps, but the prototype in this work only uses holographic elements.

The developed prototype can be seen in Fig. 5. It can display the hologram to multiple operators co-located in the AR workspace to discuss live air traffic. As main element, the holographic map is placed in the middle of the workspace on the floor. Operators are free to walk around in contrast to the work of Kiyokawa et al. [24]. Therefore, a round map is used to simplify walking around it. Accompanying the map, the menu is placed at the edge of the hologram. Everything on the map is visible in a shared view equally for all operators. Nevertheless, the menu is always headed in a private view to each operator. This applies as well to all text in the workspace to ease reading. Private and shared views are possible because each HMD renders the shared view exclusively for an operator and can add private elements.

Based on recorded ADS-B data selected flight paths are visualised on the map. They are drawn as blue semi-transparent 3D lines. A 3D object of the aircraft is placed at the current flight position (see Fig. 5b). As the application is a full functional GIS based on Mapbox [32], the map can be zoomed and dragged. Furthermore, different map layers like satellite images or roads could be chosen. On Fig. 5a the map is zoomed out to get a world view and on Fig. 5b+c the map is zoomed in to maximise the ADS-B flight path of the Airbus A380. When an aircraft is selected it is enlarged and rotates hovering in the middle of the map. Furthermore, ADS-B details are displayed in dialogs around the aircraft. Like the menu, they are always headed in a private view to each operator (see Fig. 5c).

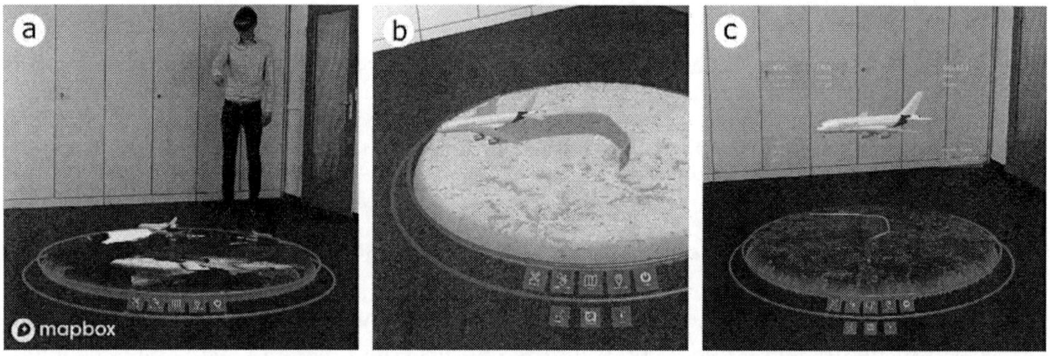

Fig. 5. Collaborative AR workspace: In (a) the collaborative AR workspace is shown with the map placed on the floor besides an operator. The map can be zoomed in (b) to maximise the ADS-B flight path of the Airbus A380. When the aircraft is selected (c) it is hovering and rotating in the middle of the map. ADS-B details are shown in dialogs around the aircraft. This application uses © Mapbox [32] and © OpenStreetMap contributors [35].

One of the next functions that should be implemented is a 3D visualisation of the flight path prediction (see Fig. 4) in 3D with altitude information. Furthermore, an analysis function could be implemented to discuss recognized pattern (see Fig. 3b). This work is related to the planned AR radar coverage analysis function proposed by Drey et al. [33] and will be merged in one software solution.

4. Conclusion and Outlook

This work shows examples where visualisation is an important part of a big data analytics process. 2D as well as 3D visualisations have their advantages and disadvantages and should be chosen depending on the task. The AR workspace prototype indicates that it eases collaboration and improves decision making. Nevertheless, a formal user study has to be conducted in the future to evaluate the concept in detail and verify this indication.

The big data air surveillance application shows how data analytics can be a useful tool for ATC and helps operators in their decision making. Further pattern and algorithms can be created to recognise even more incidents. Data analytics is useful for other domains besides ATC such as maritime surveillance as well. Pattern recognition based on automatic identification system (AIS) data and satellite images could be used to recognise illegal fishing. This could be a domain to expand the existing system in the future.

Acknowledgements

I want to thank all colleagues who supported me in writing this paper. Especially Markus Konle, Dr. Felix Opitz, Dr. Michael Rottmaier, Kaeye Daestner and Klaus Humm who developed the ADS-B based ATC system and provided me information and pictures. Furthermore, I want to thank all colleagues inside Airbus who developed together with me the AR workspace. My special thanks go to Prof. Dr. Enrico Rukzio and Jan Gugenheimer from Ulm University for their great support during my master's thesis. Map data copyrighted OpenStreetMap contributors and available from https://www.openstreetmap.org".

References

[1] T. Drey, "Investigation of the Applicability of Augmented Reality Head Mounted Displays for Geographic Information Systems," Master's Thesis (unpublished), Ulm University, 2017.

[2] Microsoft, "Microsoft HoloLens," 2018. [Online]. Available: https://www.microsoft.com/en-us/hololens. [Accessed 30 08 2018].

[3] R. S. Laramee and C. Ware, "Rivalry and Interference with a Head-mounted Display," *ACM Trans. Comput.-Hum. Interact.,* vol. 9, no. 3, pp. 238-251, 9 2002.

[4] P. Milgram and F. Kishino, "A taxonomy of mixed reality visual displays," *IEICE TRANSACTIONS on Information and Systems,* vol. 77, pp. 1321-1329, 1994.

[5] M. Tavanti and M. Lind, "2D vs 3D, implications on spatial memory," in *IEEE Symposium on Information Visualization, 2001. INFOVIS 2001.*, 2001.

[6] M. St. John, M. B. Cowen, H. S. Smallman and H. M. Oonk, "The Use of 2D and 3D Displays for Shape-Understanding versus Relative-Position Tasks," *Human Factors: The Journal of the Human Factors and Ergonomics Society,* vol. 43, pp. 79-98, 2001.

[7] J. T. Todd and J. F. Norman, "The visual perception of 3-D shape from multiple cues: Are observers capable of perceiving metric structure?," *Perception & Psychophysics,* vol. 65, pp. 31-47, 01 1 2003.

[8] M. M. Sebrechts, J. V. Cugini, S. J. Laskowski, J. Vasilakis and M. S. Miller, "Visualization of Search Results: A Comparative Evaluation of Text, 2D, and 3D Interfaces," in *Proceedings of the 22Nd Annual International ACM SIGIR Conference on Research and Development in Information Retrieval*, New York, NY, USA, 1999.

[9] J. T. Todd, "The visual perception of 3D shape," *Trends in cognitive sciences,* vol. 8, pp. 115-121, 2004.

[10] A. M. MacEachren, How maps work: representation, visualization, and design, Guilford Press, 2004.

[11] C. Tominski, G. Fuchs and H. Schumann, "Task-Driven Color Coding," in *2008 12th International Conference Information Visualisation*, 2008.

[12] S. Silva, J. Madeira and B. S. Santos, "There is More to Color Scales than Meets the Eye: A Review on the Use of Color in Visualization," in *11th International Conference Information Visualization - Supplements (IV '07)(IV)*, 2007.

[13] H. S. Smallman, M. S. John, H. M. Oonk and M. B. Cowen, "Information availability in 2D and 3D displays," *IEEE Computer Graphics and Applications,* vol. 21, pp. 51-57, 7 2001.

[14] E. Feibush, N. Gagvani and D. Williams, "Visualization for situational awareness," *IEEE Computer Graphics and Applications,* vol. 20, pp. 38-45, 9 2000.

[15] M. Lange, J. Hjalmarsson, M. Cooper, A. Ynnerman and V. Duong, "3D Visualization and 3D and Voice Interaction in Air Traffic Management," in *The Annual SIGRAD Conference. Special Theme-Real-Time Simulations. Conference Proceedings from SIGRAD2003*, 2003.

[16] C. Hendrix and W. Barfield, "Relationship between monocular and binocular depth cues for judgements of spatial information and spatial instrument design," *Displays,* vol. 16, pp. 103-113, 1995.

[17] D. N. Rapp, S. A. Culpepper, K. Kirkby and P. Morin, "Fostering Students' Comprehension of Topographic Maps," *Journal of Geoscience Education,* vol. 55, pp. 5-16, 2007.

[18] E. F. Churchill and D. Snowdon, "Collaborative virtual environments: An introductory review of issues and systems," *Virtual Reality,* vol. 3, pp. 3-15, 01 3 1998.

[19] D. A. Bowman, S. Coquillart, B. Froehlich, M. Hirose, Y. Kitamura, K. Kiyokawa and W. Stuerzlinger, "3D User Interfaces: New Directions and Perspectives," *IEEE Computer Graphics and Applications,* vol. 28, pp. 20-36, 11 2008.

[20] M. Billinghurst, S. Weghorst and T. Furness, "Shared space: An augmented reality approach for computer supported collaborative work," *Virtual Reality,* vol. 3, pp. 25-36, 01 3 1998.

[21] A. Henrysson, M. Billinghurst and M. Ollila, "Face to Face Collaborative AR on Mobile Phones," in *Proceedings of the 4th IEEE/ACM International Symposium on Mixed and Augmented Reality*, Washington, 2005.

[22] M. Krichenbauer, G. Yamamoto, T. Taketom, C. Sandor and H. Kato, "Augmented Reality versus Virtual Reality for 3D Object Manipulation," *IEEE Transactions on Visualization and Computer Graphics,* vol. 24, pp. 1038-1048, 2 2018.

[23] S. Dong, A. H. Behzadan, F. Chen and V. R. Kamat, "Collaborative Visualization of Engineering Processes Using Tabletop Augmented Reality," *Adv. Eng. Softw.,* vol. 55, pp. 45-55, 1 2013.

[24] K. Kiyokawa, M. Billinghurst, S. E. Hayes, A. Gupta, Y. Sannohe and H. Kato, "Communication Behaviors of Co-Located Users in Collaborative AR Interfaces," in *Proceedings of the 1st International Symposium on Mixed and Augmented Reality*, Washington, 2002.

[25] D. Roth, J. Lugrin, D. Galakhov, A. Hofmann, G. Bente, M. E. Latoschik and A. Fuhrmann, "Avatar realism and social interaction quality in virtual reality," in *2016 IEEE Virtual Reality (VR)*, 2016.

[26] Flightradar24 AB, "flightradar24.com," 2018. [Online]. Available: https://www.flightradar24.com. [Accessed 30 08 2018].

[27] Boeing, "Boeing: UAVs. Holograms. Wildfire.," 2016. [Online]. Available: https://www.youtube.com/watch?v=omGoz66xHU8. [Accessed 30 08 2018].

[28] Next Reality, "HoloTerrain Lets HoloLens Users Explore the Earth in 3D," 2017. [Online]. Available: https://hololens.reality.news/news/have-you-seen-this-holoterrain-lets-hololens-users-explore-earth-3d-0176044/. [Accessed 30 08 2018].

[29] TAQTILE INC, "HoloMaps by Taqtile," 2016. [Online]. Available: https://www.microsoft.com/en-us/p/holomaps-by-taqtile/9nblggh3zrtd. [Accessed 30 08 2018].

[30] Google, "Google Earth," 2018. [Online]. Available: https://www.google.com/intl/en/earth/. [Accessed 30 08 2018].

[31] UC Davis, "Augmented Reality Sandbox," 2016. [Online]. Available: https://arsandbox.ucdavis.edu/. [Accessed 31 08 2018].

[32] Mapbox, "Mapbox," 2018. [Online]. Available: https://www.mapbox.com/. [Accessed 30 08 2018].

[33] T. Drey and J. Urban, "Radar Site Studies with Automated Propagation Effect Calculations Using QSiteAnalysis," in *2018 19th International Radar Symposium (IRS)*, 2018.

[34] Bundesamt für Kartographie und Geodäsie, "GeoBasis-DE," 2018. [Online]. Available: https://www.bkg.bund.de/. [Accessed 01 09 2018].

[35] OpenStreetMap contributors, "Planet dump retrieved from https://planet.osm.org," 2018. [Online]. Available: https://www.openstreetmap.org. [Accessed 01 09 2018].

[36] V. Agafonkin, "Leaflet," 2018. [Online]. Available: https://leafletjs.com/. [Accessed 01 09 2018].

[37] Met Office, "Cartopy: a cartographic python library with a Matplotlib interface," 2018. [Online]. Available: http://scitools.org.uk/cartopy. [Accessed 03 09 2018].

Spatially resolved non-invasive strain measurement on fast rotating composite rotors

J. Lich [1], R. Kuschmierz [1], T. Wollmann [2], A. Filippatos [2], M. Gude [2], J. Czarske [1]
[1] TU Dresden, Chair of Measurement and Sensor System Technique (MST), Dresden, Germany
[2] TU Dresden, Institute of Lightweight Engineering and Polymer Technology (ILK), Dresden, Germany

Contact e-mail: julian.lich@tu-dresden.de

Abstract

Fiber reinforced plastics (FRP) have a high potential for the application in high performance rotors as they provide high specific tensile strength and stiffness as well as low density. This combination enables composite rotors to run at very high rotational speeds. However, the complex mechanical behavior of FRP impedes accurate failure prediction of the rotors. At TU Dresden, numerical simulation models are developed to describe the structural vibration and deformation behavior of FRP rotors as a function of their rotational speed and their structural health. To validate these models, measurement techniques are developed and applied to track the vibrational behavior of the FRP rotors and consequently their mechanical properties spatially and temporally. A rotor test rig accelerates the cylindrical specimens to rotational speeds up to 200 Hz and circumferential speeds above 300 m/s.

The Laser Doppler Distance Sensor, developed at MST, was already used to investigate the radial expansion of rotors as a function of their rotational speed. Next, the strain as a function of the rotor radius is measured.

To measure the strain, a holographic diffraction grating foil is applied to the rotor surface. A collimated laser beam illuminates the grating, which splits the beam into several diffraction orders. The diffracted beams are analyzed via cross correlation in order to measure the strain and the tilt of the observed spot on the surface simultaneously. By using a high-speed line scan camera, measurement rates up to 12 kHz are achievable and vibrational measurements as well as real time processing can be performed.

Contrary to conventional strain gauges, the spot of measurement does not move with the rotor. This enables an angular scan of the rotor behavior. A radial movement of the Sensor enables radial scan. Arbitrarily many spots on the surface are measurable with a spatial resolution of under 3 mm. Furthermore, no cables, slip rings or telemetric transmitters are necessary.

The robustness of the sensor towards rigid body movements, like in-plane displacement, out-of-plane displacement and tilt was demonstrated. As a next step, the system will be tested on a rotating object for the first time.

1. Motivation

Advantages of composite rotors

- High specific stiffness and tensile strength, low density
  - ➢ Large diameters
  - ➢ High rotational speeds
  - ➢ High performance and efficiency
- Advantageous fatigue behavior
- Adaptability of two component material

Fan blade of a GE90 -Turbine in Boeing 777 [GE18]

1. Motivation

Validation of models

- Models can be used to increase rotor performance and safety
- Validation and calibration by in-situ measurements
- Required: In-plane deformation and out-of-plane vibration behavior as a function of:
  - ➢ Rotational speed
  - ➢ Damage state

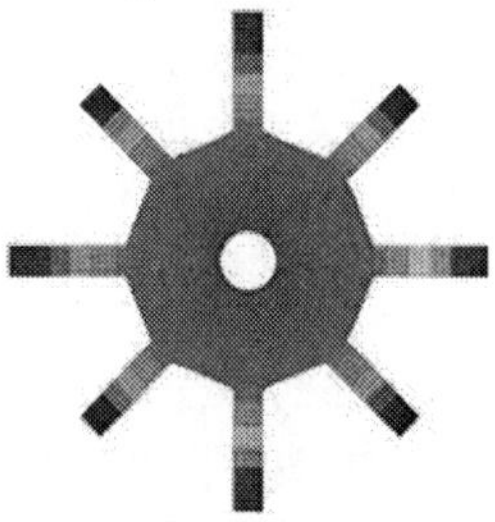

FEM Model: Out-of-Plane eigenmode of static FRP rotor [GUD15]

⟷

Rotor test rig

2. Preliminary Work

Integral in-plane deformation

- 3 Laser-Doppler Distance Sensors →
  - Surface speed 300 m/s, 120 spots measured
- Fatigue and hysteresis
- Directionally anisotropic

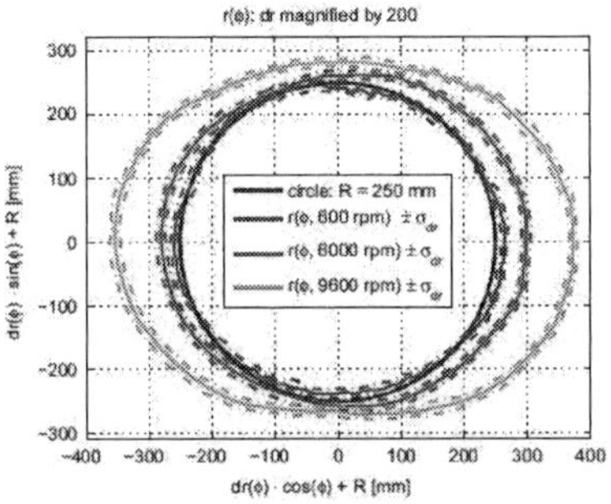

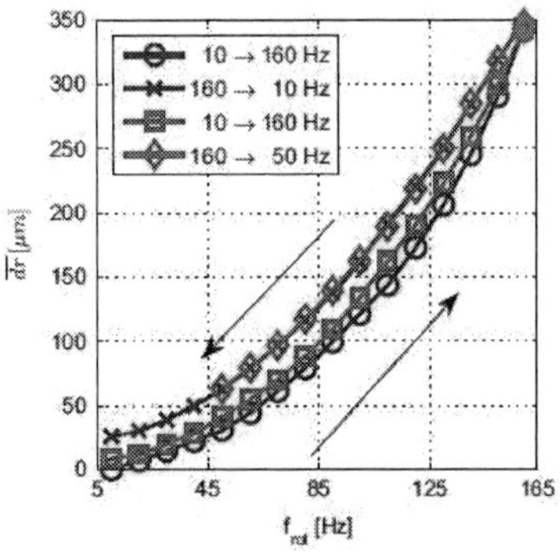

Source of Image and plots: [KUS14]

➢ High angular resolution, no radial resolution

2. Preliminary Work

In-plane vibration measurement

- Excitation with 2 piezoelectric actuators →
- 2 strain gauges
- Eigenfrequency depends on rotational speed

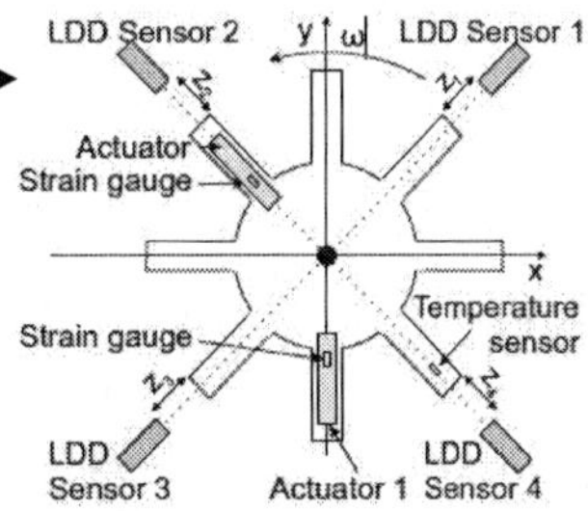

Source: [GUD15]

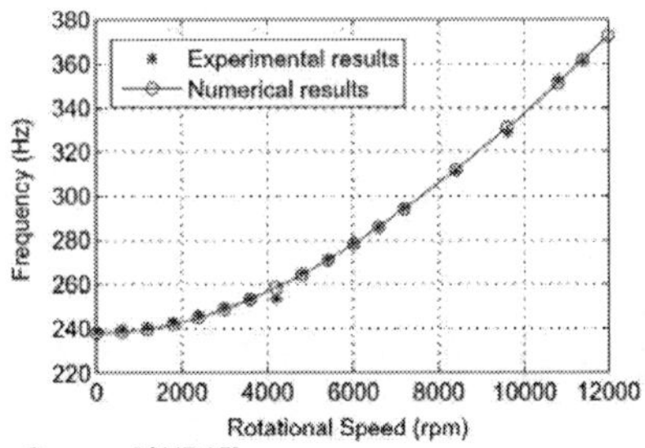

Source: [GUD15]

➢ Spatial resolution limited by number of strain gauges

3. Approach

Required

- Higher radial and angular resolution
    - In-Plane deformation
    - Tilt/ out-of-plane vibration

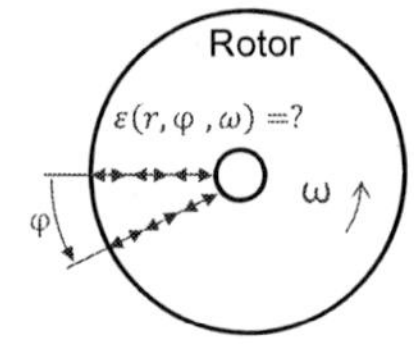

Aim: full-field measurement on moving rotor

Approach

- Optical far field measurement
- Use diffraction grating

Bragg equation:

$$\sin\alpha + \sin\beta = \pm m\lambda/d(\varepsilon)$$

➢ diffraction angle β yields deformation ε

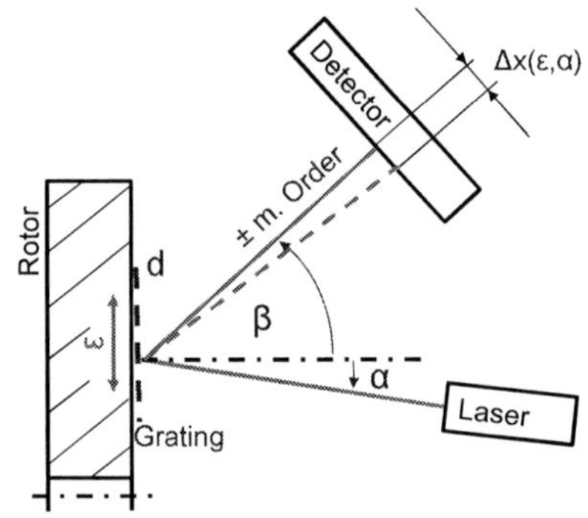

3. Approach

Strain and tilt measurement with optical grating

- Apply several gratings at different rotor angles
- Probe rotation → angular scan
- Sensor translation → radial scan

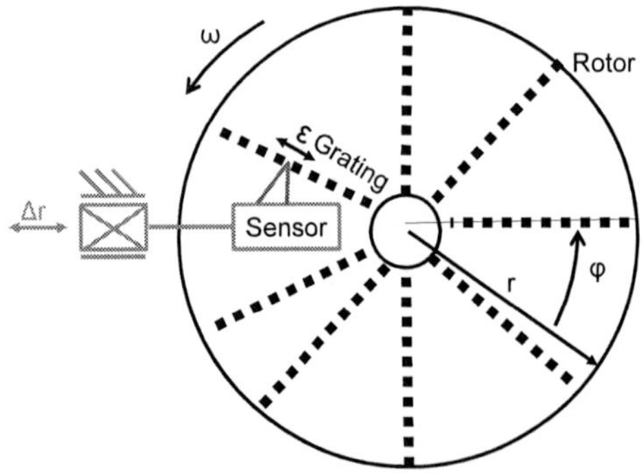

3. Approach

Strain and tilt measurement with optical grating

- Problem: Position of light spot on detector depends on axial rotor position
- Solution: Use objective lens

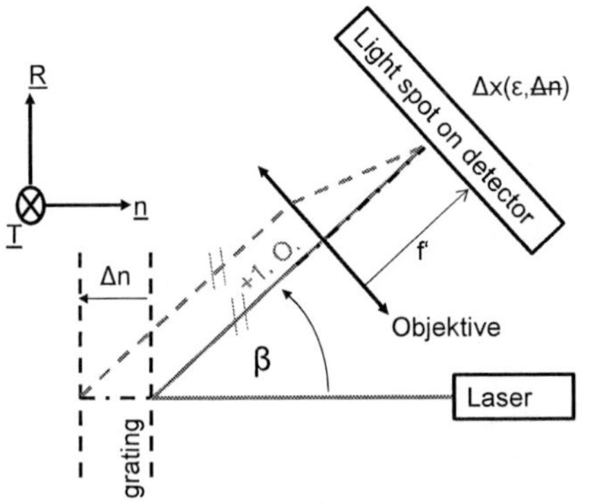

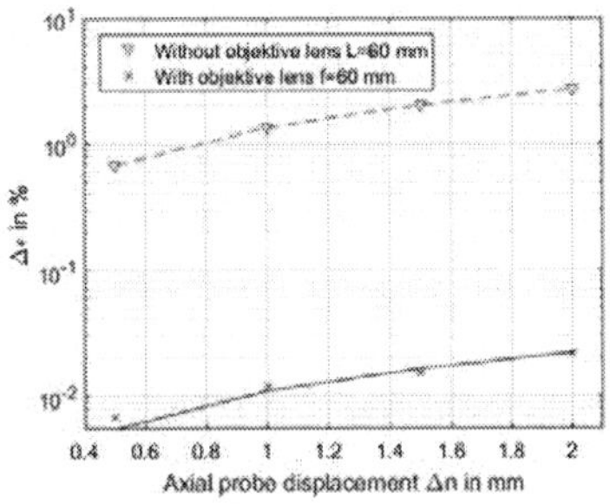

➢ Measurement independent of axial rotor position

3. Approach

Strain and tilt measurement with optical grating

- Problem: Crosstalk between strain and tilt
- Solution: Use two diffraction orders

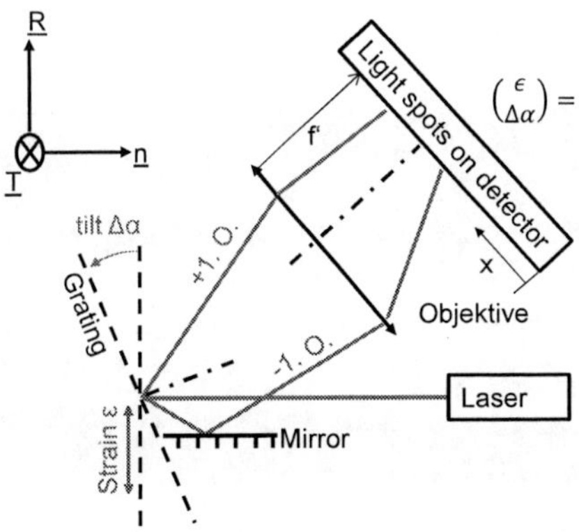

$$\begin{pmatrix} \epsilon \\ \Delta\alpha \end{pmatrix} = \begin{bmatrix} k_1 & k_2 \\ k_3 & k_4 \end{bmatrix} \begin{pmatrix} x_+ \\ x_- \end{pmatrix}$$

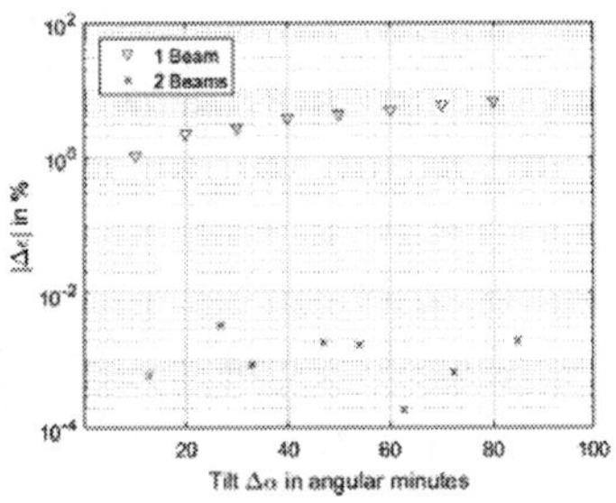

➢ Separation of strain and tilt information

4. Setup and Characterization

Measurement Setup

Static tensile testing with FRP specimen

Aluminum specimen with strain gauge and grating

Measurement Setup

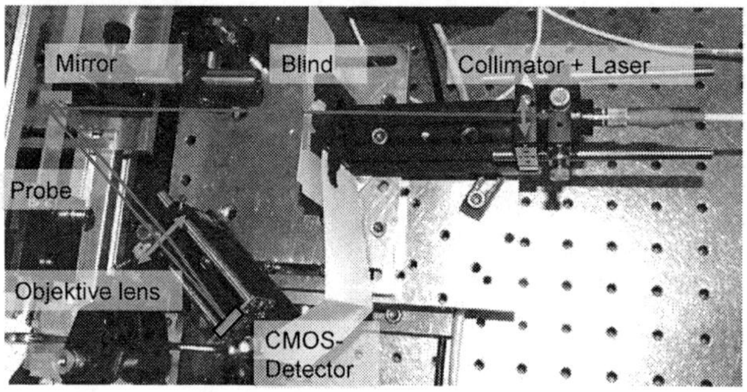

4. Setup and Characterization

Signal Processing

- Analyze movement of two diffraction orders on matrix camera
  - ➢ Determination of strain ε and surface tilt Δα

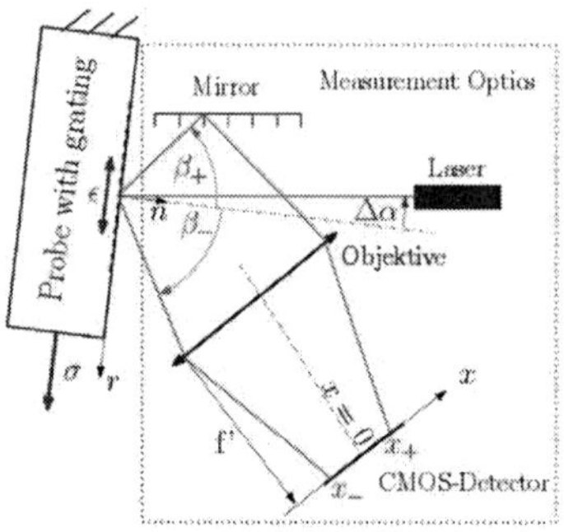

Principle

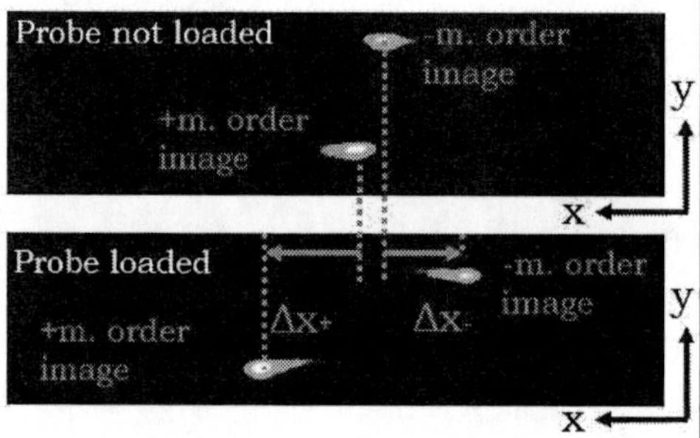

Diffraction orders ±m. on CMOS Sensor

5. Results

Static tensile test Aluminum

- Validation with strain gauges
- 8 spots every 1 mm on grating measured
- Standard deviation: 50 µε
- R^2 of mean value: 99.63%
- Systematic deviation from strain gauges: ΔE=(11…18) GPa
  - ➢ Calibration

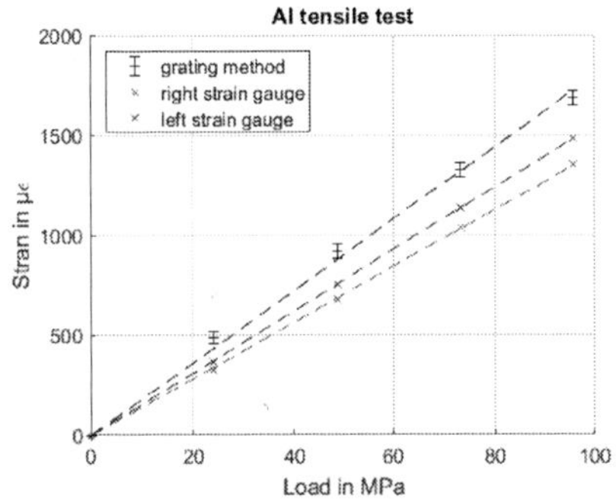

Aluminum specimen with strain gauges and grating

➢ Validated for static Al probes

5. Results

Static tensile test FRP

- 8 spots every 1 mm on grating measured
- Standard deviation: 80 µε
- R^2 of mean value: 99.84%

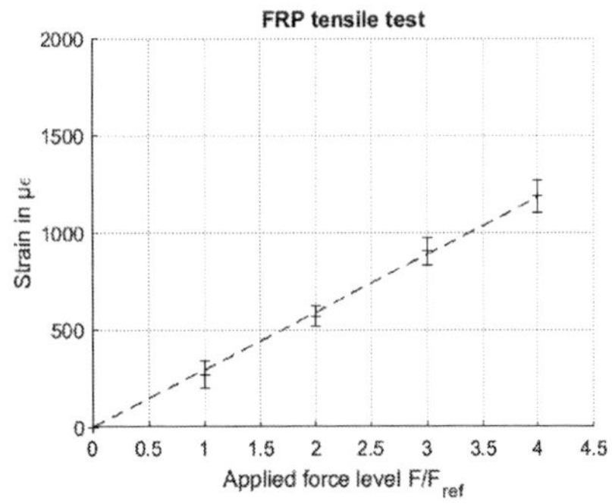

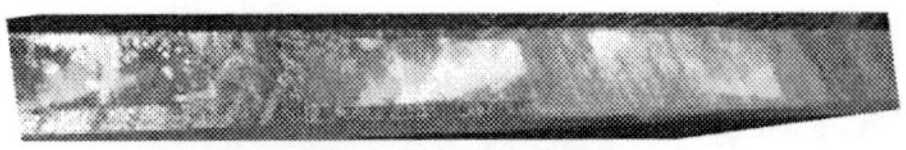
FRP specimen with gratings

➢ Validated for static FRP probes

5. Results

Dynamic test

- Compact Setup with line Camera and light barrier trigger
- Rotational speeds: (20…100) Hz
- Strain standard deviation: < 11 με
- Angle standard deviation < 2.5"

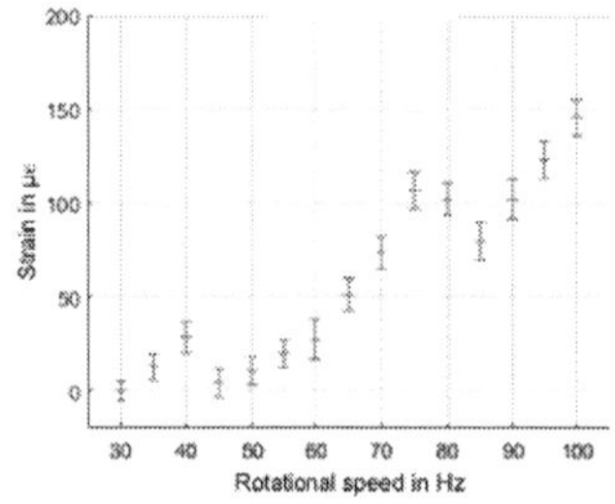

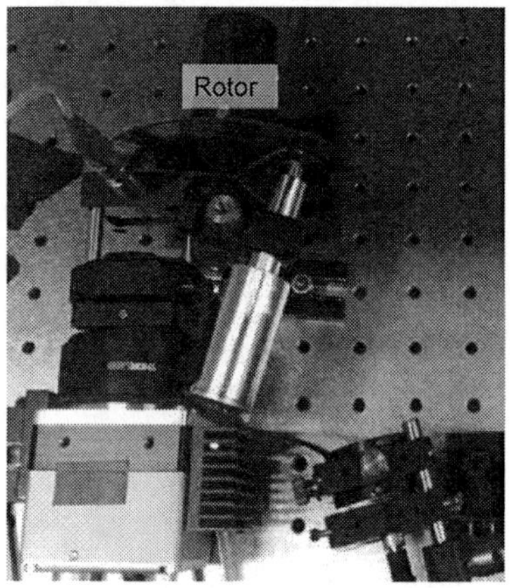

➢ Validated for moving probes

6. Conclusion

- FRP advantageous for rotors, validated models necessary
- Validation + calibration with measurements
- Strain measurement with diffraction grating method
  - ➢ Optical: fast and contactless
  - ➢ Robust towards vibrations
  - ➢ High linearity (R^2=99.63%)
  - ➢ Low uncertainties (s ≈ 10 με possible)
  - ➢ Applicable on moving rotor

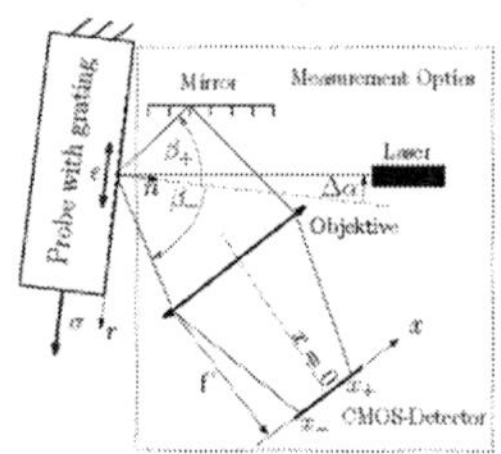

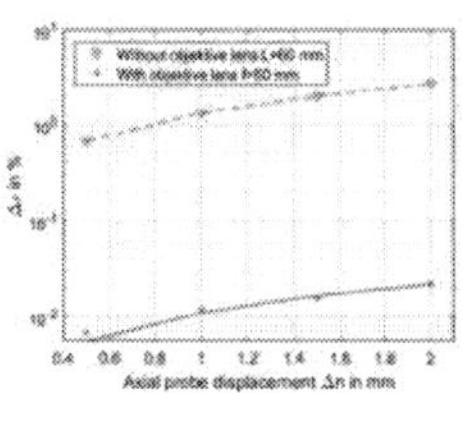

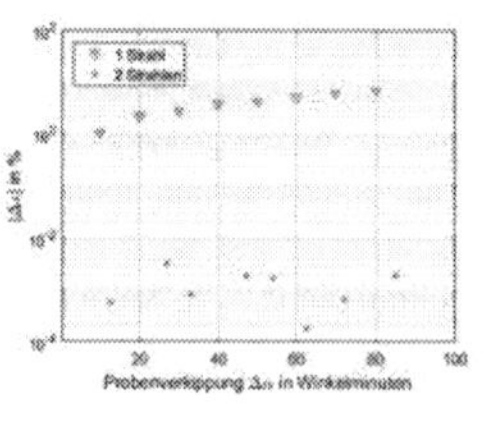

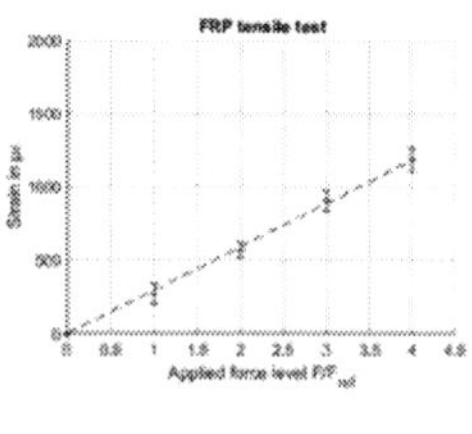

7. Outlook

Measurement in rotor test rig

- Validation and calibration of numerical simulation models

Grating application

- Currently: Glued holographic grating film
  - ➢ Cheap, high diffraction efficiency
- Surface patterning
  - ➢ Surface = grating → no strain transmission errors

Holographic diffraction grating film [ED18]

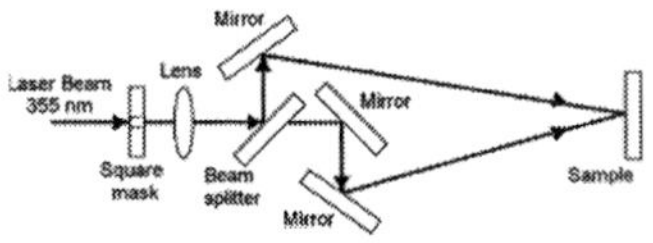

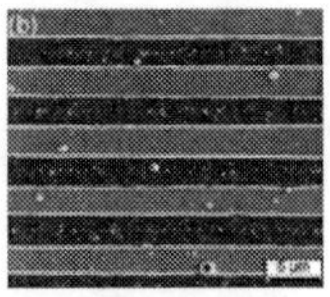

Laser Interference Patterning an Fe/Cu/Al-Schicht [LAS07]

Sources

[GUD15]	M. Gude, A. Filippatos, A. Langkamp, W. Hufenbach, A. Fischer, J. Czarske, "Model assessment of a composite mock-up bladed rotor based on its vibration response and radial expansion", Composite Structures, 2015
[KUS14]	R. Kuschmierz, A. Filippatos, P. Günther, A. Langkamp, W. Hufenbach, J. Czarske, A. Fischer, „In-process, non-destructive, dynamic testing of high-speed polymer composite rotors," Mechanical Systems and Signal Processing, 2014.
[LAS07]	A.Lasagni, C. Holzapfel, T. Weirich, F. Mückrich, "Laser interference metallurgy: A new method for periodic surface microstructure design on multilayered metallic thin films", Applied Surface Science, 2007
[SON14]	T. V. Son and A. Haché, "Thermal expansion and stress and measurement using high-order diffraction: Possibilities and theoretical limits," Optics Communications, vol. 326, pp. 134–138, 2014.
[Zha00]	B. Zhao, H. Xie, A. Asundi, "Optical strain sensor using median density grating foil: Rivaling the electric strain gauge", Review of Scientific Instruments, vol. 72, nr. 2, 2000
[GE18]	ge.com/reports/the-art-of-engineering-the-worlds-largest-jet-engine-shows-off-composite-curves/ (15.06.2018)
[LS18]	lightsoftllc.com/ (23.06.18)
[ED18]	edmundoptics.com/optics/gratings/Holographic-Diffraction-Grating-Film/ (23.06.18)

Advanced and automatic turbine blade inspection, CT resolution optimization and error estimation in reconstructed volumes

Peter WESTENBERGER [1], Tristan LOWE [2]
[1] Thermo Fisher Scientific, Düsseldorf, Germany
[2] University of Manchester, Manchester, UK

Contact e-mail: peter.westenberger@thermofisher.com

Abstract

The presentation will cover in detail the challenges in fully automatic turbine blade inspection workflows. The presented workflows have been advanced by techniques established and derived from radiography in order to optimize the spatial resolution in X-ray CT and to estimate the errors in the reconstructed volumes. We will demonstrate how such techniques help to speed up the inspection process in tomorrow's inline inspection visions where time really matters.

The blade inspection workflow itself will cover non-destructive testing standards like robust segmentation and sub-resolution precise surface determination which are the prerequisite for actual nominal comparison against CAD and/or reference models and for retrieving precise metrology and GD&T information. Furthermore it will also give attention to rather unpredictable and more challenging inspection tasks like proper and robust detection, segmentation, metrology, and analysis of air vents in turbine blades. We will present how to solve the problem of spatially unstable air vent drill patterns as well as how to properly detect foreign objects, or under- and over-drills for each individual air vent.

The inspection workflow together with the advanced optimization techniques will help to build automatic inline inspection scenarios for the future.

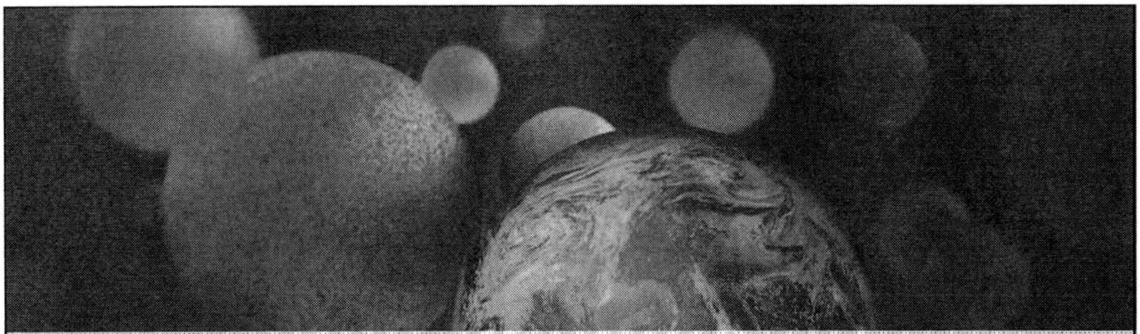

Thermo Fisher SCIENTIFIC

Automatic Turbine Blade Inspection Workflow

10th International Symposium on NDT in Aerospace

Peter Westenberger, Dr. Tristan Lowe

The world leader in serving science

Outline

- Detectability and measurability thoughts
  - Resolution
  - Blurring
  - Positioning
  - Radiographs
- Segmentation
  - Cooling system reconstruction
  - Air vent metrology
  - Defect detection

Data courtesy by diondo GmbH

2 Thermo Fisher SCIENTIFIC

General assumption of resolution

$$M = \frac{d_{so} + d_{od}}{d_{so}}$$

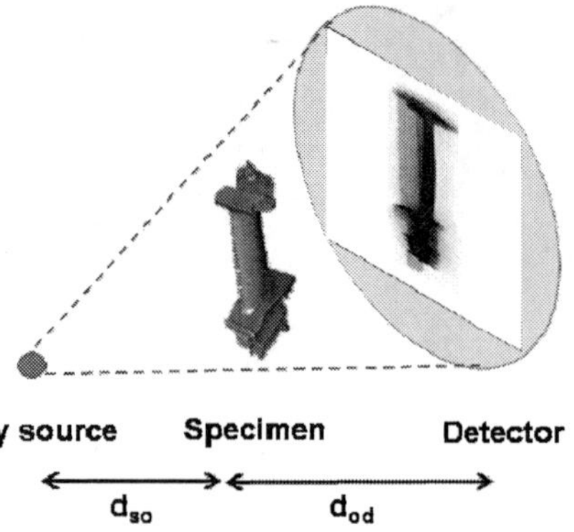

3

ThermoFisher SCIENTIFIC

Geometric blurring

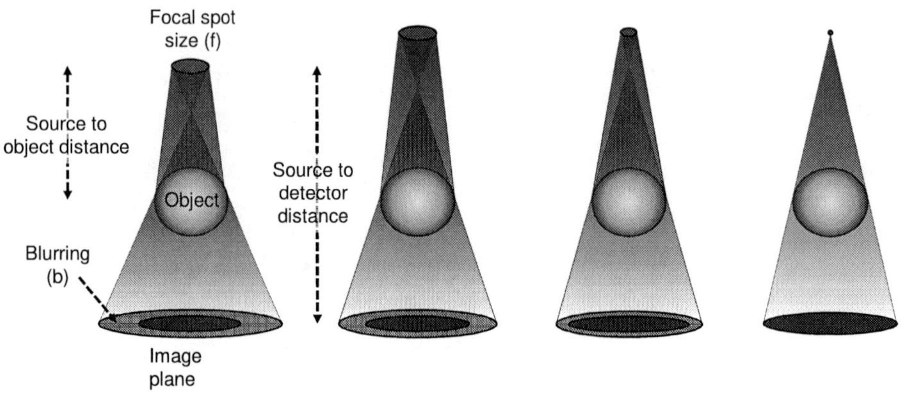

$$b = (M-1)f$$

4

ThermoFisher SCIENTIFIC

Optimising imaging positions

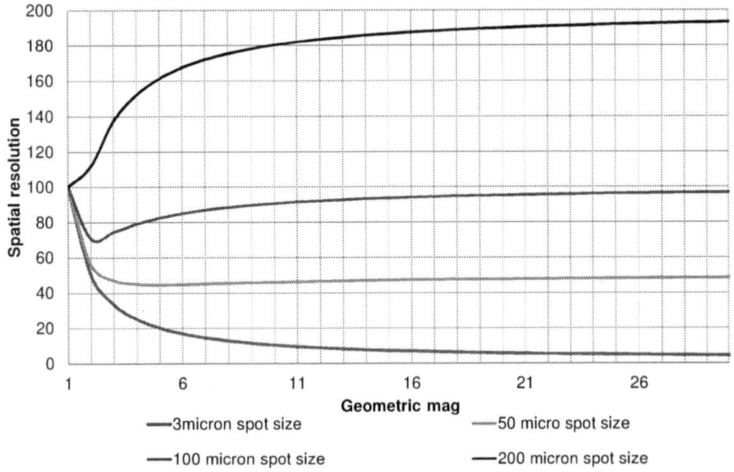

Radiograph resolution

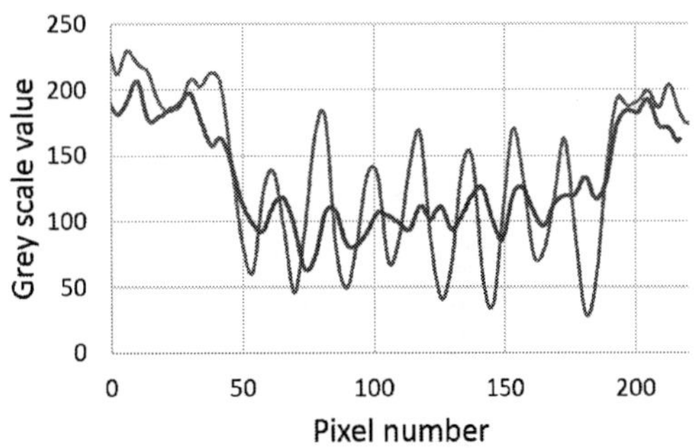

- Radiograph quality is essential for good spatial resolution
- Significantly affects post processing time, repeatability & errors
- Numerical characterisation is essential

Measuring spatial resolution

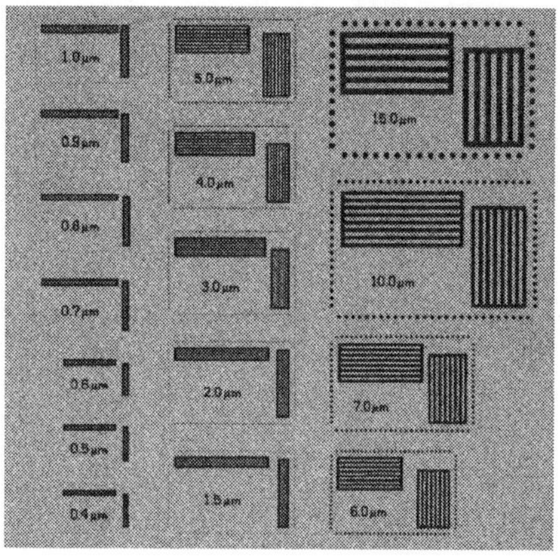

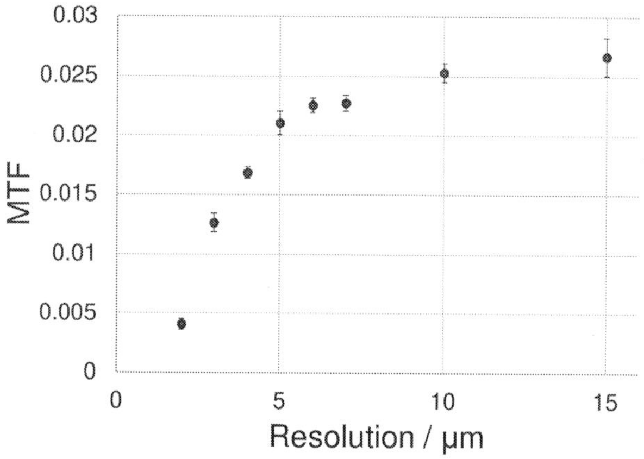

$$MTF = \frac{2\ x\ Amplitude\ (difference\ between\ crest\ and\ trough)}{Overall\ intensity\ (sum\ of\ crest\ value\ and\ trough\ value)}$$

7

ThermoFisher SCIENTIFIC

Optimising scan times

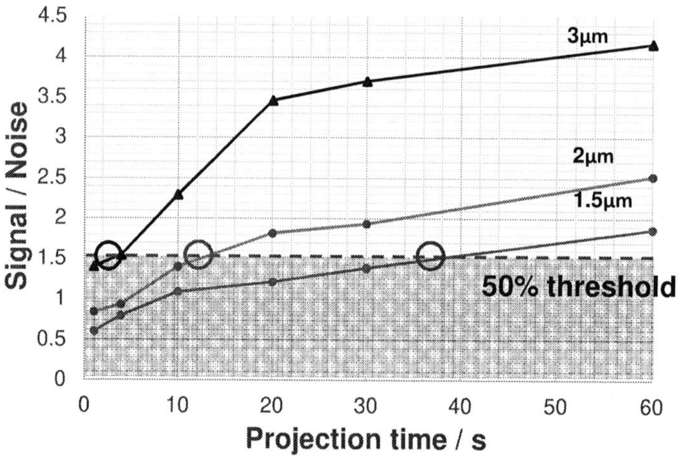

8

ThermoFisher SCIENTIFIC

Reduced analysis time

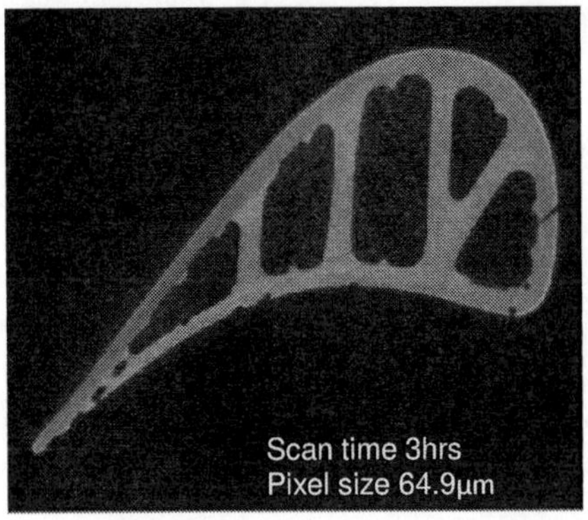

Higher resolution

Reduced scan time

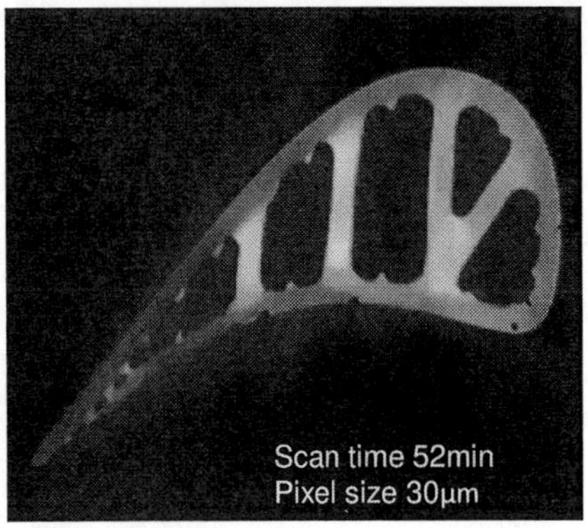

430keV, 197 W, 1500 projections, 120s exposure with LDA

320keV with Flat panel

Thermo Fisher SCIENTIFIC

Watershed

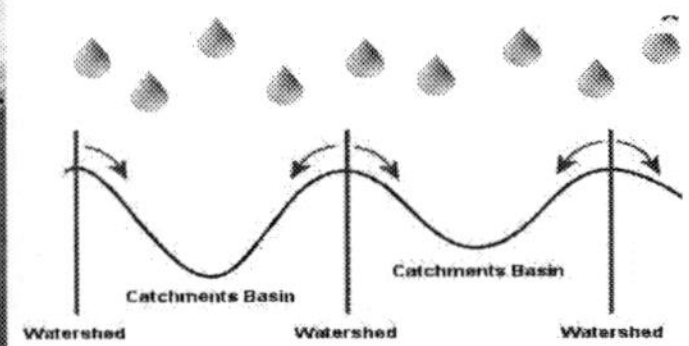

Area of land that drains into a river system

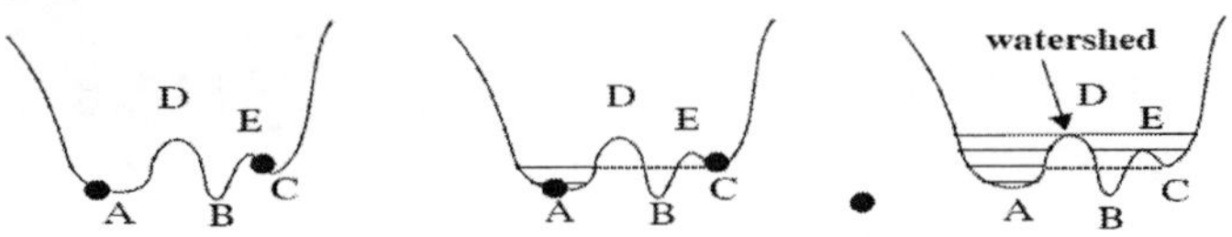

Thermo Fisher SCIENTIFIC

Watershed

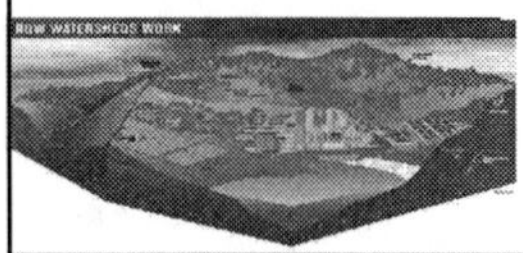

11

Thermo Fisher SCIENTIFIC

Surface extraction

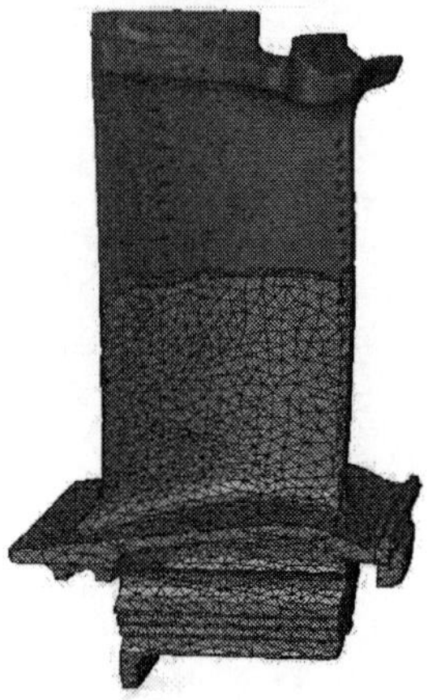

12

Thermo Fisher SCIENTIFIC

Voxel precise segmentation

Voxel precise surface

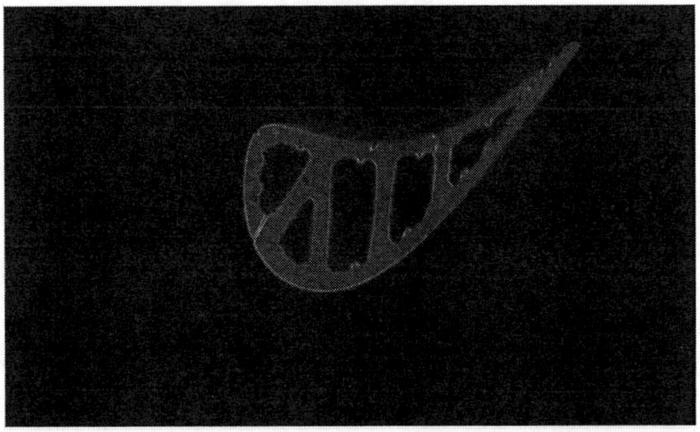

13

Thermo Fisher SCIENTIFIC

Adjusted surface vs. voxel precise surface

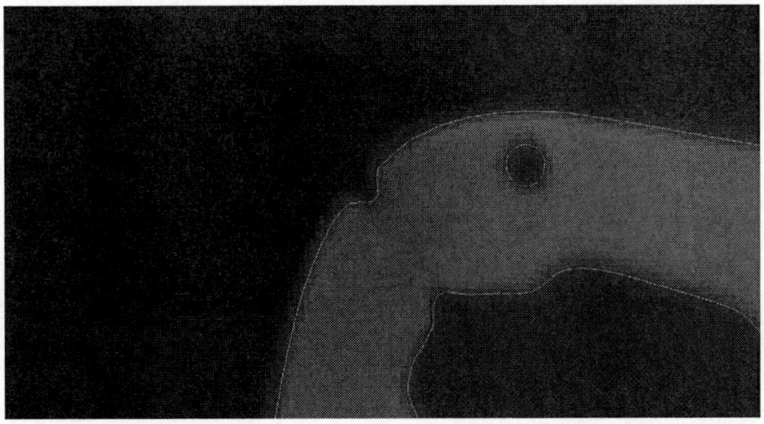

14

Thermo Fisher SCIENTIFIC

Cooling system reconstruction using ambient occlusion

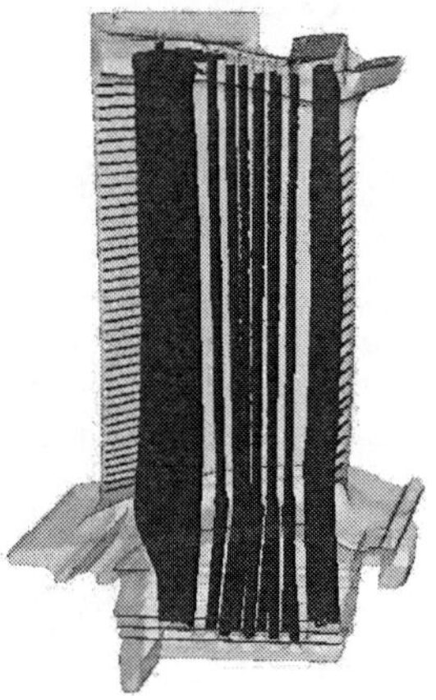

15

ThermoFisher SCIENTIFIC

Air vent segmentation

- Spatially instable air vent patterns
- Creating tolerant search mask
- Thresholding inside mask
- Center line extraction
  - Auto-Skeletonization

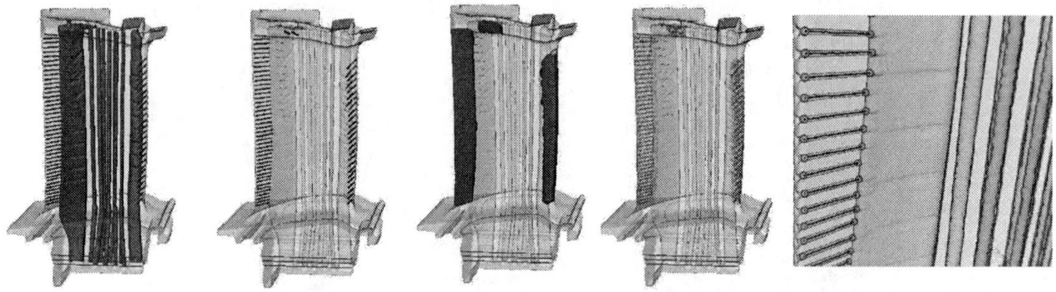

16

ThermoFisher SCIENTIFIC

Metrology: Meshing extracted air vents from extracted center lines

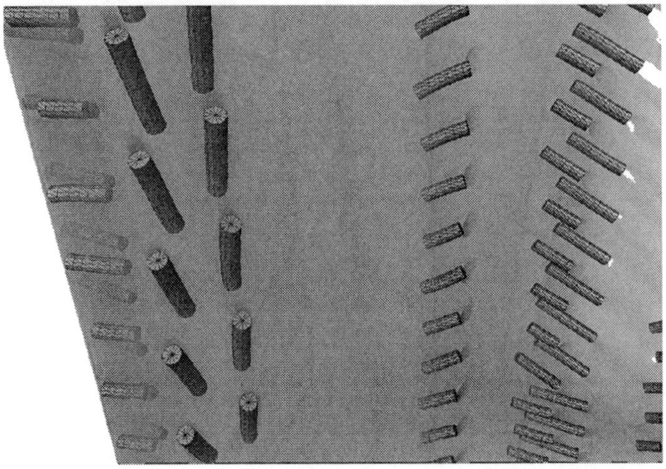

17

ThermoFisher SCIENTIFIC

Metrology: Point extraction from meshed air vents for cylinder fitting

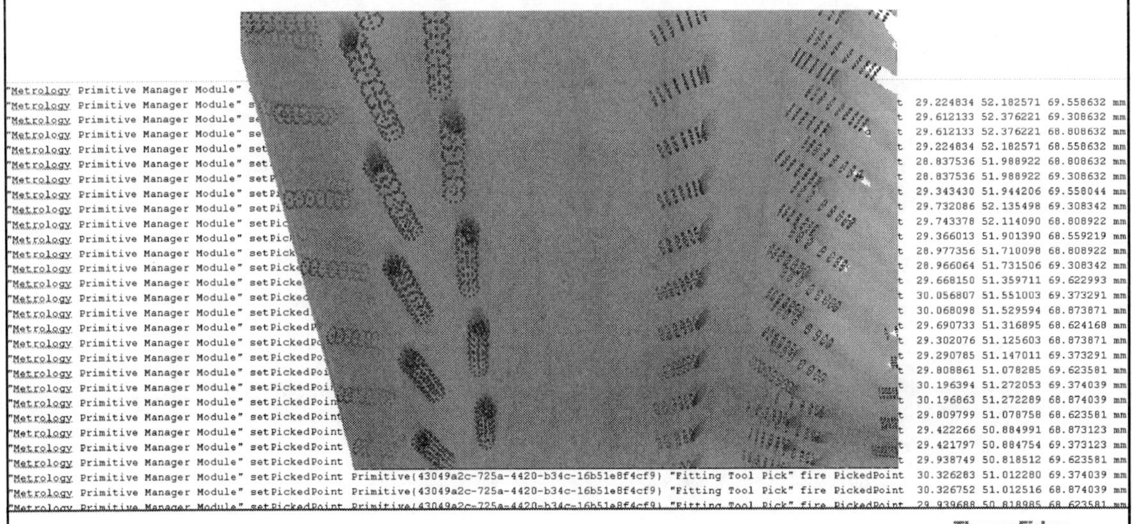

18

ThermoFisher SCIENTIFIC

Air vent metrology: Automatically fitted cylinders

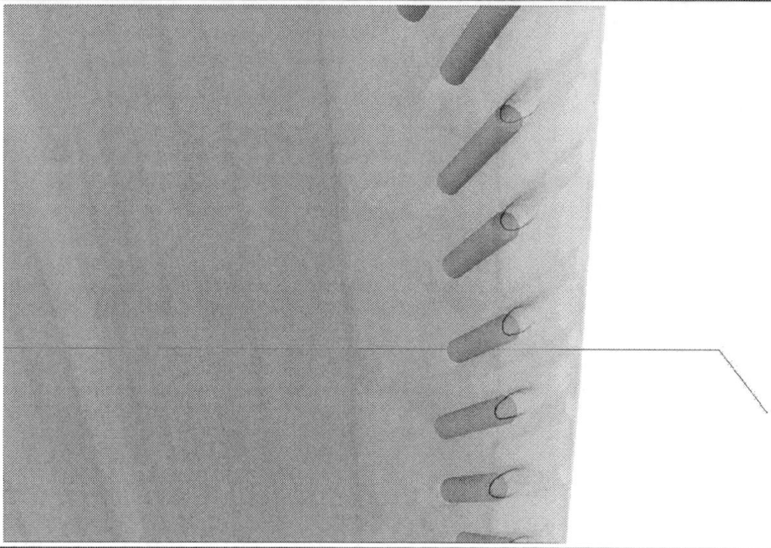

19

ThermoFisher SCIENTIFIC

Foreign object detection

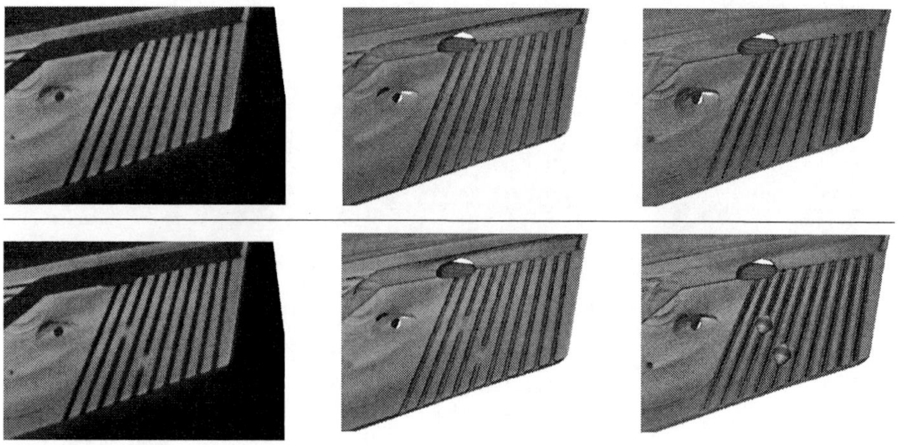

20

ThermoFisher SCIENTIFIC

Overdrill detection

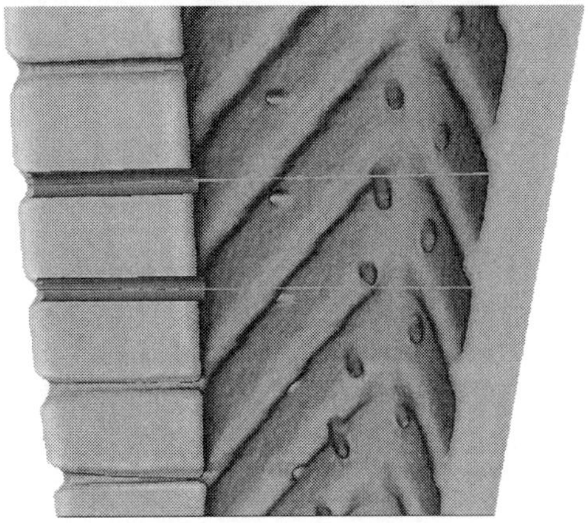

21

ThermoFisher SCIENTIFIC

Thank you!

Find out more at **thermofisher.com/amira-avizo**

22

ThermoFisher SCIENTIFIC

Virtual wave concept for thermographic porosity estimation in carbon fiber reinforced plastics

Holger PLASSER [1], Günther MAYR [2], Gregor STOCKNER [1], Günther HENDORFER [2], Peter BURGHOLZER [3]
[1] University of Applied Sciences Upper Austria, Wels, Austria
[2] University of Applied Sciences Upper Austria, School of Engineering, Wels, Austria
[3] Research Center for Non Destructive Testing GmbH, Linz, Austria

Contact e-mail: holger.plasser@fh-wels.at

Abstract

In this work the quantitative evaluation of porosity with pulsed thermography in carbon fiber reinforced plastics is shown by applying the virtual wave concept. A virtual temperature signal is calculated by applying a local transformation to the measured temperature data. This virtual temperature is a solution of the wave equation, whereby for the parameter estimation ultrasonic test methods can be used, e.g. pulse-echo method for time-of-flight measurements. The time-of-flight determined from the virtual temperature is directly related to the thermal diffusion time. Therefore, if the specimen thickness is known the effective thermal diffusivity can be carried out. The main advantage of the virtual wave concept is the possibility to use the same data algorithm for pulsed thermography measurements in reflection as well as in transmission mode for parameter estimation. The estimation of the porosity based on the virtual wave concept matches the porosity based on the well-known Parker method very well.

Effective medium theories are used to derive the porosity from the estimated thermal diffusivity. In the first step, this model-based porosity evaluation is validated on a range of different calibrated porous CFRP specimens with different number of plies and varying porosity contents. Furthermore, porosity evaluation on real aerospace parts will be demonstrated. All thermography results are validated with X-ray computed tomography.

Determination of Indication Depths at High Frequency Eddy Current Testing of CFRP

Nataliia MATVIEIEVA [1], Martin SCHULZE [2], Koichi MIZUKAMI [3],
Ievgen KHARABET [1], Henning HEUER [1,2]
[1] Technische Universität Dresden, Dresden, Germany
[2] Fraunhofer-Institut für Keramische Technologien und Systeme, IKTS, Dresden, Germany
[3] Ehime University, Matsuyama, Japan
Contact e-mail: matvieieva@avt.et.tu-dresden.de

Abstract. Nowadays, consumption and production of the carbon fiber reinforced polymer (CFRP) increase very fast, therefore a demand for the relatively simple and fast methods of quality control increases as well. Thanks to the sensitivity, ease of application and fast speed of utilization, High Frequency Eddy Current (HF EC) method is increasingly used for quality control of CFRP. One of the most important and complicated existing tasks for the method is the determination of the real depth of the indication or defect e.g. for carbon fiber ply determination.

In order to address this task experimental and numerical study has been made. The last was based on the concept of the coils combination with different operating frequencies, which provides different penetration depth into CFRP and, in the result, the possibility to quantify it. First experiments in this direction showed encouraging results, where half-transmission 2-coils sensors with different operating frequencies and penetration depth were used for the depth determination of copper elements embedded in laminated cross-ply CFRP specimen. The finite element modeling was done, compared with experiment and results are discussed in the paper.

Introduction

Every year usage of advanced composite materials increases dramatically, among them CFRP and glass fiber reinforced polymer (GFRP) are holding the dominant place [1]. This effect could be explained by their outstanding thermo-mechanical properties, superior corrosion resistance, high strength-to-weight ratio and rigidity which are required in aerospace (aircrafts, spacecrafts, micro air vehicles (MAVs)), automotive and wind industry, civil engineering (e.g. retrofitting, concrete reinforcement), sports goods and other consumer and technical applications [2].

Anisotropic materials like CFRP and GFRP consist of components, which have various physical and mechanical properties (e.g. conductive carbon yarn and dielectric epoxy matrix) and exhibit different damage mechanisms, which can occur during manufacture or in service. For evaluation of the defect's harmfulness, as well as the necessity to change or repair the part, an information about its type, location, size and depth is essential [3]. For example, one of the known defects is out-of-plane waviness (or undulations), which doesn't lead to a strong difference in inhomogeneity in comparison to the whole structure. This exacerbate a task to detect them, whilst the amplitude of the wave and its depth plays significant role in the judgment about the textile part validity [4].

The textile production of CFRP is focused on the development of various types and structures of the textile, and on the study of the innovative fiber orientations as well as binding [5]. For the assurance of woven structures' quality it is needed to define an orientation of fibers on different depth.

For these and other important tasks, an appropriate method has to be found. Conventional non-destructive methods for CFRP composites include IR thermography, X-ray tomography and a variety of ultrasonic (US) techniques. Thermography methods are portable and economic and could cover large components in a single image. The main challenge is, to determine the specimens' information precisely from the measured data, which requires many complex algorithms [6], [7]. X-ray tomography, e.g. micro-CT, is very precise and easily locates three-dimensional defects regardless of CFRP composite type [8], [9]. However, X-ray tomography scanners are very expensive and cumbersome, 3-D X-ray imaging is very time-consuming and the size of the sample is very limited, all these make X-ray unsuitable for most field applications, particularly aerospace or automotive devices. Conventional US pulse-echo techniques usually require couplants or full immersion of the sample for optimal energy transfer, which can affect overall scanning speed and greatly limit the number of applications where these techniques can be used. [10], [11]. More advanced optical methods, such as digital image correlation (DIC) and optical coherence tomography (OCT) are sometimes used, but they are usually limited to surface inspection or measurements on semi-transparent composites [12].

In recent years Eddy Current (EC) testing [13, 15, 16] becomes more widespread for CFRP investigation thank to the relatively low costs of its inspection and utilization and high operational speed. In comparison to other, EC method doesn't require additional coupling (as US), mechanical loading (as DIC) and doesn't extremely limit the inspected area (as X-ray). EC spectroscopy is used for investigation of electric properties of conductive materials. The induced primary electromagnetic field (EMF) affects conductive material, where under its influence eddy and displacement currents occur, which induce a secondary magnetic field (MF), directed opposing to the primary field and resisting its propagation. The resisting MF is dependent of the material's conductivity, where the higher the last one, the higher the opposing field. Resulting difference between generated and induced voltage (resulting induced voltage) is measured and analyzed.

Using the current EC testing systems, a depth determination is only possible to a very limited extent with use of 2D Fourier transformation. Current research is concerned with the variation of physical parameters of the EM coils, such as the measurement frequency and the diameter for receiving a new approach of the depth determination using the new HF EC method.

Methodology

HF EC tomography is based on the principle of conventional Eddy Current (EC) inspection , but operates on frequencies up to 100MHz, whereas EC works on kHz range. As well as EC introscopy, HF EC is non-invasive, relatively cheap and easily applicable for investigation of the conductivity distribution of the mechanical parts, which have various curvature and size.

Due to high operating frequencies HF EC is more sensitive to small deviations in conductivity in the material and allows evaluation of not only conductive, but dielectric properties as well. The last could be explained by the Equation 1 which shows that total currents in the material caused by the magnetic field circulation $rot\vec{H}$ are the sum of the electric current density $\vec{J}$ and displacement current $i\omega\varepsilon\vec{E_0}$, which occur in dielectric component (ε) and has the bigger influence beginning from the certain frequency ω. Presence of displacement currents is the main specific feature of the HF EC method, which differentiates it from the conventional EC spectroscopy.

$$rot\vec{H} = \frac{\partial(\varepsilon\varepsilon_0\vec{E})}{\partial t} + \vec{J} = \vec{J} + i\omega\varepsilon\overrightarrow{E_0}\,. \quad (1)$$

As CFRP is a composite structure, which consists of the inhomogeneously conductive textile and dielectric epoxy matrix, for its inspection, a HF EC method should be taken, which is sensitive to all of the components of the material.
In the paper two approaches are suggested for depth determination using the HF EC method, based on the physical principle of the attenuation of the EMF in a material .

The exponential decay of currents in thickness is basic and understandable principle of EMF behavior penetrating into conductive materials (Eq.3, Figure 1(a)). Despite the inhomogeneity, EMF in CFRP in general has the same character, as in other conductive materials, where the higher the frequency of EMF the smaller the penetration depth. The layer, where the field's concentration reaches its maximum and which depth corresponds to the value, where fields decay exponentially is known as skin layer (Eq.2).

$$\delta = \sqrt{\frac{1}{\sigma f \pi}}\ , \quad (2)$$

$$J = J_s e^{-(1+j)d/\delta}\ , \quad (3)$$

where f- EMF frequency (Hz), σ- conductivity of the material [S], J_s is the current value at the surface [A], d-depth [m].

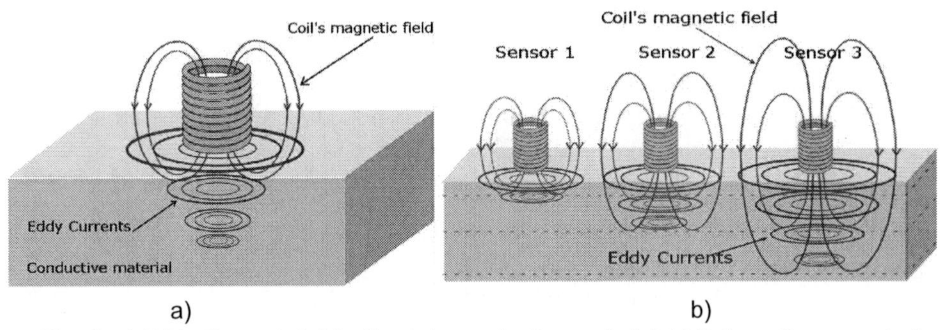

Fig. 1. a) Eddy Current distribution into conductive material; b) Schematic concept of sensors combination for depth differentiation

The first concept intends the EC image data processing using threshold for currents density on depth. From the Eq. 2 and Eq. 3 is obvious, that starting from some frequency f electromagnetic field fully penetrates the CFRP specimen. Here EMF decreases exponentially, as well as resulting currents. When investigated structure lays deeply, the EMF at that level is not strong enough to excite high density of Eddy Currents in the structure, therefore the resulting signal which corresponds to the structure's location also decreases with depth. This aspect was used in previous work for depth quantification, where for textile CFRP a voltage value threshold was set, which corresponds to the different attenuation on certain depth [14]. The deeper the defect layer was - the lower the threshold.

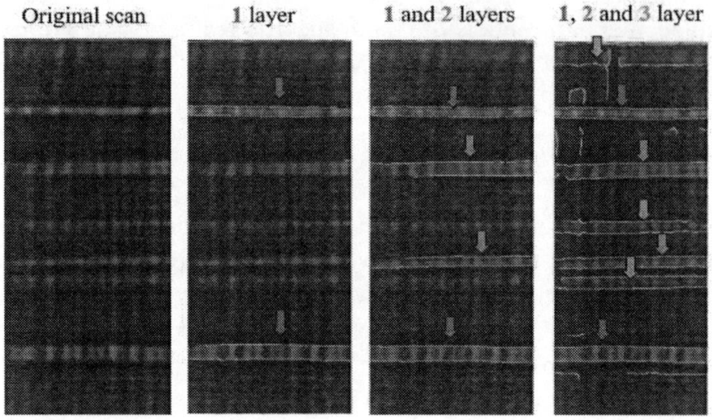

Fig. 2. Image of scanned textile CFRP with implemented defects. a) original scan, b) distinguished defects on 1st layer, threshold (thr)=0.79, c) defects on 1st and 2nd layers, thr = 0.6, d) defects on both (3) layers, thr = 0.39

The results are presented on the Figure 2. This approach could help differentiate a depth of clearly visible defects knowing the attenuation character of EMF for certain frequency for the specified material. For cases of out-of-plane waviness or CFRP plies on depth separation such approach is not appropriate. Therefore, other concept was suggested.

The second concept is based on the combination of different coils that provide different operating frequencies, depending on their parameters (diameter, amount of windings, etc.), where excitation EMF penetrates studied material on different depth. The more sensors are used – the higher depth resolution could be achieved (Figure 3). Measurement results, received in the form of images of conductivities in CFRP on different depth are combined, allowing to evaluate material's layers separately and to clearly identify defects on different depths.

To prove a concept, experiments were conducted using EddyCus MPECS® device and cylindrical multilayer coils with different diameters and operating frequencies (Table1), obtained results were compared with simulation data in ANSYS Mechanical.

Results

In order to check the validity of our concept we performed experimental study of defects' on depth in CFRP quantification by using coils with various parameters on specially fabricated specimen. The tested specimen was 16 cross-plies CFRP with stacking sequence [45/-45] with sizes of 160x75x4.5 mm length, width and high respectively, schematic representation and photo of the specimen are shown on Figure 4 (a), (b) [17]. Square copper foil elements 15x15mm were placed as conductive contrast markers with the step 1.33mm on depth from each other: 0, 1.33, 2.66 and 4 mm. In our experiment, the scanning area was 130x30 mm of the CFRP plate marked with dash-dotted contour on Figure4 (b).

A list of used sensors is presented in Table 1, where each separate coil is cylindrical multilayer coil with ferrite core and shield with different diameter and resonant frequency. In experiment seven 2-coils combinations were used (half-transmission representation), where №1 is a pick-up coil with wide pass band used in every combination and №2-8 – various driver coils which differ by the resonant frequency, diameter and amount of windings. The placing of the pick-up coil for all experiments corresponds to the x-axis (0° angle Figure 4 (a)), where the currents distribution is close to the maximum. Operating frequency for each coil was chosen near Fres on frequency, where the coils pair is most sensitive, i.e. the system has the biggest signal difference between air and CFRP.

Scanning results are presented on Figure 5 for each driver coil listed on Table 1. The underlined number on Figure is an operating frequency value for each used coil. First line is an Imaginary part of the resulting EC signal for each coil which corresponds to the currents occurred on the reactive components in the material. The Real part presented on the second line on the Figure 5 corresponds to the conductive currents.

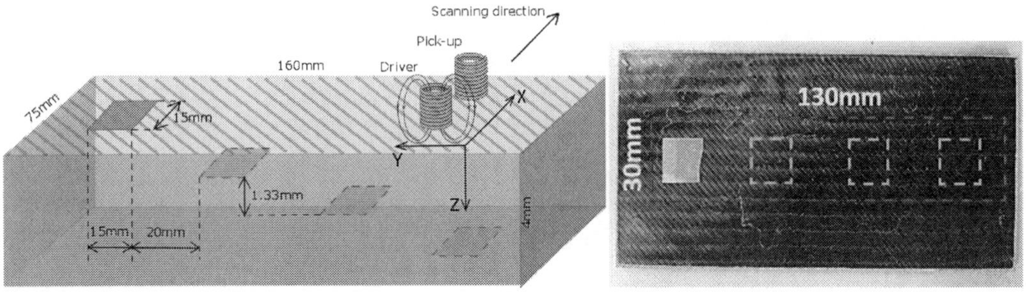

a) b)

Fig. 4. a) Schematic representation of the CFRP specimen and experimental set-up; b) Photo of the specimen with dash-lined scanned area

Table 1. List of the cylindrical coils used in the experiment

№	Type	Fres, MHz	Fact, MHz	Diam.,mm	w	dwire,mm	R, Ohm	L, uH
1.	Pick-up	8	-	2,4	50	0,06	1,367	14,24
2.	Driver	31,45	28,65	2,4	15	0,125	0,103	1,18
3.	Driver	24,1	7,1	3,3	20	0,1	0,26	2,36
4.	Driver	16,1	6,6	7	20	0,25	0,1	4,85
5.	Driver	13,6	5,9	2,4	30	0,08	0,45	5,68
6.	Driver	8	2,05	2,4	50	0,06	1,239	15,69
7.	Driver	3,45	0,55	7	80	0,1	2,136	77,1
8.	Driver	2,4	0,45	9	100	0,1	3,214	172,2

To understand behavior of the EMF in the CFRP on different frequencies, to prove experimental results, and to check the ability to predict which coils need to be created to penetrate a certain depth into the CFRP material, simulations were done and compared with experiment.

An electromagnetic field distribution in CFRP material and its dependence on the parameters of driver coils is investigated using finite-element modeling (FEM). A modelled specimen was designed as a sixteen-ply [-45/45]s CFRP, 60 x 60 x 4mm, length, width and thickness respectively. Voltage applied to the driver coil was 250 mV, longitude conductivity – 28,4 kS/m, transverse and through thickness conductivity – 4 S/m. Other parameters as coil resistance, diameter, amount of windings and resonant frequency were varied for each coil and listed in Table 1. The driver coil was placed directly above the center of the specimen with lift off 1 mm. For comparison, in Table 2 are presented an EC distribution character in accordance to the operating frequency of the coils. Here are showed Real and Imaginary parts of resulting currents (electric currents and Eddy Currents) in CFRP plate, with top and crosscut view for coil №2 on frequency 28.65 MHz and coil №6 on frequency 2.05 MHz. As it could be seen from the images, the higher the frequency, the smaller the depth and radius of the currents distribution, moreover, it is seeing, that currents propagates along the fiber and directed 45° and -45°, and overlapped on the axis.

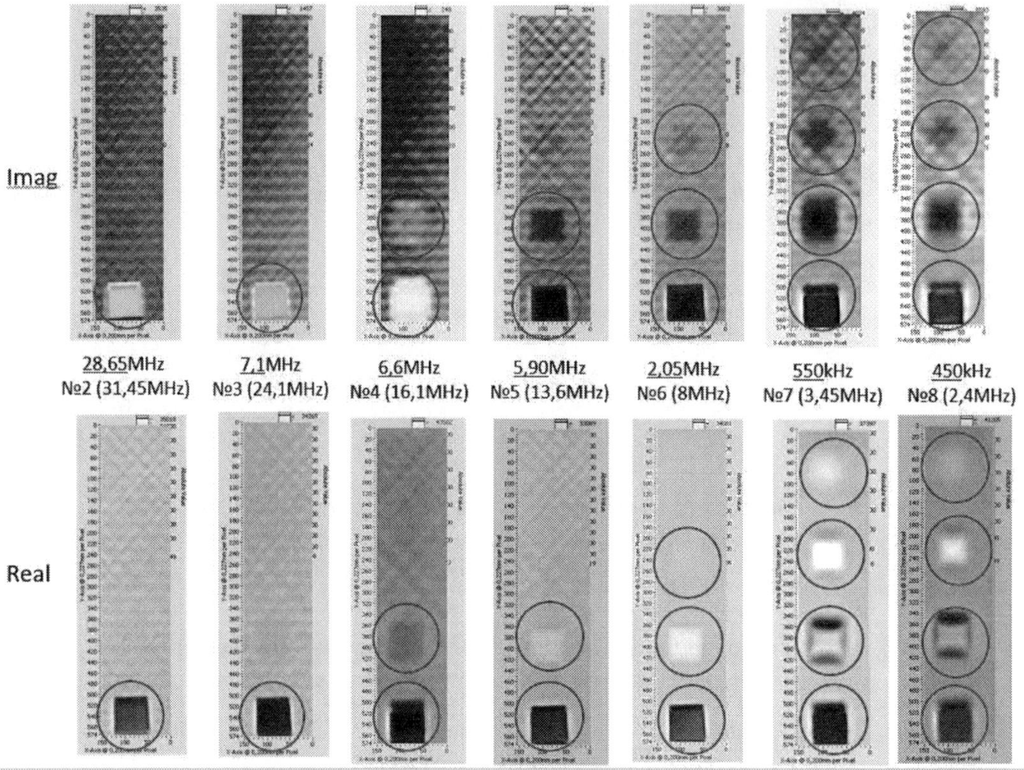

Fig.5. EC scans images of the specimen using 7 different driver coils №2 -8 listed in Table 1

For analysis of the currents' distribution in thickness, certain specimen volume under the pick-up coil was chosen. This procedure was done for each driver coil on operating frequencies used in experiment.

Table2. Current distribution in [45/-45]s CFRP for two coils with different resonant frequencies

Coil	Fres, MHz	Fact, MHz	Real Top	Imag Top	Real part, crosscut	Imag part, crosscut
№2	31.45	28.65	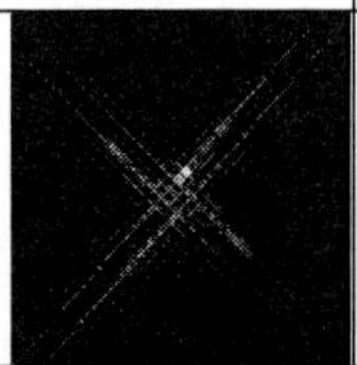	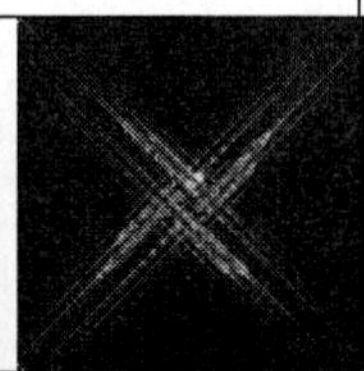	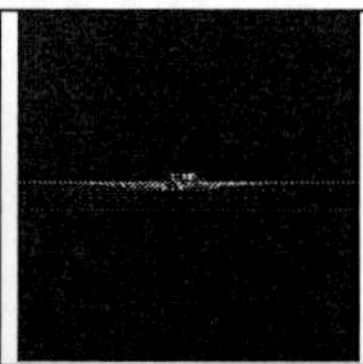	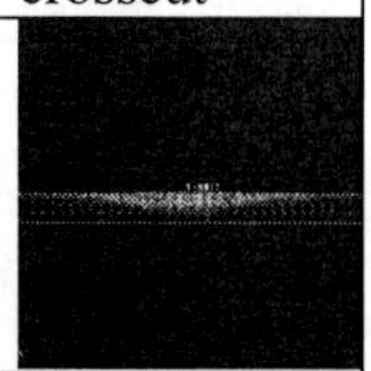
№6	8	2.05	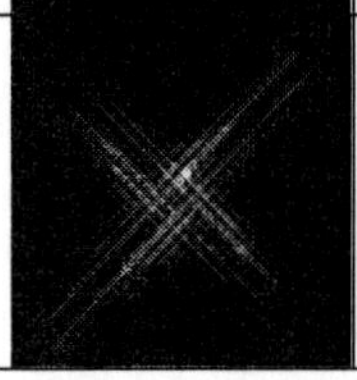	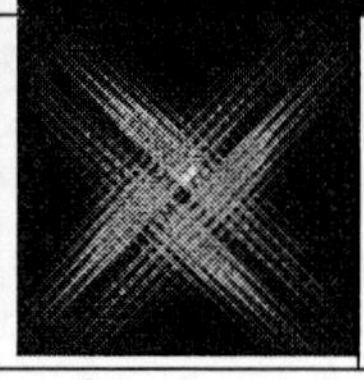	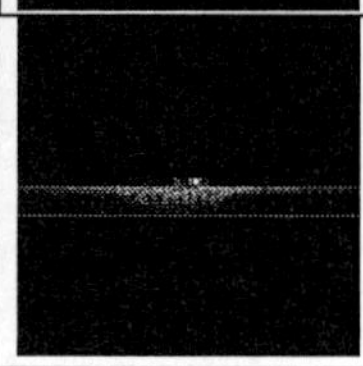	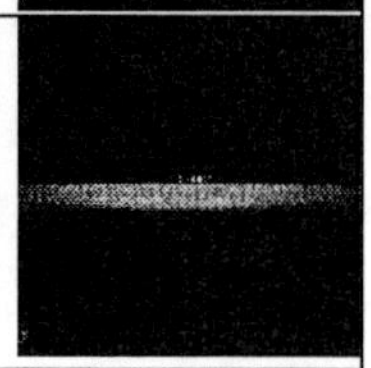

Dependencies of the normalized current density over the depth for each driver coil are presented on Figure 6. Graphs a, b, c and d show coils' curves for one, two, three and four detected layers respectively. The threshold of the highest current density, which correspond to the skin layer thickness, is shown with dash-dotted line. Simulated skin layer thickness value is compared with the experimental effective penetration depth and discussed in the paper.

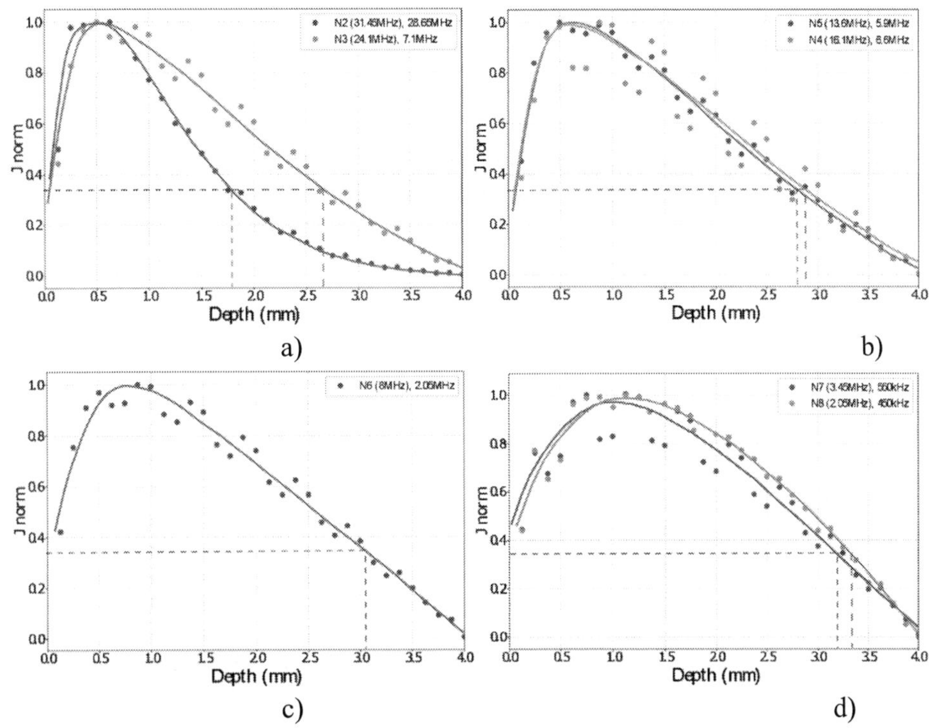

Fig. 6. Modeling results of the currents attenuation on the depth. a) detected 1 layer, coils №2 and №3; b) 1 and 2 layers, №4 and №5; c) 1- 3 layers, coil №6 ; d) all 4 layers, coils №7 and №8

Discussion

Results of the depth determination using different coils combinations are presented. The simulation and experimental results shown on Figure 5 and Figure 6 are analyzed and compared.

Estimating the penetration depth from experiment and skin layer value for driver coils from modeling, from Figure 6(a) it could be seen that for coils №2 with Fact = 28,65 MHz and №3, Fact = 7,1 MHz, modeled penetration depth is big enough to detect 2 Cu elements, but EC images for these two coils from Figure 5 shows that only the first Cu element is detected for Imaginary and Real parts. For coils №5 Fact = 5,9 MHz and №4 Fact = 6,6 MHz simulations showed a possibility to detect 2 Cu elements on depth 0 and 1.33mm, but the signal penetration depth is close enough to locate the Cu element on depth 2.66 mm. Experimental images confirm modeling results and presents only the first 2 Cu elements. Experimental results for coil №6 Fact = 2,05 MHz conform to the modeling results which declare the possibility to detect 3 Cu elements. For low frequency coils №7 Fact = 0,55 MHz and №8 Fact = 0,45 MHz the experimental EC scan showed the full penetration into the specimen and detection of 4 Cu elements, compared to modeling, where 3 Cu elements are distinguishable.

It is evident from the experiment that the contrast of the Cu elements is decreasing with depth, what correlates with the modelled curves' character. The highest image resolution was obtained with the smallest driver coils (with diameter 2.4, 3.3 mm), as well as insignificant influence of edge effects in resulting image compared to the coils with 7 and 9 mm diameter.

As it was presented, not all modeled values correspond to the experiment. That could be explained by several imprecisions. The CFRP conductivity set for modeling was the average for such type of CFRP, but not measured exactly in studied specimen. Also, mistake of the interpolation for scattered currents value which correspond to its inhomogeneous distribution in dependency of the fiber orientations, could play significant role. Cu elements embedded into the specimen could act as an effective electromagnetic conductor, therefore penetration depth differs with the model, where it was estimated in the poor CFRP. However, in authors' opinion, the energy conversion and transmission in CFRP in 2 coils' system have the most impact. Here half-transmission coils have the form of transformer where driver and pick-up coils differ by the parameters (e.g. amount of windings) and are connected partially through the CFRP, where each coils' pair has different distance in-between. All the coils parameters as well as distance in-between and properties of medium are influencing the energy transmission and have to be studied. Also, the pick-up coil plays a role of a pass band filter and could exclude the informative signal corresponded to the certain depth, therefore driver and pick-up coils system has to be spectrally adjusted.

Conclusions

In the paper was discussed the novel concept of HF EC application for depth of CFRP differentiation. It was suggested to use diverse EM coils with different resonant frequency for receiving various penetration depth following the equation of EMF in material attenuation. Validity of the method was checked experimentally and compared with the model. It was shown that the approach works and among experimental coils four could be chosen to divide a depth on 4 regions with depth resolution 1.33mm. Modeling results in the rate of 50% couldn't correlate with experiment, which could be explained by inaccuracies of the CFRP model. As results showed, this work would be a great basis for the next development of the method and its successful implementation in industry.

References

[1] Federal Aviation Administration. "Advanced Composite Materials". In: (2012), p. 6

[2] George Marsh."Greater role for composites in wind energy". In: Reinforced Plastics 58.1 (2014), pp.20–24.

[3] Hayman and Dan Zenkert, "The Influence of Defects and Damage on the Strength of FRP Sandwich Panels ", Naval Ships Brian, 2004. http://www.diva-portal.org/smash/get/diva2:466521/fulltext01.pdf

[4] Chenjun Wu, Yizhuo Gu, Liang Luo, "Influences of in-plane and out-of-plane fiber waviness on mechanical properties of carbon fiber composite laminate", Journal of reinforced plastics and composites, March 30, 2018

[5] JAN FREDERIK EVERS, MARKUS BREUER, RAPHAEL REINHOLD, "INDUSTRIALIZED MANUFACTURING OF CFRP REINFORCEMENT PROFILES"

[6] C. Meola, G.M. Carlomagno, "Impact damage in GFRP: new insights with infrared thermography", Compos. Part Appl. Sci. Manuf., 41 (2010), pp. 1839-1847

[7] Vitalij Popow, Martin Gurka" Determination of depth and size of defects in carbon-fiber-reinforced plastic with different methods of pulse thermography", spie digital library, 27 March 2018

[8] Tiseanu, E. Tsitrone, A. Kreter, et al. "X-ray micro-tomography studies on carbon based composite materials for porosity network characterization", Fusion Eng. Des., 86 (2011), pp. 1646-1651

[9] C. Goidescu, H. Welemane, *et al.* "Damage investigation in CFRP composites using full-field measurement techniques: combination of digital image stereo-correlation, infrared thermography and X-ray tomography", Compos. Part B Eng., 48 (2013), pp. 95-105

[10] G. Kim, S. Hong, et al., "NDE of low-velocity impact damages in composite laminates using ESPI, digital shearography and ultrasound C-scan techniques", Int J. Precis. Eng. Manuf., 13 (2012), pp. 869-876

[11] K.V. Steiner, "Defect classifications in composites using ultrasonic nondestructive evaluation techniques", Damage Detection in Composite Materials, 1128, ASTM STP (1992), pp. 72-84

[12] S.V. Lomov, D.S. Ivanov, et al ".Full-field strain measurements for validation of meso-FE analysis of textile composites", Compos. Part Appl. Sci. Manuf., 39 (2008), pp. 1218-1231

[13] Olympus, Introduction to Eddy Current Testing https://www.olympus-ims.com/en/eddycurrenttesting/

[14] N. Matvieieva, I Patsora, H Heuer, I Kharabet, G Bardl , "Process-Integrated Eddy Current Quality Assurance (Theoretical Simulations, Study, Investigation)", ASNT Annual Conference 2016, 104-112

[15] Mizukami, Koichi & Ogi, Keiji, "Non-contact visualization of fiber waviness distribution in carbon fiber composites using eddy current testing", Advanced Composite Materials, 2017.

[16] X Xu, H Ji, J Qiu, J Cheng, Y Wu, T Takagi , "Interlaminar contact resistivity and its influence on eddy currents in carbon fiber reinforced polymer laminates", NDT & E International 94, 79-91

[17] Specimen was produced and provided by ITM TU Dresden.

Eddy Current Array for Aircraft Engine Component Inspection

André LAMARRE [1]
[1] Olympus Scientific Solutions Americas, Quebec City, Canada
Contact e-mail: andre.lamarre@olympus-ossa.com

Abstract

Components of aircraft engines are submitted to very high stresses and temperatures during operation. Consequently, to assure the safety of the public, the integrity of these components must be ensured with reliable inspection methods. While the use of remote visual and dye penetrant inspection is widespread, the aircraft engine manufacturing and maintenance industries now benefit from recent developments in magnetic methods, namely eddy current array.

Eddy current testing (ECT) has long been considered the technique of choice to detect and size cracks on the surface of aircraft engine components. Unfortunately, inspections using single-coil eddy current probes can be slow with results that vary from operator to operator. In these and other respects, eddy current array (ECA) technology has advanced the technique significantly. The use of arrays helps to rapidly achieve full coverage of inspected zones while maintaining a high resolution. Portable eddy current electronics enable the use of large arrays and the image resulting from the software's data processing helps the operator perform a reliable analysis. Probe holders and mechanical supports are continuously being developed to address the different applications of ECA for engines.

Today, eddy current arrays are used by major aircraft engine manufacturers and maintenance companies for the assessment of engine components, such as dovetails, blade attachments, and turbine discs. This presentation will review how eddy current array technology contributes to public safety.

OLYMPUS

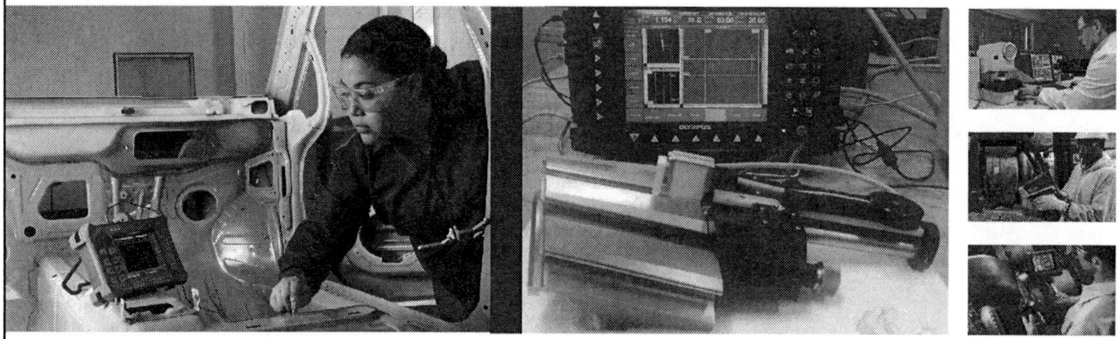

Eddy Current Array for Aircraft Engine Component Inspection

Olympus Scientific Solutions Americas | Andre LAMARRE

NDT in Aerospace 10th International Symposium on NDT in Aerospace DGZfP

OLYMPUS

Agenda

1. Eddy Current Array (ECA) Technology
2. ECA Equipment
3. Engine Applications
4. Conclusion

Eddy Current Array Technology

OLYMPUS

Eddy Current Array Principles

- ECA is a technology based on the ability to ***electronically drive several eddy current sensors*** placed side-by-side in the same probe assembly
- Data acquisition is performed by ***multiplexing the eddy current sensors***
- Most eddy current probes and techniques (absolute, reflection, etc.) for flaw detection can be used with eddy current array probes

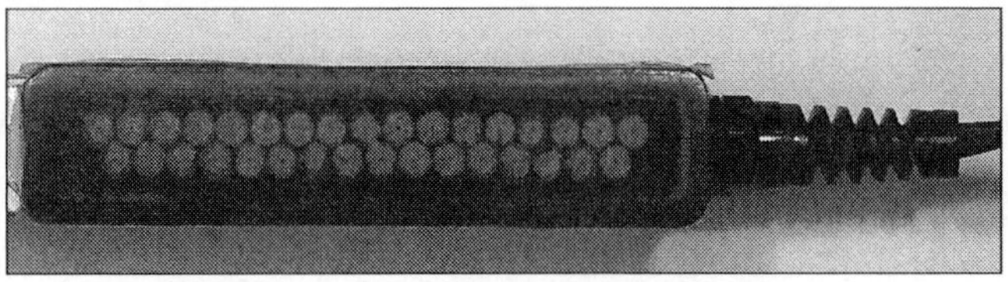

OLYMPUS

Eddy Current Array—Advantages

- Provides larger coverage in a single probe pass while maintaining high resolution
- Improves flaw detection and sizing with C-scan imaging
- Enables inspection of complex shapes with probes customized to the profile of parts

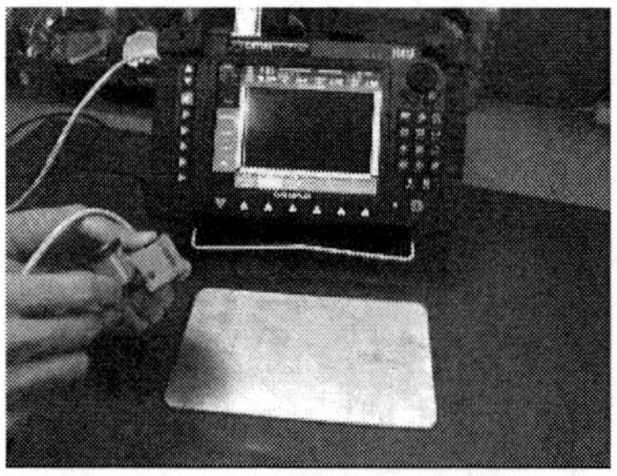

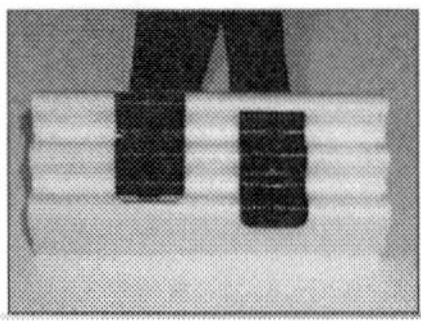

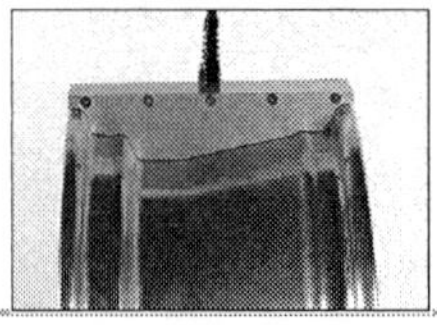

OLYMPUS

ECA – Signal Representation

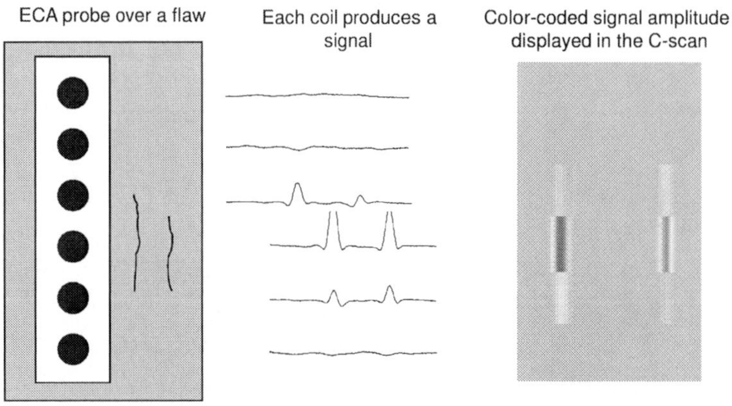

OLYMPUS

ECA – Signal Representation

- The C-scan can display either the vertical or horizontal component of the signal in the impedance plane.
- In the example below, only the vertical component of the signal (Y axis) is displayed on the C-scan, ensuring that all defects are clearly shown. The phase of the signal was adjusted so the liftoff signal is in the horizontal plane (X axis).

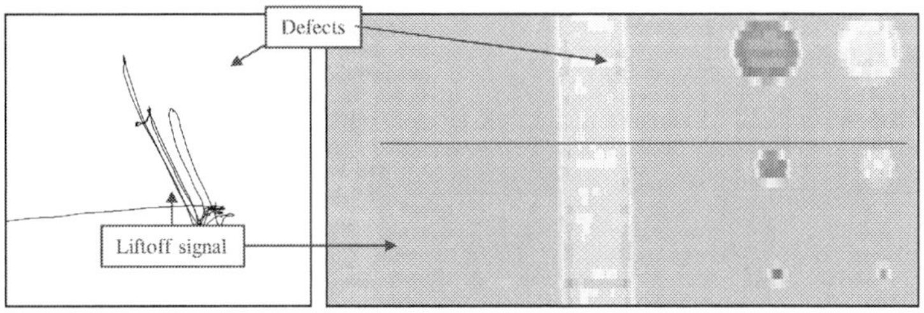

Eddy Current Array Equipment

OLYMPUS

General Hardware Features

OMNI-P-ECT4

- Conventional ECT only
- 4 input channels
- Frequency ranges from 20 Hz to 6 MHz
- Dual frequency operation
- 2 encoder inputs
- 3 alarm outputs
- 1 analog output

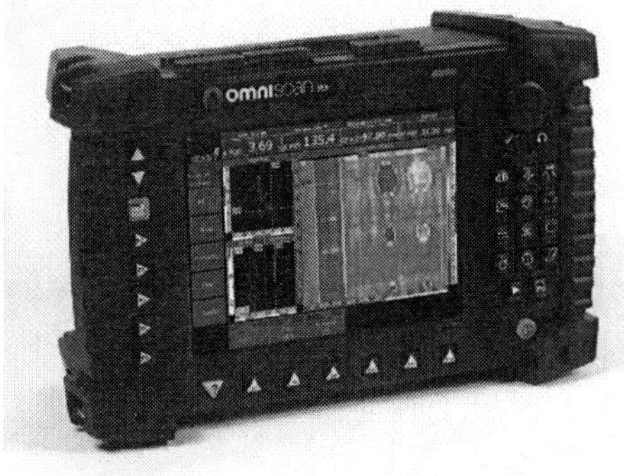

9

OLYMPUS

ECA Probes

- Can be optimized for different applications
- Can be shaped to the part to inspect

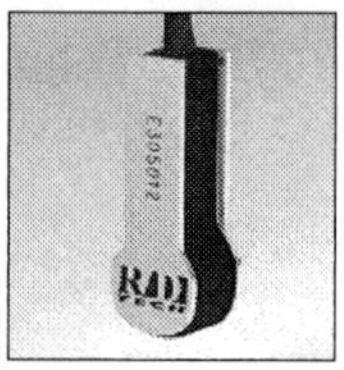

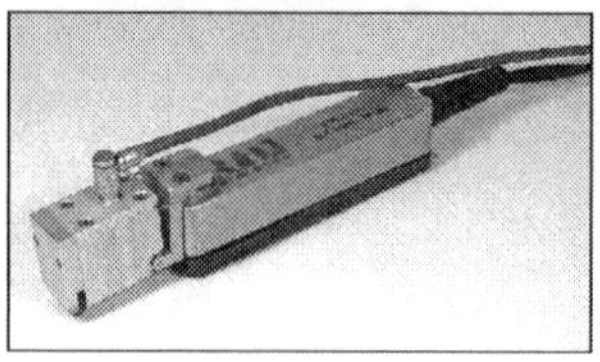

10

OLYMPUS

ECA — Probe Parameters

- Frequency (*f*)
- Number of elements (*n*)
- Resolution (*r*) (also depends on the coil configuration)
- Coverage (*C*)

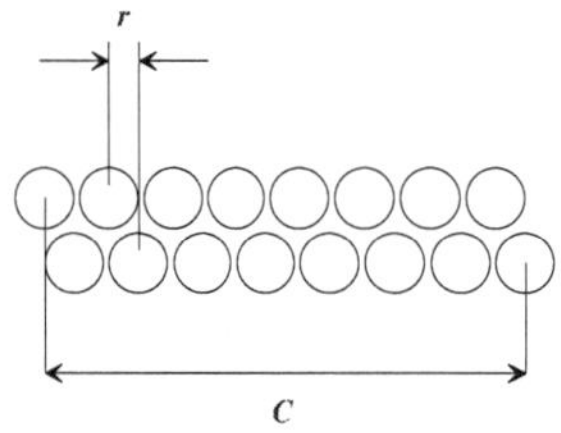

Example of an absolute bridge probe

11

OLYMPUS

Inside an ECA Probe

- Hard coils
- Many wires
- A lot of soldering SW1
- Labor intensive

12

Diapositive 12

SW1 I would call this soldering and not welding.
Sarah Williams; 2018-05-28

OLYMPUS

ECA Hard Coiled Array VS Printed Circuit Board (PCB) Array

Traditional hard coil sensors

- Pros:
  - Different type and size of coils
- Cons:
  - Time consuming
  - Repeatability and coil matching
  - Expensive

Flexible array on PCB film

- Pros:
  - Excellent repeatability and coil matching
  - Versatility, i.e., can be attached to holders of different shapes
  - Affordable price
- Cons:
  - Price/time for the initial film

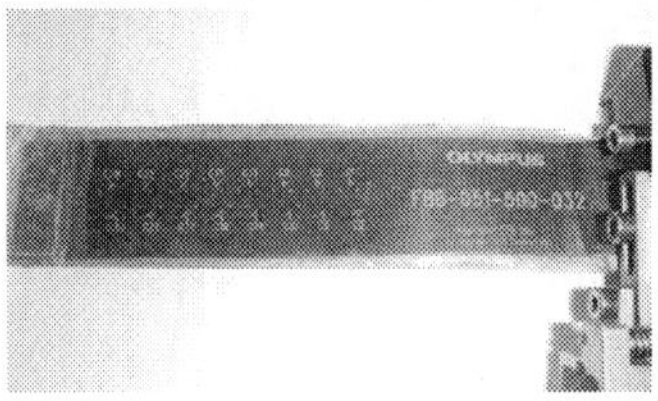

Engine Applications

OLYMPUS

Dovetail Slot Inspection — Current Method

- Manual tooling
- Scanner with eddy current pencil probe
- The operator must perform 40 line scans
- Time consuming
- Operators complain of discomfort

15

OLYMPUS

Engine Disk Dovetail Slot Semiautomated Inspection

- 32 coils cover critical zones

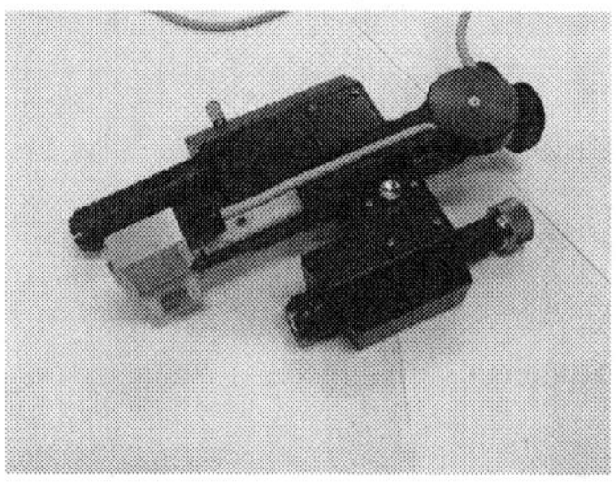

- Probe mounted on portable scanner

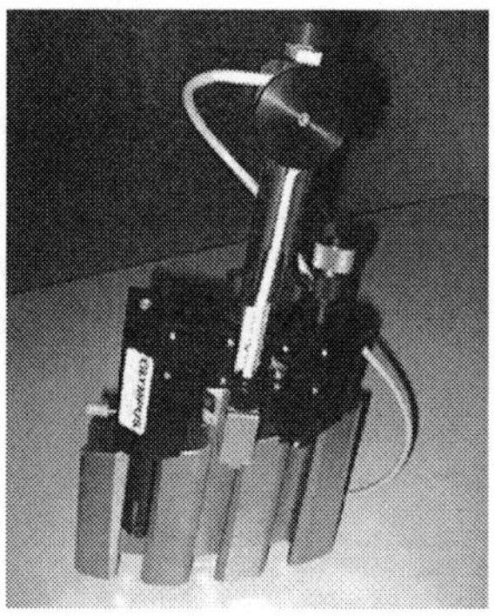

- Scanner in position

16

OLYMPUS

ECA Engine Disk Dovetail Results

- C-scan displaying EDM and calibration notches in the calibration standard

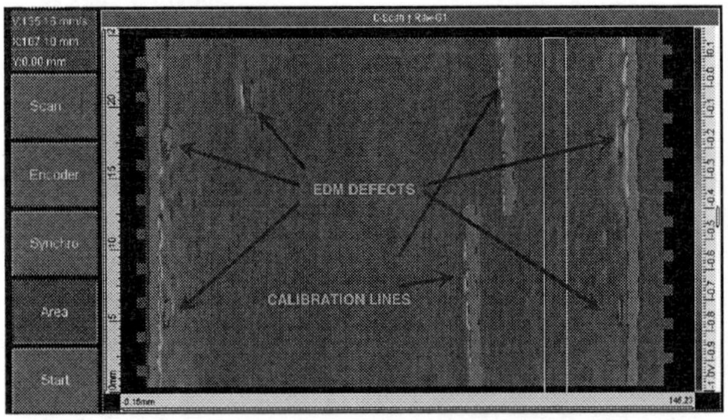

17

OLYMPUS

ECA Engine Disk Dovetail Results

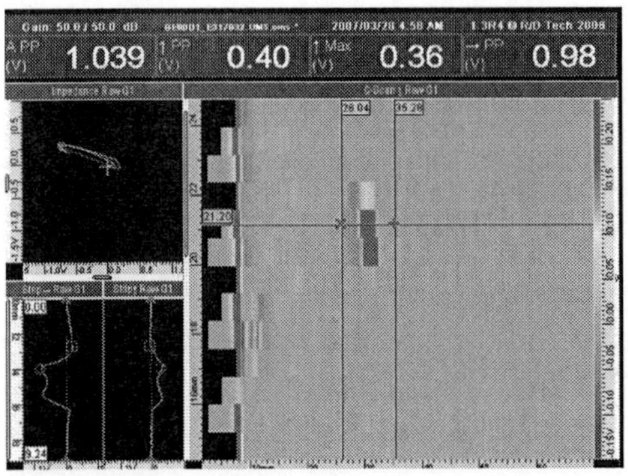

C-scan of an EDM notch: 1.5 × 0.7 × 0.1 mm

18

OLYMPUS

New Engine Disk Slot Inspection System

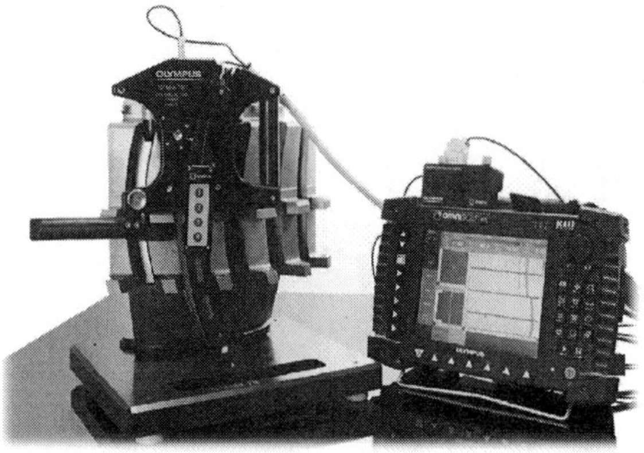

Scanner improved with OmniScan® flaw detector remote control

OLYMPUS

Engine Fan Blade Inspection

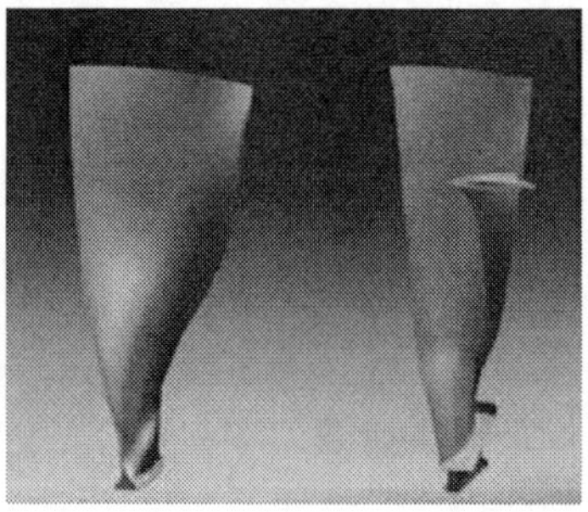

OLYMPUS

Engine Fan Blade Inspection Solution — Development Process

Stage 1: Manual inspection

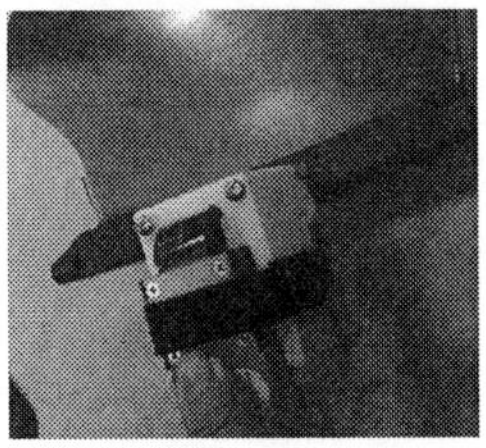

- 2 passes:
  - 1 convex side
  - 1 concave side
- Wire encoder

Stage 2: Semiautomated inspection

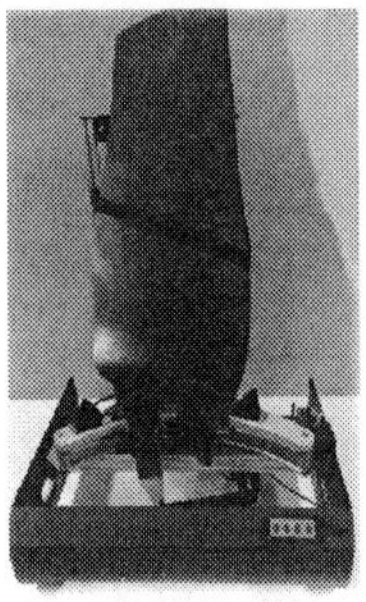

- Single pass, convex and concave sides
- Encoded scanner

OLYMPUS

Semiautomated Fan Blade Inspection Solution

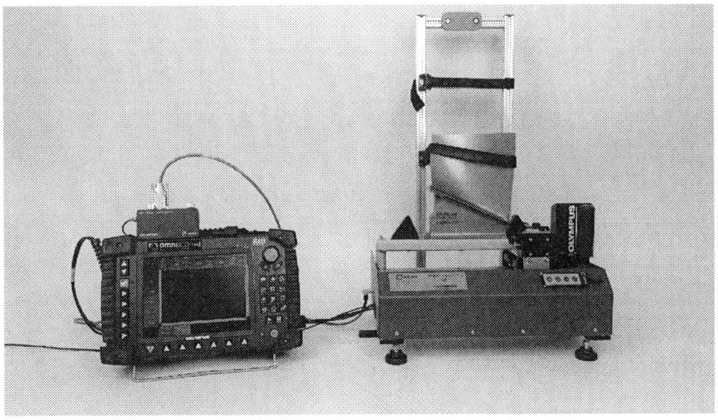

Conclusion

OLYMPUS

ECA Advantages for Aircraft Engine Inspection

- ECA probes cover large areas in a single pass
- Improved Probability of Detection due to full coverage and C-scan imaging
- Probes can be adapted to the complex shapes of components
- Flexible ECA can be used interchangeably with probe holders of different shapes
- Can be used on-site for aircraft engine component inspection during maintenance
- Referenced in procedures for aircraft and engine maintenance

OLYMPUS®

Development of a Contactless Inductive Strain Gauge for Monitoring Mechanical Load in CFRP

Ievgen KHARABET [1], Susanne HILLMANN [2], Martin H. SCHULZE [3], Nataliia MATVIEIEVA [1], Henning HEUER [3]
[1] Technische Universität Dresden, Dresden, Germany
[2] Fraunhofer-Institut für Keramische Technologien und Systeme, Institutsteil Materialdiagnostik (IKTS-MD), Dresden, Germany
[3] Fraunhofer-Institut für Keramische Technologien und Systeme IKTS, Dresden, Germany
Contact e-mail: ievgen.kharabet@tu-dresden.de

Abstract. In this paper, results of the research on the feasibility of High Frequency Eddy Current (HFEC) based inductive strain sensor are presented. Conducted studies suggest that there is a possibility to utilize HFEC as an alternative for conventional strain gauges for CFRP parts monitoring. This is based on the fact, that carbon fibers are peizoresisitive, i.e. change their conductivity due to applied mechanical strain in the same way a conductive strain gauge changes its own. The main idea is to measure the change of the conductivity of fiber bundles inside CFRP to determine the strain directly in the fiber. Advantages of the inductive strain gauge compared to classic conductive sensors are direct measurements of fibers properties' change, not influenced by glue cracking and aging, and, as a result, conceptual ability to endure exploitation without loss of the measuring capabilities, because these capabilities do not depend on the state of fixation of the sensor to the investigated part.

Results of the research indicate there is a reliable correlation between measured EC signal and an applied mechanical strain. The possibility to integrate prototype sensor design into the CFRP mesh was also investigated, and the results of the effort are represented in the paper.

Introduction

Carbon-fiber reinforced plastics (CFRPs) are prospective materials for automotive and aerospace construction, and civil engineering. This type of composites allows for creation of highly durable and at the same time light designs with increased bearing capabilities due to the decrease of the systems' intrinsic weight [1].

CFRP are also relatively cheap and uncomplicated to produce, which further increases their importance. However, as any other material, CFRP can suffer from different defects, which occur both during manufacturing process, such as missed lane, "fuzzy ball", in- and out-of-plane waviness, and due to exploitation of the mechanical part. Thus, ability to ensure high quality of production and to monitor health of parts and systems is no less desirable for CFRP production than for any other.

During previous string of studies, it was shown high frequency eddy current could be applied for monitoring of low-conductive materials, e.g. metal-epoxy pastes, graphite-based pastes for Li-ion batteries production, ceramics, and even pure epoxy. [2–9]

It is known, that CFRP possesses piezoresistivity [10–14], which means any mechanical change in the structure would translate into change of conductivity of the part, which in turn would influence eddy currents, excited in the material under the influence of the sending coil's magnetic field. This way, mechanical strain and stress, applied to CFRP part would transcend into changes or fluctuations of the measured EC signal, be it impedance or phase angle changes.

Naturally, in this regard HF EC method would become an alternative for conventional strain gauges, which measure strain indirectly through the change of the resistance of the sensor itself. It is not always possible to use strain gauges to measure decrease of the strain applied to the part, when the material relaxes. On the contrary, HF EC method allows to measure the change of material properties directly and allows to measure both increase and decrease of strain, which could be applied

Authors of this paper weren't able to find any previous attempts to use HF EC for mechanical strain monitoring in CFRP.

1. Theoretical background

When alternating magnetic field is applied to conductive material, it induces Foucault currents or eddy currents, which in turn excite secondary magnetic field, attempting to counteract the primary one. Thereupon classical Eddy Current non-destructive method is based. However, as it was shown a number of times during previous research, if primary magnetic field is alternating at higher frequencies, it is possible to excite eddy currents in low-conductive and even non-conductive materials [2, 3, 6, 7]. This is the foundation for High Frequency Eddy Current (HF EC) method.

CFRPs generally are low-conductive composite materials, where carbon fibers have anisotropic conductivity and build chaotic semi-continuous conductive mesh inside of a nonconductive polymer matrix. Thus, if alternating magnetic field is applied to CFRP material, eddy currents are induced, and their paths are defined by the location of fibers inside the material. If CFRP is undergoing a mechanical strain, its conductivity would change, the reason for that are changes, introduced in the conductive structure of CFRP, which influence both conductivity of carbon fibers, and paths and distribution of eddy currents in the material. Carbon fibers undergo changes on macrolevel, when the bundles elongate and tighten, and the surfaces of neighboring bundles press together. Also, distance between fibers inside bundles changes (fig. 1). On microlevel, carbon fibers can change their diameter to an extent, due to the specifics of carbon hexagonal structures, which could also contribute to the piezoresistive capabilities. All the changes, that occur in CFRP under mechanical strain influence eddy currents excited in the structure and thus could be detected by HF EC method, which is the basis for the idea of inductive strain gauges.

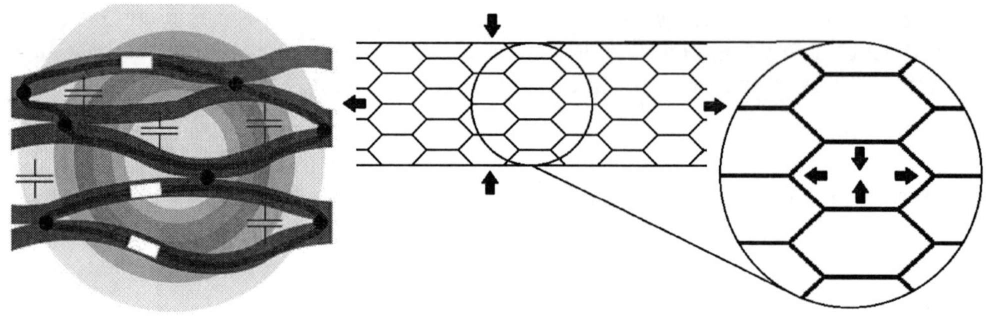

Fig. 1. Equivalent representation of carbon fibers (a), schematic representation of carbon structure on microlevel (b).

2. Experimental setup

Samples, produced for the research along the guidelines of the Type 3 specification for strain samples from DIN EN ISO 527-4, were formatted as stripes of woven and nonwoven types of CFRP, of different width and thickness, both imbued with epoxy, and dry.

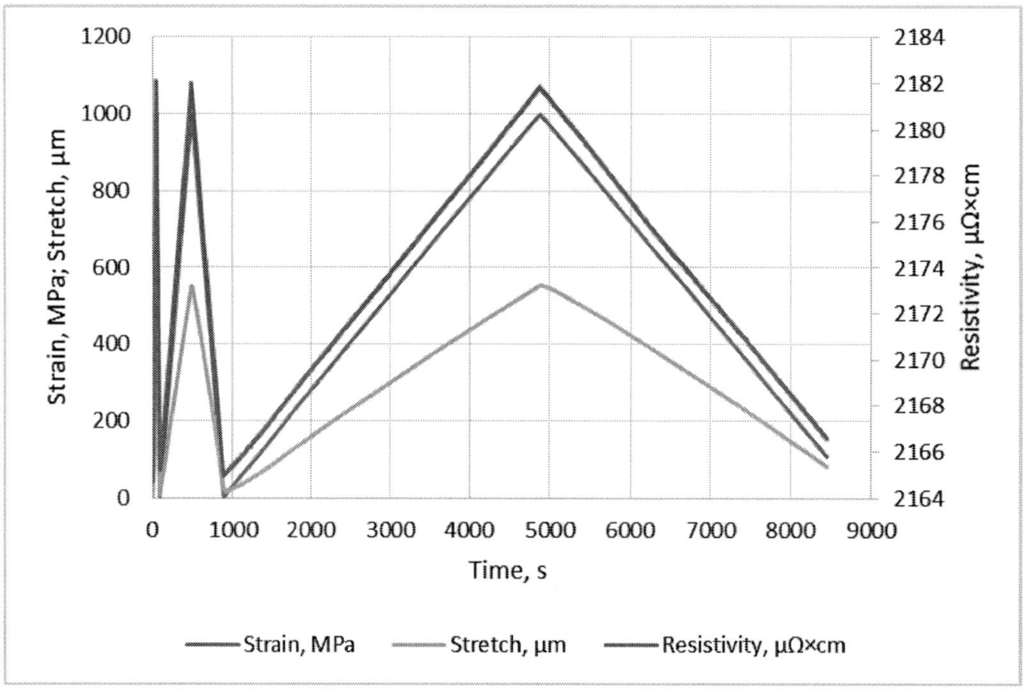

Fig. 2. Control measurements of CFRP resistance.

For the purposes of the research several experiments were devised and conducted. First of all, the piezoresistive nature of CFRP was confirmed by control measurements of CFRP plate, stretched and relaxed in the UTM (Zwick Z250, Fraunhofer IKTS). Electrical contacts to carbon fibers were established by burning away epoxy matrix at the ends of the plate and gluing copper plates onto them with silver based conductive epoxy compound. Resistance of the plate was measured with four-point method, and shows correlation to the applied mechanical strain and deformation (fig. 2).

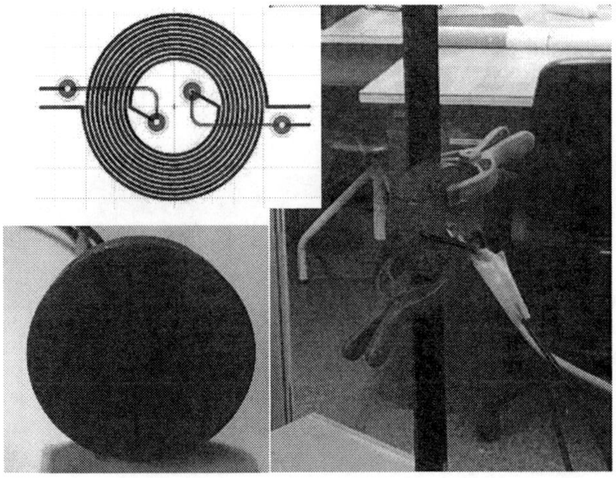

Fig. 3. Proof-of-concept design (measuring coils (upper left); sensor head (lower left); sensor, pressed to the surface of the CFRP sample (right))

During experiments, design of the sensor went through several stages. First, only the method applicability was tested, and for this purpose bulky sensor head with spiral coils was used. During experiments, this sensor was held against the surface of CFRP with spring clamps (fig. 3).

After satisfactory results were obtained, a concept of a sensor patch was developed. It was used for experiments in two configurations: glued onto the surface of CFRP with double-sided tape (fig. 4) and integrated between layers of carbon fiber textile (fig. 5).

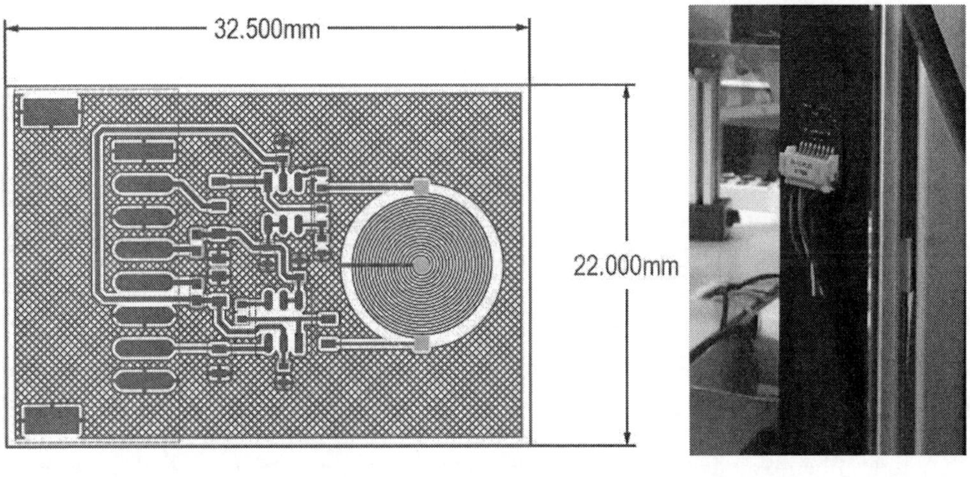

Fig. 4. Design of a EC sensor patch (left), sensor, attached to the CFRP sample during experiment (right).

The prototype design of EC-patch consisted of a pair of interwound spiral coils and a SMD-based electronic (preamplifier) on a thin flexible PCB. Preamplifier should be placed as near as possible to the coils, otherwise a distortion or dampening of measured signal could occur. However, it is possible to devise a compensation algorithm, as long as the dampening is not too extreme.

Fig. 5. X-ray of the embedded sensor (left), CFRP samples with integrated sensor inside (middle and right).

The construction form of spiral coils as a sensor body on the flexible substrate is rather well suited for integration purposes, however, this design still is only a prototype, and SMD-elements could potentially be replaced with film structures. One of the downsides of the integration is the introduction of additional potential delamination seed into material, as well as the necessity to encapsulate the sensor in an insulation pocket. The insulation pocket was created from two layers of fiberglass textile. The problem of contacting the sensor for energy and signal transfer is yet another issue, which was solved in research circumstances by opening the upper part of CFRP sample and soldering a microcoaxial cables to sensor's contacts. However, it is possible to develop a design of an autonomous sensor with either rechargeable or stationary energy source, and a wireless signal transmitter.

3. Results and discussion

After resistance measurements (fig. 2) showed there is a correlation between CFRP's resistance and mechanical strain, series of experiments with HF EC sensors were undertaken to verify the applicability of the method for CFRP monitoring. It was established, that there is a pronounced correlation between eddy current signal and a mechanical load applied to samples (fig. 6–7). It was found, that linear increase of mechanical strain leads to ever diminishing increase of EC-signal (fig.6).

The reason for that is twofold: first, there is a pizoresistive character of carbon fibers themselves, and, second, there are capacitive effects, which occur when carbon fibers move in relation to each other, become thinner, and press closer together. On microscopic level each single fiber becomes thinner and elongated. On macroscopic that translates into elongation and tightening of the whole thread and into the diminishing of distances between single threads and bundles. Capacitance changes that occur on macroscopic level influence the behavior of the magnetic fields in the material and thus the EC signal. Physical tightening of the fibers and bundles themselves also alters conductive pathways in the textile, which also impacts the flow of eddy currents, and, as a result EC signal. These considerations could be a foundation for additional research, aimed at finding of the exact interaction mechanism between the strain and the eddy currents.

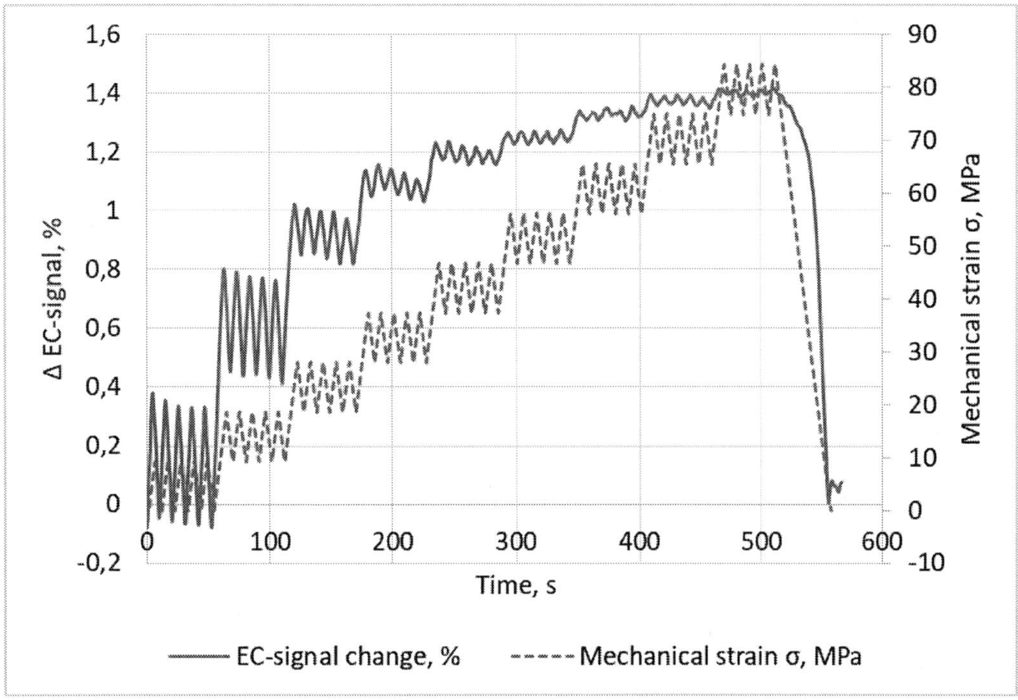

Fig. 6. Correlation between mechanical strain and measured HFEC signal (surface sensor).

It was found, that during continuous measurements, a general decrease of EC-signal over cycles appears. This occurrence could be a result of the residual plastic deformation (elongation) of carbon fibers and/or the accumulation of wear in the CFRP.

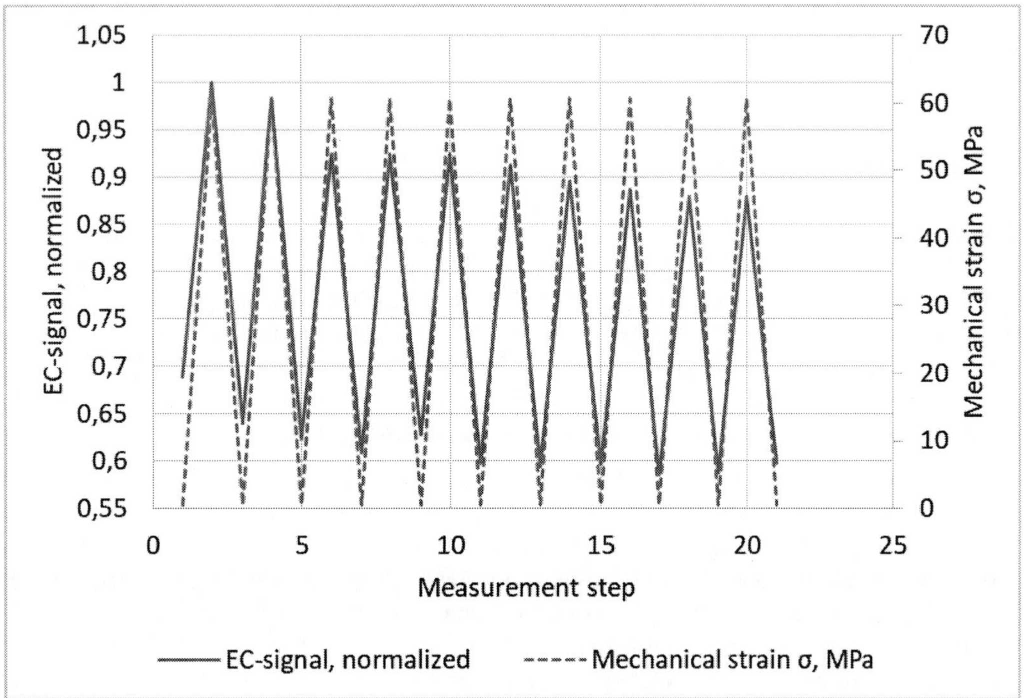

Fig. 7. Correlation between mechanical strain, applied to the CFRP and measured HFEC signal (embedded sensor)

4. Comparison with conventional strain gauges

For comparison of HF EC method to conventional strain gauges, sensitivity of the method must be evaluated. Conventional strain gauges are usually characterized by their k-factor (usually it is 1,9...3,0, and for most sensitive Pt gauges it is as high as 6,0). This factor is obtained as a ratio between sensor's resistance and length changes:

$$\frac{\Delta R}{R} = k \cdot \frac{\Delta l}{l},$$

where $\frac{\Delta R}{R}$ is a change of sensor's resistance, $\frac{\Delta l}{l}$ is a change of sensor's length.

To evaluate sensitivity of HF EC method, similar approach was adapted. The k_{EC}-factor was calculated as a ratio between the change of EC-signal and an elongation of CFRP-sample:

$$\frac{\Delta EC}{EC} = k_{EC} \cdot \frac{\Delta l}{l},$$

where $\frac{\Delta EC}{EC}$ is a change of EC-signal, $\frac{\Delta l}{l}$ is a change of specimen's length.

As an alternative index of sensitivity, a relative EC-signal's change per MPa of mechanical strain:

$$\frac{\Delta EC}{EC} = G_{EC} \cdot \Delta\sigma,$$

where $\frac{\Delta EC}{EC}$ is a change of EC-signal, $\Delta\sigma$ is a strain increase.

For surface sensor the k_{EC}-factor turned out to be 2,2...3,86, which is comparable to that of most common strain gauges. For embedded sensor, k_{EC}-factor was as high as 16. For embedded sensors the growth index G_{EC} was as high as 1,5 $\frac{\%}{MPa}$.

It is also worth noting, that HF EC method allows to measure relaxation of previously elastically strained carbon fibers, which is not always possible with conventional strain gauges.

5. Conclusions

Obtained results show, that there is a strong correlation between mechanical strain, applied to the sample, and an EC-signal measured simultaneously, which corresponds to the correlation between strain and resistivity. Possible mechanisms of such occurrence are microscopic changes in carbon fibers and macroscopic alterations of fibers' and bundles' relative positions, distances between them, tightening of bundles and treads. Ideas for future studies of exact mechanisms, which lead to change in CFRP resistivity, as well as a research and development of an autonomous sensor design were proposed.

It was shown from obtained results, that HF EC method is applicable for mechanical strain and wear monitoring of CFRP parts and is comparable to conventional strain gauges in its sensitivity and even surpasses them completely with use of embedded sensor, and is capable to measure relaxation of elastically strained carbon fibers, and its working principle is based on direct measurement of material's properties' changes. It is also worth noting, that HF EC method allows to measure relaxation of previously elastically strained carbon fibers, which is not always possible with conventional strain gauges.

The advantages of the proposed use of HFEC method include the ability to directly measure changes of material's properties, caused by mechanical load and wear, the ability to measure both increase and decrease of mechanical load applied to the part, high sensitivity of the method towards measured effects.

References

[1] R. Bilgram, W.Zimmermann; Processing and Uses of Carbon Fibre Reinforced Plastics; International Publisher Service, Incorporated; S. 67-87; 1981

[2] S. Hillmann, H. Heuer, J.G. Calzada, A.T. Cooney, B.C. Foos, N. Meyendorf, "Validation of potentials and limitations of a high-frequency eddy current system in frequency sweep mode", AIP Conference Proceedings 1430 (1), pp. 433-440, 2012

[3] Xin Li; Eddy Current Techniques for Non-destructive Testing of Carbon Fibre Reinforced Plastic (CFRP); A thesis submitted to The University of Manchester for the degree of Doctor of Philosophy in the Faculty of Engineering and Physical Science; School of Electrical and Electronic Engineering; 2012

[4] I Patsora, S Hillmann, H Heuer, BC Foos, JG Calzada, "High-Frequency Eddy Current System for Analyzing Wet Conductive Coatings during Processing", ASNT Annual Conference 2014, pp. 88-95, 2014

[5] Patsora, I.; Kharabet I.; Heuer, H.; Schulze, M.; Fritsch, M.; Tatarchuk, D.; Didenko, J. High Frequancy Eddy Current Spectroscopy for In-Line Control of Li-Ion Batteries' Quality 20th International Workshop on Electromagnetic NonDestructive Evaluation, ENDE2015

[6] S. Gäbler, H. Heuer, G. Heinrich, R. Kupke, "Quantitatively analyzing dielectrical properties of resins and mapping permittivity variations in CFRP with high-frequency eddy current device technology", AIP Conference Proceedings 1650 (1), pp. 336-344, 2015

[7] S Gäbler, H Heuer, G Heinrich, "Measuring and Imaging Permittivity of Insulators Using High-Frequency Eddy-Current Devices", IEEE Transactions on Instrumentation and Measurement 64 (8), 2227-2238, 2015

[8] Matvieieva, N.; Patsora, I.; Kharabet, I.; Heuer, H.; Bardl, G. Process-integrated eddy current quality assurance (theoretical simulations, study, investigation) ASNT Annual Conference 2016

[9] Iryna Patsora, Henning Heuer, Susanne Hillmann, Dmitry Tatarchuk, "Study of the particle based films' cure process by high frequency eddy current spectroscopy", Coatings 2017, vol. 7(1), 3, 2017

[10] R. Lange, G. Mook; Structural analysis of CFRP using eddy current methods, NDT&E International; Vol. 27; Nr. 5:241; 1994

[11] Sihai Wen, D.D.L. Chung; Electrical-resistance-based damage self-sensing in carbon fiber reinforced cement; Carbon #45 (2007), p.p. 710 – 716

[12] Sihai Wen, D.D.L. Chung; Uniaxial tension in carbon fiber reinforced cement, sensed by electrical resistivity measurement in longitudinal and transverse directions; Cement and Concrete Research #30 (2000), pp. 1289 – 1294

[13] Xiaojun Wang and D.D.L. Chung; Piezoresistive behavior of carbon fiber in epoxy Carbon Vol. 35, No. 10-I 1, pp. 1649-1679, 1997

[14] Sihai Wen, Shoukai Wang, D. D. L. Chung; Piezoresistivity in continuous carbon fiber polymer-matrix and cement-matrix composites; Journal of Materials Science #35 (2000); pp. 3669 – 3675

Funding information

The IGF research project 18901BR/1 of the research association "Forschungskuratorium Textil (FKT) e. V." is funded through the AiF within the program for supporting the industrial collective research („Industrielle Gemeinschaftsforschung", IGF) from funds of the Federal Ministry of Economics and Energy (BMWi) by a resolution of the German Bundestag. The providing of funds is gratefully acknowledged.

NASA Applications of Robotics for NDT of Aerospace Vehicles

William PROSSER [1], George STUDOR [1], Michael ROLLINS [2], Eric BURKE [1], Ken HODGES [1], Bence BARTHA[3], Elliott CRAMER [1]
[1] NASA Langley Research Center, Hampton, VA USA
[2] NASA Johnson Space Center, Houston, TX USA
[3] NASA Kennedy Space Center, FL USA

Contact e-mail: william.h.prosser@nasa.gov

Abstract

NASA has employed robotic methods to deploy NDT sensors and techniques for a number of critical inspection applications on aerospace vehicles and will continue to enhance robotic capability to support future missions. These robotic systems include both ground-based applications for inspection of aerospace components and vehicle assemblies as well as in flight inspections of spacecraft. Examples to be discussed include robotic x-ray systems that were used for the ground based inspection of a variety of Space Shuttle components such as the aft-skirt and carry structure assemblies, as well as the Orion spacecraft heatshield that flew on the Exploration Flight Test – 1 (EFT-1). Robotic NDT systems are also being employed for ultrasonic inspections of welds for the gigantic Space Launch System (SLS) propellant and oxidizer tanks. Additionally, NASA research efforts are underway to develop and demonstrate additional robotic NDT systems. A cutting edge coordinated multi-robotic inspection system to deploy infrared thermographic NDT systems on a full scale aircraft fuselage has been demonstrated, which can be expanded to accommodate other NDT sensor modalities. NASA is integrating in-situ NDE inspection techniques into large scale fiber placement robots to provide real-time NDE feedback for large scale composite structure construction. Further, more advanced deployment systems such as robotic "snake" technologies are being explored for NASA applications where access is an issue.

The need for in-space NDT was highlighted by the loss of the Space Shuttle Columbia due to impact damage to the thermal protection system (TPS). In response, the Shuttle Remote Manipulator System (SRMS), provided by the Canadian Space Agency, was augmented with a 50 ft. boom and the Orbiter Boom Sensor System (OBSS). The OBSS inspection sensor package provided optical and LIDAR measurement capabilities that were used to inspect the Shuttle TPS for impact damage. The results of these inspections were critical input for the decision to proceed with re-entry of the Shuttle and the safe return of its crew. In-space inspections are also important for the International Space Station (ISS). The ISS and visiting crewed vehicles are exposed to hypervelocity impacts from micrometeoroids and orbital debris (MMOD), which can cause damage to aging structures on the ISS as well as the TPS of vehicles docked there. NASA currently conducts surveys of the ISS exterior with visual cameras on the Space Station Remote Manipulator System (SSRMS) augmented by Dextre (Special Purpose Dextrous Manipulator or SPDM). NASA's Image Science and Analysis Team have improved on their Shuttle-era techniques to inspect the current and future visiting vehicles to ISS. New robotic inspection tools are also being developed to enhance the inspection capability on ISS. The Visual Inspection Poseable Invertebrate Robot (VIPIR) experiment for NASA's Satellite Repair Mission has a flexible-deployable videoscope with an articulating tip and the Canadian Dextre Deployable Vision System (DDVS) will provide high resolution imager, LIDAR and infrared imager for much improved effectiveness. It is hoped that the humanoid-like R2 Robot, flown aboard ISS, can be developed to accomplish inspections inside and eventually be adapted to be used outside the ISS to do at least some inspections that an astronaut might do with less human and mission risk. NASA has been testing robotic free-flying platforms (SPHERES and Astro-bee) to perform various tasks, including inspection, with the goal of little or no astronaut involvement. Also, new robotic gripping tools (Gecko gripper) have been tested on ISS that enable temporary adhesion for inspection tools, robotic crawlers and other applications in zero-g, including satellite capture!

Such robotic NDT systems will become increasingly important and enabling as humans travel deeper into the solar system, where return to Earth for vehicle inspection and repair is no longer a viable scenario. Robotic in-space inspection will require more autonomy to reduce or eliminate the need for ground communication or astronaut involvement. These robotic tools will need to perform over longer lifetimes in more severe environments. They will also need to support a wider variety of NDT sensor modalities to include penetrating imaging capabilities to examine structures under thermal blankets or MMOD shielding.

Development of a method for the non-destructive evaluation of fiber orientation in multilayer 3D carbon fiber preforms and CFRP with robot-guided high-frequency eddy current testing technology

Johannes MERSCH [1], Georg BARDL [1], Andreas NOCKE [1], Chokri CHERIF [1], Martin H. SCHULZE [2], Henning HEUER [2]

[1] Institute of Textile Machinery and High Performance Material Technology (ITM), TU Dresden, Dresden, Germany

[2] Fraunhofer Institute for Ceramic Technology and Systems, Material Diagnostic (IKTS-MD), Dresden, Germany

Contact e-mail: johannes.mersch@tu-dresden.de

Contact address: Technische Universität Dresden, Fakultät Maschinenwesen, Institut für Textilmaschinen und Textile Hochleistungswerkstofftechnik (ITM), 01062 Dresden

Abstract. This paper presents results of recent cooperative works at the ITM and IKTS on the 3D fiber orientation measurement with eddy current technique, which are the result of a recent dissertation. For the development of CFRP with load-adapted fiber orientations, a process is required that is able to nondestructively measure the fiber orientations in multi-layer, 3D preforms and CFRP structures.

Eddy-current testing (ECT) with frequencies in the megahertz range is able to reveal the filament structure in multilayer CF structures. To further expand its applicability a method is developed that can automatically generate the robotic paths for robot-guided ECT from an arbitrary 3D surface. From the obtained EC images, the local fiber orientation is determined on a grid of evaluation points. For this purpose the surface in near each point is projected onto a plane and converted into a pixel image. Fourier transformation is used to calculate the main fiber directions which are assigned to the individual layers by means of a robust algorithm and strategies are developed to exclude false information caused by faulty measurements or the misinterpretation of edges from the fiber orientation measurement.

The method is tested on a 3D geometry, a complex four-layer CFRP component. The fiber orientations can be determined in all four investigated layers. The results can be used to improve forming strategies and draping simulations or as inputs to structural simulations of CFRP components to verify the structural integrity under expected loads. This allows to further optimize CFRP design and manufacturing processes, exploiting CFRP's full potential in order to minimize costs and resource consumption while maximizing durability and robustness in aerospace applications.

1. Introduction

This paper presents results of recent cooperative works at the ITM and IKTS on the 3D fiber orientation measurement with eddy current technique, which are the result of a recent dissertation [1]. Preliminary results, which are not included here, have also been published in a number of articles [2, 3].The outstanding mechanical properties of carbon fiber reinforced plastics (CFRP) are achieved by adjusting the orientation of the load-bearing carbon fibers exactly to the type and direction of the external loads acting on the structure. If the property potential of the fibers is fully exploited in this way, CFRP can be used to realize load-bearing structures that are up to 70% lighter than steel and up to 30% lighter than aluminum with the same load-bearing capacity [4]. Additionally to the weight saving, composite structures offer the further benefit of being able to integrate smart measurement systems [6] or functional components [7].

On the other hand, any deviation of the yarn orientation from the load direction must be compensated by additional material. A deviation of the fiber orientation reduces the stiffness by up to 5 %, the strength by up to 20 % . Each fibre angle deviation thus increases the weight, increases the material and manufacturing costs and in many cases makes production with traditional light metals more economical [5].

The prerequisite for exploiting the property potential of CFRP to reduce weight is precise knowledge of the fiber orientation in the component. Without this knowledge, manufacturing processes cannot be systematically optimized, draping simulations cannot be validated and component design cannot be carried out reliably. Significant research resources are currently dedicated to mapping the entire CFRP development process in a virtual process chain and designing the semi-finished products and manufacturing processes on the basis of the desired fiber orientations in the component. Without a process that confirms the validity of the statements made about the fiber orientations after the manufacturing process, enormous uncertainties must always be accepted in these virtual process chains. It is not sufficient to analyze only the fiber orientation of the top layer or the deformation behavior of each individual layer, because the fiber orientation is essentially dependent on how many layers are formed together and in which order they are stacked.

There is currently no method that can non-destructively measure the yarn orientation in multilayer, three-dimensionally formed CFRP components. Optical methods are highly precise when measuring the fiber orientation, but can only examine the fiber orientation in the uppermost layer and provide neither the information required for component design nor for process optimization [8 – 12]. Micro X-ray CT investigations, on the other hand, are able to measure the fiber orientation in the inner layers of a CFRP component, but are complex, cost-intensive and limited to small samples [13 – 17]. High-frequency eddy current testing is a method that permits the measurement of fiber orientation in multilayer carbon fiber structures [18 – 22]. In contrast to traditional eddy current testing, measurement frequencies in the megahertz range are used to make the course of the weakly conductive carbon fibers in the component visible with the aid of eddy currents. As shown in Fig. 1, not only the fiber orientations of the top layer are visible, but also those of deeper layers.

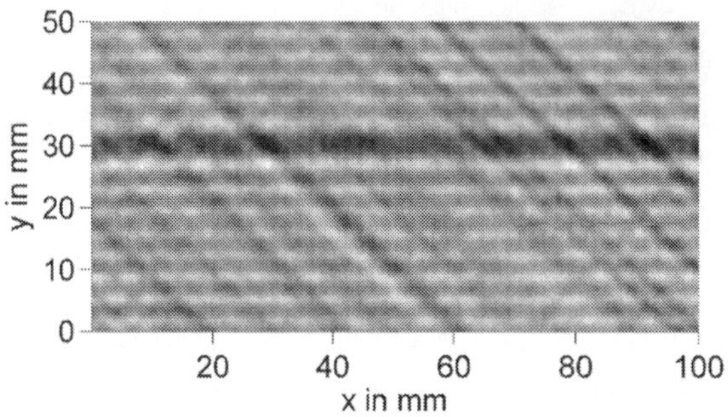

Fig. 1. Eddy current image of a four-ply non-crimp fabric with a missing thread in the fourth layer (dark stripe)

While the measurement of the fiber orientation in two-dimensional textile semi-finished products and semi-finished product stacks using the eddy current method is state-of-the-art, there is currently no method for full-surface measurement of the fiber orientation in multi-curved components as required for process development and quality assurance in CFRP production [23 – 26]. The aim of this contribution is therefore to develop a method based on high-frequency eddy current testing technology for measuring fiber orientation in three-dimensional, multi-layer carbon fiber preforms and CFRP. Since eddy current testing is based solely on the conductivity of the carbon fibers, such a method, in contrast to ultrasonic testing, can be used both for dry preforms - three-dimensionally formed and fixed stacks of semi-finished products that are passed on to the infusion process - and for the cured composite components themselves.

Eddy current testing of carbon fiber materials uses imaging eddy current testing and rotational eddy current testing as measuring principles. In eddy current imaging, the sensor is moved over the component surface in several paths. From the measured impedance at the individual measuring points an intensity image of the surface (eddy current image) is generated, comparable to a C-scan from ultrasonic testing. Since the conductivity of each fiber bundle is higher at the center, the highest eddy current densities are generated when the transmitting coil is located centrally above a fiber bundle of this layer. When scanning the surface, a superimposed striped pattern is created, which indicates the thread orientation of the individual layers. In addition to the fiber orientation, local defects can also be detected in the eddy current image due to the different conductivity, which is not possible when using rotational eddy current testing. Commercial 2D inspection systems for eddy current imaging of carbon fiber materials are distributed by Suragus GmbH, Germany and Fraunhofer IKTS.

The determination of the fiber orientation from 2D eddy current images was carried out in all previous work by evaluating the Fourier transformation, which converts the image to the frequency domain [23, 27 – 31]. Fig. 2 shows an evaluation software based on the Fourier transformation from Suragus for fiber orientation measurement on 2D eddy current images. In addition to the fiber directions, the software allows the automatic detection and quantification of paths, undulations and local defects.

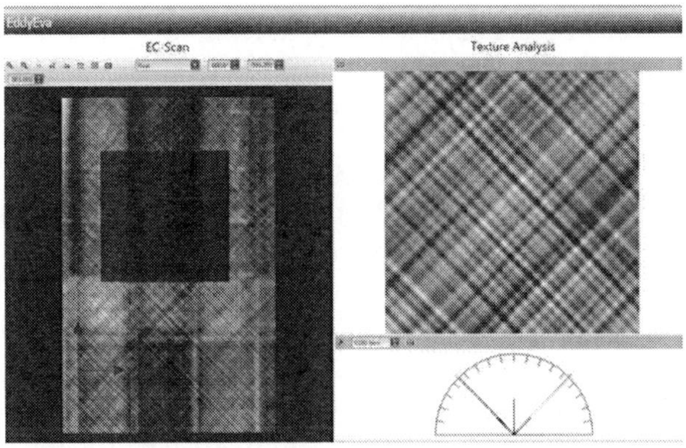

Fig. 2. EddyEva evaluation software from Suragus *[32]*

Various test systems were developed for testing single curved, so-called 2.5-dimensional test specimens. Salski [33] introduces an array scanner that is equipped with a flexible sensor arm and can therefore detect simply curved surfaces (Fig. 3 left). Printed planar coils are used as sensors. Since the individual coils have a different angle of tilt to the surface with double curved surfaces, the measuring signals are subsequently adjusted by averaging. A 2.5D scanner (Fig. 3 right) distributed by Suragus in cooperation with Fraunhofer IKTS has a tilting mechanism of the z-axis, so that slightly curved components can also be inspected. Both systems extend the testability for 2.5-dimensional test specimens, but do not allow testing of multiple curved surfaces. Since no shear occurs during the draping of textile semi-finished products to 2.5-dimensional test specimens, thus maintaining the original thread orientation, these are not relevant for fiber orientation measurement.

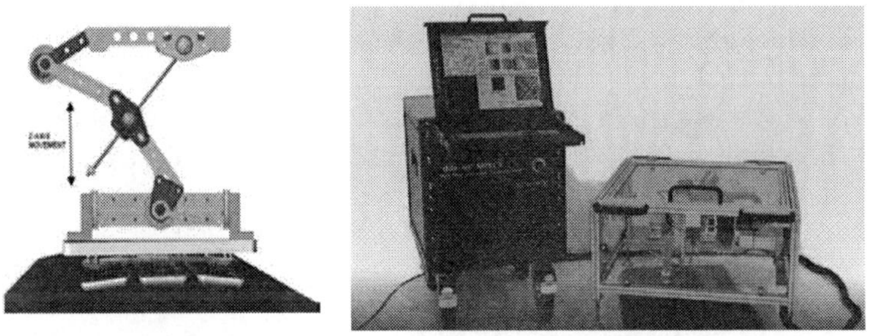

Fig. 3. 2.5D scanner: Salski 2016 [33], Fraunhofer IKTS [34]

In addition to these 2.5-dimensional systems, Fraunhofer IKTS has developed a system that enables eddy current testing of multiple curved surfaces. For this purpose, an eddy current sensor was integrated into an industrial robot. With the aid of a strip light projection, the surface and the position of the test specimen are recorded, followed by path planning with the aid of dedicated software, in which parallel test paths are projected onto the surface. After a collision check in a robot simulation, the test procedure can be carried out. An evaluation software is available for visualizing the results.

For the first time, an inspection system is available that is able to detect even multiple curved 3D surfaces and to visualize the yarn gradients in three-dimensionally formed carbon fiber materials. The advantages are a fully integrated, easy-to-use path planning and inspection software. However, a major limitation of this work is seen in the fact that the path planning is carried out by a point projection. This method requires little computing effort and can be controlled very intuitively, but in the case of complex components with inclined or vertical side surfaces it results in fewer or no paths being placed on the side surfaces, so that the measurement must be carried out in several sub-steps. Furthermore, although the three-dimensional yarn gradients are made visible, it is not possible to measure the yarn directions on the 3D data. Consequently a method is developed to calculate robotic paths with a defined maximum distance on any three-dimensional surface. In addition a method to robustly measure fiber orientations in multiple layers in three-dimensional preforms and CFRP components.

2. Materials and Methods

2.1 Robotic Path Planning

As an input to the path planning software CAD data of the object to be investigated is used. The representation of the 3D surface is an approximated triangulation mesh, which is derived from the given CAD data. This allows the representation of non-unique (i.e., not as a function z = f(x,y) representable) surfaces with vertical edges and undercuts, on the one hand, and a fast calculation of basic geometric operations (intersection line/surface, normal vector, progress by defined arc length on the surface), on the other hand. By specifying a sufficiently small triangle size, even strongly curved 3D surfaces can be modelled with sufficient accuracy.

There are two strategies for calculating the paths on the 3D surface: firstly, the starting points of the individual paths can be located with the specified path distance. The paths are then calculated so that they follow a given direction on the surface (method 1, Fig. 4 left). On the curved 3D surface, however, the paths have, depending on the surface curvature, a distance different from the specified path distance, resulting in additional paths being inserted to such an extent that the specified path distance is not exceeded.

The calculation of additional paths between the calculated paths can be avoided by a second strategy, in which the first path is given on the surface, and the points of the following path are calculated in such a way that they have a constant distance to the points of the first path (method 2, Fig. 4 right). A curved surface still requires additional paths. In contrast to method 1, however, the filling paths are placed at the end of the calculated paths.

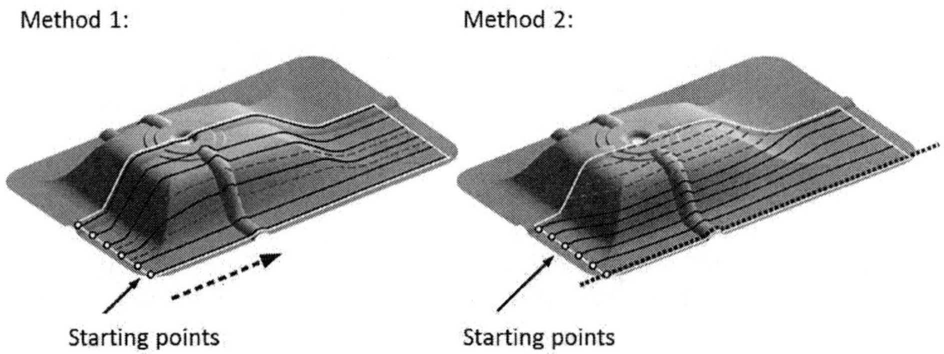

Fig. 4. Strategies for path planning: Starting points and path vector (left) or starting points and first path (right)

Both methods are mathematically similarly complex, but the first method has the advantage that the start and end points of all paths lie on one line and all paths run similarly, which simplifies the collision and accessibility check. Therefore, a method for surface detection based on method 1 is developed.

1. The starting point for the calculation of each path is the starting point and the same path vector for all paths. For a uniform interpolation point distance, equal distances on the surface in the direction of the path vector must be determined, which is done by the following algorithm (see Fig. 5).

2. Find the corresponding triangle for the starting point. Project the path vector into this triangle and add it to the first starting point. Determine the intersection of the direction vector with the triangle edge to which the direction vector points.
3. Find the next triangle that contains this intersection edge. Project the path vector into this triangle and add it to the calculated intersection. Find the intersection with the next triangle edge.
4. Check if the travelled path has exceeded the specified path distance. If yes, the next interpolation point on the way section lies in the current triangle and is calculated by dividing the distance.

Repeat steps 1 to 4 until the limit of the defined measuring range or surface or the specified number of paths is reached. The limit of the surface is reached if no second triangle containing the same triangle edge is found in step 2. The current triangle then lies at the edge of the surface.

If the specified maximum path distance is exceeded for at least one of the support points, a number of fill lines are inserted between the two paths so that the maximum path distance is guaranteed for all support points.

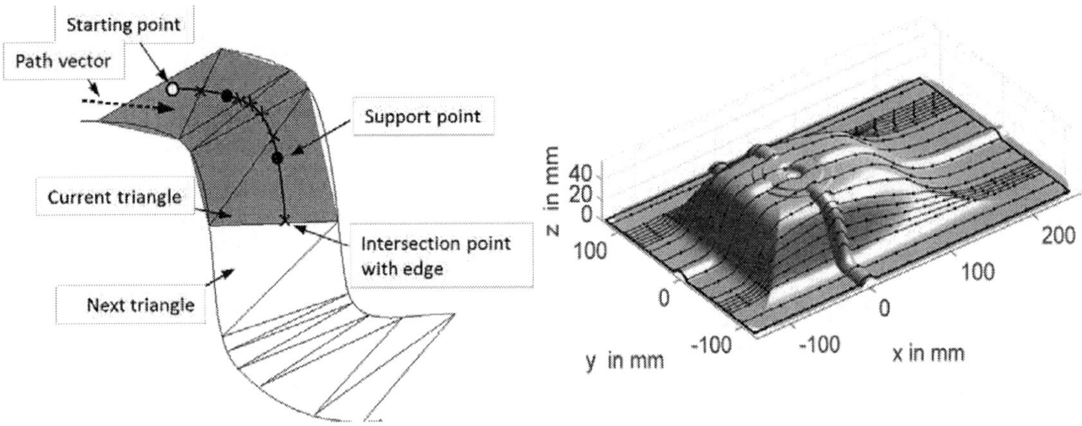

Fig. 5. Path calculation on triangulated surface and result for a complex 3D-geometry

The sensor orientation for each point is calculated from the normal vectors of the surface and converted to the euler angles (positive rotation around Z-, Y- and X-axis) used in the resulting robotic program. Since small radii (< 15 mm) cannot be scanned without a lift-off of either transmitter or receiver coil, it is acceptable to restrain the euler angles at these points in order to avoid collisions and improve inspection speed.

2.2 Eddy current measuring and evaluation system

For eddy current testing an EddyCus Integration Kit (developed by Fraunhofer IKTS) is employed. To avoid lift-off effects due to a large sensor geometry a special 3D-printed probe with a small coil diameter of 3.3 mm is used. It is attached to an adapter plate via a spring-loaded linear guidance, which is mounted on an industrial robot Kuka KR6 R900 sixx, Agilus (Kuka AG, Germany) (Fig. 6).

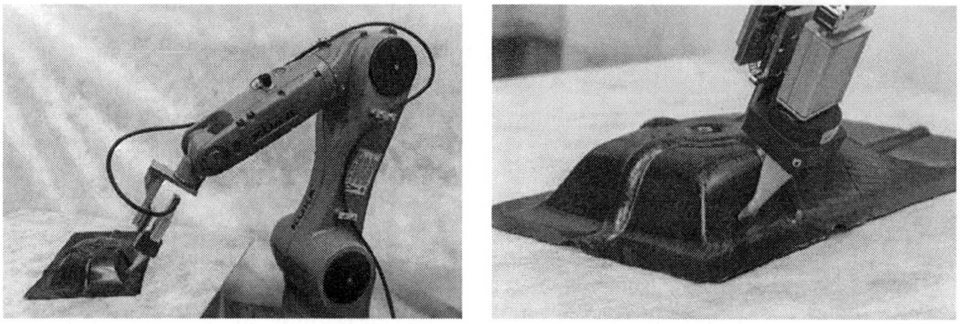

Fig. 6. Robot guided eddy current testing of CFRP component

The system uses a time-trigger approach. After receiving a trigger input, the measurement data is recorded at a constant measuring rate until the trigger signal is set to OFF. At the end of each path (trigger signal = OFF), the control software stores all recorded measured values together with a time stamp. To determine the x,y,z coordinates of the measured values of the individual paths, the eddy current software expects a file for each measuring path, which contains the actual movement of the robot (time stamp + x,y,z coordinates). The coordinates for the individual measured values can subsequently be determined by temporal interpolation.

The result of the 3D eddy current test is a measurement data set in which each x,y,z measuring point is assigned a complex impedance value (real and imaginary part). If all real and imaginary values of a measurement are displayed in the complex plane, the result is a point cloud in the complex plane. The eddy current signal is affected by several effects, especially differences in conductivity, lift-off and thickness. A separation of the different effects is possible if they have different directions of action in the complex impedance or voltage level. Consequently, the point cloud must be rotated in such a way that the interference effects are minimized when projecting the rotated point cloud on the x-axis. The optimal phase rotation, which maximizes the contrast of the carbon filaments, can be determined by a principal component analysis.

2.3 Measuring the fiber orientation

For the measurement of the local fiber orientation at an evaluation point, the stripe pattern of the carbon fibers in the vicinity of this point must be observed. All known image analysis methods (see e.g. [35]) require a 2D image with a constant x and y point grid (pixel grid/bitmap), so that the 3D surface in the vicinity of the point must be smoothed, i.e. projected into a plane.

For this projection, the normal vector of the tangential plane to the surface, X_n, is determined at the respective evaluation point X_0 (for the definition of the evaluation points see below). Then all points X of the environment are projected into this tangential plane:

$$X_{proj} = X - \big((X - X_0) \cdot X_n\big)X_n \tag{1}$$

However, this procedure must be extended by a correction: in the vicinity of strong curvatures, the projection would lead to distortions, which is corrected by extending the distance X_0X_{proj} so that it corresponds to the original distance of the point X to the evaluation point, X_0X. To further enhance the image quality a high-pass filter is applied that accentuates the carbon fiber bundles. Subsequently, noise is reduced by applying a Gaussian filter.

As a next step the 2D-FFT is used to measure the fiber orientations which are then mapped onto the 3D-surface. If there is not enough image information to determine all fiber orientations in an evaluation section, e.g. if it is located on the edge of the scanned surface, the information will be calculated by interpolating between adjacent evaluation points.

3. Results

Fig. 7 shows the obtained 3D eddy current image for a complex CFRP component and sensor orientation 0°. The component has a layer structure of [+45°/-45°/0°/90°] and both the upper +/-45° layers and the lower 0°/90° layers are clearly visible as a stripe pattern in the eddy current image.

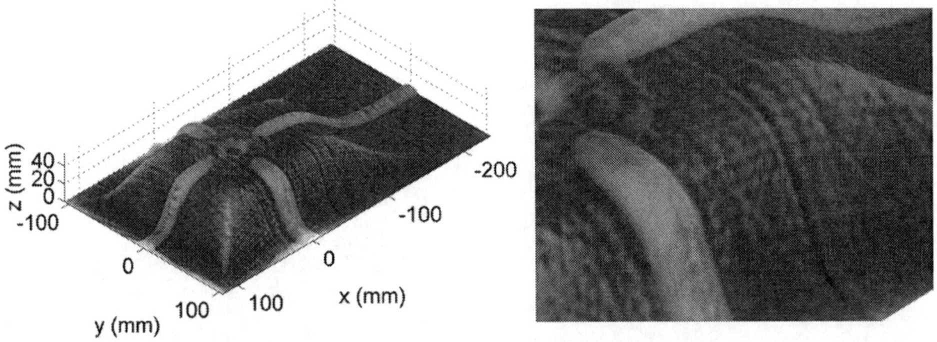

Fig. 7. Eddy current image of complex CFRP component

Lift-off effects are unavoidable at the longitudinal and transverse ribs as well as at the edges (bright areas): due to the narrow geometry radius, either the receiver coil or the transmitter coil of the sensor loses contact with the surface. The phase rotation can reduce these interferences, but not completely eliminate them, so that the ribs are excluded from the fiber orientation measurement in order to avoid false detections. This is not necessary for the edges, since the fiber orientation measurement algorithm automatically detects edges in the image section and prevents a false detection of an edge as the fiber direction.

Fig. 8 shows the reconstructed fiber orientations for all four layers according to the presented method. The results are validated qualitatively by comparing the course of single rovings in the eddy current image to the detected fiber courses. As can be seen in Fig. 8 the measured fiber courses in the plane as well as in the curved areas correspond very well with the stripe patterns recognizable in the eddy current image.

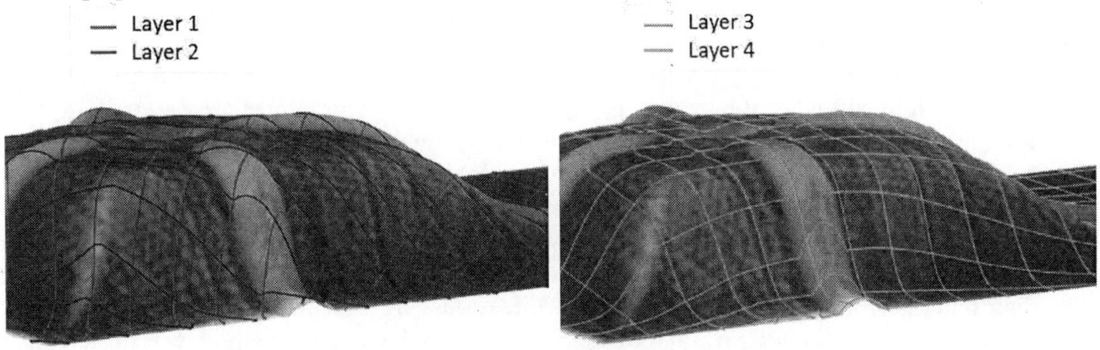

Fig. 8. Fiber courses detected by Fourier transformation based algorithm

For an optical validation of the measured fiber orientations in the upper layer, the component was photographed in vertical perspective and the 2D photo was mapped onto the 3D surface of the measurement data (Figure 9). The fiber orientations determined with the eddy current inspection follow the optically recognizable fiber orientations very well.

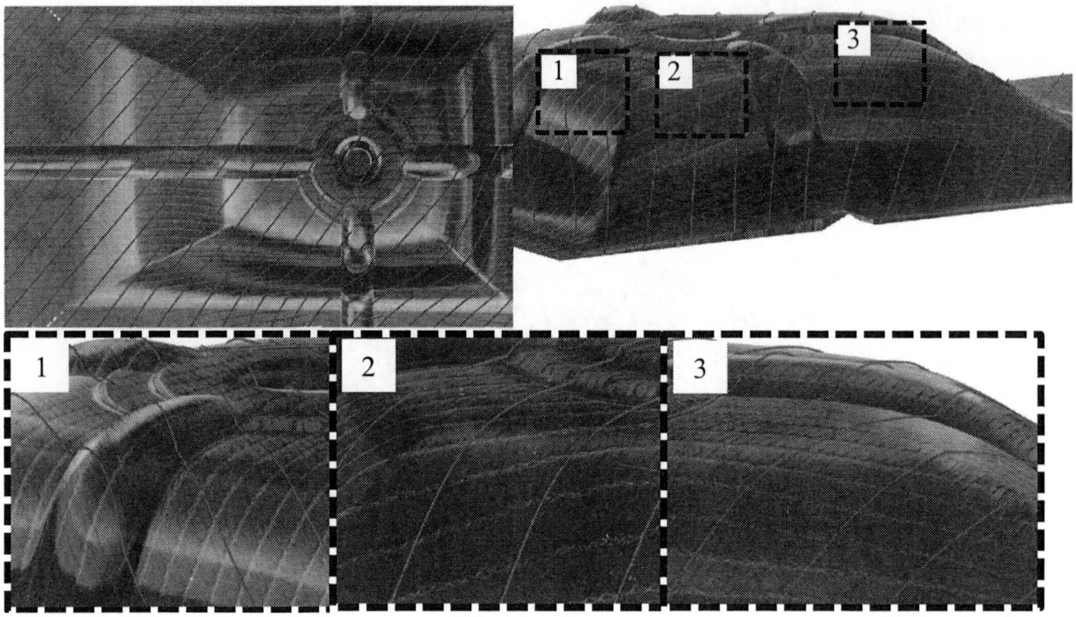

Fig. 9. Optical validation

Nevertheless only the top layer can be validated quanlitatively due to the lack of alternative methods to measure fiber orientations of deeper layers.

4. Conclusion

The development of carbon fiber-reinforced plastics with load-bearing fiber courses requires a process capable of non-destructively measuring the fiber orientations in multilayer, three-dimensionally formed preforms and CFRP structures. Previous work has shown that imaging eddy current testing with frequencies in the megahertz range is capable of imaging the yarn structure in multi-layer two- and three-dimensional carbon fiber structures. In this article a method is presented that can fully capture any 3D surface, even in the presence of strongly curved and vertical areas with uniform path spacing, and thus ensure a defined distance between the measuring points. This procedure serves as a basis for the development of robot programs for eddy current inspection of three-dimensional surfaces. From the measured eddy current images, the local fiber orientation is determined on a grid of evaluation points. In a first step, the surface in the vicinity of each evaluation point is smoothed and converted into a pixel image, which is pre-processed to increase the yarn contrast. The Fourier transformation is used to calculate the local fiber orientation distribution, from which the main fiber directions are calculated and assigned to the individual fiber layers using a robust method. The developed method was tested on a complex CFRP component and validated optically.

The developed process is the first to provide a tool that allows the non-destructive measurement of fiber orientation in multilayer, three-dimensional carbon fiber preforms and CFRP and can be used for the systematic optimization of 3D preforming processes. The resulting CFRP structures will have fiber orientations precisely adapted to the external loads, which exploit the lightweight construction potential provided by the CFRP material.

References

[1] Bardl, Georg: Entwicklung eines Verfahrens zur zerstörungsfreien Messung der Faserorientierung in mehrlagigen 3D-Carbonfaserpreforms und CFK mit robotergeführter Hochfrequenz-Wirbelstromprüftechnik. Dresden, Technische Universität Dresden, Fakultät Maschinenwesen, Dissertation, 2018, Dresden: 2018.

[2] Bardl, Georg; Nocke, Andreas; Hübner, Matthias; Gereke, Thomas; Pooch, Matthias; Schulze, Martin; Heuer, Henning; Schiller, Marko; Kupke, Richard; Klein, Marcus; Cherif, Chokri: Analysis of the 3D draping behavior of carbon fiber non-crimp fabrics with eddy current technique. Composites Part B: Engineering 132(2018), S. 49–60.

[3] Bardl, Georg; Nocke, Andreas; Cherif, Chokri; Pooch, Matthias; Schulze, Martin; Heuer, Henning; Schiller, Marko; Kupke, Richard; Klein, Marcus: Automated detection of yarn orientation in 3D-draped carbon fiber fabrics and preforms from eddy current data. Composites Part B: Engineering 96(2016), S. 312–324.

[4] Cherif, Chokri: Textile material for lightweight constructions, Berlin, Heidelberg: Springer, 2016. - ISBN 978-3-662-46341-3.

[5] Assmann, Wolfgang; Witten, Elmar: Handbuch Faserverbundkunststoffe - Composites. 4 Auflage, Wiesbaden: Springer Fachmedien, 2013. - ISBN 978-3-658-02755-1.

[6] Haentzsche, Eric; Mueller, Ralf; Ruder, Tristan; Nocke, Andreas; Cherif, Chokri: Integrative Manufacturing of Textile-Based Sensors for Spatially Resolved Structural Health Monitoring Tasks of Large-Scaled Composite Components. Materials Science Forum 825-826(2015), S. 571–578.

[7] Ashir, M.; Hindahl, J.; Nocke, Andreas; Cherif, Chokri: Time dependent deformation behavior of adaptive fiber reinforced plastics. In: 21st international conference on composite materials. Xi'an, China, August 2017.

[8] Lomov, S. V.; Boisse, Ph.; Deluycker, E.; Morestin, F.; Vanclooster, K.; Vandepitte, D.; Verpoest, I.; Willems, A.: Full-field strain measurements in textile deformability studies. Composites Part A: Applied Science and Manufacturing 39(2008)8, S. 1232–1244.

[9] Mersmann, Christoph: Industrialisierende Machine-Vision-Integration im Faserverbundleichtbau. Band Bd. 8/2012, Aachen: Apprimus Verlag, 2012. - ISBN 9783863590628.

[10] Long, A. C.; Rudd, C. D.; Blagdon, M.; Smith, P.: Characterizing the processing and performance of aligned reinforcements during preform manufacture. Composites Part A: Applied Science and Manufacturing 27(1996)4, S. 247–253.

[11] Haanappel, S. P.; Thije, R.H.W. ten; Sachs, U.; Rietman, B.; Akkerman, R.: Formability analyses of uni-directional and textile reinforced thermoplastics. Composites Part A: Applied Science and Manufacturing 56(2014), S. 80–92.

[12] APODIUS GMBH: Produkte | Apodius 3D. http://apodius.de/de/produkte#apodius3d, (17.9.2017).

[13] Sutcliffe, M.P.F.; Lemanski, S. L.; Scott, A. E.: Measurement of fibre waviness in industrial composite components. Composites Science and Technology 72(2012)16, S. 2016–2023.

[14] CZABAJ, MICHAEL W.; RICCIO, MARK L.; WHITACRE, WILLIAM W.: Numerical reconstruction of graphite/epoxy composite microstructure based on sub-micron resolution X-ray computed tomography. Composites Science and Technology 105(2014), S. 174–182.

[15] SKET, F.; ENFEDAQUE, A.; ALTON, C.; GONZÁLEZ, C.; MOLINA-ALDAREGUIA, J. M.; LLORCA, J.: Automatic quantification of matrix cracking and fiber rotation by X-ray computed tomography in shear-deformed carbon fiber-reinforced laminates. Composites Science and Technology 90(2014), S. 129–138.

[16] GOMMER, F.; WEDGWOOD, K.C.A.; BROWN, L. P.: Stochastic reconstruction of filament paths in fibre bundles based on two-dimensional input data. Composites Part A: Applied Science and Manufacturing 76(2015), S. 262–271.

[17] EMERSON, MONICA J.; JESPERSEN, KRISTINE M.; DAHL, ANDERS B.; CONRADSEN, KNUT; MIKKELSEN, LARS P.: Individual fibre segmentation from 3D X-ray computed tomography for characterising the fibre orientation in unidirectional composite materials. Composites Part A: Applied Science and Manufacturing 97(2017), S. 83–92.

[18] GARCÍA-MARTÍN, JAVIER; GÓMEZ-GIL, JAIME; VÁZQUEZ-SÁNCHEZ, ERNESTO: Non-destructive techniques based on eddy current testing. Sensors (Basel, Switzerland) 11(2011)3, S. 2525–2565.

[19] MOTTL, Z.: The quantitative relations between true and standard depth of penetration for air-cored probe coils in eddy current testing. NDT International 23(1990)1, S. 11–18.

[20] SHULL, PETER J.: Nondestructive Evaluation. Band v. Vol. 142, New York, USA: CRC Press, 2016. - ISBN 9780203911068.

[21] VAN DRUNEN, G.; CECCO, V. S.: Recognizing limitations in eddy-current testing. NDT International 17(1984)1, S. 9–17.

[22] BLITZ, JACK: Electrical and Magnetic Methods of Non-destructive Testing. Band 3, Dordrecht: Springer Netherlands; Imprint; Springer, 1997. - ISBN 9789401158183.

[23] SCHULZE, MARTIN; HEUER, HENNING: Textural analyses of carbon fiber materials by 2D-FFT of complex images obtained by high frequency eddy current imaging (HF-ECI). In: SPIE Smart Structures and Materials + Nondestructive Evaluation and Health Monitoring, San Diego, California, Sunday 11 March 2012, S. 83470S.

[24] MIZUKAMI, KOICHI; MIZUTANI, YOSHIHIRO; TODOROKI, AKIRA; SUZUKI, YOSHIRO: Detection of in-plane and out-of-plane fiber waviness in unidirectional carbon fiber reinforced composites using eddy current testing. Composites Part B: Engineering 86(2016), S. 84–94.

[25] MIZUKAMI, KOICHI; MIZUTANI, YOSHIHIRO; KIMURA, KENSHI; SATO, AKIYOSHI; TODOROKI, AKIRA; SUZUKI, YOSHIRO: Detection of in-plane fiber waviness in cross-ply CFRP laminates using layer selectable eddy current method. Composites Part A: Applied Science and Manufacturing 82(2016), S. 108–118.

[26] POURDEYHIMI, B.; KIM, H. S.: Measuring Fiber Orientation in Nonwovens: The Hough Transform. Textile Research Journal 72(2016)9, S. 803–809.

[27] MOOK, GERHARD; LANGE, ROLF; KOESER, OLE: Non-destructive characterisation of carbon-fibre-reinforced plastics by means of eddy-currents. Composites Science and Technology 61(2001)6, S. 865–873.

[28] GRIMBERG, R.; SAVIN, ADRIANA; ROTUNDU, C.R: Eddy current microscopy applied to graphite-epoxy composite. Sensors and Actuators A: Physical 91(2001)1-2, S. 73–75.

[29] HEUER, H.; SCHULZE, MARTIN; POOCH, M.; GÄBLER, S.; NOCKE, ANDREAS; BARDL, G.; CHERIF, CHOKRI; KLEIN, M.; KUPKE, R.; VETTER, R.; LENZ, F.; KLIEM, M.; BÜLOW, C.; GOYVAERTS, J.; MAYER, T.; PETRENZ, S.: Review on quality assurance along the CFRP value chain – Non-destructive testing of fabrics, preforms and CFRP by HF radio wave techniques. Composites Part B: Engineering 77(2015), S. 494–501.

[30] CHENG, JUN; QIU, JINHAO; JI, HONGLI; WANG, ENRONG; TAKAGI, TOSHIYUKI; UCHIMOTO, TETSUYA: Application of low frequency ECT method in noncontact detection and visualization of CFRP material. Composites Part B: Engineering 110(2017), S. 141–152.

[31] KOSEKUGAWA, H.; YOSHIKAWA, Y.; URAYAMA, R.; UCHIMOTO, TETSUYA; TAKAGI, T.: Non-Destructive Assessment of Fiber Alignment in CFRP using Eddy Current Testing with Differential Type Probe, München, (2016).

[32] SURAGUS GMBH: SURAGUS Carbonfaser Prüfung - automatisierte Auswertung von Strukturen und Defekten. https://www.suragus.com/ de/produkte/karbonfaser-pruefung/struktur-und-defekte/auswertungssoftware-eddyeva/, (05.10.2017).

[33] SALSKI, B.; GWAREK, W.; KOPYT, P.; THEODORAKEAS, P.; HATZIIOANNIDIS, I.; KOUI, M.; Chong, A. Y. B.; TAN, S. M.; KAPPATOS, V.; SELCUK, C.; GAN, T. H.: Portable Automated Radio-Frequency Scanner for Non-destructive Testing of Carbon-Fibre-Reinforced Polymer Composites. Journal of Nondestructive Evaluation 35(2016)2, S. 130.

[34] FRAUNHOFER IKTS DRESDEN: Wirbelstromscanner EddyCus (R) MPECS. https://www.ikts.fraunhofer.de/ content/dam/ikts/forschung/ el-ektronik_mikrosystemtechnik/pruef_und_analysesysteme/IKTS_EddyCus_MPECS_2014.pdf, (5.10.2017).

[35] JÄHNE, BERND: Digitale Bildverarbeitung, Berlin, Heidelberg: Springer Berlin Heidelberg, 1997. - ISBN 9783662067338.

Automated single view 3D Texture Mapping and Defect Localisation of Thermography Measurement on Fuselage utilising an industrial robot and laser system

Somen DUTTA[1], Klaus DRECHSLER[2], Michael KUPKE[1], Alfons SCHUSTER[1]
[1]Deutsches Zentrum für Luft- und Raumfahrt, [2] TU München, Garching, Germany
Contact e-mail: somen.dutta@dlr.de

Abstract. Thermography measurements of large objects are always difficult and very expensive to automate. The main reason is that most large objects are bigger than measuring field. Furthermore, continuous measurements are prevented due to lateral thermal diffusion. Another unhandy and less intuitive method is that certified examiners evaluate every 2D thermography image per measuring field one by one. Defect analysis from 2D images restricts to demine the accurate defect size and 3D defect propagation. 3D evaluation is complex but gives complete 3D defect propagation knowledge. Moreover absolute robot accuracy is major challenge for robot based automated thermography. Robot Tool-Centre-Point (TCP) positional accuracy depends mostly on robot pose and their orientation. Many factors affect the absolute positioning accuracy of created robot TCP positions. 0.1° TCP inaccuracy can cause for 1 meter working distance up to 3mm position error.

In order to overcome these limitations and retrieve a 3D thermography texture for the observed object, DLR Augsburg has developed both hardware and software components. The essential information required to generate 3D thermography are objects geometry, location and orientation of object with reference to Camera. DLR-AQP department is working on two different methods to calculate 2D-3D camera projection parameters. Method one projects predefined 3D points per measuring field by using laser projector and applies 2D-3D point correspondence algorithm. The second Method uses the robot pose and a kinematic-coordinate-relationships algorithm is applied to the images. For both methods single thermal image is captured by using thermography camera, which is also used for NDT. The first method is suitable for research and development as well as for production inspection with or without using a robot for camera positioning while the second method requires a robot. The first method enables also contactless localisation of detected defects. About accuracy, reproducibility and flexibility of both methods have been compared and discussed in this paper. The 3D results are to be imported into the CAD environment for further structure analysis. Furthermore thermography results are visualised in accordance to their depth in 3D-Thermo-Thermography model.

Introduction

Large object measurements are always difficult and very expensive to automate. The main reason is that most large objects are bigger than the working envelope of measuring machines. Furthermore, modulated thermal waves during Lock-In Thermography diffuses laterally to the surrounding surfaces of measurement filed, which prevents continuous measuring of the very next measurement field. In addition, certified examiners have to evaluate every 2D thermography image per measuring field one by one which is unhandy and less intuitive method for industrial applications. These 2D images are projections of measurement objects, which prevent to demine the accurate defect dimension and 3D defect propagation [1]. Evaluation in 3D components coordinate, gives complex but complete 3D defect propagation knowledge. Another challenge is robot's absolute positioning inaccuracy. Robot

Tool-Centre Point (TCP) positional accuracy depends mostly on Hand-Eye calibration results, robot pose and their orientation. Many other factors like gravitational or deflecting forces may affect the absolute positional accuracy of robot TCP positions. For example, 0.1° absolute robot inaccuracy (TCP) can cause for 1 meter working distance up to 3 mm positional error.
In order to overcome these limitations and retrieve a 3D thermography evaluation for the observed object, DLR Augsburg has developed robot based measuring system for industrial thermography applications and 3D-Thermo-Tomography visualisation technology (see previous papers [2, 3, 4 and 5]). To address limitations of existing 3D thermography texture mapping solutions, DLR Augsburg proposes a new 3D visualisation (see Figure 1) method using thermography camera, which is often used for NDT, and a laser system mounted above of the robotic cell. This approach eliminates the uncertainty of robot's position accuracy. With an easy calibration process within a few minutes the laser system is calibrated to the component coordinate system and projects predefined 3D points for every measuring field. These 3D points are captured by the thermography camera; afterwards a 2D-3D point correspondence algorithm is applied to calculate 2D-3D camera projection parameters. The second Method (Figure 2) uses the robot pose and thereby a kinematic-coordinate-relationships algorithm is applied to the images. For both methods single thermal image is captured using a thermography camera. The first method is suitable for research and development as well as for production inspection with or without using a robot for camera positioning while the second method requires a robot. The first method enables also contactless localisation of detected defects. Accuracy, reproducibility and flexibility of both methods are compared and discussed in this paper. The 3D results from above mentioned methods are imported directly into the CAD environment for further structural analysis.

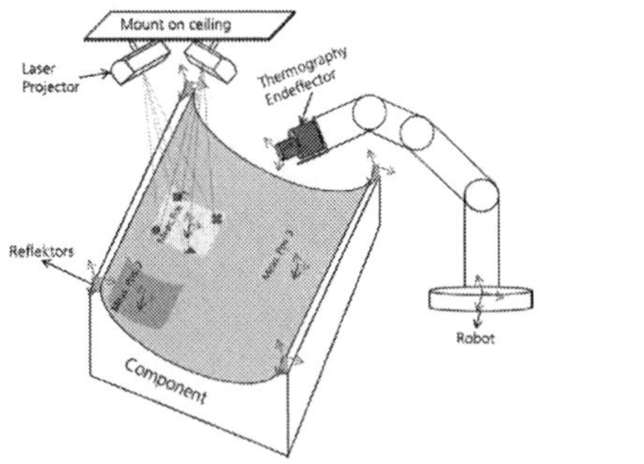

Figure *1: 2D-3D point correspondence approach*

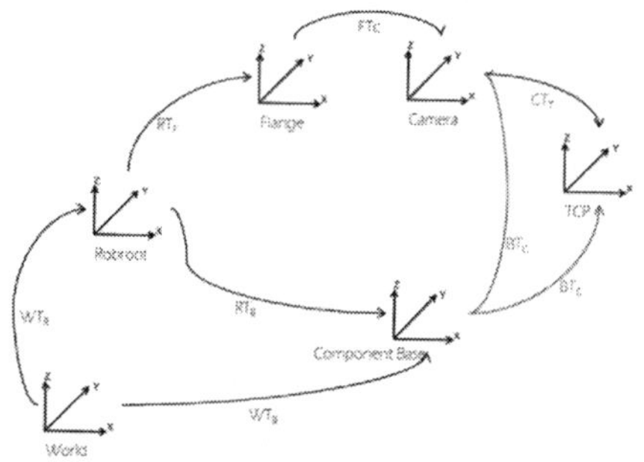

Figure 2: Kinematic coordinate chain approach

Feasibility Study of new concept

For the new concept (see above), the provided and respectively projected laser points, which are visible by thermography camera due to local surface temperature difference, have to be recognised automatically. By using a similar template like laser spot in 32 x 32 Pixel (Figure 3b) and several filters to minimise noise from thermal image, software can detect the laser points automatically (Figure 3a). Experiment has been conducted to find influencing parameters like point size, laser projector distance, projection time, surface reflection and materials temperature absorption degree, which may influence the detection. All test results were reproducible and showed no significant influence of the above mentioned parameters. Details of experimental results have been presented in paper [5].

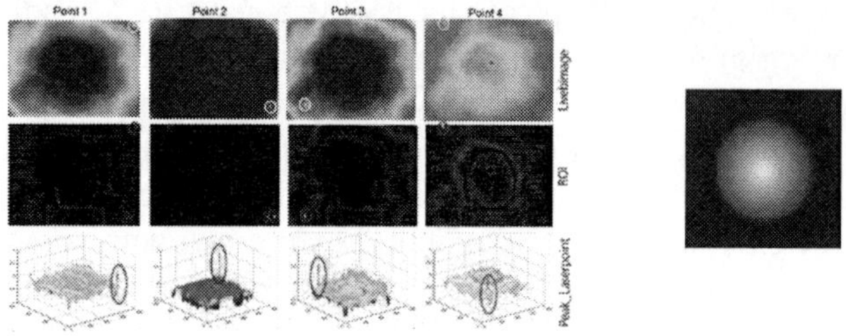

(a) Laser point detection (b) Correlation template

Figure 3: (a) Laser point detection, (b) Correlation template

Geometric and Hand-Eye camera calibration

Importance of accurate both geometric and hand-eye calibration along with their results have been published in conference papers [3, 4, 11 and 12]. The paper [4] has claimed max 0.22 mm translational and 0.03° rotational standard deviation of repeatability accuracy. However, a maximum magnitude pointing error was 1.052°, which was caused by cable carrier. This pointing error has considerable influence in X and Y direction. Furthermore, any robot velocity influence has been ascertained. The achieved hand-eye accuracy for a 400 mm working distance was 10.31 mm and 2.87 mm. The Author has emphasized poor absolute accuracy caused by cable carrier as main reason for such error. In [5] paper the hand-eye calibration results with normal objective has been presented. But no accuracy analysis has been performed, as experiment was focused on development of robot based 3D thermography. ±0.15 re-projection error from geometric calibration was achieved. Therefore, a further experiment has been conducted. The aim of the experiment is to find out whether there is any influence of cable carrier at all and how much that affect on absolute accuracy at the measuring position. During experiment Cable carrier was disassembled from axis 6 and wrapped around axis 3 in the hope of achieving a better result than yet. So that cable carrier effect can be avoided. In general, intrinsic camera parameters are independent of the position and orientation of the camera. The MATLAB Camera calibration toolbox, mostly based on Zhang model, calculates intrinsic parameter with Pinhole camera model. Since the principle point is directly connected to the projection centre (for Pinhole model), the exterior and interior orientation are highly correlated. In general, there are three different kinds of distortions, which are expressed as radial distortion, dicentric distortion and thin prism distortion. Both dicentric and prism distortions contain radial and tangential distortion coefficients. As first two radial coefficients are predominant, default values in MATLAB calibration toolbox were changed for skew (i.e. α_c=0) and last three radial tangential coefficient ($k_c(3)$=0, k_c (4)=0 and k_c (5)=0), for the first geometric calibration experiment. Hereby, it has been assumed that pixels are rectangular and used standard 12 mm objective has no 4th order radial and tangential distortions. The results once with assumed distortions coefficient as in default setting and once without default setting are shown in Tables 1 and 2 respectively. Comparing both tables one can see, that both tangential and skew coefficients are also almost zero with default setting (without changing distortions coefficient). This means presumptions were correct. Since skew coefficient zero was set, the aspect ratio fc(1)/fc(2) has become 1. Which means the focal length in X and Y direction are in tables 1 and 2 are same. It is to note, that radial distortions coefficient is highly correlated with focal length, as radial distortion is a function of focal length. Therefore, different focal length values can be observed between these two table. The reprojection error with changed distortions coefficients is 0.01 pixel worse than without changed distortions. The result is within tolerance range and leads to an unmeasurable influence in 3D thermography and therefore can be ignored. Hand-eye calibration was continued with geometric calibration parameters, which were generated with default values (with 2nd order of radial and tangential coefficient). Although the radial component of lens distortion is predominant, it is coupled with tangential one. Therefore image distortion with radial distortion without considering tangential part is not enough. In addition, tangential coefficients are almost zero (in our experiment). Therefore further experiment has been conducted with both distortions coefficient to generate 3D thermography results.. In this experiment, it is to analyse, why TCP is shifted from beam centre, as claimed in paper [4].

Table 1: Intrinsic camera matrix with changed distortion coefficients

Focal length	[1962.02836 1962.02836] ± [7.34006 17.34006]
Principal point	[315.60483 222.51200] ± [5.69753 4.96031
Skew	[0.00000] ± [0.00000]
Distortion	[-0.19156 -3.89772 0.000 0.000 0.000]
Pixel Error	[0.12853 0.11233]

Table 2: Intrinsic camera matrix without changing distortion coefficients

Focal length	[1981.79812 1979.75644] ± [16.86725 16.78202]
Principal point	[316.25993 225.41139] ± [5.34492 4.60791]
Skew	[-0.00020] ± [0.00016]
Distortion	[-0.20327 -4.20002 0.00092 0.00009 0.00000]
Pixel Error	[0.11801 0.10161]

In this experiment, it is to analyse, why TCP is shifted from beam centre, as claimed in paper [4]. According to pinhole camera model, for ideal case the 3D position of a spot on the plane is derived via the principle of intersection of a line and a plane in 3D space. Derived from Zhang camera model, it is to assume that the calculated camera TCP lies on principle point. Therefore the calculated hand-eye calibration result (new TCP) has been entered in KUKA software and parallel to that offline robot program was generated with the same results. This program moves the robot to a position where the TCP is perpendicular to the calibration pattern base centre at a distance of 550 mm in z direction. At this position one live image has been saved to analyse image centre relative to TCP position and second amplitude image has been taken to analyse, if TCP position is tilted to pattern base. According to detector size, image has 640 x 512 Pixel, so that image centre is at 340 x 256 pixel, beginning from upper left of image. In ideal experimental result, image centre and pattern base position will be congruent. By analysing figure 4 one can observe that image centre is sifted from pattern base position. But principle point is close to pattern base position. By enlarging the image and comparing with principle point in table 1, it is easily recognisable (in figure 5) that the calculated TCP is at Principle point, which is close to pattern base.

It is to note, that the line has been drawn on live image manually and pattern base was also measured manually with robot. In addition, robot absolute position differs from target position. These are main reasons wherefore the principle point is not exact at pattern base position. This proves that the calculated TCP is not randomly offset from the centre of the image, but always at the principle point. The Amplitude image has been analysed with toolbox implemented in MATLAB tool box by C. Frommel. This toolbox calculate, with the new image, centre of black circles and fits a rectangle. Sides of the rectangle passes through the circle centres and connect to each other. In ideal case, the angle between two connected lines will be 90°. The angle deviations have been presented. The average angel deviations are -0.034° in upper left, -0.11° in upper right, -0.12° in lower leftand -0.13° in lower right. The pointing error for 550 mm working distance and 0.13°average TCP tilt angel, is 1.24 mm. As these deviations are extraordinary small and within the measurement equipment tolerance range, it is difficult to identify absolute tilt angel of TCP.

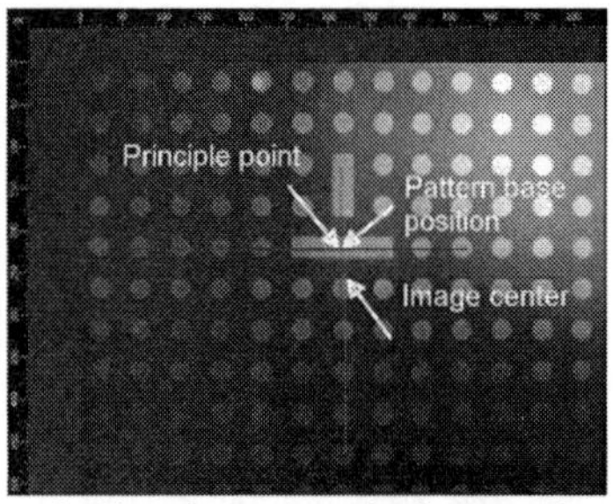

Figure 4: Live thermography image of pattern

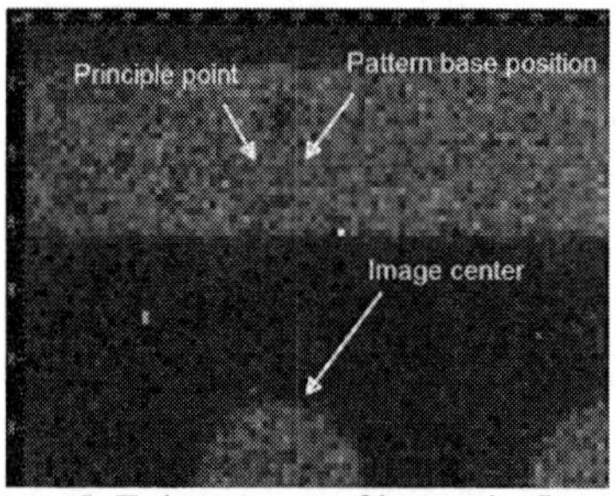

Figure 5: Enlargement of image in figure 4

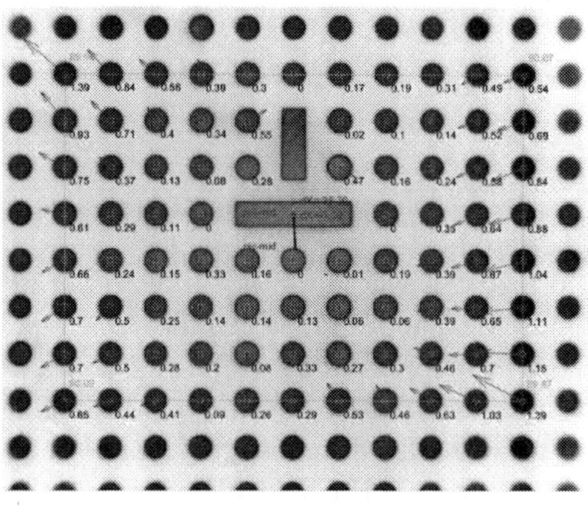

Figure 6: Validation of Hand-Eye calibration

3D Visualisation

2D-3D Point correspondence based texture mapping

In order to establish 2D-3D correspondence relationship, most of the approaches use either geometric features of the object or previously applied marker points on the object. This concept follows the approach of using laser pointer as contactless marker point. The developed automated 3D thermography process has been demonstrated on complex geometry along with process chain. Figure 7 represents the automated 3D thermography visualisation and defect localisation process. The automation process starts with geometric camera calibration and the following steps are shown with red arrows. After measuring the real geometry, the corresponding 3D points were constructed in CATIA. There are nine such points per measurement field. The first three measurement fields are highlighted in colour. The defined nine correspondent points per measurement field are projected onto the component by the laser projector. Those constructed corresponding points are used to generate offline robot program. Thus, the robot moves with the thermography end effector to the respective measurement field and the thermal camera captures the projected laser points. The image recorded by the measurement position was then evaluated in Matlab and displayed together with the nine reference points in image coordinate. Hence, each image receives the reference points for the respective position. To be able to reconstruct in 3D, the projection matrix, which operates in homogenous coordinate system, has to be calculated from 2D-3D data set. Projection matrix has the dimension of 3x4 and contains intrinsic and extrinsic information. This matrix can be calculated by using mathematical camera model of Tsai, Zhang and DLT, which is built on the principles of the pinhole camera. The generated intrinsic parameters $K \in \mathbb{R}^{3\times3}$ and extrinsic parameters $[R \mid t] \in \mathbb{R}^{3\times4}$ represents altogether corresponding mapping in homogeneous coordinate, 3D point $P \in \mathbb{R}^{4}$ to 2D $p \in \mathbb{R}^{3}$ image point, which is defined by $\mathbf{p} = \boldsymbol{K}\,[\mathbf{R}\,|\mathbf{t}]\,\mathbf{P}$. Due to less noisy image projection, matrices are calculated per measurement field by using a mathematical approach (DLT - Discrete Linear Transformation and SVD Single Value Decomposition). The projection matrix contains the relationship between the 2D image and the 3D part because the reference points are related to the part coordinate system. These generated projection matrix per measurement field are then taken to calculate an image position for each model vertex. All images are taken with overlap at their boundary areas. To stich images with each other without overlap by using graphic card, one single texture is generated from all images in a post-processing step.

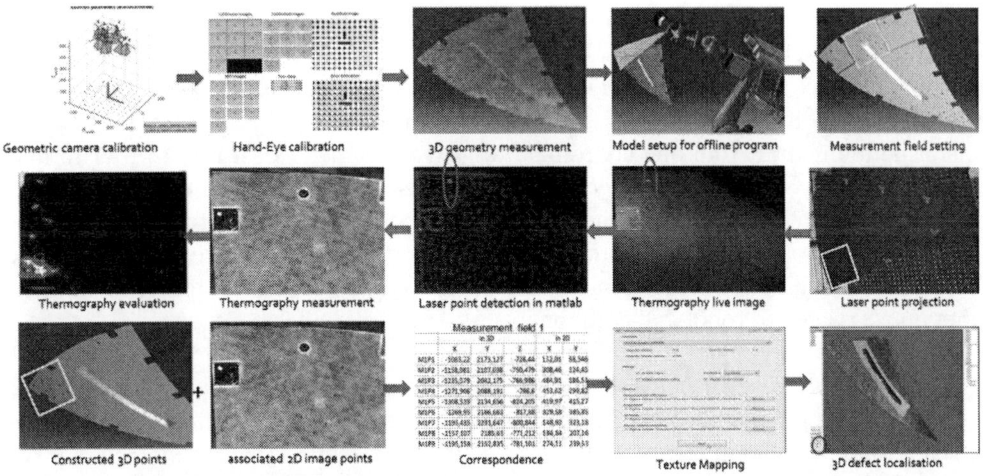

Figure 7: automated 3D thermography visualisation and defect localisation process

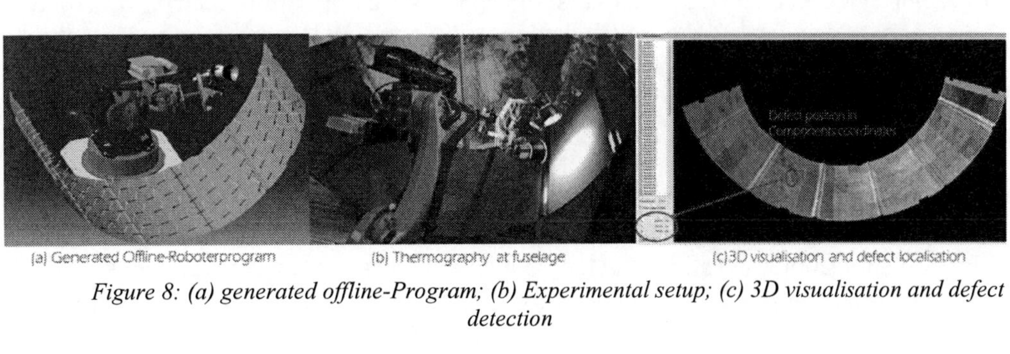

Figure 8: (a) generated offline-Program; (b) Experimental setup; (c) 3D visualisation and defect detection

This texture mapping and post processing method is described in detail in the master thesis of Pascal Keller "A new application for texture mapping using 2D-3D Correspondences". In Addition to 3D visualisation, 3D defect localisation method has been implemented. By clicking on defect location in the software application, 3D coordinates in components coordinate are displayed. This information can be provided to laser projector to show the desired defect location on components surface. A further experiment (figure 8b) has been conducted to demonstrate the complete process chain on large component. For that a fuselage with the dimensions of 4 m in diameter and 2 m width are taken for the experiment. In order to get high 3D mapping accuracy, the real surface of the fuselage has been measured and total number of 1640555 vertices generated. Due to robots reachability, measurement distances for thermography were between 1 and 1.2 m. Pixel resolution for approx. 1.2 m camera distance was 0.6 Pixel / mm. Due to manufacturing and engineering edge, 50 mm from border side has not been measured. In total 105 measurement fields were required to cover the complete surface. Offline Robot program was generated after measuring component's base position (see figure 8a). Due to large surface area two laser projectors were calibrated for each half of the surface area and thermography measurement has been executed with 0.02 Hz frequency. To avoid lateral thermal diffusion during thermography measurement execution, measurement field has been chosen arbitrary. Total measurement time was with 0,02Hz (EP = 1; MP = 3): 200 (s) x 105 = 5.83 hrs, which is compared to conventional water coupled ultrasonic method very less. It took less than 2 minutes to visualise 3D thermography (figure 5c). By a right click on the surface area (figure 8c) 3D coordinates of defect can be localised.

Robot based 3D texture mapping

While 3D texture mapping, using robot kinematic coordinates relationship, camera pose has to be calculated related to component. In previous sections and papers, robot's drawbacks for automated 3D thermography have been analysed. The proposed approach to calculate projection matrix in this paper has been illustrated in figure 2. The arrows represent the 4x4 transformation matrix from one coordinate to next coordinate. Last row of transformations matrix represents scale and perspective coefficients, which can be neglected. Therefore, transformation matrix provides at the end 3x4 projections matrix. In order to implement the algorithm (based on kinematic coordinate relationship) in the new software, 3x4 projection matrix is required. This projection matrix contains the extrinsic parameters in terms of camera pose related to robot and component.
The attached thermography end-effector at flange is rigid joint. Therefore transformation matrix FT_C (see figure 2) between flange and camera plane needs only to be calculated once. These calibrations have been performed in section 2. The RT_B transformations matrix between robot and component needs to calculate whenever the component's positions change. This WT_R transformations matrix can be read from KUKA machine.dat file. Furthermore, KUKA software delivers the entire transformation matrix RT_F between robroot and flange for all the new positions. RT_B and WT_B transformation matrix are generated through measuring the component's position by using 4 point method, provided by KUKA. The BT_C transformation matrix can be calculated by multiplying the other matrices. $BT_C = RT_B \cdot RT_F\, FT_C$. This has to be done for every measuring field. To verify the system, a new experimental setup has been prepared (see figure 9). The component was 1000 mm long and 500 mm wide. With a measurement distance of 500 mm 30 measurement fields were required to cover the complete surface. After doing the hand-eye calibration and geometric calibration the component's base was measured by using 4 point method, provided by KUKA. Offline robot program has been generated using measured TCP and base position. Robot program was generated in base coordinate system. After that,

the component was measured with thermography camera by using 0.4 Hz, 0.1 Hz and 0.05 Hz parameters. 30 images per frequency, intrinsic parameters, dat file of robot program, which contains all measured robot positions, and rigid transformation matrices are provided to software to calculate the projection matrix. The results have been illustrated in figure 11.

Figure 9: experimental setup for kinematic relationship based algorithm

3D Thermography in CATIA

Since examiner has to evaluate every 2D thermography image one by one, a possible solution has been developed to evaluate in 3D. Thus, for the measurement 0.4 Hz, 0.1 Hz and 0.05 Hz frequency has been chosen, the penetration depth can be calculated using $\mu = \sqrt{(\alpha/\pi * f \text{ Hz})}$ equation. The penetration depths for the chosen frequencies (with 0.004 cm^2/s thermal conductivity) are 0.56mm, 1.2 mm and 1.6 mm. For the three thermal depth three offset surfaces with appropriate parameters (0.56mm, 1.2 mm and 1.6 mm) have been generated (see figure 10f) and exported as stl file, so that the thermography results can be visualised in accordance to their depth. Thermography images are then mapped with their belonging offset surfaces in the software and exported as VRML file format. This file contains all information of image colours associated with 3D vertices with their corresponding orientation and positions. At the next step the VRML file together with a merged image will be imported again in CATIA. This allows to illustrate all thermography images together corresponding to their depth. Figure 11a shows the plies and implanted kapton foil in different depth. Figure 10b, 10c and 10d illustrates the 3D thermography results in three different depths. In these figures, it can be recognised, that with increasing thermal depth, more kapton foil and real manufacturing defects are getting visible. Figure 10e represents the 3D thermo-tomography model. This 3D thermo-tomography model can now easily be analysed by clicking the CATIA's show option to visualise the results of desired thermal depth. In this way other manufacturing tolerances and coherent influences (for example ultrasonic welding during preform process) can be evaluated. Furthermore, structural analysis can be compared with thermography results as both 3D thermos-tomography and structural analysis are using the same software CATIA. This thermo-tomography model represents a simplified version of real computer tomography. Evaluation of 3D damage propagation is now easier. After mapping all the 90 thermography images, the 3D location and size of those kapton foils have been exported from CATIA to laser projector to project directly on components surface. This allows to avoid marking the damage area and robot or employee can take required step. For an example, only desired damages, which need to be repaired, can be projected. In figure 11a few of damages are marked with yellow line. As results, only these damages (dimension and location) are exported and projected (see figure 11c).

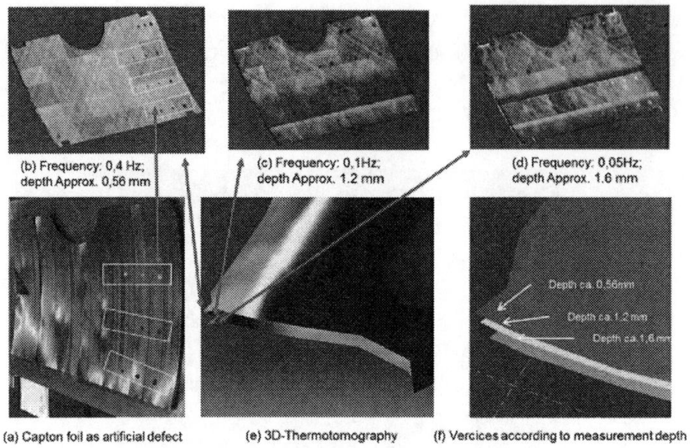

Figure 10 (a) Preforming process; (b) 3D thermography results in 0.56 mm depth; (c) 3D thermography results in 1.2 mm depth; (c) 3D thermography results in 1.6 mm depth; (d) 3D-Thermo-tomography model; (e) vertices with different offset

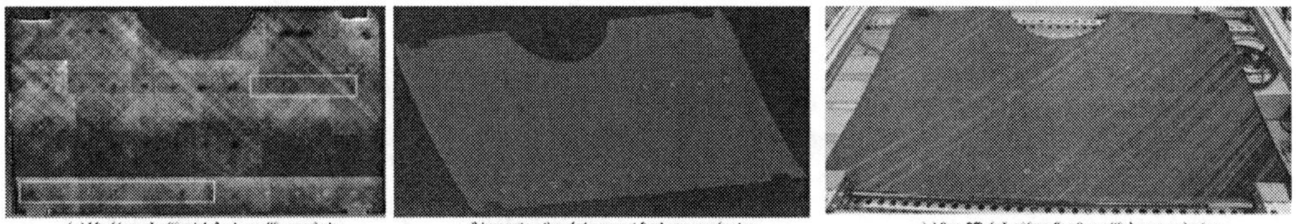

Figure 11: (a) Critical defects; (b) exporting defect location and dimensions from catia; (c) projection of defects with their dimensions

Texture mapping accuracy

After developing of two methods another experiment has been conducted to determine the accuracy. Results have been illustrated in figure 12. Deviations are measured in 3D between reference point and corresponding textured point. For this experiment PnP method has shown better results than robot based method. Details of this experiment results will be published in coming conference.

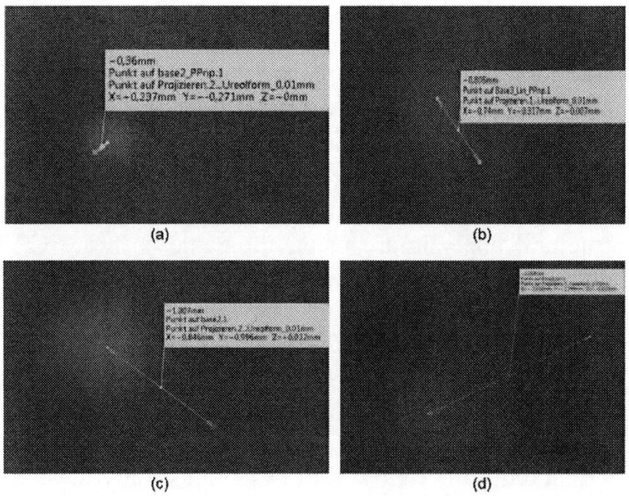

Figure 12: (a) and (b) Min. and Max. Error at reference point with PnP method; (c) and (d) Min. and Max. Error at reference point with Robot based method

Conclusions

In this series of experiments, thermographic images were 3D visualized using the 2D-3D point correspondence method and robot based kinematic relationship chain method. At first, to minimize the drawbacks of robot based automated inline thermography, concept of 2D-3D corresponding method is described. Secondly feasibility of this concept is experimentally verified. In this experiment, it could be shown that the projected contour size and the camera distance to the component are dependent to each other, while detecting the laser point. For a reliable laser point detection smaller circles with 0.1 mm diameter and the largest possible camera distance is recommended. The correlation value in this test series was between 70% and 80%. Afterwards, DLT method has been applied to calculate projection matrix out of 2D-3D corresponding data set. Using two experimental setups, one on large single curved component (fuselage) and second on a double curved component, said method has been validated. The results confirm the feasibility of the new concept. An automated thermography process chain and visualisation of the results in 3D were presented. The individual process steps of the automated inline thermography were reported as well. After that, a system accuracy analysis overview was carried out and the possible source of error is shown. This 2D-3D corresponding method can also be applied without any manipulator like robot. Only laser projector and component should be calibrated to each other. This simplification allows this system to applicable in laboratory. Furthermore, robot based 3D visualisation method has been implemented. Results have been validated through experiment. Besides that a new approach for 3D evaluation by means of 3D-thermo-tomography has been showed. This new approach helps examiner to analyse coherent influences of manufacturing process over the surface and in their depth. Another future goal is to implement alternative non-linear 2D-3D corresponding method, like PnP, POSIT etc.

REFERENCES

[1] Volker Carl, Konstantin Eisler; „3D-Formerfassung im Infraroten und Sichtbaren sowie Techniken zum Visualisieren"; Thermografie-Kolloquium 2011

[2] Thomas Schmidt, Somen Dutta; "Automation in production integrated NDT using thermography". NDT in Aerospace 2012.

[3] Somen Dutta, Thomas Schmidt; "New concept for higher Robot position accuracy during thermography measurement to be implemented with the existing prototype automated thermography end-effector utilising an industrial robot and laser system". QIRT 2014.

[4] Thomas Schmidt, Christoph Frommel; "Geometric Calibration for thermography cameras", 7th International Symposium on NDT in Aerospace, 2015

[5] Somen Dutta, Prof. Dr.-Ing. Klaus Drechsler, Prof. Dr.-Ing. Michael Kupke, Dr. Alfons Schuster, Jens-Peter Tuppatsch; Automated single view 3D Texture Mapping and Defect Localisation of Thermography Measurements on large Components utilising an industrial robot and a laser system; QIRT 2018 Berlin

[6] Thomas Schmidt, Somen Dutta "CFRP manufacturing process chain observation by means of automated thermography," 5th International Symposium on NDT in Aerospace, Singapore, 2013.

[7] Thomas Schmidt, "Hand-eye-calibration of a robotized thermography end-effector", 6th International Symposium on NDT in Aerospace, 12-14th November 2014, Madrid, Spain

Autonomous Systems Imaging of Aerospace Structures

Shakeb DEANE [1*], Nicolas AVDELIDIS [1,2], Clemente IBARRA-CASTANEDO [2], Xavier MALDAGUE[2], Antonios TSOURDOS [1], Alex A. WILLIAMSON [3], Hamed YAZDANI-NEZHAD [1], Hai ZHANG [2]

[1] School of Aerospace, Transport and Manufacturing, Cranfield University, United Kingdom.

[2] Universite Laval, Computer Vision and Systems Laboratory, Department of Electrical and Computer Engineering, Quebec City, G1V 0A6, Canada.

[3] Mapair Thermography Ltd, Melbourn, South Cambridgeshire, United Kingdom.

Contact e-mail: shakeb-deane@hotmail.co.uk

Abstract. Manufacturers are constantly looking for more cost-efficient means to produce aircraft components. An effective way to do this is to reduce the weight, which results in less fuel required to power the aircraft. This has led to an increased use of composite materials. Carbon fibre reinforced polymer (CFRP) composite is used in industries where high strength and rigidity are required in relation to weight. e.g. in aviation – transport. The fibre-reinforced matrix systems are extremely strong (i.e. have excellent mechanical properties and high resistance to corrosion). However, because of the nature of the CFRP, it does not dint or bend, as aluminium would do when damaged, it makes it difficult to locate structural damage, especially subsurface. Non-Destructive Testing (NDT) is a wide group of analysis techniques used to evaluate the properties of a material, component or system without causing damage to the operator or material. Active Thermography is one of the NDT risk-free methods used successfully in the evaluation of composite materials. This approach has the ability to provide both qualitative and quantitative information about hidden defects or features in a composite structure. Aircraft must undergo routine maintenance – inspection to asses for any critical damage and thus to ensure its safety. This work aims to address the challenge of NDT automated inspection and improve the defects' detection by performing automated aerial inspection using an Unmanned Aerial Vehicle (UAV) thermographic imaging system. The concept of active thermography is discussed and presented in the inspection of aircraft's CFRP panels along with the mission planning for aerial inspection using the UAV for real time inspection. Results indicate that this inspection approach could significantly reduce the inspection time, cost, and workload, whilst potentially increasing the probability of detection.

Introduction

Technology is currently thriving in industry, therefore companies are competing with one another and are all looking for innovate ways to meet the demands of the global economy [3]. In this global economy, companies must turn out the highest quality of products in an efficient manner with minimal manpower and remain profitable [3]. For Aircraft to remain the safest way to travel the structure and parts need to be manufactured to an impeccable standard and then maintained regularly. Non-destructive testing (NDT) is a range of different inspection methods which can be used to evaluate the health of materials and components without interfering with the serviceability [1]. Infrared thermography (IRT) is a popular NDT technique that has risen due to the increased use in varied applications and the rapidly decreasing price of thermal technology in today's market. Infrared Thermography is predominately being used to detect subsurface defects in a plethora of structural elements and mechanical systems [1]. It is one of the newest non-destructive technologies and has proven to provide substantial improvements in performance as well as good cost savings, allowing more frequent maintenance with minimal manpower [16].

Airbus and Boeing are considered the most elite commercial aircraft manufacturers of the past few decades. The demand in the market is putting pressure on companies to develop new solutions for larger aircraft [12]. Since the successful release of the A380 and Boeing 787 Dreamliner, commercial Aircraft manufacturing has seen a rapid increase in the use of composites materials. In 1985, around 5 percent of the A310-300 consisted of composite materials, 18 years later and more than half (53%) of the materials used to manufacture the A350XWB consists of composites [2]. Composites are being used more and more for primary structures in many industries such as; rail, wind power and marine. Composites offer greater strength characteristics compared to the common structural metallic materials [9]. The materials of an aircraft structure have immense responsibilities, there are required to resist many intense loads repeatedly and be free from damage, delamination, fatigue and corrosion after each cycle. IRT is suitable for use on a wide range of materials, including glass and carbon fibre composites, natural fibres, ceramics and metallic materials. Due to the increased popularity in composite materials there is a demand for reliable and fast NDT methods to ensure the safety of the material.

The idea of inspecting aircraft structures came from an ongoing project at Cranfield University in collaboration with Airbus, where a commercially available UAV is performing a non-destructive liquid penetrant test. A liquid paint substance is added to a metallic wing panel and a UAV equipped with a UV light inspected the aircraft in a dark room, the primary objective is to execute the algorithms in order to display, classify and log defects [6]. Since there is such a rise in the use of composite materials it seems necessary to discover a fast and cost-efficient inspection method. IRT is a fast-non-contact method and is well-suited for testing large areas of complex geometry [7]. Nevertheless, the traditional thermographic systems are bulky and inconvenient for use on many applications. Therefore, it is necessary to develop a more versatile thermographic system, with the help of UAV technology.

1. Composites

Composite materials have significant beneficial properties for IRT inspection, including low reflectivity (high emissivity), meaning that they absorb energy very well. Once the energy is absorbed, the material will retain the energy and release it much slower than traditional metallic materials, such as aluminium, therefore using infrared NDT is very attractive for inspecting composite materials. Composites have some limitations, perhaps the most noteworthy limitation is the poor response to impact loading [17], which can be caused easily by runway debris, a dropped tool or any other impact. It's been reported that 80% of in-service damages to composites is caused by impact strikes [18]. Composite materials are known to have poor impact resistant, the material responds differently than metals in terms of impact loading and its ability to dissipate the incident kinetic energy of the projectile [17]. For minor incident energies, metals absorb energy through elastic and plastic deformation, this will usually result in permanent structural deformation however, metals can carry the load and it results in consequences that are less-significant compared to composites. Plastic/Elastic deformation for a composite is very limited as the resultant energy is frequently absorbed in creating large areas of fracture, which will affect the performance of the material due to the reduction in both strength and stiffness [11]. During composite manufacturing, it is possible for the material to be damaged due to the complexity of the manufacturing process.

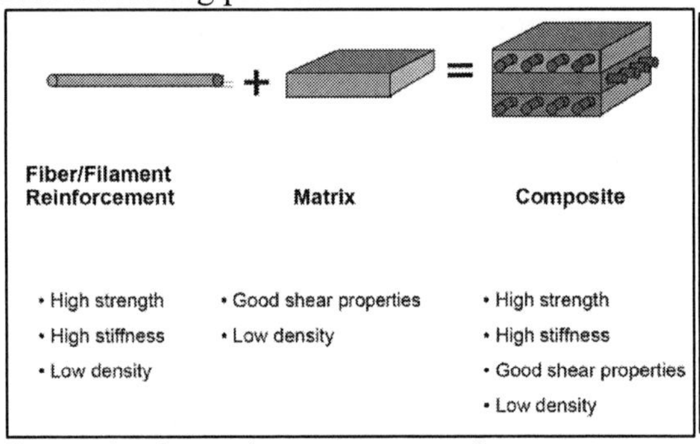

Figure 1: shows the composition of composite material [9].

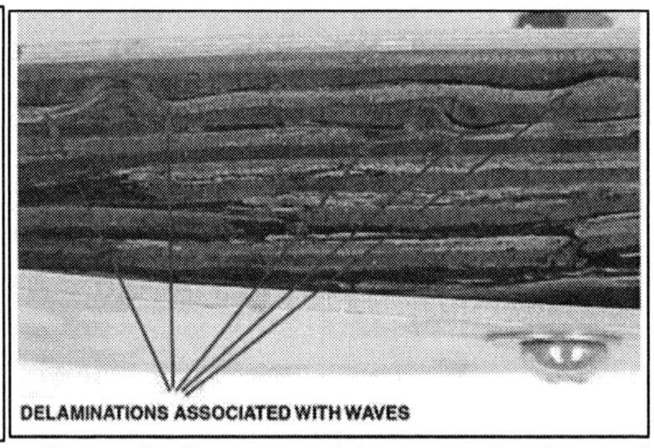

Figure 2. Composite delamination's, this is the consequence of misalignments in large aircraft structures [13].

1.1 Demand for NDT inspection on composites

Due to the increased demand of CFRP, it has subsequently caused rapid large-scale manufacturing. Airbus estimates its Carbon Fiber demand to soar to around 20,000 tons by the year 2020 (see figure 3, right) [2].

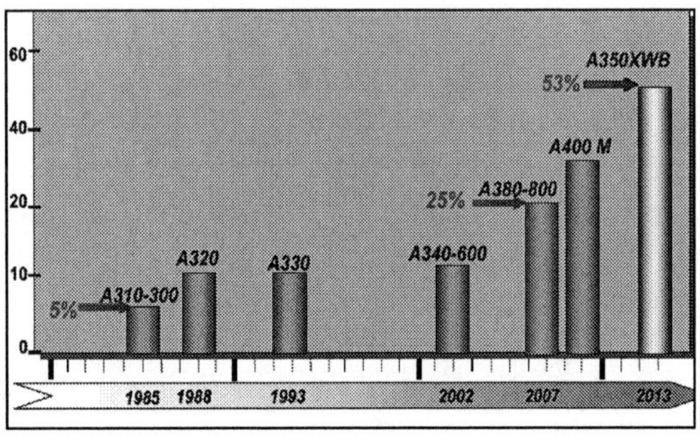

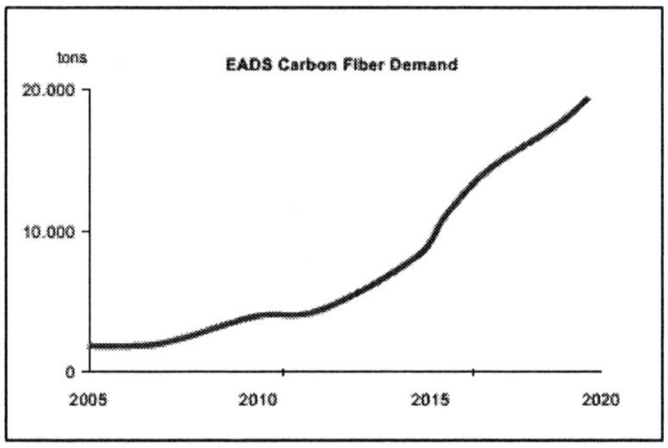

Figure 3: The graph on the left shows the use of composites in Airbus aircraft since 1985. The line graph represents the demand by tons from 2005 up to 2020 [2].

The current manufacturing and machining process are novel compared to the manufacturing of traditional metallic structures, in addition, composites have different specific mechanical properties. As a consequence, its proven difficult to control manufacturing defects, due to the complexity of the manufacturing. Composites are layered pieces of laminar with alternating orientation typically 45 or 90°. Composite impact damage can result in damage in many different forms such as matrix cracking, fibre/matric debonding, surface microbuckling, delamination and fibre breakage [11]. Such impacts can cause barely visible impact damage (BVID), this is where the damage will still exist subsurface and extend way beyond the impacted area. This can leave aircraft vulnerable by jeopardising the safety of the structure. It is clear that the damage zone of a composite is of a more complex nature and can be very difficult to characterise. The complexity of the problem multiplies due to the lack of impact damage testing techniques on a composite material [1]. This calls for an adequate way to search for defects within a composite material.

2. Methodology

The specimens were made of Carbon Fibre Reinforced polymer (CFRP) (composite laminate). Sample A4 was manufactured via manual woven and the other two samples where unidirectional. The composites consist of laying up pre-preg piles to form a laminate stack, the material is then autoclave processed at 180C and 7 bars for a few hours suitable for the thermoset resin cure, this is according to the supplier's specification (Hexcel). There were three specimens, one that was undamaged, and two that had been impacted with a force of 15J and 20J of energy.

2.1 Experiment

The infrared camera was a FLIR Phoenix, with inSb sensor material, 3-5 mm, 640x512 pixels and allows data acquisition at 50Hz.

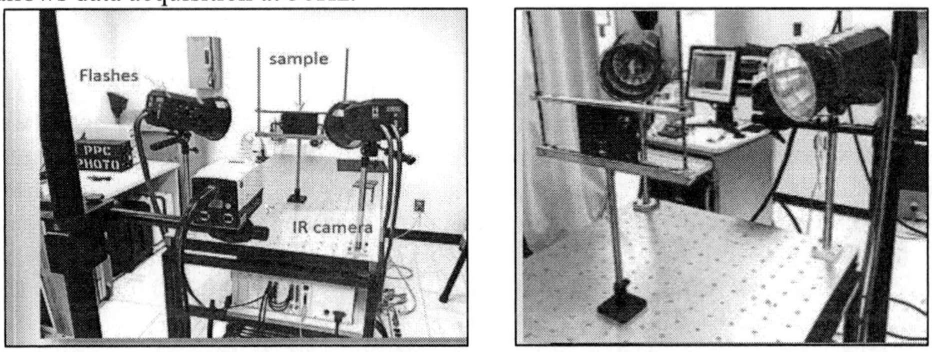

Figure 4: Images showing the NDT testing equipment and the experimental set up.

These camera needs to be cooled to 77 Kelvin (i.e. -196.15 °c) during testing. Cooled cameras work by collecting photons of infrared energy that pass through the optic. The photons are converted into electrons that are stored in an integration capacitor. After a certain period, called integration time, the charge is read out to a digital count calibrated to temperature, assigned a colour or greyscale value and then presented as a viewable infrared image [5].
The surface of the specimens where positioned parallel to the camera lens. The data was acquired for 40 seconds with a 1.5 millisecond integration time, that includes 10 frames before the flashes, plus 1990 frames during cooling a total of 2000 frames where recorded. The software used to acquire the data was RDac from FLIR. For signal processing MATLAB and Ir_view from Visiooimage inc were employed. Two advanced processing techniques where used; PCT (principal component thermography) and PPT (pulsed phase thermography). The specimen was excited by two Balcar Xenon flash lamps, which provide 6.400 J per flash, the pulse duration was 2 ms at FWHM (full width half maximum).
The specimens were heated using the flash lamps, and then the infrared camera has captured its thermal response. There are two main types of thermography, active and passive as depicted in Figure 5. Passive thermography is an inspection without the use of an external

excitation source, its commonly used for inspecting materials having a temperature naturally different than its surrounding [10]. Active thermography uses external heating dissipation to create thermal contrast, once excited the thermal waves propagate through the heated specimen, and when they reach inhomogeneities caused by a medium with different thermal properties, the diffusivity coefficient changes. These inhomogeneities could be defects, such as cracks, surface cracks, air/water/material inclusions, delaminations and debonding damages. Mathematical processing techniques are the key to gather the necessary data from the raw thermogram and help visualise the damages within the material [10].

[4] 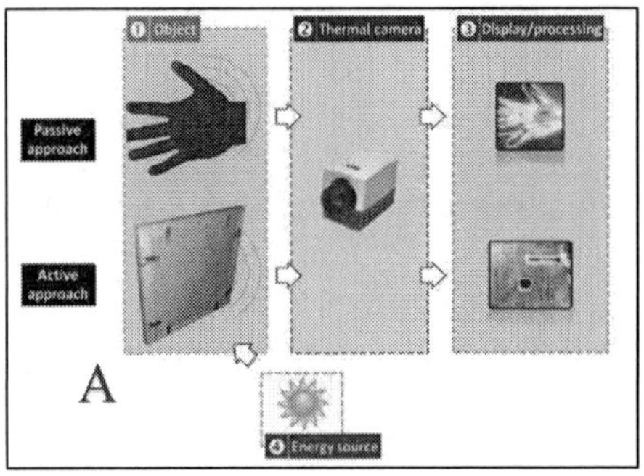

[15] 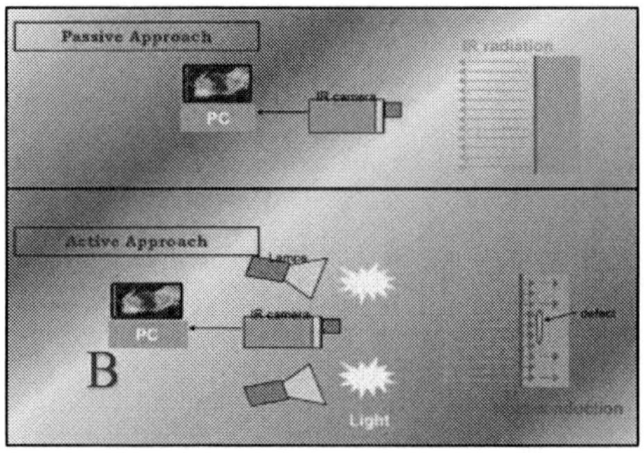

Figure 5: Show the experimental set up for both passive and active thermography.

3. Results

The figures below (Figures 6-12) are all the postprocessed results. A reflection test means that the camera and the excitation source are on the same side. Generally, reflection is more suited to detect defects closer to heated surface. PPT (pulsed phase thermography) was one of the methods used, with the acquired temperature data it varies with time and is transferred to phase data which varies with frequency, the frequencies which show adequate data are below each image in hertz. PCT (principal component thermography) was the other, method used, the data is re-ordered by variance, this method finds both time series and spatial patterns, most of the variability is contained in the first component: EOF1 (empirical orthogonal function). Usually, EOF1 shows thermal variations related to surface heating. Later EOFs are typically related to surface and subsurface damage. Comparing both techniques, for this specific data set in figure 6 it seems that the PCT technique displays the damage clearer.

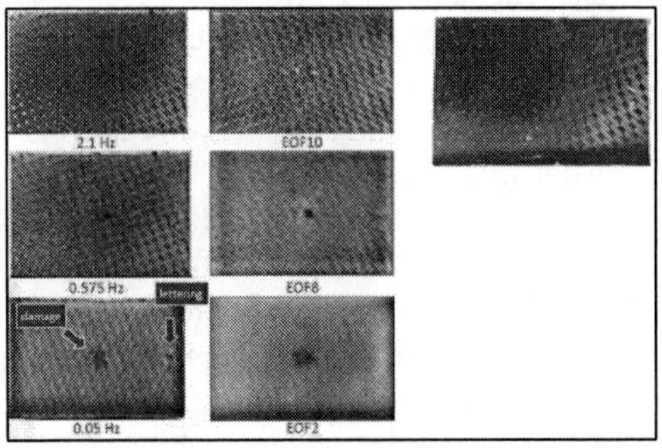

Figure 6: Sample A4 (15J woven), front face (impacted side), PPT and PCT. *The damage is barely visible in the original image of the composite (Far right image), the white marks are lettering from labelling the specimen. The 0.05Hz PPT image, shows clear subsurface damage.*

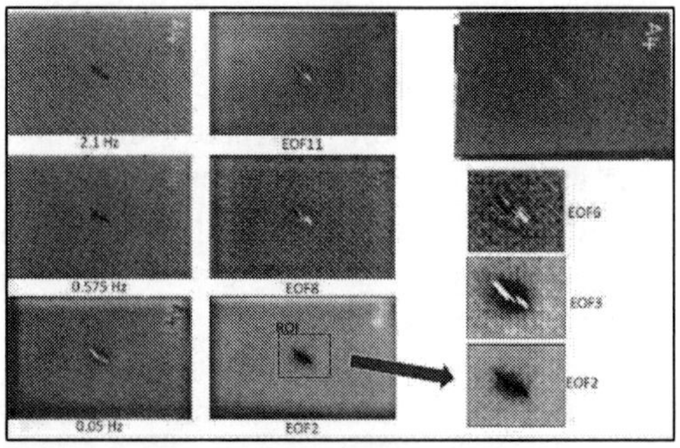

Figure 7: Sample A4 (15J), Reflection rear face PPT and PCT. *The rear surface is cracked, this is interesting because this side is significantly more damaged than the impacted side. This is a prime example of barely visible impact damage.*

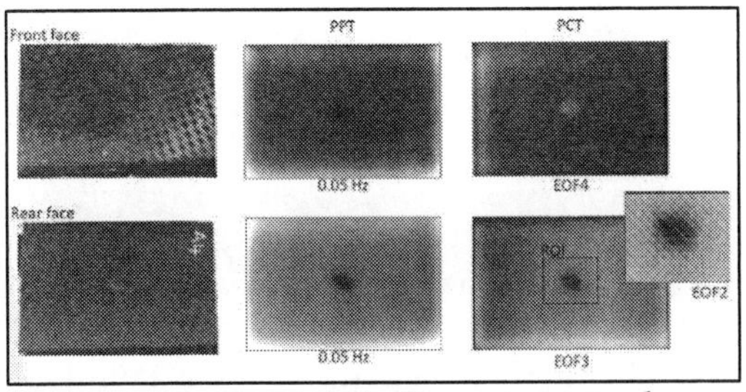

Figure 8: Sample A4 (15J). *Transmission is where the one side of the composite is excited, and the other side is observed by the camera and captures the data.*

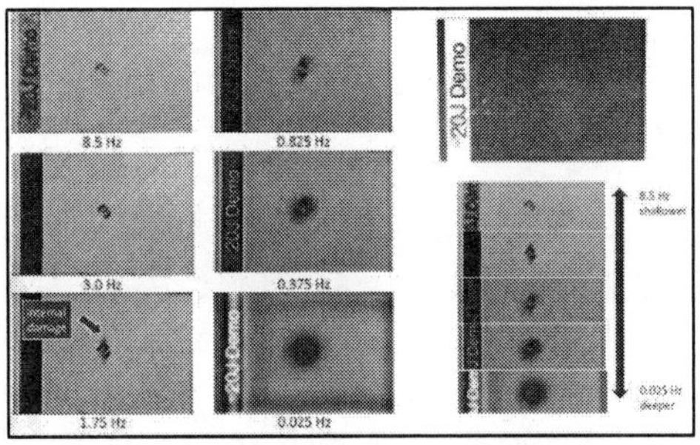

Figure 9: Sample 20J Demo, Reflection front face, PPT. *The damage is visible on the front face, how, it's clear that the damage is much worse subsurface and spreads further than the impacted area.*

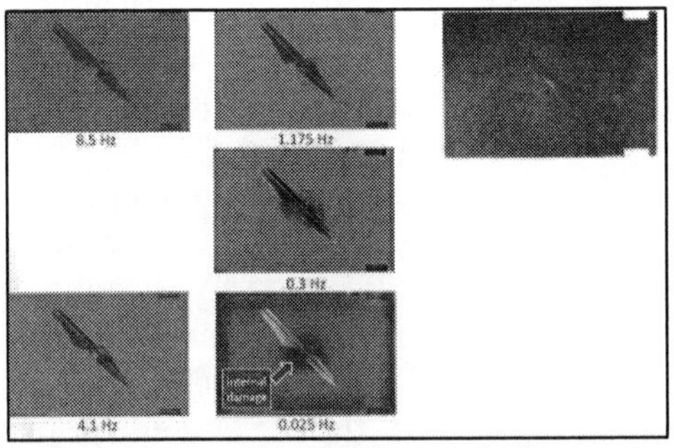

Figure 10: Sample 20J Demo, Reflection rear face*. PPT. The rear face has some substantial damage. The PPT locates the surrounding internal damage.*

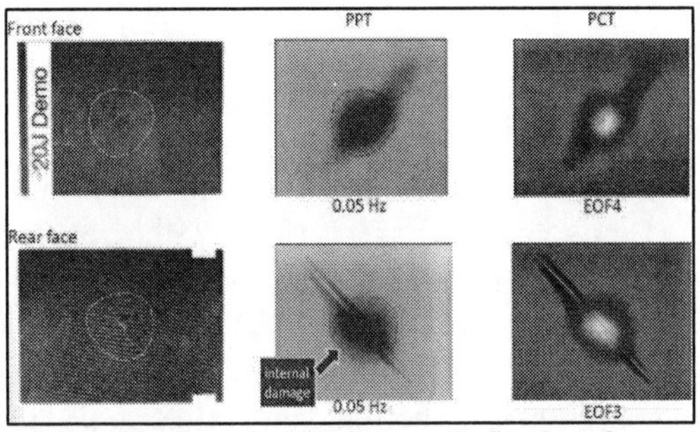

Figure 11: Sample 20J Demo, Reflection front face, Transmission. *The test has adequately located the internal damage.*

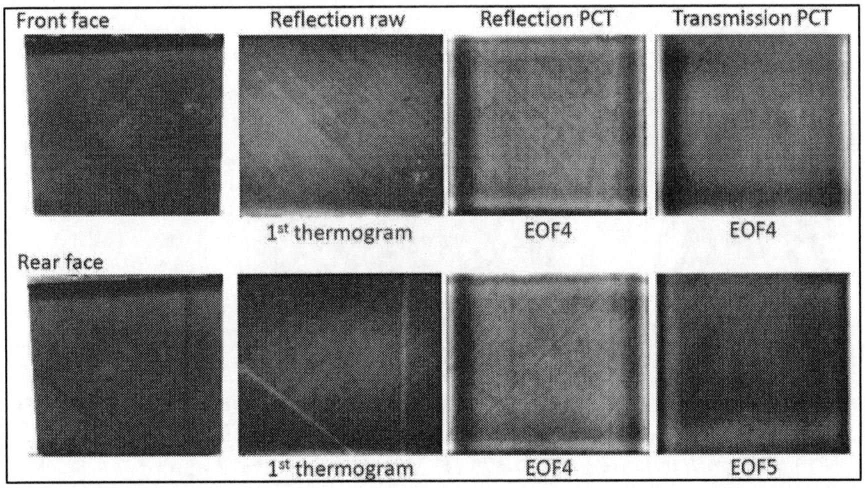

Figure 12: Sample No Name, Reflection front, back, transmission PCT. *Undamaged composite sample, which was manufactured the exact same way as the 20J sample, the different NDT test prove that this sample is damage-free.*

4. Unmanned Aerial Vehicles

4.1 Proposed UAV

Although a cooling camera was used in the beforementioned NDT test, this type of camera will not be feasible for UAV thermographic inspection. Cooling cameras usually require cryogenic cooling, which is an excess of weight. The equipment we currently have is a DJI M210, which carries an RGB and thermal camera. The thermal camera is 640 x 512 pixels which is the exact same pixels as the FLIR Phoenix that was used. It has a 13mm lens and is radiometric, therefore it uses a set of techniques to measure electromagnetic radiation, including visible light, this means that the acquired data can be extensively analysed after the survey in order to visualise and understand the data such as the temperature variations [8].

UAV dynamic inspection has the potential to provide a complete portrait of the aircraft structure much faster than static inspections. For a UAV inspection an external excitation source will be employed to heat the surface. The thickness of the composite panel may make it difficult to gather adequate results, therefore many excitation sources need to be tested other than flash lamps. For instance, halogen lamps (heating the surface for longer), Heat Blankets or a Laser. Along with the liquid penetrant NDT method, a UAV thermographic inspection was tested for civil engineering applications (see Figure 14). The UAV is equipped with a FLIR A320, this IR camera is connected to a computer via a cable, one is a power cable and the other provides real time radiometric thermal images which will reveal heat patterns and thermal anomalies [7]. Transient thermography with flash and step heating where the methods used in this inspection. An inspection took place and gathered data on difference distances between the camera and the samples (2, and 6 metres). Along with the UAV equipped with the IR camera, there was a fixed IR camera, therefore two sets of data could be acquired and compared. A thermally uniform area of a sample is delimited, and the IR cameras captured the data [7].

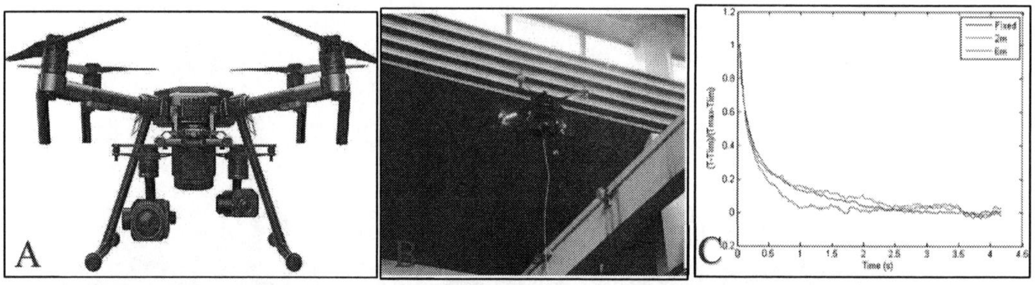

FIGURE 13 A: DJI M210 equipped with thermal camera [8]. FIGURE 13 B: UAV equipped with thermal camera [7]. FIGURE 13 C: Comparison of front face temperature responses to flash heating in the case of a large uniform irradiation with a fixed camera and a stabilized camera with UAV at two distances (2 m and 6 m) [7].

4.2 Performing an NDT inspection

The purpose of this paper is to prove that active thermography can locate defects within composite materials, then to propose the integration of this methodology into a UAV. There are several things that need to be taken into consideration. The idea is to inspect composite panels for aircraft structures straight from the manufacturing line, to verify their worthiness and quality assurance. There are some limitations for this proposed idea. Due to the fact the inspection is inside it could provide problems for the UAVs GPS system, which will result in an unstable flight and affect the data as seen. There are a few ways to combat this issue, one of them is to set up a localisation scheme by distributing some ultrasonic sensors around a specific geographic area, the UAV communicates with these sensors to understand its location therefore which will result in stabilisation. Once the flight is ready, due to the size of aircraft parts, the subject will need to be segregated into separate regions of interest, and separated images will be taken and then stitched together as a mosaic image, this segregation will ensure the subject is excited uniformly. Once the data is acquired it will then need to be postprocessed to improve results, when post processing a lot of properties need to be considered;

- Thermal Properties: conductivity, diffusivity, effusivity, specific heat
- Spectral Properties: emissivity, absorption, reflection, transmission
- Material's properties-characteristics: density, porosity, thickness, geometry [10].

With these properties understood, data process can be performed, for instance; when a material presents voids or pores in its structure, then its thermal conductivity and density decreases, its thermal diffusivity is altered and so the conduction of heat transfer within the material is affected. For instance, materials with low-effusivity values will present higher temperatures. With all this information taken into consideration, a better damage assessment can be done [10].

5. Conclusion

This study contributes to reducing the time and cost of NDT composite inspections. The research shows that NDT is an effective inspection method, whilst using PPT and PCT allows to visualise the extent of the damage and the surrounding areas. When acquiring data, the value of each pixel will represent the temperature at a specific point. The temperature profile of the point corresponding to the healthy area is different from that of the point corresponding to the defective area. Using digital image processing it is possible to enhance the images to reveal the location and geometry of the defects [1]. These methods can potentially become beneficial to the aerospace industry if integrated onto a UAV. The raw data that is acquired often contains noise caused by the environment, the reflections, the emissivity variations of the specimen or if the heating was non-uniform, it is important to take this into consideration and to employ specific algorithms for image processing in order to produce accurate results. The results from the UAV inspection indicate an excess of noise from the UAV, since this flight was done without the use of GPS, this affected the results and the noise was due to the stabilisation problem. It is necessary to perform pixel by pixel treatment in addition to the standard average signal study [7]. To do this the IR camera needs to be as static as possible, GPS or other localisation techniques can be used to compensate for any stability problems. The data acquired has shown that different techniques are better suited for different defects. Pulsed thermography can accurately locate delamination's. Delamination is where the material fails due to the layers separating which result in significant loss of mechanical toughness. In the future, a UAV inspection will be performed, and the fidelity and performance of the image processing algorithms will be further evaluated. An excitation source that is small, light and produces enough energy to penetrate the material, can also be equipped on the UAV soon, which will offer an all in one inspection product and minimise time as there will be no need to set up an external excitation source. There are a few halogen lamps on the market which offer sufficient weight to energy ratio.

This research was part supported by the British Engineering and Physics Sciences Research Council.

References

[1] A P Chrysafi, N Athanasopoulos, N J Siakavellas. 2017. "Damage detection on composite materials with active thermography and digital image processing." International journal of thermal sciences 242-253.

[2] Airbus. 2018. Composites in Airbus. Airbus, 10 September.

[3]2014. What is Thermography (thermal imaging). Directed by Bin Industrial Training. Produced by Training, Bin Industrial. Performed by Bin Industrial Training.

[4] Clemente Ibarra-Castanedo, Jose Ricardo Tarpani, Xavier P V Maldague. 2013. "Non destructive testing with thermography ." European Journal of Physics s91-s109.

[5] 2016. What's the difference between cooled & uncooled thermal detectors? Directed by FLIR Systems. Produced by FLIR. Performed by FLIR.

[6] Kostantinos Malandrakis, Al Savvaris, Jose Angel Gonzalez Domingo, Nico Avdelidis, Panagiotis Tsilivis, Florence Plumacker, Luca Zanotti Fragonara, Antonios Tsourdos. 2018. "Inspection of Aircraft Wing Panels Using Unmanned Aerial Vehicles." 2018 IEEE MetroAeroSpace. Rome.

[7] L E Mavromatidis, J -L Dauvergne, R Saleri, J-C Batsale. N.D. "First experiments for the diagnosis and thermophysical sampling using impulse IR." L. E. Mavromatidis, J.-L. Dauvergne, R. Saleri, J.-C. Batsale.

[8] Mapair. 2018. Frequently Asked Questions. Accessed September 09, 2018. https://www.mapair.co.uk/faqs.

[9] Mrazova, Maria. 2013. "Advanced composite materials of the future in aerospace industry." Incas Bulletin 5 (3): 139-150.

[10] N P Avdelidis, D Almond, A Dobbinson, B Hawtinb, C, Ibarra-Castanedo, X Maldagued. 2004. "Aircraft composites assessment by means of transient thermal NDT." Progress in Aerospace Sciences 40 (3): 143-162.

[11] N Razali, M T H Sultan, F Mustapha, N Yidris, M R Ishak. 2014. "Impact Damage on Composite Structures – A Review." The International Journal Of Engineering And Science 08-20.

[12] Pora, Jerome. n.d. Composite materials in the Airbus A380 - from history to future. Blagnac Cedex: Airbus.

[13] S, Cairns Douglas. 2009. "Composite Materials for Aircraft Structures." Montana State University. Accessed September 09, 2018. www.montana.edu/dcairns/documents/composites/MSUComposites2009.pdf.

[14] Seung Yeob Nam, Gyanendra Prasad Joshi. 2017. "Unmanned aerial vehicle localization using distributed sensors." International Journal of Distributed Sensor Network 13 (9).

[15] Theodarakeas, P. 2018. Thermal NDT & E of Composites. Athens, 12 09.

[16] Ullmann, Thomas, Thomas Schmidt, Severin Hofmann, and Raouf Jemmali. 2010. "In-line Quality Assurance for the Manufacturing of Carbon Fiber Reinforced Aircraft Structures." 2nd International Symposium on NDT in Aerospace.

[17] W J Cantwell, J Morton. 1991. "The impact resistance of composite materials — a review." Composites 347-362.

[18] Yifan Zhao, Jörn Mehnen, Wei Xu, Mosab Alrashed, Stephen Abineri, Rajkumar Roy. 2015. "Degradation Assessment of Industrial Composites Using Thermography." Procedia CIRP 147-152.

Accubot - a precision Robot NDT Kinematics for several modalities

Wolfgang HAASE [1], Thomas GRAMBERGER [1]
[1] Fill Gesellschaft m.b.H., Gurten, Austria
Contact e-mail: wolfgang.haase@fill.co.at

Abstract

In Aerospace NDT applications especially for composite the demands for flexible NDT Solutions are increasing.

Linear kinematics was the standard approach for UT inspection systems for Through Transmission or Pulse Echo inspection due to the achievable accuracy of positioning required for the correct inspection paths. Especially for Through Transmission (TTU) using Squirter a high positional accuracy was required to be compliant to the specifications of Aerospace companies.

Robot based systems are nowadays an interesting alternative as possible accuracy enhancements allow to have the same or even better path accuracy as linear kinematics. The challenge is that the repeatability of robotics solutions is excellent while the dynamic positional accuracy along a path is not perfect or only achievable with reduced inspection speed and low data acquisition rates based on the position feedback. Additionally standard robots need to be protected against water splashes especially in Squirter Through Transmission technique.

This presentation shows the realized system based on standard Stäubli TX200 L robots with industrial standard Numerical Control SIEMENS Sinumerik 840 D and additional encoders to achieve high precision on complex scan paths still keeping a high inspection speed.

The robot kinematics is fully integrated in the Sinumerik control chain and based on that, flexible for any additional requirements as active tools, tool changers or multi-modality applications. Challenges to inspect narrow cavities with the robot hand are overcome by active extenders fully integrated in the system kinematics.

The inspection system is calibrated in the entire inspection volume in different arm configurations during the installation by a laser system. More than 1000 positions are taken into account to perform a mathematical correction higher order (not only the standard DH parameters) and introduced in the NC control system. During the inspection run the compensation parameters are online integrated in the path calculations.

By applying the compensation absolute positional TCP accuracies of 0.2 mm with additional encoders on all robot axes are achieved.

ACCUBOT

- Motivation to develop a precision robot
- Cooperation STÄUBLI, SIEMENS & FILL
- Designed for NDT and CFC processing
- No limitation in flexibility
- One system for multiple applications

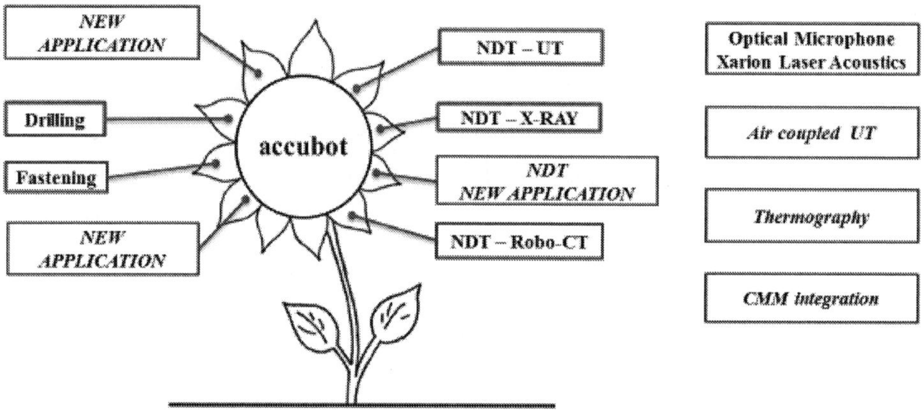

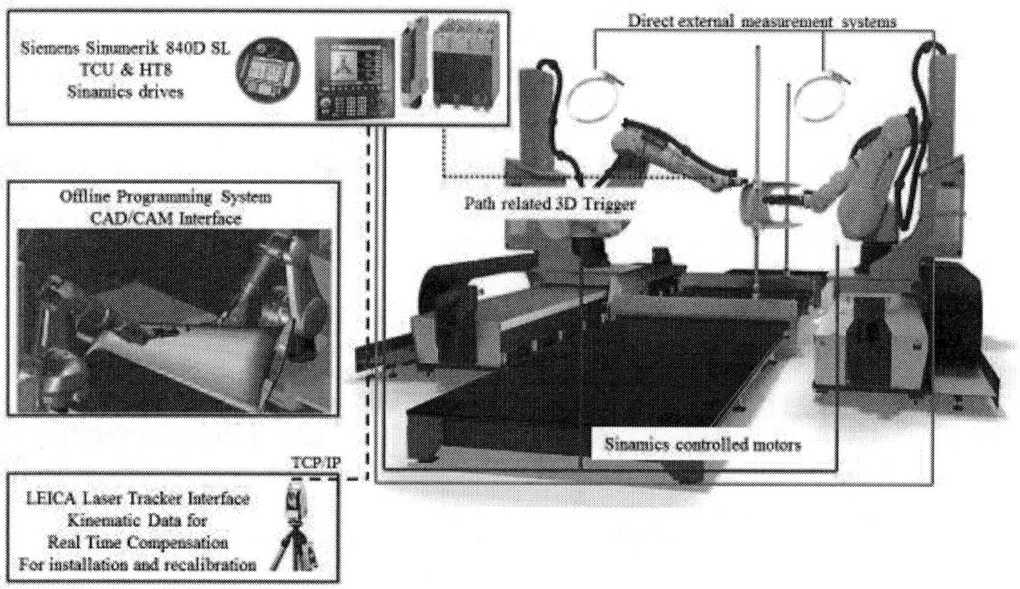

Accubot functional diagram of the system

Accubot Dual Robot system during setup at FILL

Accubot **X-Ray and Robo-CT** (courtesy EZRT 2017 DGZfP)

Accubot **UT with Phased Array**

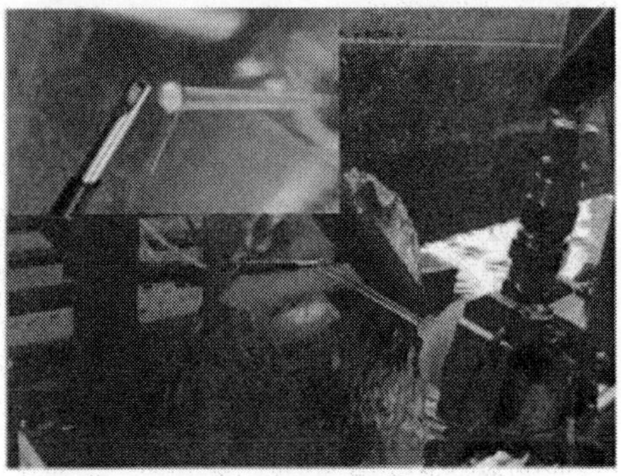

Accubot new application **optical microphone**

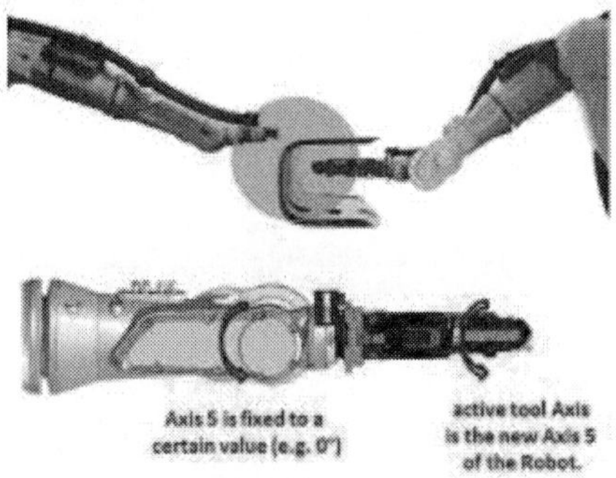

Accubot active tool for **UT TTU** in cavities

Microstructure and residual stresses in AM metallic parts: Do we know what we do not know?

Tobias THIEDE, Tatiana MISHUROVA, Naresh NADAMMAL, Arne KROMM, Johannes BODE, Sandra CABEZA, Giovanni BRUNO
BAM Bundesanstalt für Materialforschung und -prüfung, Berlin, Germany

Contact e-mail: giovanni.bruno@bam.de

Abstract

The freeform and the revolutionary design possibilities offered by additive manufacturing have skyrocketed the amount of optimization studies in the realm of engineering, and metallic additive manufactured parts are becoming a reality in industry. Not surprisingly, this has not been paralleled by a similar enthusiastic wave in the realm of materials science, and still very little is known about AM materials properties. This has the consequence that, typically, classic materials properties are still used in design and even in simulations.

In this talk, I will give a few examples of how necessary it is to dig a lot deeper than at present, in order to understand these new materials classes, and in particular their microstructure and their internal stresses, largely different from their cast or wrought companions.

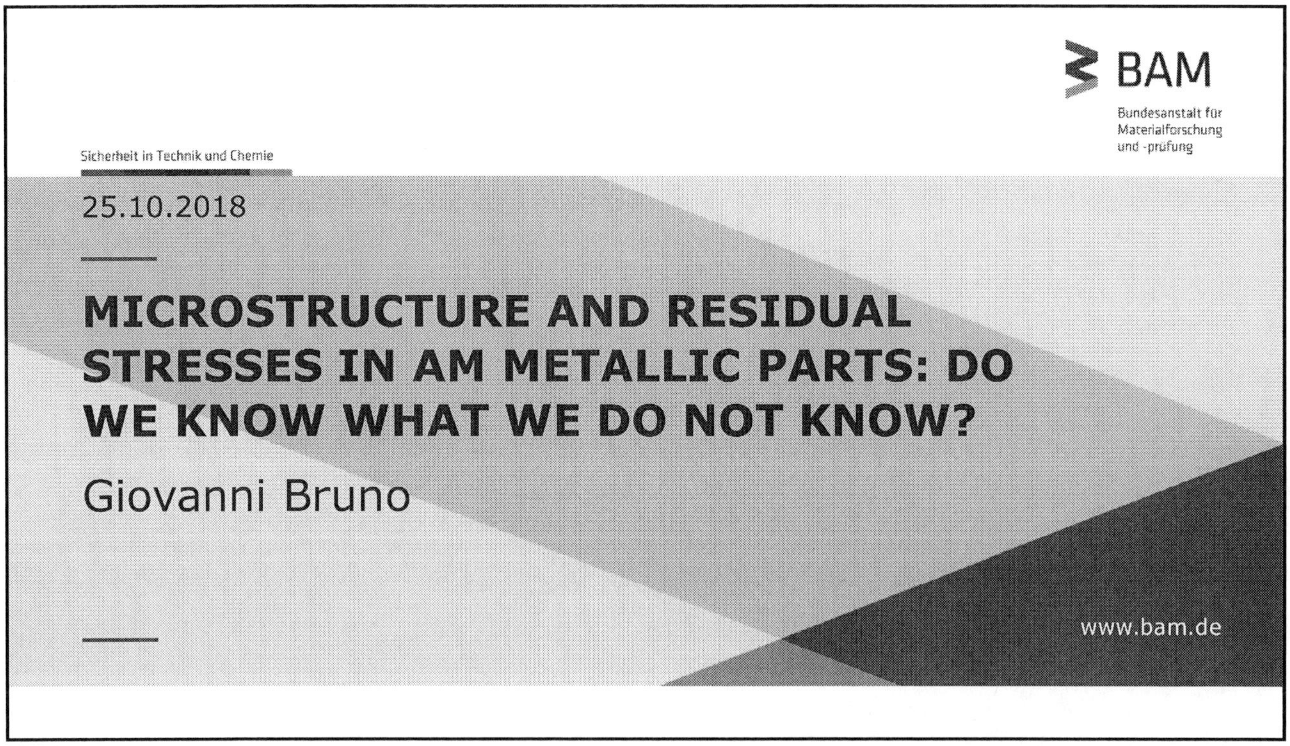
BAM
Bundesanstalt für Materialforschung und -prüfung
Sicherheit in Technik und Chemie
25.10.2018
MICROSTRUCTURE AND RESIDUAL STRESSES IN AM METALLIC PARTS: DO WE KNOW WHAT WE DO NOT KNOW?
Giovanni Bruno
www.bam.de

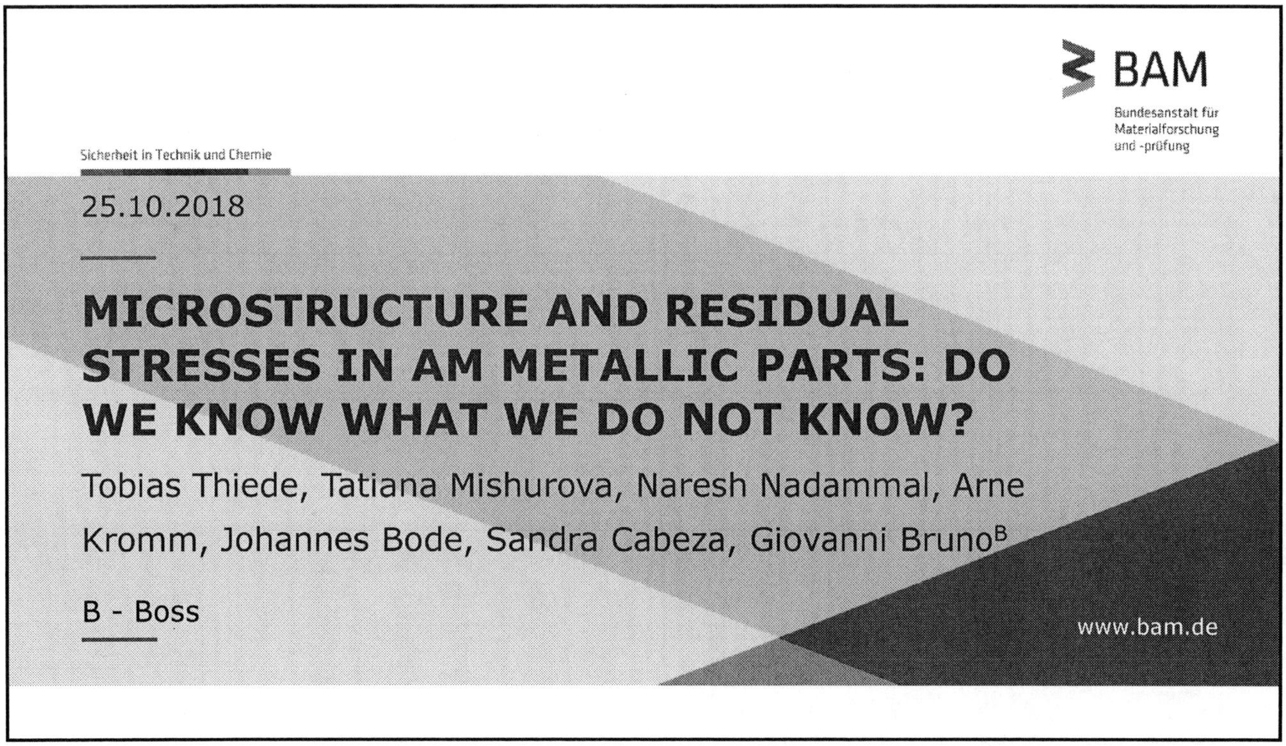
BAM
Bundesanstalt für Materialforschung und -prüfung
Sicherheit in Technik und Chemie
25.10.2018
MICROSTRUCTURE AND RESIDUAL STRESSES IN AM METALLIC PARTS: DO WE KNOW WHAT WE DO NOT KNOW?
Tobias Thiede, Tatiana Mishurova, Naresh Nadammal, Arne Kromm, Johannes Bode, Sandra Cabeza, Giovanni Bruno[B]
B - Boss
www.bam.de

Simplifying the Universe

BAM

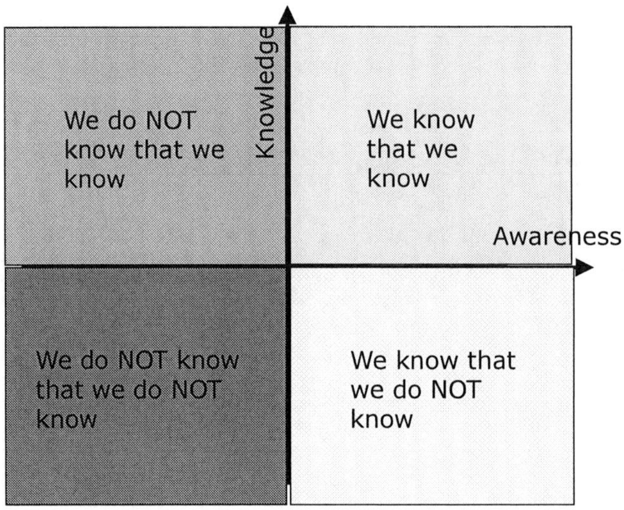

Avoid living dangerously

BAM

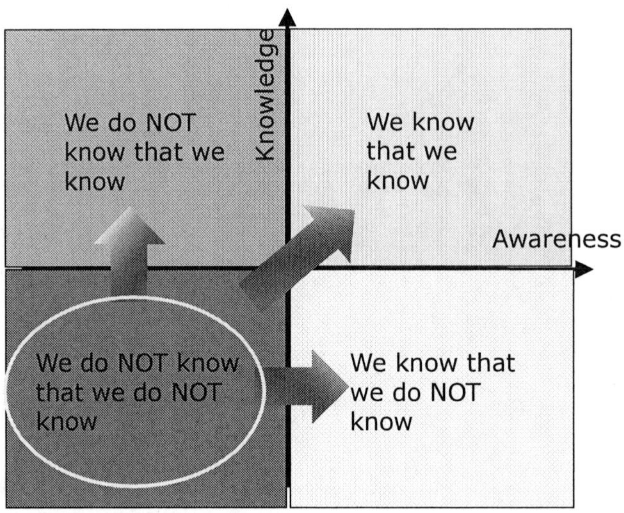

Example

BAM

In Additive Manufacturing everybody is talking about

- Free Form
- Unconventional Design
- Re-thinking Components
- **Think out of the box**

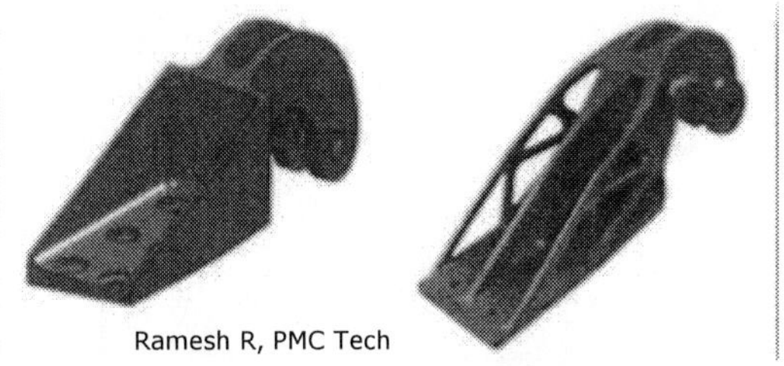

Ramesh R, PMC Tech

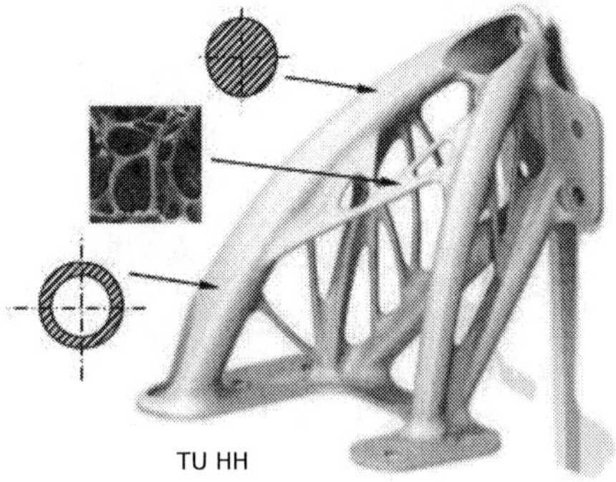

TU HH

25.10.2018 NDT in Aerospace- Dresden 5

FEM Simulations

BAM

However, when it comes to FEM Simulations

If we ask the question:
Which material properties are we using?
The answer is:
Literature values, for Conventional Materials...

If we ask the question:
What about residual stress?
The answer is:
We know they are there,
We heat treat...with conventional HT

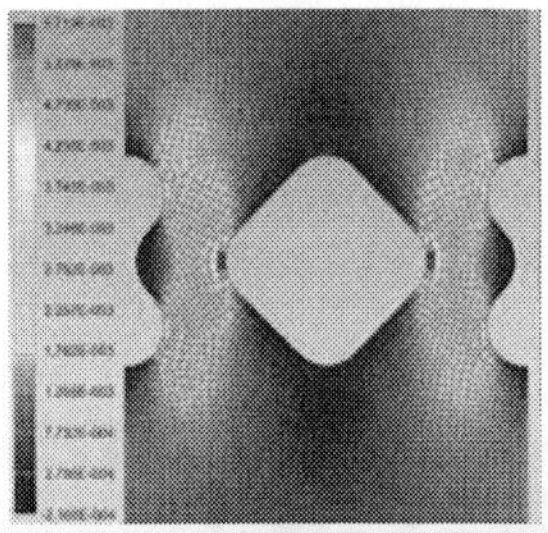

Examples from KU Leuven (polymer)

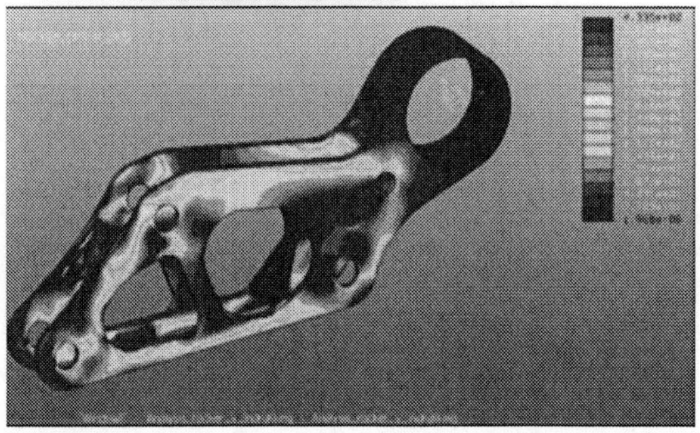

25.10.2018 NDT in Aerospace- Dresden 6

Where do you think we are?

BAM

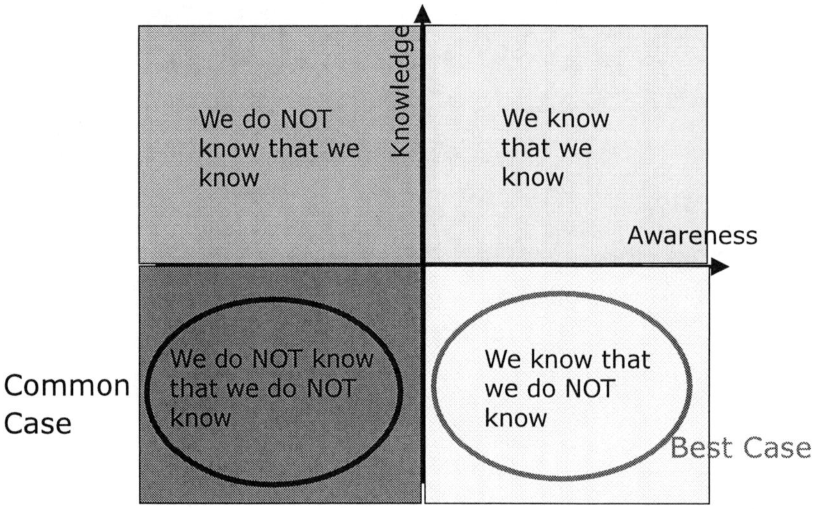

What we should do (our mission at BAM)

BAM

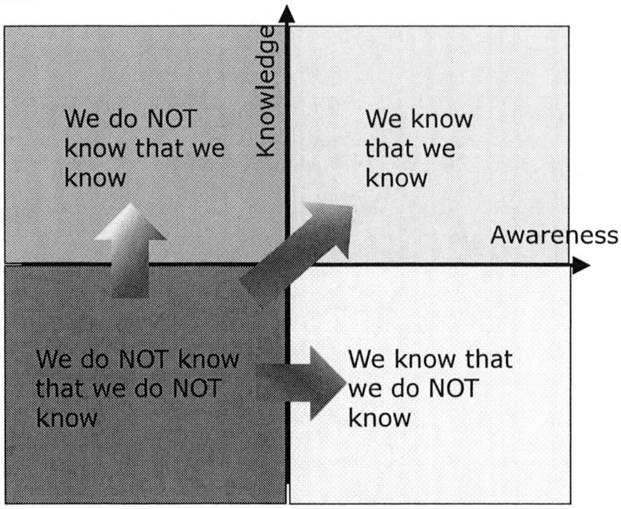

Case Study: SLM IN718 parts
Lots of modeling, little data

BAM

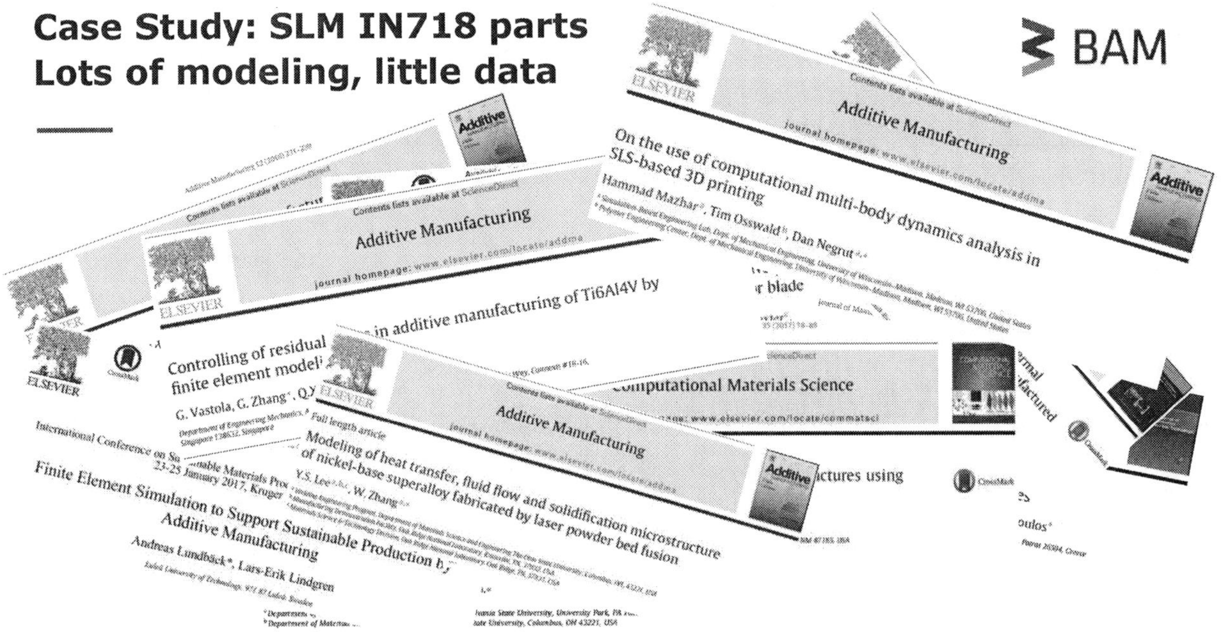

Macro and Microstructure

BAM

Optical and Electron Microscopy

Electron Backscattering Diffraction (EBSD)

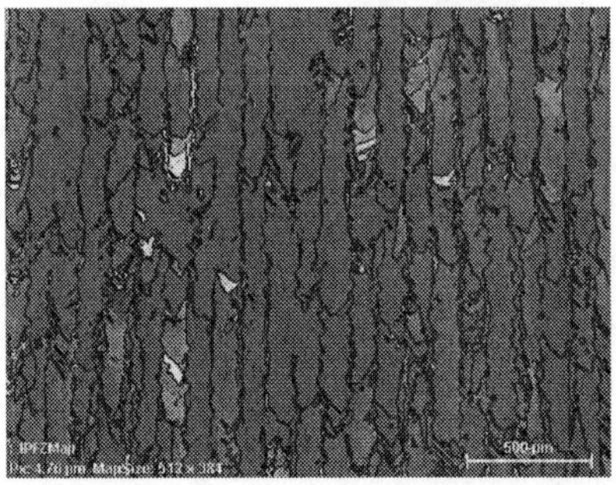

Coordinate Measuring Machine (CMM)

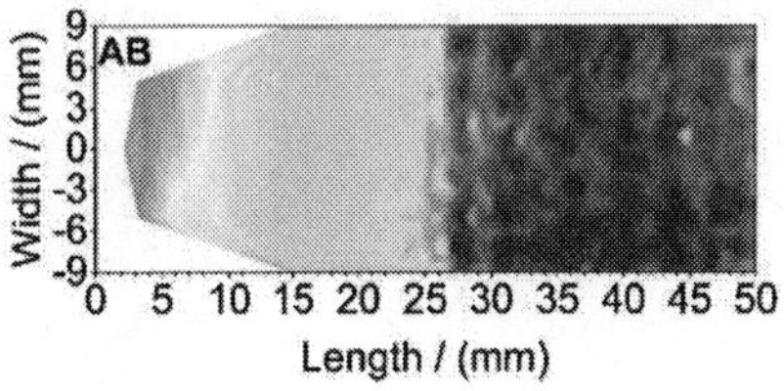

Advanced Characterization Methods

BAM

Neutron Diffraction (ND)

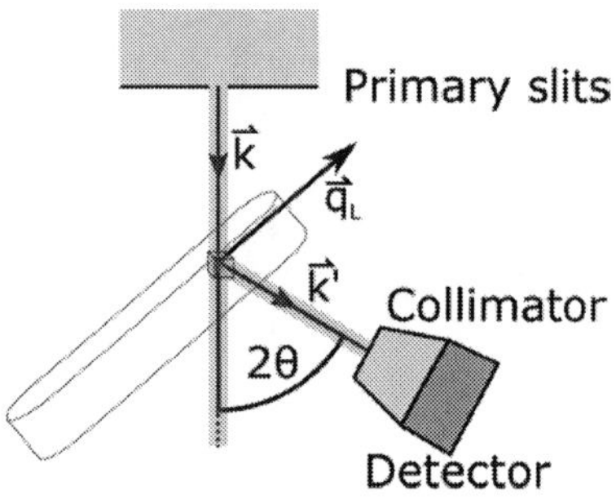

Bragg's law → d (Lattice parameter) → d_0 Reference → ε (Strain) → Hooke's law → σ (Stress)

Stress Analysis by Diffraction

BAM

- Non-destructive Method
- The lattice is our gauge length

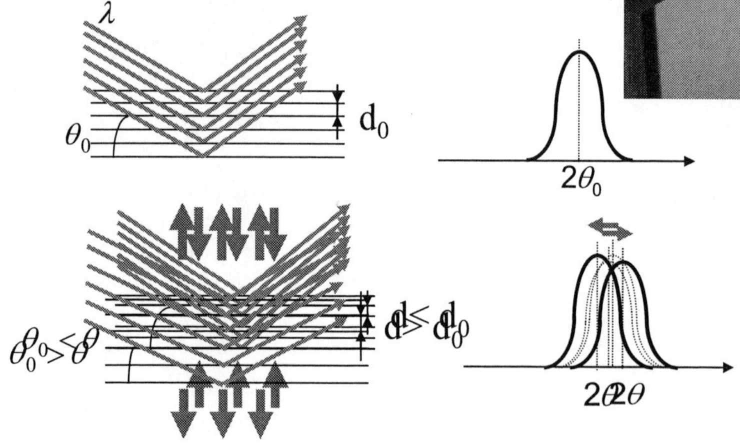

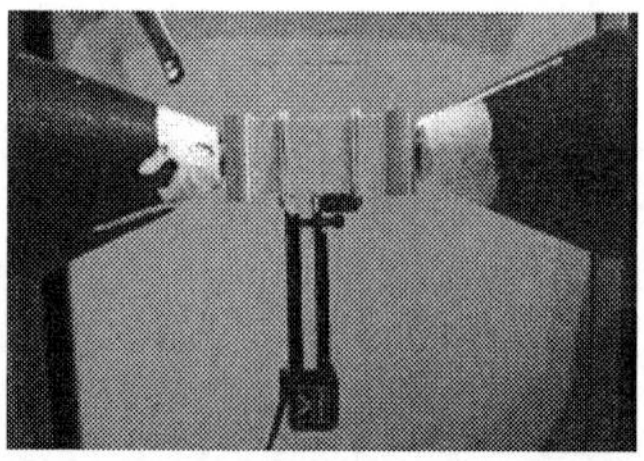

$$\varepsilon = \frac{d - d_0}{d_0}$$

Strain Scanning by diffraction

BAM

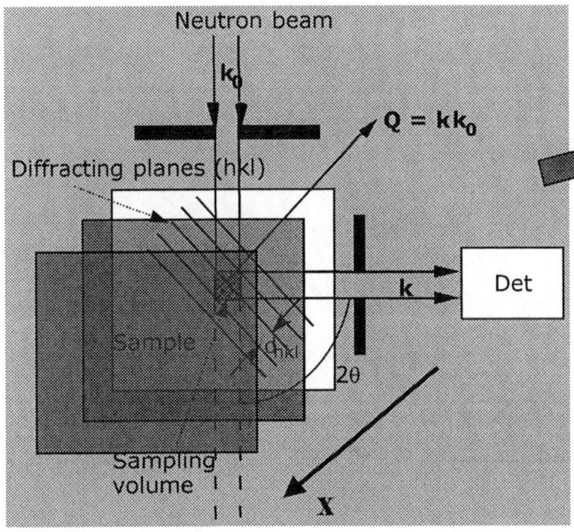

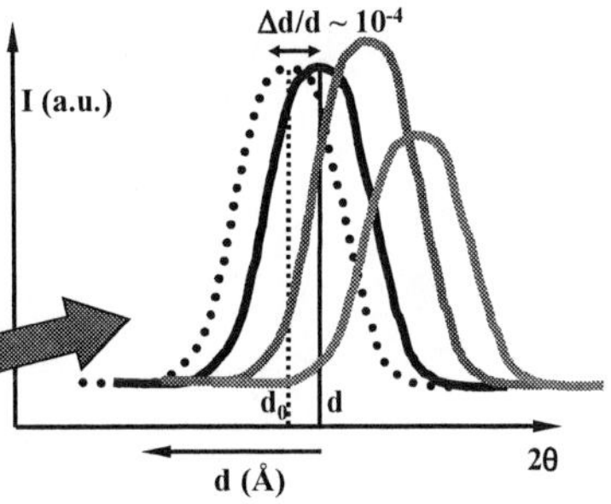

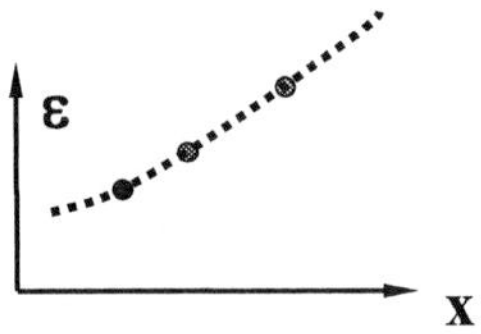

3D Hooke´s Law: Strain → Stress (Phase Specific)

$$\sigma = C\,\varepsilon$$

Samples – SLM IN718

SIEMENS BAM

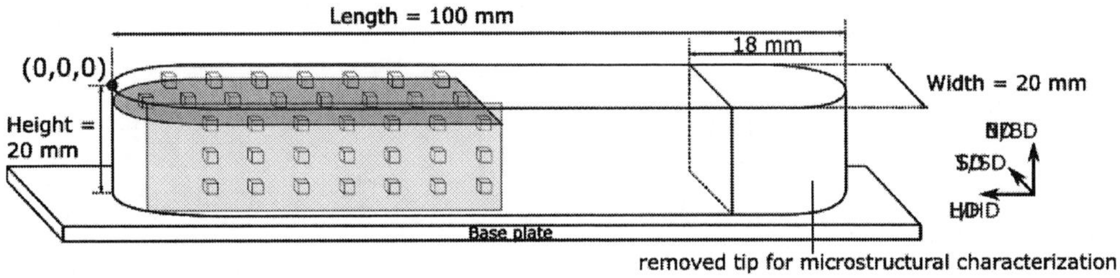

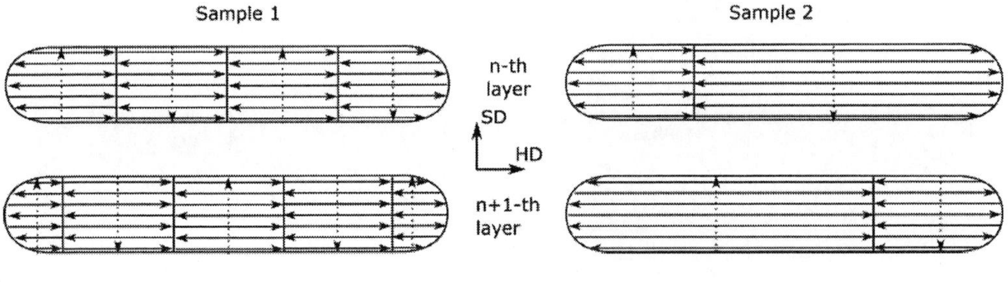

Microstructure

BAM
SIEMENS

Influence of hatching

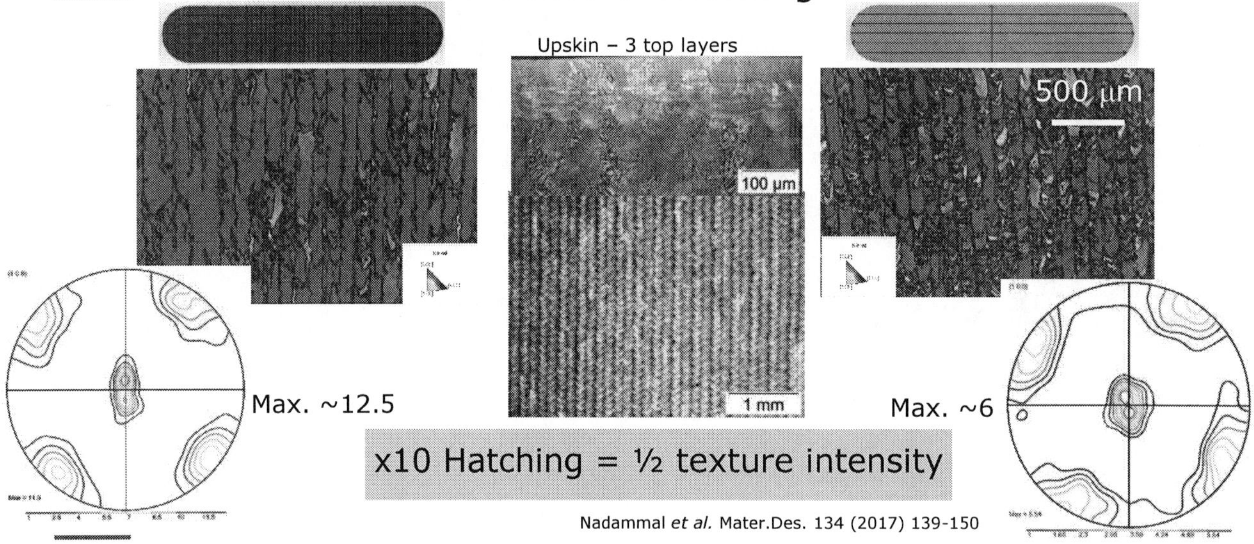

x10 Hatching = ½ texture intensity

Nadammal *et al.* Mater.Des. 134 (2017) 139-150

Residual Stress

BAM

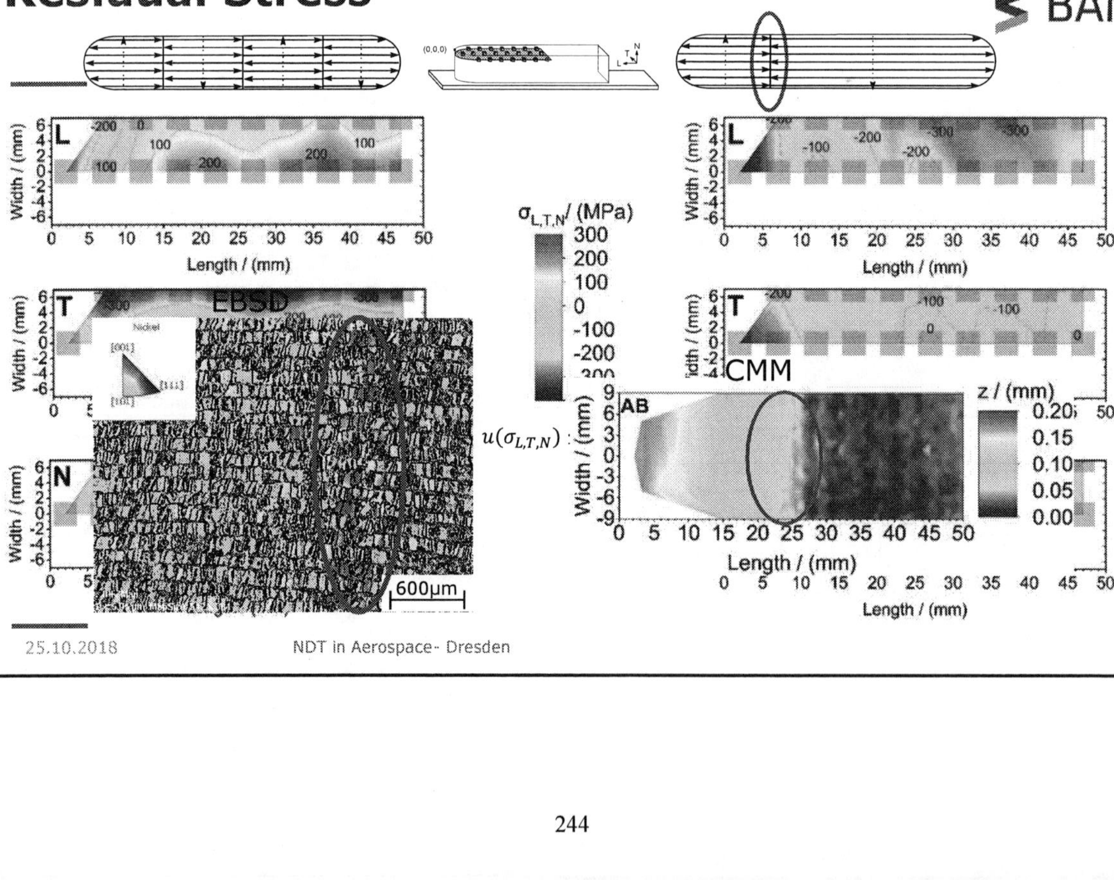

Influence of Hatch length

BAM

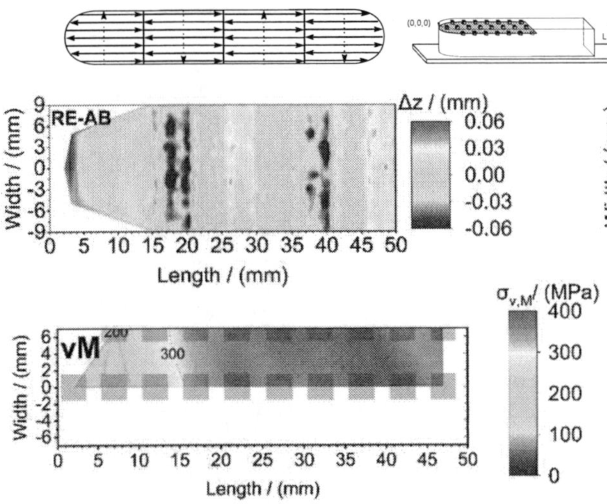

25.10.2018 NDT in Aerospace- Dresden Thiede *et al.* MPC 7 (2018) 20170119 17

Possible Scenario

BAM

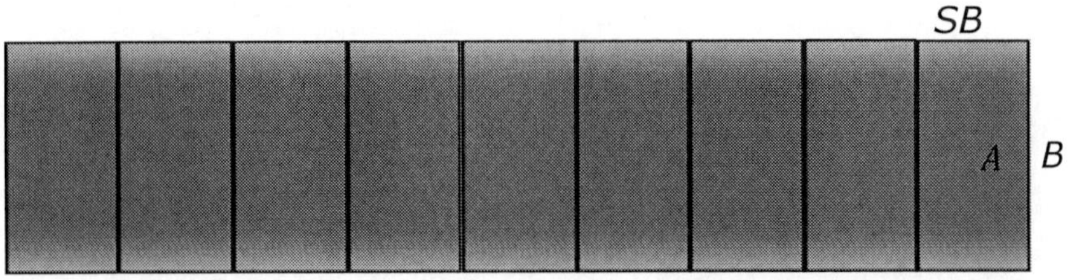

Heat input:

Heat output:

⇨ $A = SW \cdot W = 10SW \cdot W/10$

p minimum for $W = SW$

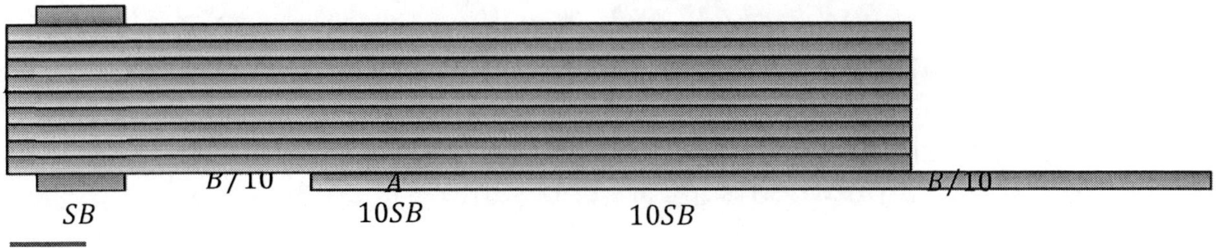

25.10.2018 NDT in Aerospace- Dresden 18

Influence of the support structure

BAM

Why support structures are important?

- Preventing cracking and compensate distortion
- Necessary for overhanging features
- Facilitating heat flux
- Easier and more precise removal from base plate

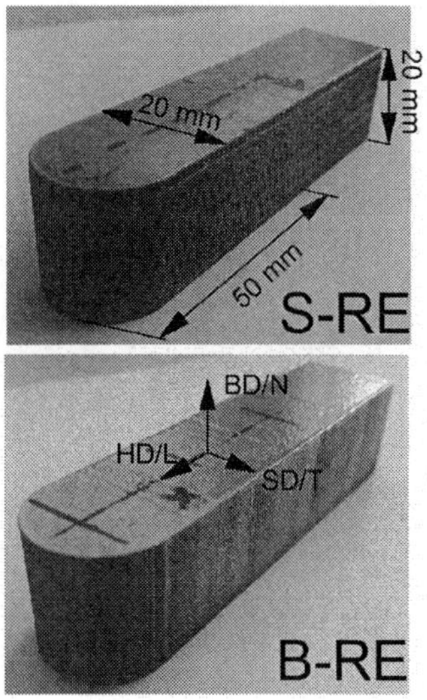

Microstructure

BAM

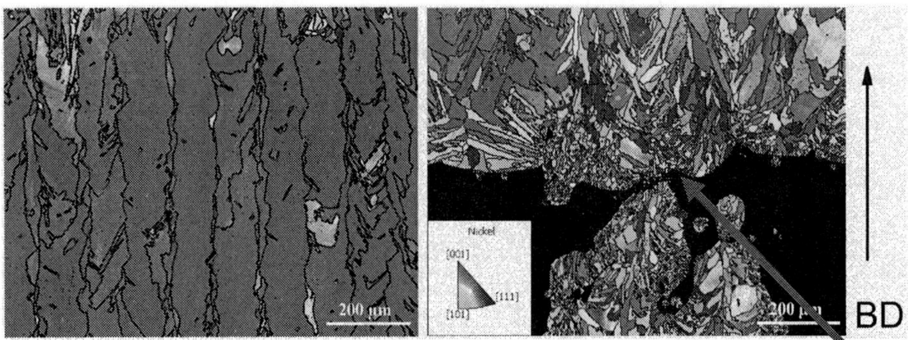

- Columnar shaped grains with preferred texture (<001>)
- Contact area between support and sample is small

Mishurova *et al.* Met.Mat.Trans. 49A (2018) 3038-3046

Synchrotron X-ray diffraction

BAM

Subsurface residual stress (penetration around 100µm)

EDDI beamline

BESSY II, HZB, Berlin

RS mapping- von Mises stress

BAM

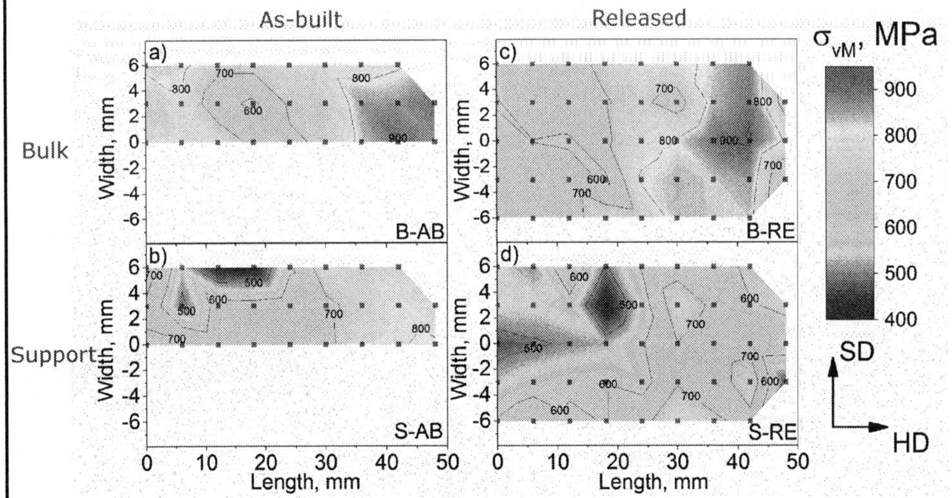

- High tensile stresses (up to "*yield*") near the surface
- RS redistribution and relaxation after removal
- Support structure leads to *reduction* of RS → HT cannot be avoided

Distortion

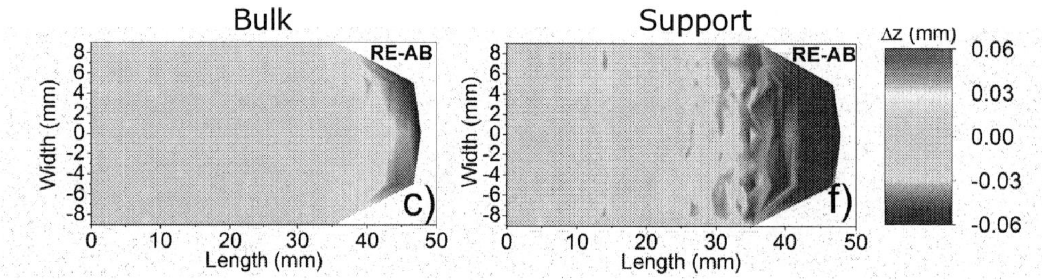

The support structure gives more compliance to the sample and results in larger distortion.

The stripe-like pattern correlates with period of the support structure.

Lower residual stress corresponds to larger distortion.

Insights- 1- d_0 reference

Requirement: **quantitative** assessment of stress

Calculation of Strains by diffraction $$\varepsilon = \frac{d - d_0}{d_0}$$

Which reference state needs to be used?
1- Initial powder? → Stress-free, but it did not undergo the same HT as the AM part
2- Small cubes extracted from the sample? → Not completely stress-free (e.g., cutting)
3- Powders extracted from the sample? → Possibly plastically deformed…(filing)

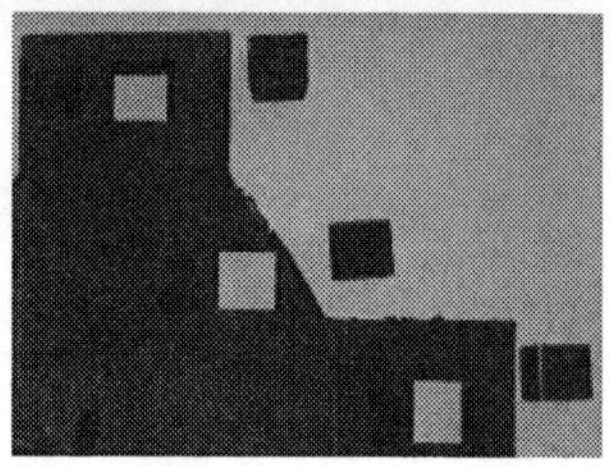

Insights- 1- d_0 reference

Lab XRD results:
d_0 scan on the surface

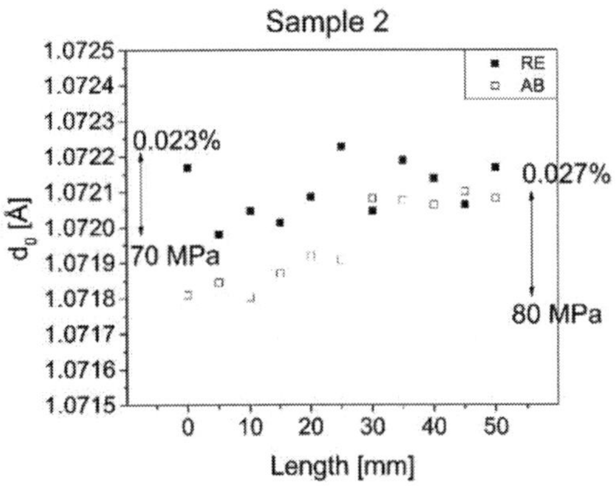

ND results on cubes (L, T, N), raw powder (RP), Sample powders (SP)

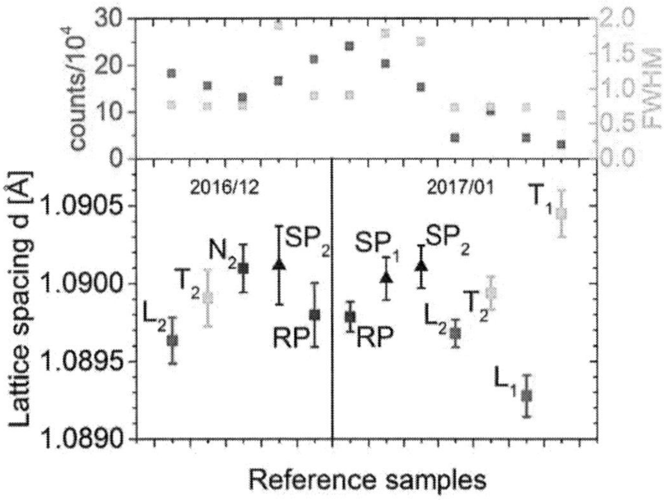

Takeaways- 1- d_0 reference

1- Initial powder? → Not suitable: it did not undergo the same HT as the AM part, the chemistry is not the same

2- Small cubes extracted from the sample? → Not suitable: Not stress-free

3- Powders extracted from the sample? → Plastically deformed, but reproducible and macro-stress-free. OK

Deeper analysis is required → BAM

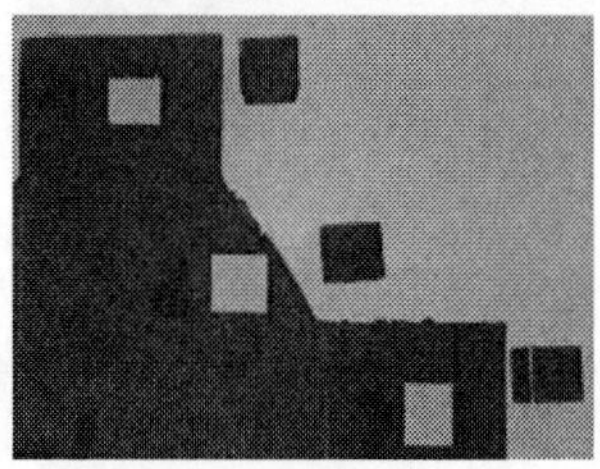

Insights- 2- Principal stress axes // Texture

BAM

In AM the deposition strategy strongly influences the microstructure, through columnar growth

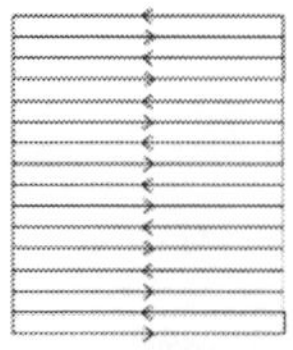

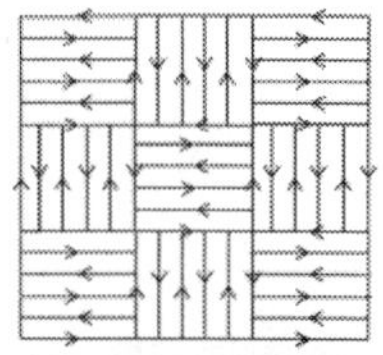

1- What happens to the stress axes, if the hatching has a different geometry? Are the principal axes rotated (like in the case of a weld)?

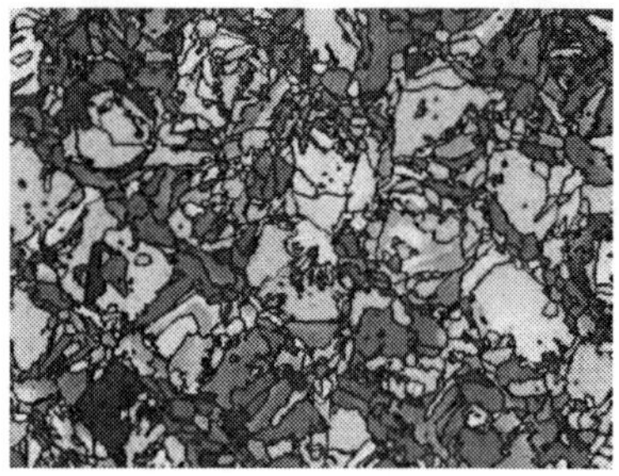

2- Texture can be very strong. Does it influence the assessment of residual stress?

The $\sin^2\psi$ technique

BAM

From laboratory XRD we borrow a useful technique that allows determining the influence of texture and the principal axes

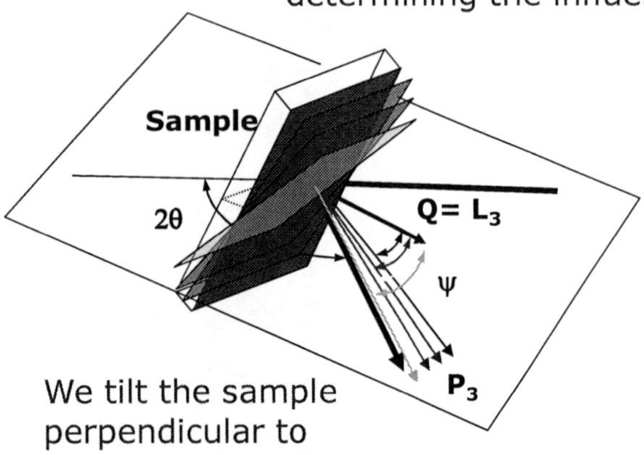

We tilt the sample perpendicular to the scattering plane

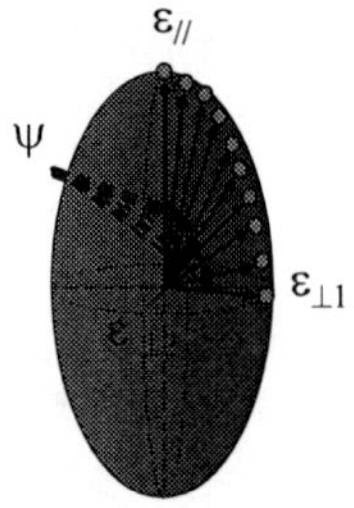

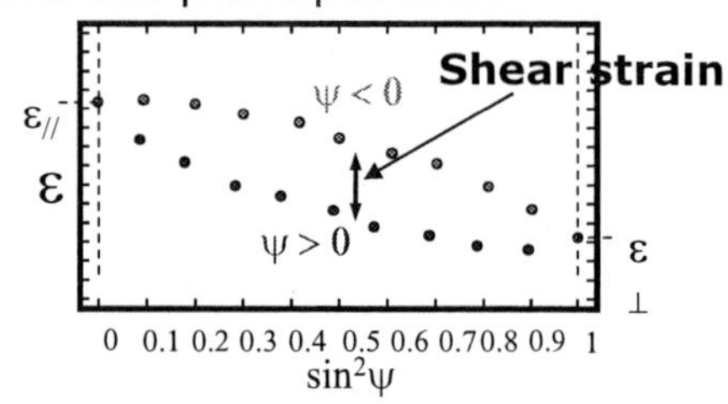

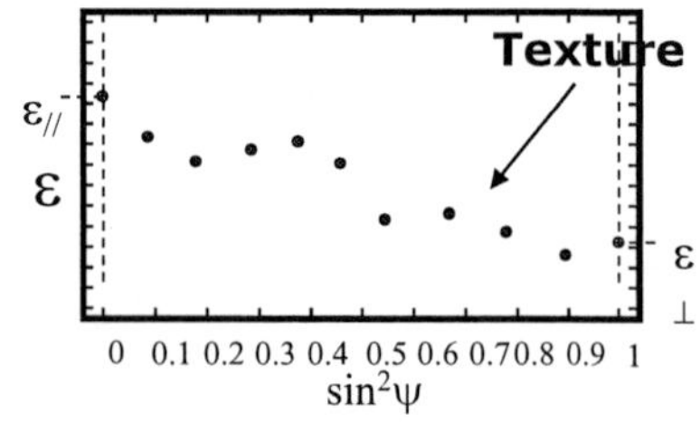

Insights- 2- Influence of Texture

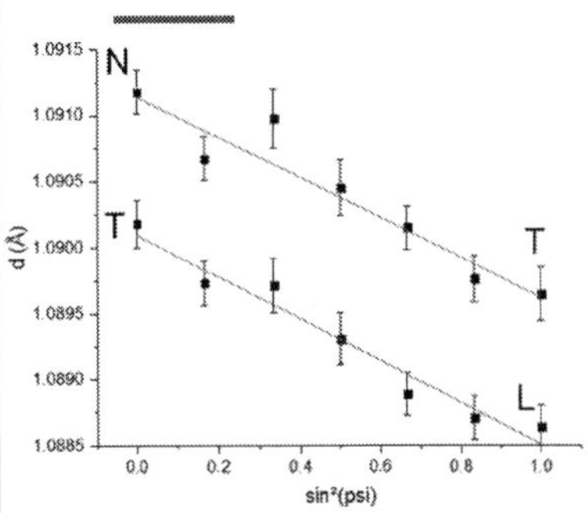

Neutron Diffraction (Bulk)

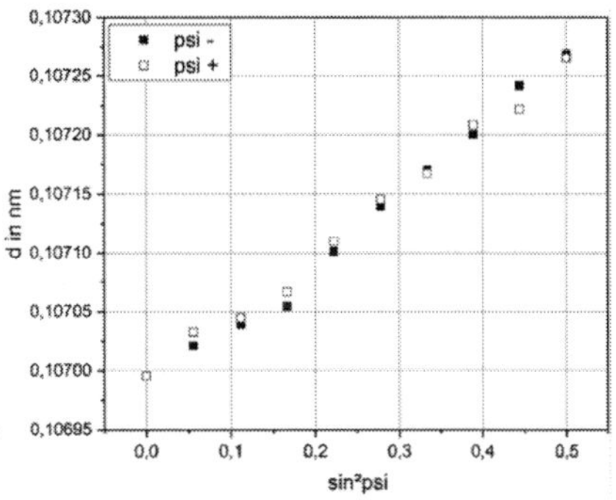

Lab X-ray Diffraction (Surface)

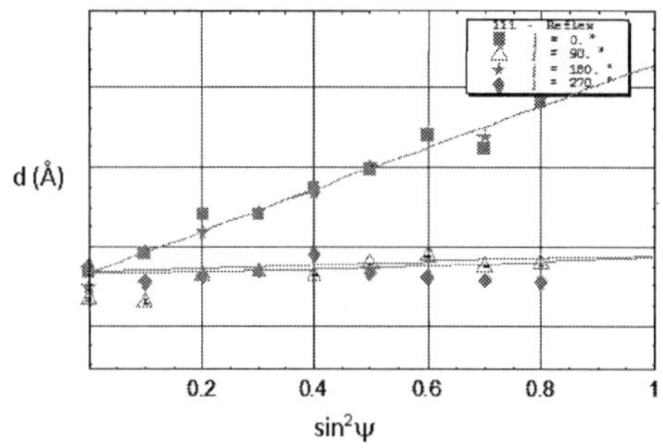

Synchrotron X-ray Diffraction (Sub-Surface)

All show linear plots :
No large influence of texture
All show no difference for pos. and neg. tilts :
geometrical directions are principal

In this case, the classic RS analysis is valid

Takeaways 2- Principal stress axes // Texture

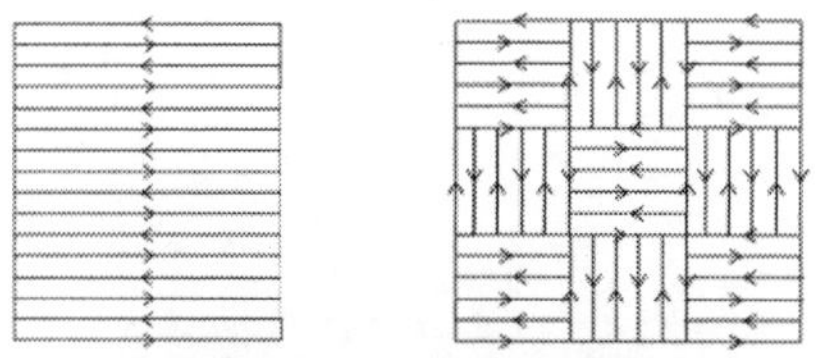

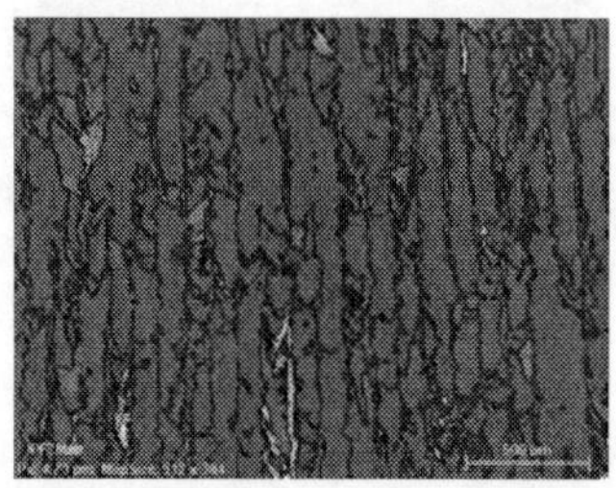

In AM the deposition strategy strongly influences the microstructure, through columnar growth

1- In the case of IN Ni alloys principal axes are not rotated (unlike in the case of a weld)

2- Texture can be very strong. However, it does not influence the assessment of residual stress by diffraction methods.

What happens to other materials/ alloys? → BAM

Conclusions and Takeaways

BAM

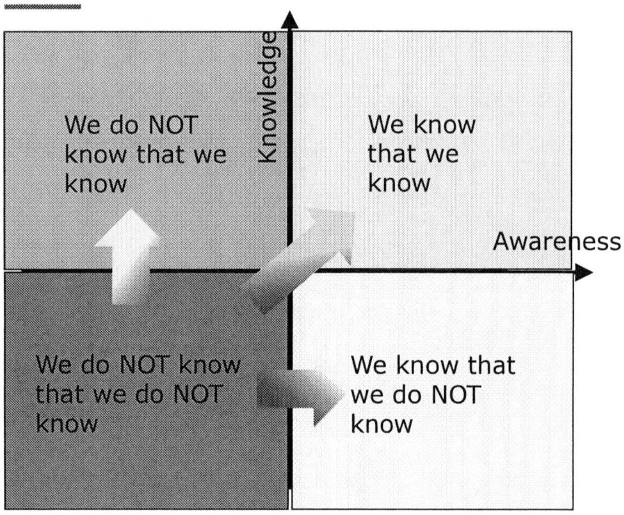

The large texture does not strongly influence Residual Stress Analysis

Residual Stress and distortions in SLM IN 718 strongly reflect the hatch pattern and depend on the peculiar thermal history

Powders extracted from the AM are the best macro-stress-free reference

25.10.2018 NDT in Aerospace- Dresden 31

BAM

Competence Centre
Safety in Technology and Chemistry

Safety creates markets.

ONLY Safety creates SUSTAINABLE markets.

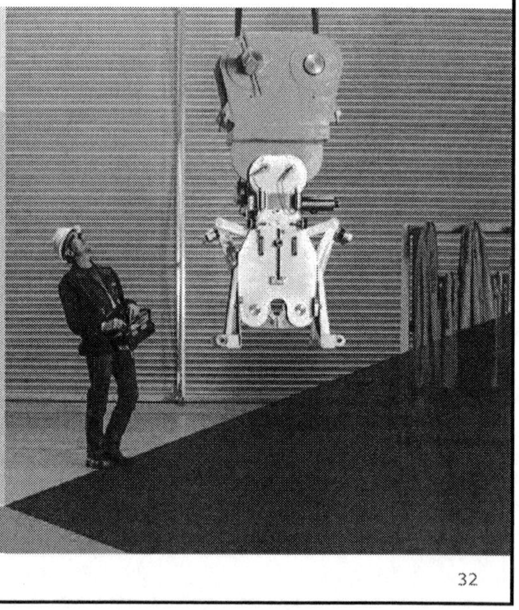

25.10.2018 NDT in Aerospace- Dresden 32

BAM

Thank you for your attention.

Contact:

Giovanni Bruno
Head of Division 8.5

Phone: +49 30 8104-1850
Email: giovanni.bruno@bam.de

www.bam.de

Quality assurance of additively manufactured alloys for aerospace industry by non-destructive testing and numerical modeling

Mustafa AWD[1], Shafaqat SIDDIQUE[1], Jan JOHANNSEN[2], Tillmann WIEGOLD[3], Sandra KLINGE[3], Claus EMMELMANN[2], Frank WALTHER[1]

[1]Department of Materials Test Engineering, TU Dortmund University, Dortmund, Germany
[2]Fraunhofer Research Institute for Additive Manufacturing Technologies, Hamburg, Germany
[3]Institute of Mechanics, TU Dortmund University, Dortmund, Germany

Contact e-mail: mustafa.awd@tu-dortmund.de

Abstract. Development of the layer-wise build-up of metallic materials offered significant technical and economic advantages for modern structural components. Airliners are now losing much weight which is due to efficient designs which were made possible by additive manufacturing technologies. In addition to topology optimization, selective laser melting (SLM) induces enhanced microstructural features and mechanical properties. Highly dynamic melt pools result in refined microstructures which increase the strength of the conventionally manufactured titanium and aluminum alloys. At the scanning electron microscope (SEM), it was revealed that SLM produces a very fine cellular dendritic microstructure in aluminum alloys of the Al-Si system. In the recently developed Scalmalloy®, the ultra-fine ceramic precipitation of $Al\text{-}Sc_3$ enables the tensile strength to reach 490 MPa and improves fracture strain by 100%. Along with improved microstructures due to high cooling rates, unstable melt pools induce spherical porosity as the release of gases from the powder bed is entrapped by the high solidification rate.

Platform heating (PH) which slows down cooling and enhances energy density is proved to stabilize melt pools and improve the degassing mechanism. Al-Si alloys are well-known cast alloys in the automotive industry; however, with the application of SLM technology, the tensile strength of these alloys is now 200% of the cast counterparts. The interest of aerospace industry in a high-strength aluminum alloy encouraged the development of Scalmalloy® which has a similar tensile strength to the wrought EN AW 7075. The latter is sensitive to hot cracking during SLM. For assurance of quality of additively manufactured alloys for the aerospace industry, the non-destructive testing (NDT) technology of X-ray computed tomography (CT) is applied in this study to develop finite element (FE) models which will be used in a statistical simulation to deduce a quality-controlled fatigue strength in high- and very high-cycle fatigue (HCF to VHCF) regimes. Modeling of fatigue strength will be based on the quasistatic and cyclic deformation behavior of SLM aluminum alloys.

1. Introduction

The flexibility of powder-bed-fusion (PBF) processes, such as selective laser melting (SLM), earned them a great interest in industry and academia. SLM fuses microlayers of powder in a layer-wise fashion so arbitrary complex components can be manufactured without special tooling. Initial SLM of aluminum (Al) alloys proved difficult due to high thermal conductivity. Only when high laser power SLM systems became operational that processing of aluminum alloys was possible [1]. SLM of AlSi10Mg is comparable or better than the cast alloy in terms of mechanical properties due to significantly refined microstructure resulting from high cooling rates. Significant anisotropy is observed in properties [2]. The constituents of the microstructure are an Al matrix and eutectic lamella of Si along the cell boundaries. The microstructure is homogenized, and anisotropy is reduced after T6 heat treatment. Application of platform heating (PH) reduces fatigue resistance due to the coarser microstructure [3]. The steep cooling gradients cause Si particles to supersaturate in the Al matrix which contributes to increased hardness. Due to scan tracks overlap, a coarser microstructure is induced due to remelting [4]. The supersaturation of the Al matrix by Si particles in SLM AlSi12 leads to the tensile strength which is double the cast alloy [5].

The remnant porosity in SLM Al structures is a detrimental factor to their mechanical properties. Research work has shown that scanning parameters have a significant impact on the porosity. Larger overlaps of scan tracks allow slower cooling rates and enhance degassing mechanisms [6]. Statistical methods were introduced to optimize build density of SLM AlSi10Mg [7]. Additionally, control of cooling rate through PH allowed reducing remnant porosity. On the other hand, tensile and fatigue strength are slightly reduced [8]. To quantitatively evaluate the effect of porosity on structural integrity, X-ray microcomputed tomography (μ-CT) was used to build finite element (FE) models of the internal structure of a fatigue specimen. In this analysis, it was deduced that the combined pore characteristics of size and proximity to the free surface are the influencing factors [9]. Manufacturing of high-strength precipitation hardening Al alloy such as EN AW 7075 was not possible using the same standard wrought composition due to the formation of hot cracks which significantly reduced the structural integrity [10]. Modification of composition by addition of 4% Si improved melt viscosity and eliminated hot cracks as well as resulted in a reasonable hardness [11].

High-strength alloying systems of Al alloys, different from Al-Zn, were successfully processed by SLM. An Al-Cu-Mg alloy with a relative density > 99.8% could reach a tensile strength of 402 MPa [12]. Scalmalloy® which is a Sc/Zr modified Al alloy that reaches density > 99% possesses a fine precipitation hardening mechanism which enables to reach mechanical strength higher than the traditional SLM Al alloy, AlSi10Mg [13]. The composition allows the mush to reach very high undercooling rate before solidification which makes this alloy particularly suitable for SLM processing [14]. Additionally, hardness properties become directionally independent and tensile strength can reach up to 490 MPa [15]. In this study, a comparison is made between structural and cyclic deformation behavior of AlSi10Mg, and it's most comparable in composition Scalmalloy®. The comparative study is realized through application of mechanical testing in quasistatic and cyclic loading modes. The load increase test (LIT) is applied as an initial mean to compare the fatigue strength of the studied batches. Evolution of the dynamic stiffness, as well as damage accumulation, is monitored online during constant amplitude testing (CAT). Finite element modeling techniques are applied to calculate statistically reliable Woehler curves. The results are correlated with the process and structural features of the two alloying systems.

2. Materials and methods

2.1 Experimental work

The study consists of two batches of SLM AlSi10Mg (A, B) where details about manufacturing can be found in [16]. The comparison of the two batches highlights the influence of PH on mechanical properties which has been discussed earlier by several researchers [3, 8]. The two AlSi10Mg batches are compared with two batches of Scalmalloy® (C, D), where details about manufacturing procedure and parameters can be found in [15]. The investigation of these two batches highlights the process differences between SLM and laser metal deposition (LMD). The final objective is to compare the fatigue strength of AlSi10Mg and Scalmalloy® under similar loading conditions. The study also aims at the use of non-destructive testing techniques as well as numerical simulations to qualify additively manufacturing materials for reliable application in structural components.

A tensile test was applied for batches A-D at an Instron 3369 system with 50 kN load cell. The extension rate was controlled by 1 mm/min. In Fig. 1a, a comparison of the stress-strain behavior of batches A-D can be seen. Batches A and B have a slightly higher yield strength (YS) than batch C. The plastic flow of batch C was disturbed by the extensive porosity as it will be shown by μ-CT analysis. Hence, batches A and B showed significant plastic hardening which resulted in ultimate tensile strength (UTS) of 426 and 452 MPa, respectively. Batch D showed a YS > 450 MPa and good nearly isotropic plastic behavior with a fracture strain > 8%.

With respect to cyclic deformation, a load increase test (LIT) at stress ratio of R = -1 and frequency of f = 20 Hz was applied to assess material's plastic damage reaction at continuously different amplitudes of stress where the stress amplitude σ_a was changed at a rate of 10 MPa/10^4 cycles. The experiment was applied with an Instron 8872 system that has a 10 kN load cell. In Fig. 1b, the plastic strain amplitude is shown with respect to the applied stress amplitude. The comparison between batches deduces that batch C endured the least magnitude of loading and failed at σ_a = 149 MPa. Followed by batch A at 152 MPa, batch B at 174 MPa, and batch D at 203 MPa. This result complies qualitatively well with the findings of the tensile test. The monitoring of the evolution of the plastic strain amplitude $\varepsilon_{a,p}$ allows tracking the onset of plastic damage marking the transition between high-cycle fatigue (HCF) and very high-cycle (VHCF). This point was detected at a stress amplitude of 100 MPa for batch B and D.

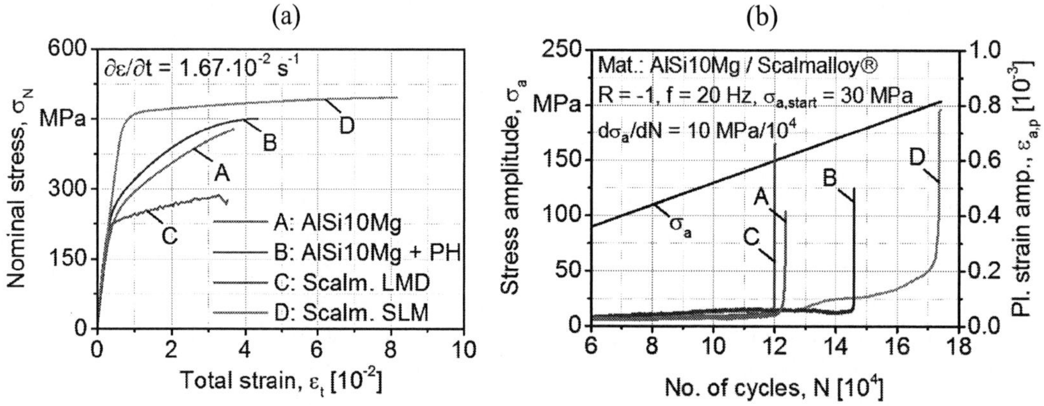

Fig. 1. Deformation behavior of batches A-D: a) Quasistatic tensile test; b) Continuous cyclic load increase test.

For the understanding of the mechanical behavior of these batches, it is necessary to correlate their mechanical behavior with the process-induced structure which is possible through μ-CT. In this experiment, where the setup is shown in Fig. 2, fatigue specimen (Fig. 3a) is coaxially clamped to the table of the μ-CT and is set in front of X-ray gun such that the cone of X-ray fully covers the gauge section. The X-rays which go through the specimen are reduced in intensity due to the partial absorption of the internal features of the specimens such as gas porosity or heterogeneities. The specimen geometry which is shown in Fig. 3b is used for validation tests at a mechanical resonance testing system of Rumul with a frequency f = 68 Hz.

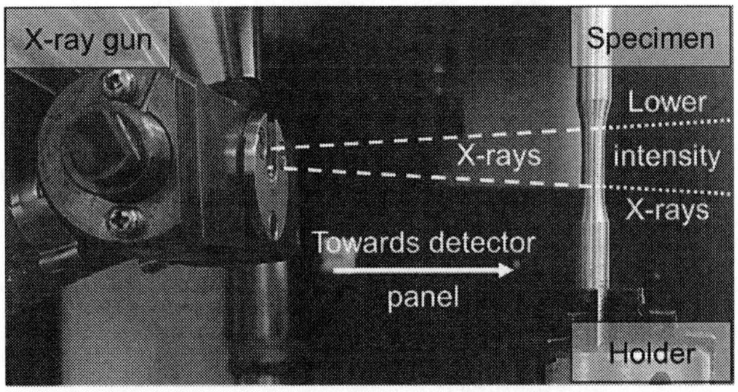

Fig. 2. Experimental setup of X-ray μ-CT scanning of fatigue specimens.

During μ-CT scanning, the specimen rotates, and images are acquired which includes the X-ray signature of the specimen. Afterward, the images are used to reconstruct the 3D volume and to conduct a defect analysis of the remnant porosity. In Fig. 3c, the volumetric defect distribution can be seen for batches A-D which had relative densities of 99.96% and 99.94% for batch A and B, respectively. The latter observation agrees with earlier research work [3, 9]. Also, batch D which is manufactured by SLM had a relative density of 99.96%, which is in contrast to LMD batch C, with extensive porosity, that had a relative density of 99.65%.

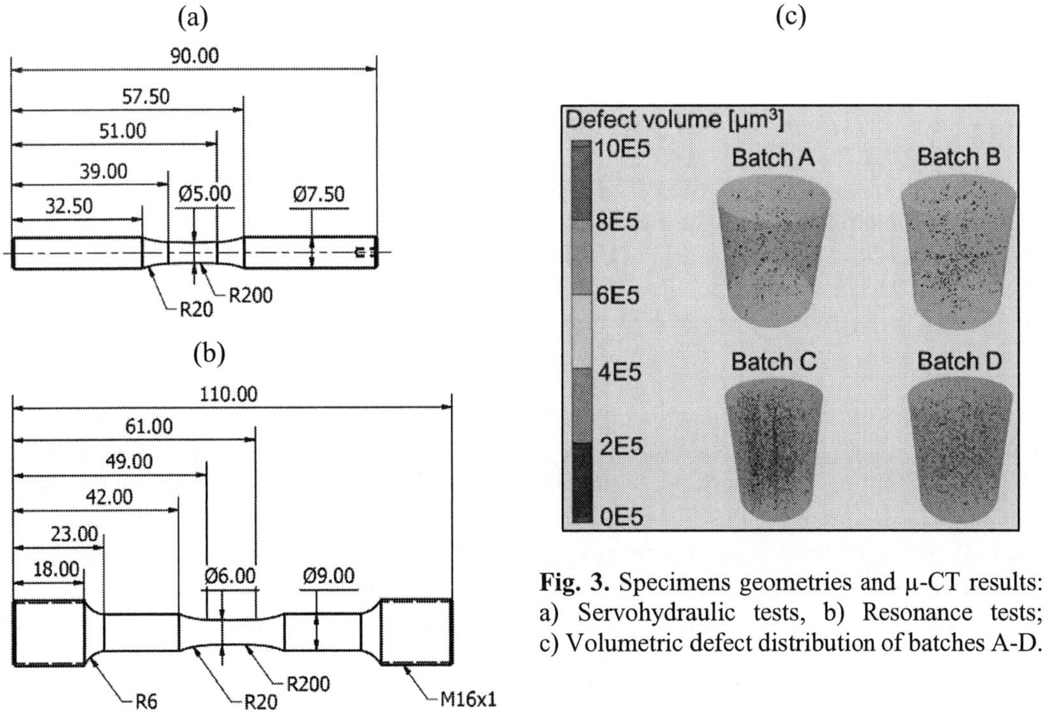

Fig. 3. Specimens geometries and μ-CT results: a) Servohydraulic tests, b) Resonance tests; c) Volumetric defect distribution of batches A-D.

2.2 Modeling and simulation

Finite element (FE) models of the internal structures were constructed based on μ-CT scans to quantify the effect of porosity on mechanical deformation. The tensile loading was simulated in Abaqus [17] using built-in isotropic and kinematic hardening models. A comparison of flow curves is shown in Fig. 4 for batches A and B. The curve of kinematic hardening shows earlier yielding than the curve for isotropic hardening with a higher strain hardening modulus. The nodal stresses and strains were exported to nCode fatigue lifetime evaluation software which was used to calculate the expected lifetime for each node according to its stress state. The advantage of this procedure is that it accounts for stress triaxiality at internal porosity which is contrary to the remote loading state that is uniaxial.

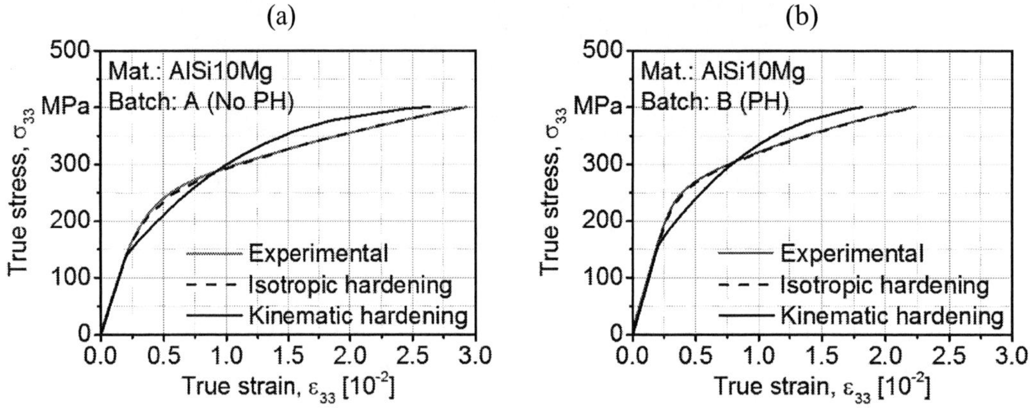

Fig. 4. Simulation of quasistatic tensile deformation using built-in Abaqus [17] models of isotropic and kinematic hardening for: a) batch A; b) batch B.

The phase-field method is used in order to model the crack propagation. This method can be implemented in a straightforward manner while avoiding the modeling of discontinuities [18]. According to this approach, an additional field variable d is introduced for the damage. It describes the current state of the structure by taking values in the range [0, 1] where $d = 1$ describes the fully broken and $d = 0$ the unbroken material. The basis of the model is the free energy

$$\psi(\varepsilon^{\mathrm{e}}, d) = \frac{1}{2}\,[(1-d)^2 + k_{\mathrm{n}}]\ \varepsilon^{\mathrm{e}} : \mathbb{C} : \varepsilon^{\mathrm{e}}, \qquad \varepsilon = \varepsilon^{\mathrm{e}} + \varepsilon^{\mathrm{p}}, \tag{1}$$

where ε^{e} are elastic strains, $\mathbb{C}$ the stiffness matrix and k_{n} a small numerical constant required to achieve numerical stability in the case of fully broken material. Typically of small strains theory, the additive decomposition of total strains ε into an elastic ε^{e} and an inelastic ε^{p} part is assumed. The dissipation potential for this approach reads

$$D = \frac{g_c}{l} d\dot{d} + g_c l \nabla d \cdot \nabla \dot{d} + k_{\mathrm{y}} \|\dot{\varepsilon}^{\mathrm{p}}\|, \qquad k_{\mathrm{y}} = k_{\mathrm{y0}} + k_{\mathrm{h}}, \qquad \dot{k}_{\mathrm{h}} = H \|\dot{\varepsilon}^{\mathrm{p}}\|, \tag{2}$$

where g_c is a constitutive threshold value related to the critical Griffith-type fracture, l is the length-scale parameter and k_{y} the current yield limit with the evolution depending on the initial yield limit k_{y0} and isotropic hardening H. The full formulation additionally requires a Lagrange term stipulating the non-negativity of the damage rate. Equation (1) corresponds to the case that both, tension and compression, affect the evolution of the damage which is not consistent with the real physics of the process where solely the tension contributes to damage. This drawback can be eliminated by assuming an additive decomposition of the total free energy into a tension part (($\psi^{+}(\varepsilon^{\mathrm{e}})$) and a compression part ($\psi^{-}(\varepsilon^{\mathrm{e}})$). The corresponding anisotropic energy storage function reads

$$\psi(\varepsilon^e, d) = [(1-d)^2 + k_n]\,\psi^+(\varepsilon^e) + \psi^-(\varepsilon^e). \quad (3)$$

3. Results and discussion

3.1 Mechanical strength in relation to material characteristics

During constant amplitude testing (CAT), the cyclic damage evolution was monitored based on dynamic stiffness and the total mean strain. In Fig. 5a, the evolution of the dynamic stiffness for batch A is shown at σ_a = 120 and 140 MPa. It can be observed that at 120 MPa that after initial stiffening of the specimen the response saturates at ~1,500 cycles. At a higher stress amplitude of 140 MPa, the stiffening phase is too short followed by softening.

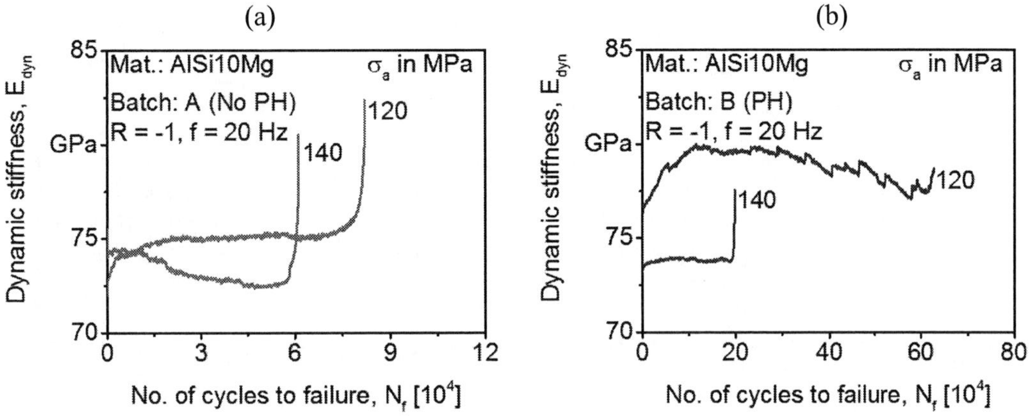

Fig. 5. Dynamic stiffness monitoring during constant amplitude testing for: a) batch A; b) batch B.

For batch B (Fig. 5b), extensive stiffening is observed at the beginning at 120 MPa, which is followed by softening that begins around half the life of the specimen. At 140 MPa no such degradation exists after specimen reaches stable saturation.

In Fig. 6a, the change of resonance frequency is derived based on the change in dynamic stiffness for batch D.

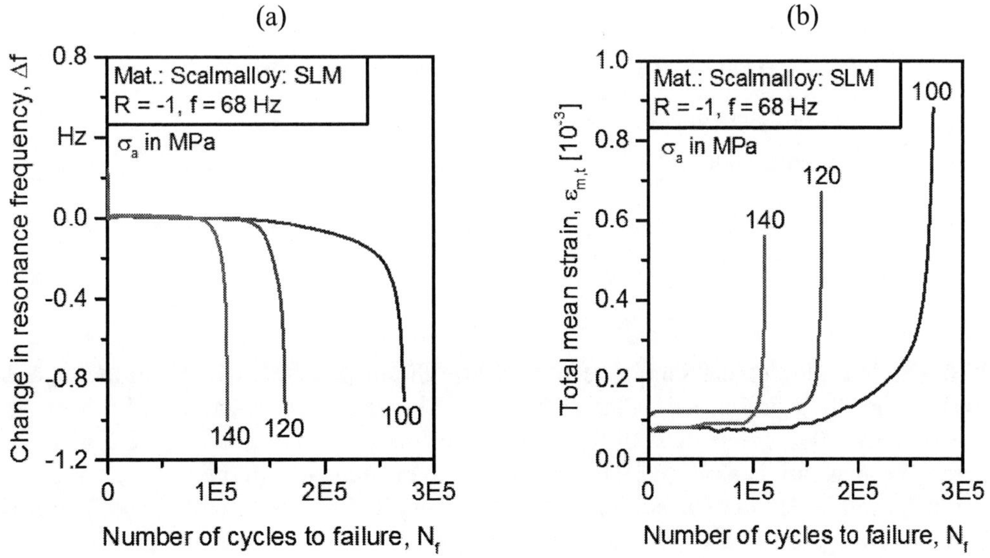

Fig. 6. Damage evolution during constant amplitude testing for batch D: a) Change in resonance frequency Δf; b) Total mean strain $\varepsilon_{m,t}$.

The comparison of the three stress amplitudes shows that at 100 MPa where it was at the critical level of stress amplitude identified by the LIT. The damage evolution was significantly delayed, and the transformation from a damaged state to the other was exponentially smoother. The same observation was made with respect to the evolution of cyclic creeping ("ratcheting") and the total mean strain as identified in Fig. 6b.

3.2 Modeling and NDT for reliability assurance

In Fig. 7a, the distribution of nodal lifetimes is given according to the Weibull distribution function. The intersection between the reference line and Weibull percentiles agrees well with experimental data. The shift of Weibull distribution to higher fatigue life is observable from Fig. 7a at 140 MPa to Fig. 7b at 120 MPa.

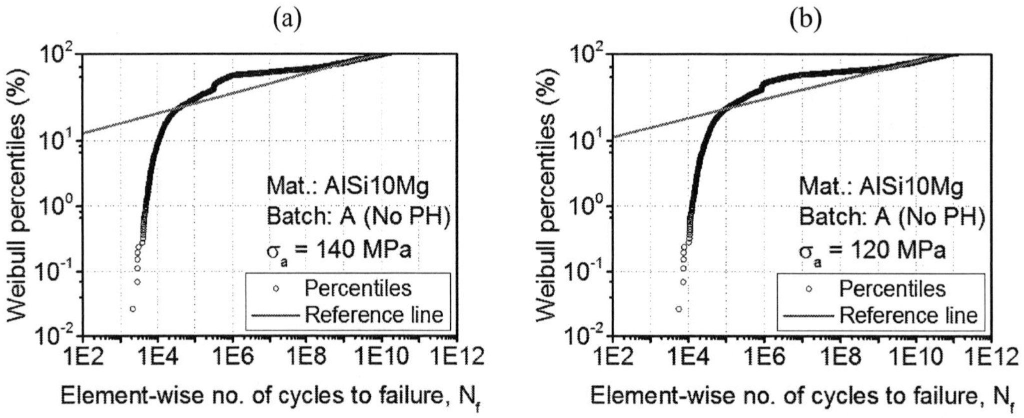

Fig. 7. Weibull distribution of nodal fatigue lifetimes of a specimen of batch A at simulated constant stress amplitude: a) 140 MPa; b) 120 MPa.

The selected numerical example for the phase-field model is concerned with the simulation of the crack propagation within a square sample with two pores, shifted relative to each other. The dimensions of the sample are 100×100 μm² and applied material parameters correspond to AlSi10Mg of batch A. An overview of the material parameters used in the simulation is presented in Table 1. The results of the numerical simulations are presented in Fig. 8.

Table 1. Material parameters used in the simulation.

Parameters of AlSi10Mg			
Young's modulus	E	68.55	GPa
Poisson's ratio	υ	0.3	-
Yield limit	k_{yo}	250	MPa
Isotropic hardening	H	166.41	GPa

The left edge of the sample is fixed while the displacement with the increment of 0.0001 μm/cycle is prescribed at the right edge. The sample is discretized by a mesh of approximately 7,000 elements such that a finer mesh is generated in the critical area where the crack propagation is expected. The characteristic element size in this area is set to 1 μm. The cracks initiate at the surface of pores at the applied displacements of 4.22 μm and thereafter propagate rapidly. The sample totally collapses at a prescribed displacement of 4.23 μm. Note that red color corresponds to the fully broken material where the value of damage parameter d is close to one.

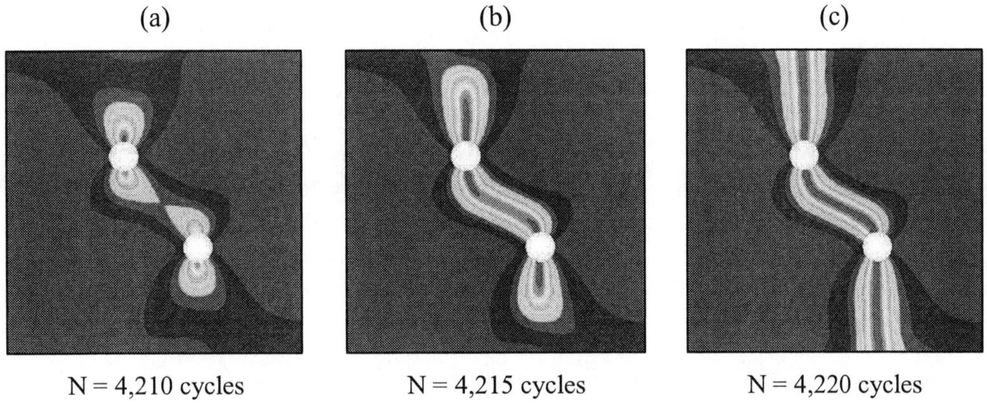

Fig. 8. Crack propagation in a sample of AlSi10Mg with two circular pores at prescribed displacements: a) 4.220 μm; b) 4.225 μm; c) 4.230 μm.

The evolution of the damage parameter by monitoring the ε_{11}^{p} component of the plastic strain tensor and the σ_{11} component of the stress tensor is presented in Fig. 9. The values correspond to a node at the top off the right hole where one of the cracks initiates. The following conclusions can be drawn here. The slope of the stress decreases after the initiation of the plastic strain. The damage parameter rapidly increases once a certain value is exceeded. The kink of the stress/cycle curve goes back to the fact that hardening in the current model is not coupled to the damage. An extension of the model in order to incorporate the kinematic hardening and influence of the damage on material constants will be concerned in our future work.

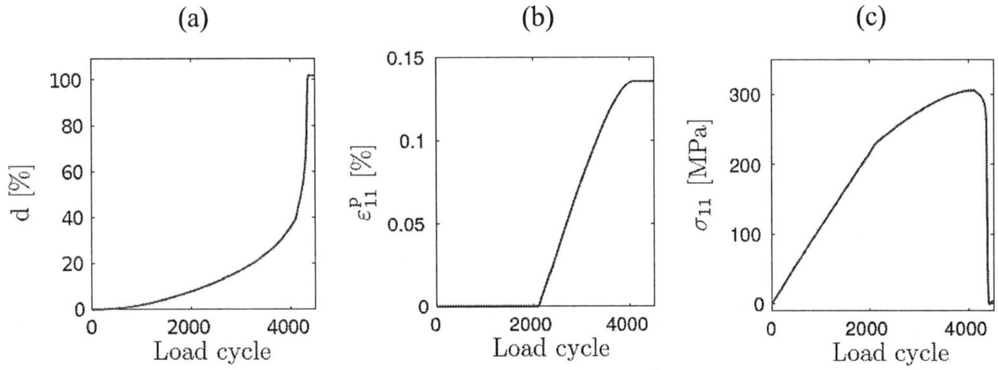

Fig. 9. Evolution of the damage parameter *d*, the first component of the plastic strain tensor ε_{11}^{p} and the first component of the stress tensor σ_{11}.

4. Conclusions and outlook

In this study, a comparison between mechanical properties of AlSi10Mg and Scalmalloy® manufactured through SLM and LMD was introduced. It was shown that Scalmalloy® possesses a significantly higher yield strength and fracture strain than AlSi10Mg due to significant grain refinement by addition of Sc and Zr. In cyclic loading, the enhanced microstructure contributed to delay the onset of plastic damage which results in Scalmalloy® reaching significantly higher stress amplitudes. This was not absolutely true because of LMD process-induced extensive porosity as it was revealed by μ-CT which resulted in premature failure of the specimen at significantly lower remote stress and fracture strain.

Simulation of tensile deformation was carried out on FE models which were constructed based on the μ-CT scans. The application of Weibull distribution to the calculated nodal lifetimes proved successful in the estimation of fatigue lifetime. Additionally, the newly introduced phase-field model for simulation of fatigue crack initiation and propagation based on the energy function proved interaction of closely located pores possible. The coalescence of multiple cracks initiating from nearly located pores has a detrimental effect on fatigue lifetime. To quantitatively evaluate this effect, it is envisaged that the model will be upgraded to include kinematic hardening and the influence of the damage on the material constants in FE models which reflects the realistic porosity states of the investigated materials.

Acknowledgment: The authors thank the Deutsche Forschungsgemeinschaft (DFG, German Research Foundation) for its financial support within the research project "Mechanism-based understanding of functional grading focused on fatigue behavior of additively processed Ti-6Al-4V and Al-12Si alloys" (WA 1672/25-1).

References

[1] Louvis, E.; Fox, P.; Sutcliffe, CJ.: Selective laser melting of aluminum components. Journal of Materials Processing Technology 211 (2011) 275-284.

[2] Kempen, K.; Thijs, L.; Humbeeck, J.; Kruth, J.: Mechanical properties of AlSi10Mg produced by selective laser melting. Physics Procedia 39 (2012) 439-446.

[3] Brandl, E.; Heckenberger, U.; Holzinger, V.; Buchbinder, D.: Additive manufactured AlSi10Mg samples using selective laser melting (SLM): Microstructure, high cycle fatigue, and fracture behavior. Materials and Design 34 (2012) 159-169.

[4] Thijs, L.; Kempen, K.; Kruth, J.; Humbeeck, J.: Fine-structured aluminium products with controllable texture by selective laser melting of pre-alloyed AlSi10Mg powder. Acta Materialia 61 (2013) 1809-1819.

[5] Prashanth, KG.; Scudino, S.; Klauss, HJ.; Surreddi, KB.; Löber, L.; Wang, Z.; Chaubey, AK.; Kühn, U.; Eckert, J.: Microstructure and mechanical properties of Al-12Si produced by selective laser melting: Effect of heat treatment. Materials Science and Engineering A 590 (2014) 153-160.

[6] Aboulkhair, NT.; Everitt, NM.; Ashcroft, I.; Tuck, C.: Reducing porosity in AlSi10Mg parts processed by selective laser melting. Additive Manufacturing 1-4 (2014) 77-86.

[7] Read, N.; Wang, W.; Essa, K.; Attallah, M.: Selective laser melting of AlSi10Mg alloy: Process optimisation and mechanical properties development. Materials and Design 65 (2015) 417-424.

[8] Siddique, S.; Imran, M.; Wycisk, E.; Emmelmann, C.; Walther, F.: Influence of process-induced microstructure and imperfections on mechanical properties of AlSi12 processed by selective laser melting. Journal of Materials Processing Technology 221 (2015) 205-213.

[9] Siddique, S.; Imran, M.; Rauer, M.; Kaloudis, M.; Wycisk, E.; Emmelmann, C.; Walther, F.: Computed tomography for characterization of fatigue performance of selective laser melted parts. Materials and Design 83 (2015) 661-669.

[10] Kaufmann, N.; Imran, M.; Wischeropp, TM.; Emmelmann, C.; Siddique, S.; Walther, F.: Influence of process parameters on the quality of aluminium alloy EN AW 7075 using selective laser melting (SLM). Physics Procedia 83 (2016) 918-926.

[11] Sistiaga, ML.; Mertens, R.; Vrancken, B.; Wang, X.; van Hooreweder, B.; Kruth, J-P.; van Humbeeck, J.: Changing the alloy composition of Al7075 for better processability by selective laser melting. Journal of Materials Processing Technology 238 (2016) 437-445.

[12] Zhang, H.; Zhu, H.; Qi, T.; Hu, Z.; Zeng, X.: Selective laser melting of high strength Al-Cu-Mg alloys: Processing, microstructure and mechanical properties. Materials Science and Engineering: A 656 (2016) 47-54.

[13] Spierings, A.; Dawson, K.; Voegtlin, M.; Palm, F.; Uggowitzer, P.: Microstructure and mechanical properties of as-processed scandium-modified aluminium using selective laser melting. CIRP Annals 65 (2016) 213-216.

[14] Spierings, AB.; Dawson, K.; Heeling, T.; Uggowitzer, PJ.; Schäublin, R.; Palm, F.; Wegener, K.: Microstructural features of Sc- and Zr-modified Al-Mg alloys processed by selective laser melting. Materials and Design 115 (2017) 52-63.

[15] Awd, M.; Tenkamp, J.; Hirtler, M.; Siddique, S.; Bambach, M.; Walther, F.: Comparison of microstructure and mechanical properties of Scalmalloy® produced by selective laser melting and laser metal deposition. Materials 11, 1 (2018) 1-17.

[16] Awd, M.; Johannsen, J.; Siddique, S.; Emmelmann, C.; Walther, F.: Qualification of selective laser-melted Al alloys against fatigue damage by means of measurement and modeling techniques. Fatigue 2018, Matec Web of Conferences 165, 02001 (2018) 1-7.

[17] ABAQUS 6.7, User documentation, Dessault systems, 2007.

[18] T. Wiegold, 2018, On phase-field modeling of ductile damage, Master thesis, TU-Dortmund, Germany.

Utilising Advanced Computed Tomography as NDT Technique in Additive Manufacturing

Jens LÜBBEHÜSEN, Key Account Manager Aerospace Europe [1],
Gerhard ZACHER, Application Engineer & Additive Manufacturing Specialist [1]
[1] GE Sensing & Inspection Technologies GmbH, Niels-Bohr-Str. 7, D-31515 Wunstorf, Germany

Contact e-mail: Jens.Lubbehusen@bhge.com, Gerhard.Zacher@bhge.com

Abstract

Additive Manufacturing (AM) is presently revolutionising industrial production in many sectors (Aerospace, Automotive, Power Generation, Oil&Gas, Medical, etc.) and is expected to substantially substitute traditional molding and casting processes within the next decade to come.
For many workpieces made of plastic and metal based materials, AM tends to be the more convenient manufacturing technique due to their complexity and geometry.

Keywords: Additive Manufacturing, AM, Computed Tomography, CT, 3D Visualization, Metrology, Standardisation ISO/ASTM.

1. Introduction

Apart from other known non-destructive techniques to assure quality of AM products, computed tomography (CT) turns out to be one of the most efficient NDT methods. Not only does CT enable the three-dimensional volumetric visualization of indications and internal geometries that traditional NDT methods can scarcely or not access, it also allows to be used for metrology application tasks, such as dimensional measurement of features, variance analyses (CAD data to part, part to part).

2. Content

In view of new types of defects and flaws to be detected in AM parts, CT can meet requirements for increased resolution, higher penetrability and X-ray scatter-corrected volume data sets of workpieces printed of various plastic or metal materials, even if they are manufactured from dense materials (e.g. Ti6Al4V, Inconel 625/718, CoCr, etc.) allowing for reliable and reproducible measurement results.

When irradiated, dense materials tend to produce physical X-ray backscatter, in particular for cone beam CT (i.e. using flat panel detectors) that is used to achieve fast CT scan times. This backscatter is negatively influencing the CT image and volume quality, as displayed by the example of a high-pressure turbine blade [1].

The proprietary «scatter|correct» method allows to measure the backscatter and deduct it from the CT volume data set to achieve a nearly scatter-free CT volume enabling to do analysis such as wall thickness measurements, as shown in the HPT blade example [2].

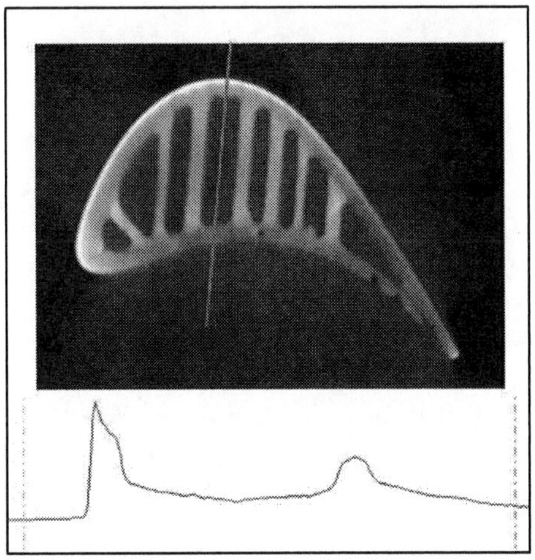

Figure 1. HPT blade CT image slice with physical backscatter [1]

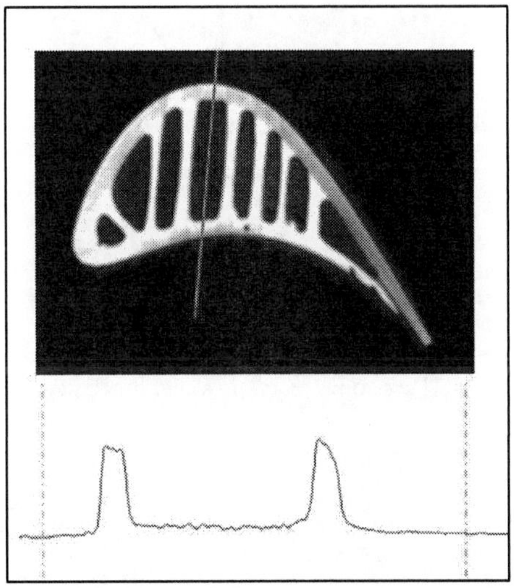

Figure 2. HPT blade CT image slice without physical backscatter [2]

Achieving a high level of CT resolution is essential when AM parts are inspected, defects such as powder entrapments, porosities and cracks are usually smaller than in traditionally casted parts. Cone beam CT allows for both, full part scanning and region of interest (ROI) part scanning, dependent on which level of resolution is required to detect the expected defects [3]

Computed tomography can complement dimensional analyses of coordinate measurement machines since it allows for measuring both, internal and external features at high precision and reproducibility.

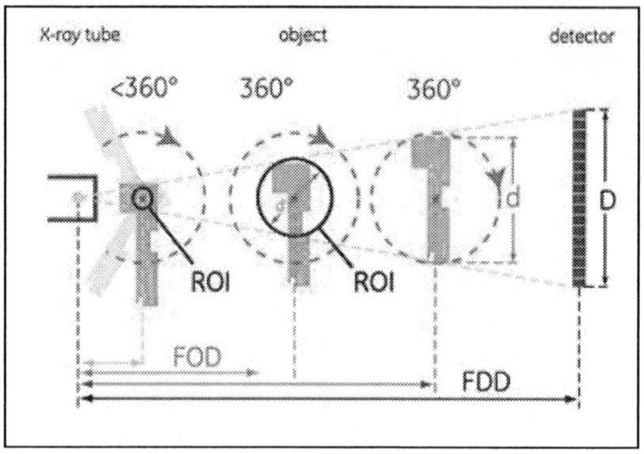

Figure 3. CT principles – full and ROI scan [3]

Some examples demonstrating the functionality range that CT can provide both, in failure analysis and in dimensional measuring are shown here below [4], [5].

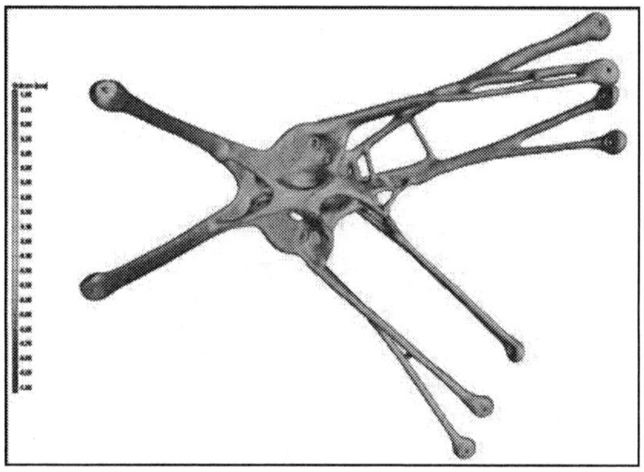

Figure 4. CT example reaction wheel bracket[1], CAD analysis [4]

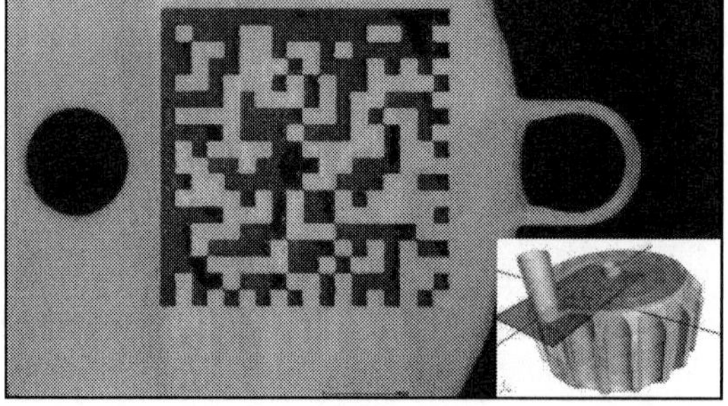

Figure 5. CT example rotary carousel, powder entrapments [5]

In addition to NDT tasks that concern the entirely printed workpiece, high-resolution CT allows for incoming inspection of the essential raw material required for 3D printing – the powder. Granulometry, particle size distribution and sphericity of AM metal powders (newly manufactured and recycled compounds) can be inspected [6], [7].

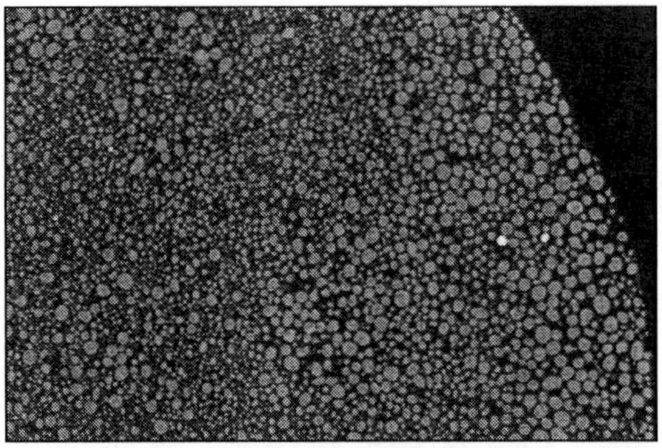

Figure 6. CT example powder – structural analysis [6]

Even material homogeneity inside the powder grains may be analysed by CT, subject to sufficient resolution and contrast.

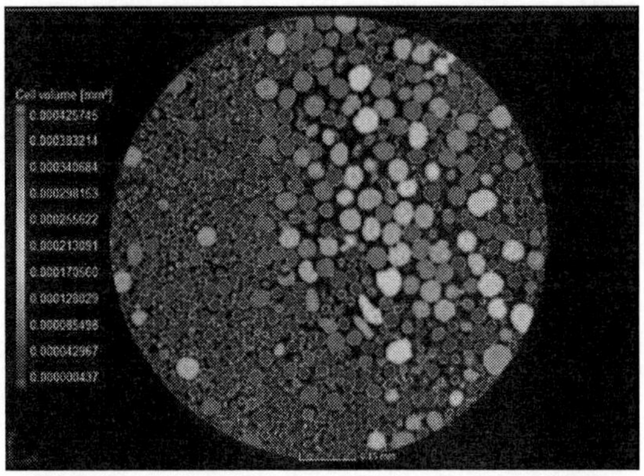

Figure 7. CT example powder – particle size analysis [7]

Being complementary to other NDT techniques such as in-situ monitoring during the AM process or on or near surface based eddy current or ultrasonic inspection, CT can help setting guidelines for volumetric quality assurance of AM parts [2]. Hence, standardisation organisations such as ISO (ref. TC 261) and ASTM (ref. F 42) rely on the recommendations and the expertise of CT system suppliers and CT users among the AM community members.

Thanks to improved automatic workflows, CT continues its move from being a traditional expert R&D device towards the use as an automatic measurement technique on or close to the production floor. Further automation steps such as the use of robot based manipulation units for unmanned part handling coupled with assisted defect recognition (ADR) have been developed and will find their way into the shop floors [8, 9].

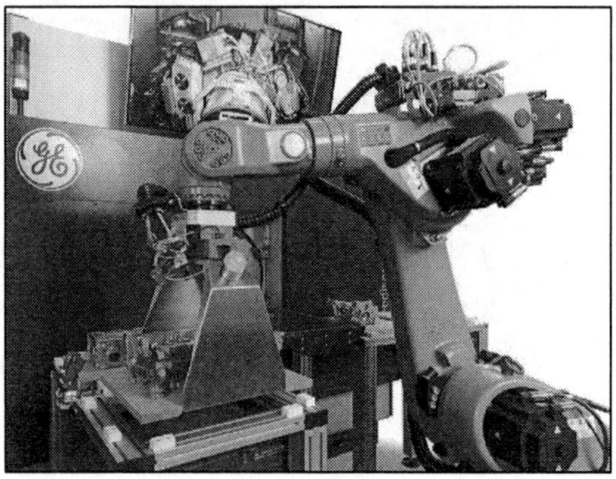

Figure 8. CT solution “speed|scan” with robotic part handling [8]

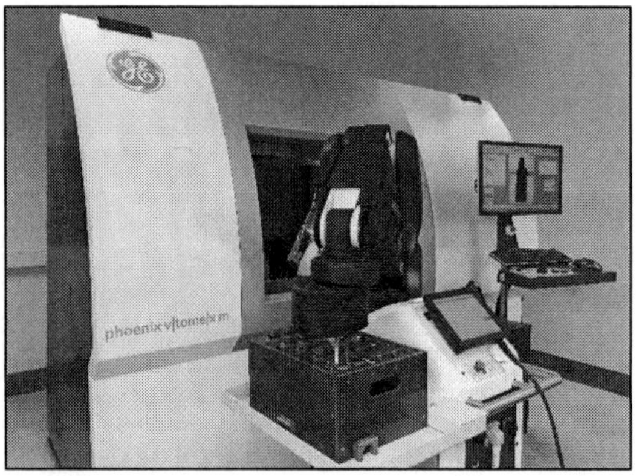

Figure 9. CT solution “v|tome|x m 300” with robotic part handling [9]

3. Conclusion

Computed Tomography is a leading technique for non-destructive testing of additively manufactured parts and complementary to other NDT techniques already in use. It fits the AM industry’s needs for quality assurance today and in future, if it delivers enough X-ray energy, resolution, contrast and speed.

Key enablers for future AM production based control by CT are a high degree of automation as well as an efficient-communication between the different manufacturing and quality checks within the AM supply chain.

References

[1] Reiher T., Koch R., "FE-optimization and design of additive manufactured structural metallic parts for telecommunication satellites", Paris Space Week, Paris, Feb. 2015.

[2] Herderick E.D., “Additive manufacturing applications and development: the past, the present and an exciting future”, IQPC Additive Manufacturing Conference, London, Feb. 2017.

Computed Tomography in NDT and Metrology for Additively Manufactured Aerospace Components

Nils ACHILLES [1], Gabriele MÄURER [1], Jason ROBBINS [1]
Julia MOLL [2], Silvia LÓPEZ-VIDAL [3], Ignacio ZAMORA [3]
[1] YXLON International GmbH, Hamburg, Germany
[2] TRUMPF Laser- und Systemtechnik GmbH, Ditzingen, Germany
[3] RAMEM S.A, Madrid, Spain

Contact e-mail: nils.achilles@hbg.yxlon.com

Combining computed tomography and additive manufacturing to optimize production cost and lead time of engine rakes

This study conducted by RAMEM S.A, TRUMPF Laser- und Systemtechnik GmbH and YXLON International GmbH investigates the joint impact of additive manufacturing (AM) in combination with computed tomography (CT) on the production process of engine rakes. Rakes are the support parts for a series of tubing that are connected to sensors which analyse the engine behaviour. Their high required load capacity as well as their complex geometry made them a particularly interesting case study. Rakes conventionally are produced by using subtractive manufacturing, i.e. they undergo a series of different machining steps, assembly and welding along with dimensional inspection and NDT on the sub-assemblies. This manufacturing complexity with its numerous production steps results in costly parts with long processing times which are a hurdle for the aerospace industry.

In contrast to the conventional manufacturing technique, the process steps for the rake can be significantly reduced by using additive manufacturing and combining dimensional measurements and NDT in one CT-scan. When comparing the times, conventional processing required 70h while AM and CT only needed 5h. The recurring process time could be reduced from 55h to 38h. The overall throughput could therefore be increased by 66%. Using advanced processes resulted in a cost reduction of 76% in non-recurring processes and a cost reduction of 66% in recurring processes. The overall manufacturing costs therefore could be reduced by 71%.

Being able to substitute most of the machining and all assembly steps by additive manufacturing resulted in an efficiency increase on the manufacturing side. By conducting only one CT-scan all dimensions within the part as well as the build quality could be checked semi-automatically in only one dataset. Applying computed tomography substituted several NDT and dimensional inspection steps on sub-assemblies by only one process. This not only reduced the amount of time needed for the inspection but also decreased complexity significantly.

**numbers are partly based on assumptions, are non-committable and represent the results achieved only at the time of publication. All numbers can be subject to modification.*

Local Acoustic Resonance Spectroscopy: An Escalation Approach for Fast Non-Destructive Testing

Philipp JATZLAU [1], Christian U. GROSSE [1]
[1] Technical University of Munich, Chair of Non-destructive Testing, Munich, Germany
Contact e-mail: philipp.jatzlau@tum.de

Abstract. Local acoustic resonance spectroscopy (LARS) is a non-destructive technique suitable for fast defect detection in large planar parts, especially in fiber composite structures. Applied in the acoustic frequency spectrum LARS is closely related to the coin tap test. In recent years, several applications in mechanical engineering have been established. These include the inspection of GFRP rotor blades as well as defect detection in CFRP parts in the automotive and aeronautic sectors. In addition, there are similarities to the impact-echo method, which is used in civil engineering for the detection of delaminations in concrete pavements or for thickness measurements of concrete slabs and structures. In all cases LARS is suitable for fast and therefore inexpensive measurements, also because it can be automated easily. Potential defect locations can then be more closely inspected by more complex non-destructive or destructive methods. The integration of LARS into such an inspection approach is subject of this paper. Moreover, an overview is given of the various applications of LARS and the respective measurement and analysis methods, where excitation technique and sensor technology are of major importance. While excitation is normally performed using an impulse hammer, the response signal is typically recorded by several microphones (capacitive or MEMS). Alongside experimental data, numerical simulation of the impulse response within the part is particularly useful to optimize the configuration on both the transmitting and receiving side for early identification of possible disturbances.

Introduction

In aerospace applications, non-destructive testing is of greatest importance for safety and reliability of parts and structures. Thus, full volume testing of all parts is performed according to specific standards and certified procedures. For CFRP structures, the NDT methods applied include ultrasonic testing and computed tomography, which are typically able to detect and locate defects relevant for safety with high precision. However, these methods are expensive and time-consuming due to complex testing devices and/or lengthy deployment of personnel. For certain defect types, e. g. delaminations due to compression after impact ("tool drop" scenario), it can be feasible to obtain information on size and location of those defects before examining the complete part using complex methods. The authors suggest that local acoustic resonance spectroscopy (LARS) can help reduce cost and time by narrowing down defective areas to be examined by more complex methods whenever suitable, e. g. on large planar parts such as fuselage fairing. In contrast to the conventional coin-tap test and commercially available tapping test devices, LARS uses both hammer force pulse and microphone data to create objective information, aiming at making the testing procedure less dependent on the personal experience of the tester. Thus, LARS has already been proven to be suitable for large plate-shaped parts made of GFRP [1] [2].

LARS is part of the holistic concept for an efficient selection of NDT techniques suggested by Grosse [3] as shown in Figure 1, where selected inspection methods are listed in the center.

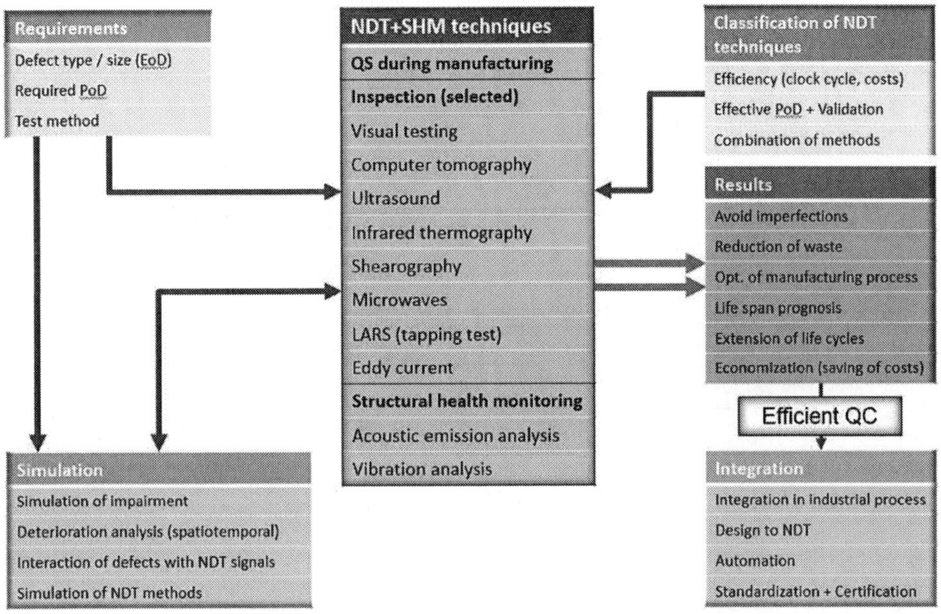

Figure 1: Holistic concept for an efficient selection of NDT and SHM techniques in the context of quality control [3]

Considering criteria like e. g. equipment and personnel costs, test duration and capability, the different methods may be compared and divided into different complexity levels. This leads to the idea of an escalation approach, in which only certain areas previously defined as "potentially defective" by the lower-level methods need to be inspected by the higher-level (and thus more costly) method. This principle is schematically presented in Figure 2.

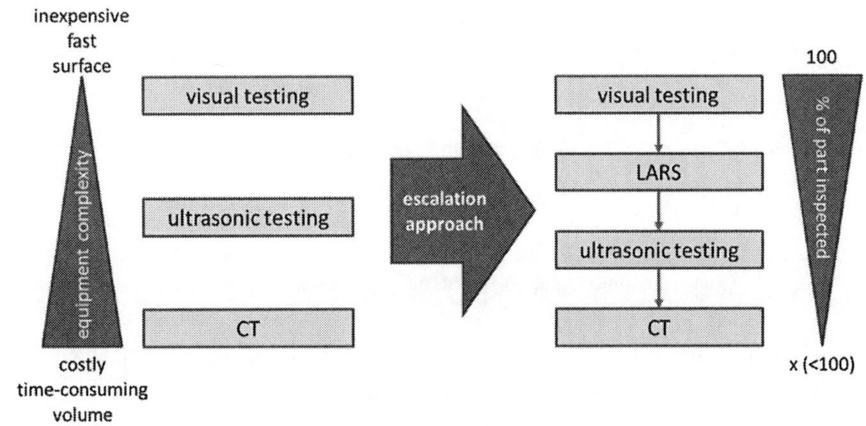

Figure 2: Simple example for the escalation approach suggested by the authors using VT, LARS, UT and CT

On the left, visual and ultrasonic testing (VT, UT) as well as computed tomography (CT) are sorted according to their complexity level. An inspection of the entire part using UT or CT is time-consuming and leads to high costs. If the suggested escalation approach including LARS is applied, the area of the part to be inspected by complex methods is decreased, resulting in fewer costs.

In practice, the measurements could look like presented in Figure 3 and Figure 4. Ultrasonic testing requires meandering one hundred percent of the surface with a sensor head using contact or immersion technique. In contrast, LARS requires a measuring grid of certain spacing, thus reducing the amount of time required for inspection. Using the method currently developed, a quasi-real time colored 2D map is visible on the inspectors screen, similar to other imaging techniques (see section 2).

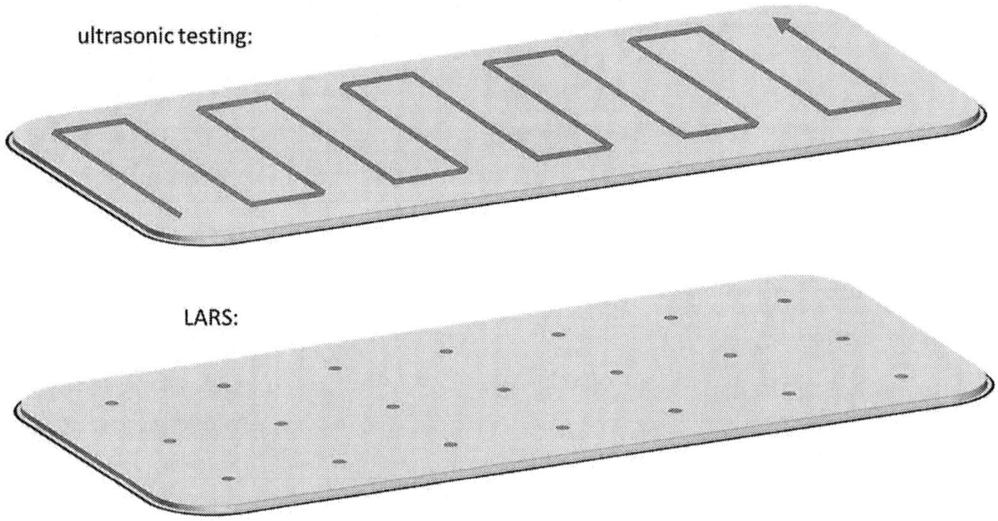

Figure 3: Schema for meandering a surface during ultrasonic testing of a large planar part (above) and a measuring grid for LARS (below)

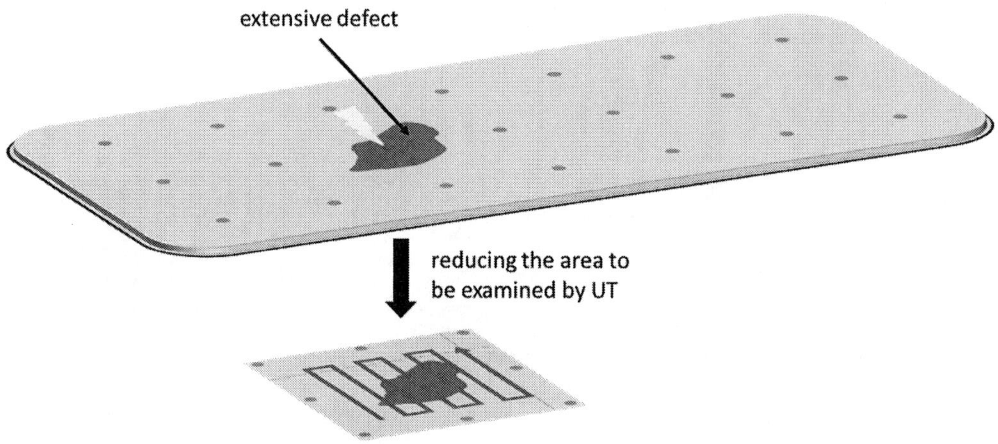

Figure 4: Reducing the area to be inspected by UT after application of LARS

1 Basic Principle and Physical Background

Like the conventional coin tap test, LARS is based on hitting a structure with a hard object. However, instead of the human ear a microphone is used as a sensor for the acoustic waves emitted by the impact and the structure in the audible range (<20 kHz). The basic principle of the coin tap test is depicted in Figure 5, whereas Figure 6 shows the sensor equipment used for LARS, floating above a CFRP part.

Figure 5: Detection of disbonded areas using the coin tap test

Figure 6: LARS: Using an instrumented hammer and a microphone instead of a coin and a human ear [4]

Both methods are based on the physics of colliding solids first described mathematically by Hertz [5]. An important physical quantity described by Hertz is the contact stiffness, which specifies the contact between two solids according to their elastic properties and the excitation force. In later investigations, this concept was extended to the effective stiffness, which includes a defect stiffness altering contact duration and frequency content of the received signal [6]. Thus, defects like delaminations can be detected by analyzing frequency shifts of the audio signal as well as the full width at half maximum of the hammer signal over the entire area of the measuring grid. Examples of this operating principle are described in the following section.

2 Experimental Work

First experimental work has been performed on CFRP parts with delaminations due to impact damage. For thin plates, the defect location can be detected above a certain impact energy [7]. Further experiments were conducted on a thicker plate with defined features such as flat bottom holes (FBH), a stepped wedge and a reference area. The geometry of the specimen is shown in Figure 7.

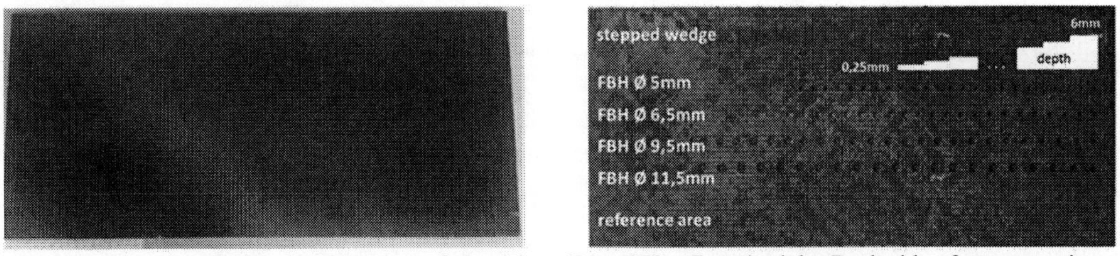

Figure 7: left: Front side of FBH test specimen (size: 620 x 270 x 7 mm); right: Back side of same specimen with rows of flat bottom holes of different depths, an "undamaged" reference area at the top and a stepped wedge at the bottom [4]

For the investigation, two different measuring grids were used. Firstly, the hammer impact locations were chosen to be directly above the FBH and a certain number of steps on the wedge (Figure 8 top left). Secondly, a regular spaced grid was used as in a real inspection situation. The results are shown in Figure 8 (center and bottom).

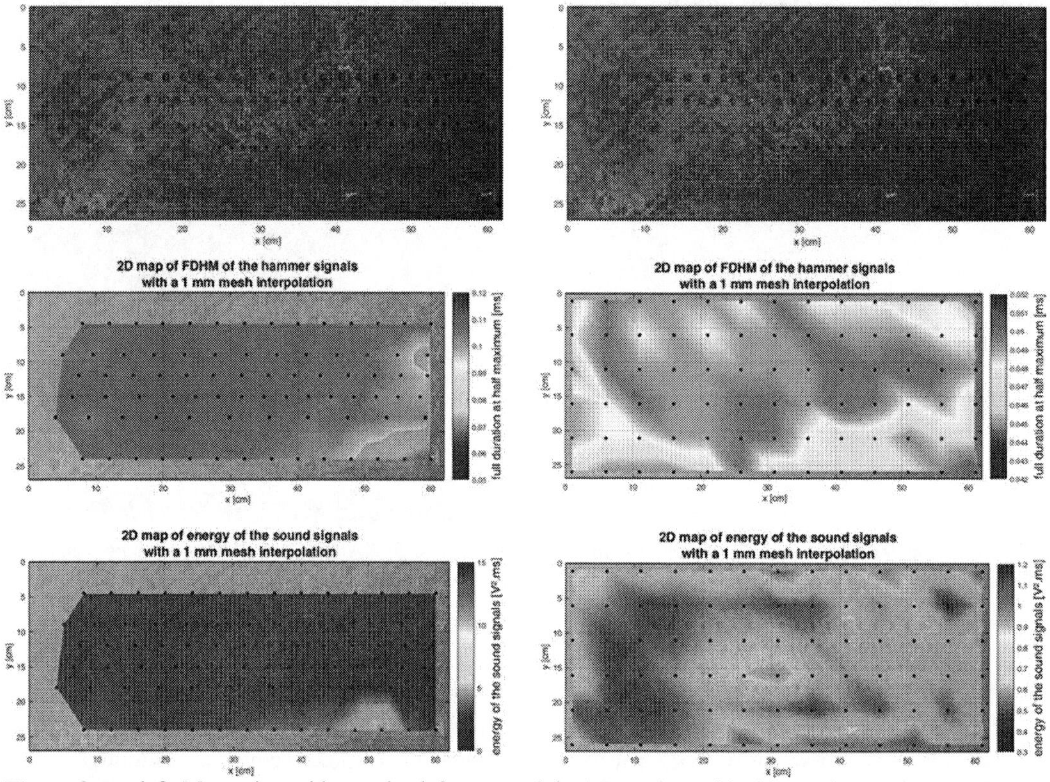

Figure 8: top left: Measuring grid over the defects; top right: Measuring grid with regular spacing; tapping on opposite side; center left: Map of FDHM distribution on defect grid; center right: Map of FDHM distribution on regular grid; bottom left: Map of time domain signal energy distribution on defect grid. Bottom right: map of time domain signal energy distribution on regular grid [8]

In this case, the full duration at half maximum (FDHM) of the hammer signal and the energy of the sound signals were used for the analysis. As expected, the results show that significantly thinner areas such as the deepest steps of the wedge or the deepest FBH can be identified under the given conditions. This can be a hint for the detectability of defects depending on their size and depth.

In other studies, LARS was applied extensively on GFRP materials. More information can be found in [1] [2] and [8].

3 Simulation

Numerical simulation helps improving the understanding of wave propagation within the structure as well as in the surrounding air thereby widening the range of application of LARS. For example, Figure 9 shows a simulation result of a plate impacted by a hammer. The propagation of surface and body waves can be observed and analyzed. Moreover, the waves emitted into the air can be identified as Lamb waves if the lateral dimensions of the plate are large compared to the thickness. If a delamination is present in the area of influence of the hammer impact, the frequency of such a wave is shifted dependent on the depth of the defect [9].

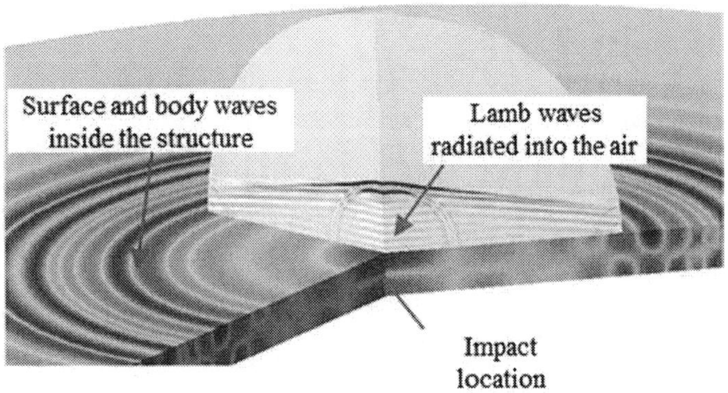

Figure 9: Numerical simulation of wave propagation in a plate after impact and emission of waves into the air [9]

4 Outlook

In further studies, possibilities and limits of LARS as a fast NDT method for CFRP parts are to be determined. For this, criteria like detectable defect type, size and location as well as the minimum part size are examined experimentally and supported by numerical simulation. Thus, the constraints for the application of LARS within an escalation approach are established.

5 References

[1] Juengert, A., C. Grosse, and M. Krueger, *Local Acoustic Resonance Spectroscopy (LARS) for Glass Fiber-Reinforced Polymer Applications.* Journal of Nondestructive Evaluation, 2013. **33**(1): p. 23-33.

[2] Hornfeck, C., et al., *Comparative Study of State of the Art Nondestructive Testing Methods with the Local Acoustic Resonance Spectroscopy to Detect Damages in GFRP.* Journal of Nondestructive Evaluation, 2015. **34**(2): p. 1-14.

[3] Grosse, C.U. *Bausteine für eine moderne Zerstörungsfreie Prüfung von Faserverbundwerkstoffen.* in *4. Würzburger Tagung Innovative Zerstörungsfreie Prüfverfahren (ZfP) für moderne Kunststoffe.* 2014. Süddeutsches Kunststoff-Zentrum SKZ Würzburg.

[4] Vandewalle, N., *Local Acoustic Resonance Spectroscopy - Non-destructive defect detection for carbon fiber reinforced composites*, Internship report, 2017. Chair of Non-destructive Testing. Technical University of Munich. 43 p.

[5] Hertz, H., *Über die Berührung fester elastischer Körper. Journal für die reine und angewandte Mathematik 92 , S. 156–171.* 1881.

[6] Cawley, P. and R.D. Adams, *The mechanics of the coin-tap method of non-destructive testing.* Journal of Sound and Vibration, 1988. **122**(2): p. 299-316.

[7] Mueller, M., P. Jatzlau, and C. Grosse. *Identification of Flawed CFRP Samples Using Local Acoustic Resonance Spectroscopy (LARS).* in *Proc. 19th World Conference on Non-Destructive Testing.* 2016. Munich: DGZfP.

[8] Grosse, C.U., A. Jüngert, and P. Jatzlau, *Local Acoustic Resonance Spectroscopy*, in *Handbook of Advanced Non-Destructive Evaluation*, N. Ida and N. Meyendorf, Editors. 2018, Springer International Publishing: Cham. p. 1-24.

[9] Groschup, R. and C.U. Grosse, *MEMS Microphone Array Sensor for Air-Coupled Impact-Echo.* Sensors (Basel, Switzerland), 2015. **15**(7): p. 14932-14945.

Simulation Based Key Performance Indicator Determination in Guided Wave Monitored Metallic Patched Repairs

Aadhik ASOKKUMAR [1], Ramanan SRIDARAN VENKAT [2], Christian BOLLER [2]
[1] Dresden International University (DIU),01067, Dresden, Germany
[2] Chair of NDT & Quality Assurance (LZfPQ), Saarland Univ.
66125, Saarbrücken, Germany
Contact e-mail: aadhikasok1994@gmail.com

Abstract. Monitoring of structures based on structural health monitoring (SHM) provides a variety of significant challenges. One of them is related to simulation, which is a necessity with respect to determining an optimum transducer configuration for a SHM system. The example to be considered in the end is a metallic patched repair of an aircraft fuselage component considering different damage scenarios for which guided wave simulations have been performed such that an SHM system can be developed. In this context the sensor pattern, actuation pattern and the input signal have been varied for reasons of optimization. The guided waves generated have been used to analyze the signals obtained with artificial intelligence (AI) algorithms trying to identify key performance indicators (KPI) with respect to SHM. Further methods considered with respect to AI are principal component analysis (PCA), artificial neural networks (ANN), time reversal methods or the Gaussian mixture model. The results obtained are presented and discussed in light of concluding the significance of simulation in SHM.

1 Introduction

The primary aim of repairing aircraft structures is to ensure the integrity of the aircraft structure by restoring the damaged parts to their original condition. In an aircraft fuselage one of the ways to repair is to add a patch over the damaged area using the same or even a different material. Attachment is often done by riveting and sometimes also by bonding and material combinations between metals, and polymer-based composites has also become increasingly popular. Repairs can often also become rather complex and non-standard, which may result in some unique piece of engineering work. As a consequence, there is a risk of those repairs to get damaged again in service and as a consequence structural health monitoring (SHM) has become an interesting option to safeguard the integrity of those repairs specifically when they have even been designed damage tolerant. With this, the condition of the repair can be evaluated continuously.

An SHM system is a monitoring system where sensors and possibly also actuators and hence the non-destructive testing (NDT) technology become an integral part of the structure being monitored. As such the number and location of sensors and possibly also actuators allowing a tolerable damage to be detected reliably have to be decided in advance. A strong option in that regard is numerical simulation when being available, since this allows an understanding of the physical processes involved with respect to reliably detecting a tolerable damage defined.

Among the various NDT techniques being available acoustic wave propagation analysis and vibrational analysis are two techniques being very much considered in SHM [1]. Both techniques can be in a passive and an active mode where in the passive mode signals are generated by damage (acoustic emission) or operational loads (modal analysis) while in the active mode actuators are generating the respective acoustic or vibration signal. Lamb waves (also called plate or guided waves) are dispersive waves, which are guided between two parallel preferably free surfaces such that the acoustic impedance between the structure considered and its environment becomes significantly different. As a consequence, the waves being generated by an actuator can comparatively travel long distances with limited energy loss. Considering this phenomenon, it is used in tubular structures such as pipelines. Currently within the context of SHM it is also very much discussed with respect to aircraft structures and also other applications. The two modes being characteristic for Lamb waves are the symmetric and the antisymmetric mode, where each of the modes is characterized by its fundamental (mode 0) and its higher harmonics (modes 1,2,3, ...). Since Lamb waves are dispersive, each of the modes propagates at different speeds depending on the excitation frequency applied.

In case damage occurs in a structure the constant shape of the structure will be disturbed and the Lamb wave travelling as well. As a consequence, the difference of the Lamb wave signals between the pristine and the damaged condition becomes a clear indicator regarding the damage's severity and location. Lamb waves can be generated by different methods such as non-contact techniques by using a laser [2], air-coupled transducers [3] and laser-induced plasma shock waves [4]. However, contact methods like using Giant Magnetostrictive Actuators (GMS effect) [5], electromagnetic transducers (EMAT) [6] and piezoelectric transducers (PZT) are the options more frequently used. PZT, especially piezoelectric wafer active sensors (PWAS) are the most popular ones used because of their low weight and volume and their ease in being integrated into structures. Also, with the PWAS, Lamb waves can be generated even with very few micro volts.

SHM systems being integrated into a structure can be exposed to a variety of environmental conditions while being in operation. This will further enhance the complexity of the SHM system and specifically the Lamb wave signals being generated including their interpretation. As a consequence, there is a need for a standardized process along which the phenomena monitored can be determined by excluding the external influences on purpose. Finite Element (FE) modeling and analysis is a cost and time effective solution in that regard and the analysis can be done in a very distinct way. Some of the models may even allow the environmental factors to be simulated. In order to increase the probability of detection (POD), artificial intelligence such Artificial Neural Networks (ANN) can be developed and trained to get a desired optimal solution [7] possibly represented by Key Performance Indicators (KPI).

In this paper, Lamb wave signals are generated by simulating different models with the FE software *COMSOL Multiphysics* and different kinds of analysis are performed with the signals in view of determining KPIs. These KPIs can be used as an input for the ANN, which can help the network to enhance the results with respect to a higher POD.

2 Theory

Lamb waves have been named after Sir Horace Lamb, one of the men who discovered guided waves. Lamb waves as shown in Fig 1 are traction-free acoustic waves that are guided between two parallel free surfaces of a thin plate having a significantly different acoustic impedance when compared to its environment. They are dispersive in nature, which means their phase and group velocities depend on the excitation frequency [8]. The reader is referred to [9] for transformation of Rayleigh–Lamb frequency equations into an equation with only real values of wavenumbers k and also to [10] for the numerical solution of the equation. The dispersion curves are the numerical solution for the Rayleigh-Lamb equations. Within the *Matlab software, the Lamb toolbox* [11] was used to generate the dispersion curves. Fig 1 shows the group velocities versus the frequency for the material '*Aluminium 2024 T3 of 2mm thickness*'. In order to make the analysis simple, the excitation frequency is kept at a lower range so that only the fundamental frequencies do occur in the plate.

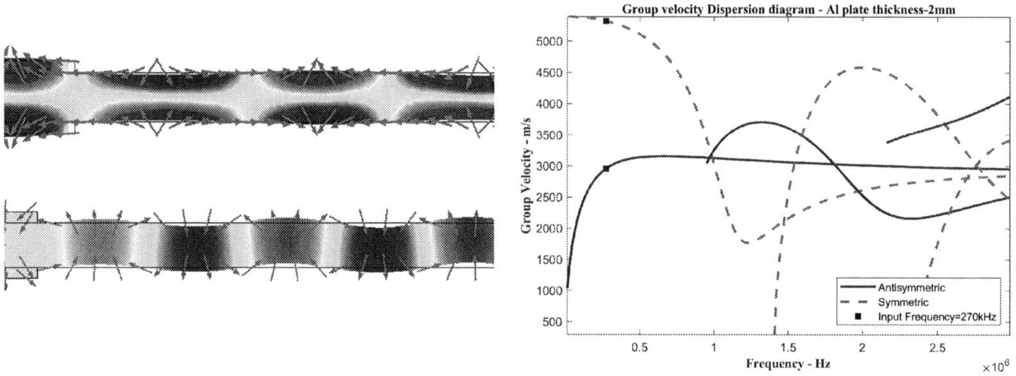

Fig 1 : Lamb waves travelling in a plate. The arrows represent the particle movement. Symmetric wave (S0) is out of phase from the mid plane (left-top) and anti-symmetric wave (A0) is in phase from the mid plane (left-bottom). Group velocity dispersion diagram and the respective mode velocities for 270 kHz of excitation frequency(Right).

3 Simulation

The material used for analysis is *Aluminium 2024 T3 with 2mm thickness, density 2700 kg/m³, Youngs modulus of 70 GPa and Poisson's ratio 0.33.* The material of the PZT is *Lead Zirconate Titanate 5A with Density 7750 kg/m³ and Poisson's ratio 0.33. The elasticity matrix (GPa) being* $E_{11}, E_{22}, E_{33}, E_{44}, E_{55}, E_{66}, E_{12}, E_{13}$ = *120.4, 120.4, 110.9, 21.1, 21.1, 22.6, 75.2, 75.1 respectively, coupling matrix (C/m²) being* $C_{15}, C_{24}, C_{33}, C_{31}, C_{32}$ = *12.3, 12.3, 15.8, -5.4, -5.4 respectively and the relative permittivity matrix being* $\varepsilon 11, \varepsilon 22, \varepsilon 33$ = *919.1, 919.1, 826.6 respectively. The diameter and thickness of PZT is 8 mm and 0.48 mm excited with a voltage of 1 V. A 5 cycles Hanning window function was used for excitation.*

COMSOL Multiphysics is a Finite Element software in which there is a possibility to combine different physical phenomena. For generating Lamb waves, the structural mechanics and the piezo electrics modules in COMSOL are combined. Analyzing the signals from a patched repair is an interesting but complicated task because of various mode conversions occurring due to waves travelling through varied thicknesses as shown in Fig 2.

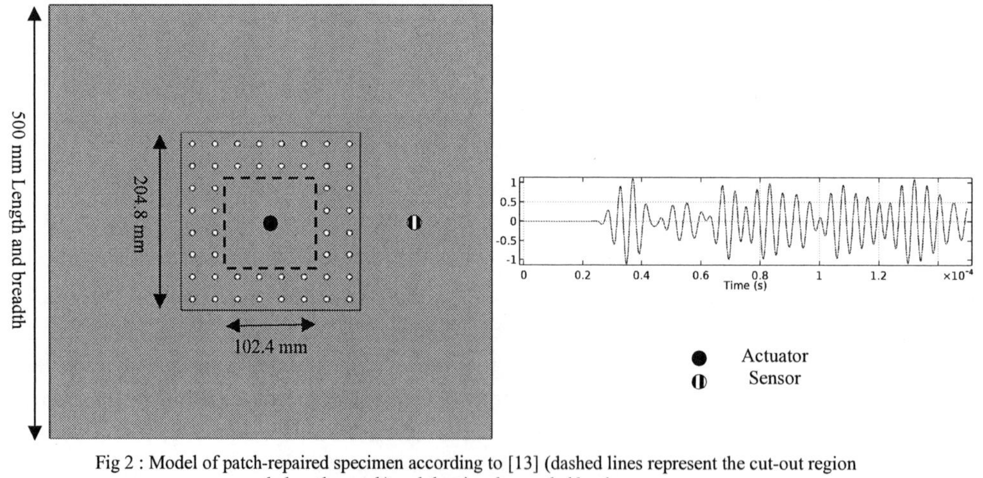

Fig 2 : Model of patch-repaired specimen according to [13] (dashed lines represent the cut-out region below the patch) and the signal recorded by the sensor.

As a consequence, a stepwise approach is considered. For analysis purposes just in view of generating signals, a 2D simulation is sufficient in the first place because of the limited number of degrees of freedom and the simulation speed when compared to a 3D simulation. Following an enhancement in structural complexity a single plate, a lap joint and a patched repair have been considered for simulation where, the model of the patched repair has been taken from [12]. Tasks to be performed while simulating have been to identify the locations where transducers should be placed best, where the method of differential imaging [13] has been applied that can be even combined with damage parameters [14].

For now, the single plate configuration is shown in Fig 3. Selective mode excitation is preferred in order to reduce the complications in signal processing. Either the S0 mode or the A0 mode can be generated by placing two actuators on the top and bottom of the plate. The pure S0 mode is generated by actuating the two PZT wafers with the same polarity while the pure A0 mode is generated by actuating the two PZT wafers with opposite polarities. The simulation results obtained are shown in Fig 4 for the Al 2024 plate of thickness 2 mm. All results were obtained using a 270 kHz excitation frequency with the time domain signals being measured with a step of 0.1 μs.

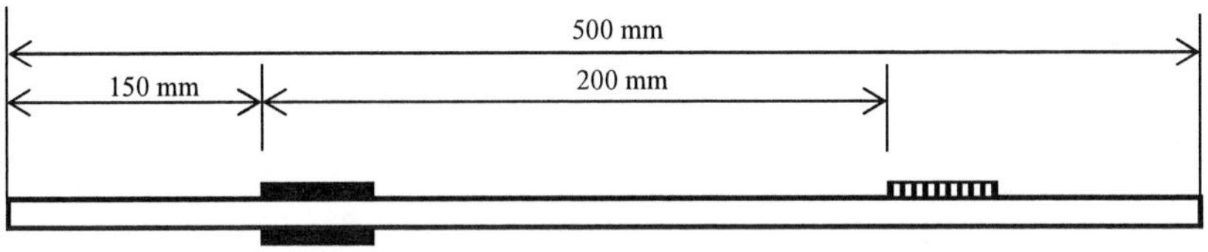

Fig 3 : Sketch of 2D plate, solid box being actuators and the patterned box being the sensor

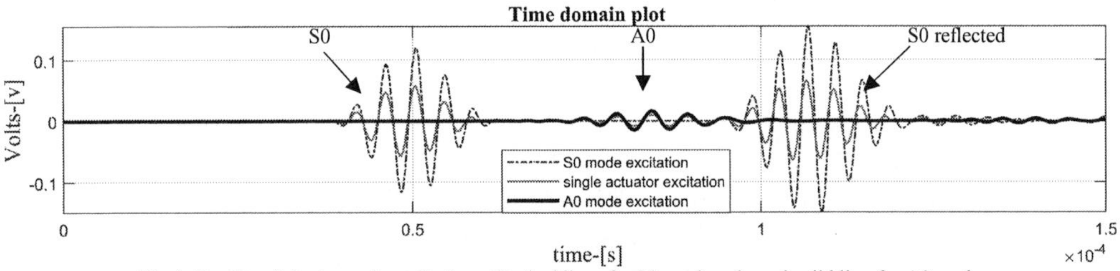

Fig 4: Results of single mode excitations. Dashed lines for S0 mode only and solid line for A0 mode only (black) compared to normal mode of excitation in which both modes exist (red)

In order to get the highest amplitude in the signal, a frequency sweep has been carried out with frequencies ranging from 100 kHz to 500 kHz with a step of 10 kHz. The frequency sweep was carried out for modes S0 and A0 separately and their normalized amplitude distribution is shown in Fig 5. These amplitudes are compared to some theoretical analytical result obtained from the software *WAVEFORM REVEALER* [15].

In simulation, the velocity and amplitude of the Lamb waves can be very sensitive to the mesh size because it is dependent on the wavelength [9]. This is true in the case of A0, which can be seen from Fig 5. Smaller mesh sizes yield better results but at the price of increased computation effort and time. As to be seen from Fig 5, frequencies below 270 kHz look to be a good frequency-mesh size tradeoff for the available resources. The excitation frequencies of 270 kHz for the S0 mode and 150 kHz for the A0 mode look to be optimum for aluminium T2024 2 mm thick plates.

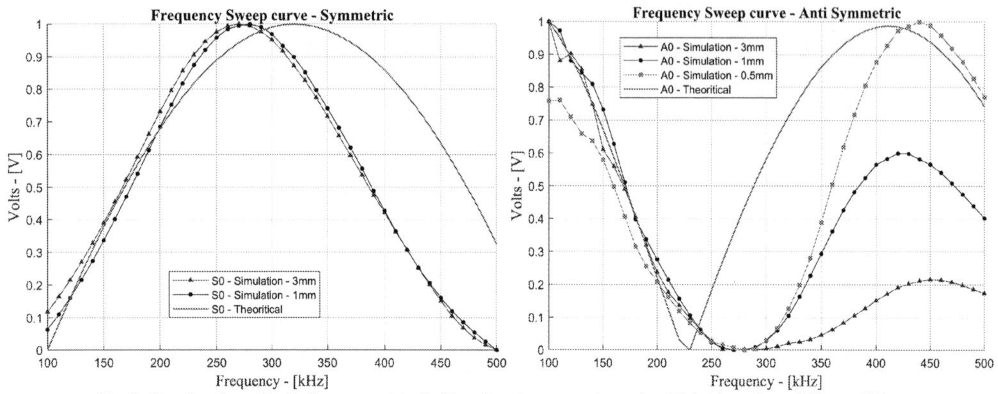

Fig 5: Simulated amplitude for symmetric (left) and anti-symmetric modes (right) based on different FE mesh sizes.

4 Analysis

First simulation studies have been performed on the plate specimen shown for the pristine condition in Fig 3 and the artificially with a surface slit 'damaged' condition in Fig 6. Different KPIs have been selected for which the signals were captured only up to 92 µs such that boundary reflections of the guided waves could be avoided. The different KPIs having been considered are briefly described below.

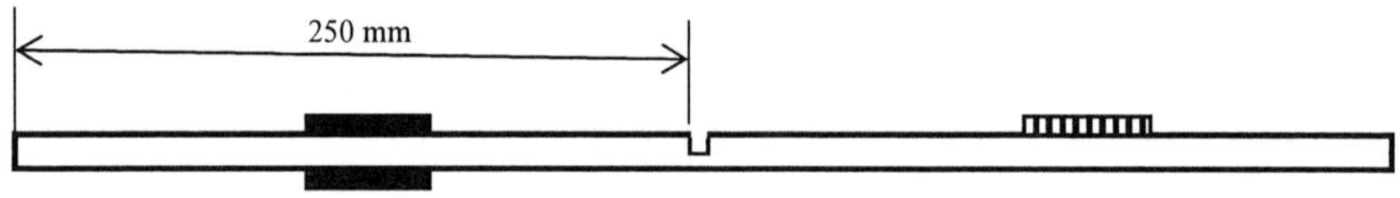

Fig 6: Plate with a crack at 250 mm from the corner. Crack width is 0.5 mm and depth is 1.5 mm

4.1 Amplitude and phase shift

This technique is the classic way of analysing signals and detecting damage. If a baseline signal is compared to the signal from the damaged specimen, a change in amplitude and phase can be observed. Fig 7 shows the peak amplitude pattern for a change in width and depth for the S0 mode. The S0 mode is very sensitive in detecting through the thickness cracks recorded in an amplitude and phase shift. The difference in signal between the pristine and damaged condition is an indicator of crack severity and hence a KPI.

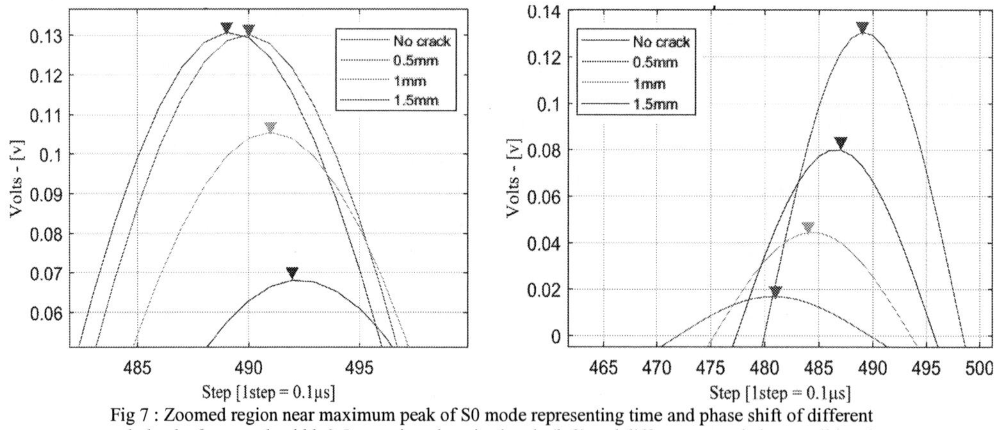

Fig 7 : Zoomed region near maximum peak of S0 mode representing time and phase shift of different crack depths for a crack width 0.5 mm; time domain signals (left) and difference to pristine condition for cracked conditions (right)

4.2 *Hilbert Transform*

Hilbert transform $H(t)$ is a signal transformation technique which is used to canvas the energy distribution over time [16]. The real part of the signal is phase shifted to 90° for constructing the analytical signal (Eq. (2)). $\mathrm{e}(t)$ is a module of $F_A(t)$ from which the envelope of the signal is obtained to reveal the energy distribution of $f(t)$ and from the imaginary part, the phase angle information $\phi(t)$ of the signal is determined as shown in Fig 8. The Hilbert transform is defined as

$$H(t)=\frac{1}{\pi}\cdot\int_{-\infty}^{+\infty}\frac{f(\tau)}{t-\tau}\cdot\mathrm{d}\tau \tag{1}$$

$$F_A(t)=f(t)+\mathrm{i}H(t)=\mathrm{e}(t)\mathrm{e}^{\mathrm{i}\phi(t)} \tag{2}$$

where, $$\mathrm{e}(t)=\sqrt{f^2(t)+H^2(t)}\ ;\ \phi(t)=\frac{1}{2\pi}\cdot\frac{\mathrm{d}}{\mathrm{d}t}\arctan\left(\frac{H(t)}{f(t)}\right) \tag{3}$$

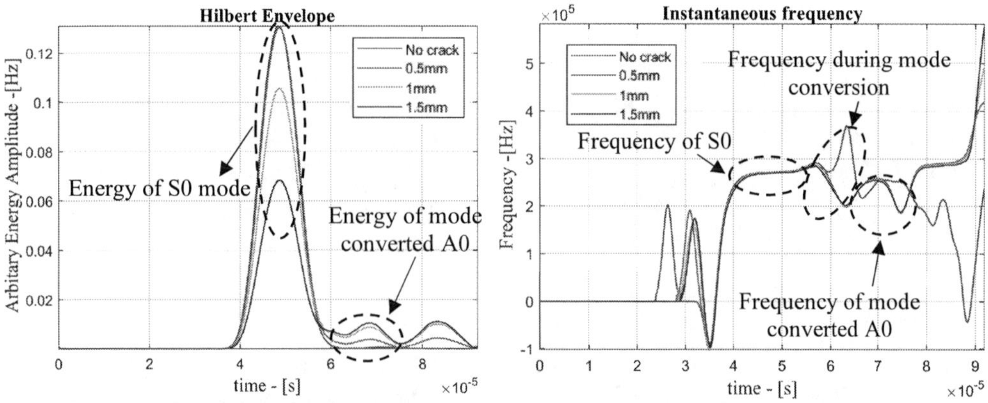

Fig 8 : Hilbert envelope of the time domain signals (left) and respective smoothened, unwrapped phase angle (right) both at instantaneous frequency.

The Hilbert envelope shows the energy distribution for different crack lengths. The amplitude decreases as the crack depth increases. The instantaneous frequency plot indicates the unwrapped phase angle of the Lamb wave signal which in turn gives the information about different frequencies at different times.

4.3 *Short time Fourier transform (STFT)*

The time frequency representation (TFR) is a transform done to visualize the frequency components occurring at different time instances. Short time Fourier transform is a technique in which the signal in time domain is chopped into a series of small overlapping pieces [17]. The STFT $S(\omega,t)$ is defined as

$$S(\omega,t)=\frac{1}{2\pi}\int_{-\infty}^{\infty}\mathrm{e}^{-\mathrm{i}\omega\tau}s(\tau)h(\tau-t)\mathrm{d}\tau \tag{4}$$

where, $h(t)$ is the window function. It is to be noted that in STFT, both time and frequency for a signal cannot be resolved at the same time. Therefore, the window length has to be changed appropriately for different applications. For analysis, the Hanning window with a length of 151 μs and an overlap of 98% was used. The power spectral density of STFT is the absolute of $S(\omega,t)$. It was converted to dB scale and the frequencies below -35 dB were suppressed. The magnitude can be used as a KPI. The result of STFT is shown in Fig 9.

4.4 Smoothed Pseudo Wigner-Ville Distribution (SPWVD)

Wigner-Ville distribution (WVD) is a type of time-frequency representation which overcomes the limitation of STFT with localization in both time and frequency. The major issue with WVD is the interreference due to cross terms with non-zero values. Hence the modified version of WVD, the SPWVD is used to overcome the cross terms [18]. SPWVD is given by

$$S(t,f)=\iint h(\tau)g(s-t)x\left(s+\frac{\tau}{2}\right)x^{*}\left(s-\frac{\tau}{2}\right)e^{-j2\pi f\tau}ds\,d\tau \tag{5}$$

where x^* is the complex conjugate of x, h and g are the smooth window function for time and frequency respectively. In order to compare SPWVD with STFT, the same window size of 151 μs with Kaiser (Kaiser was better for this TFR) window was used for both h and g. The end result is the power spectrum that had been processed the same way as STFT and is shown in Fig 9.

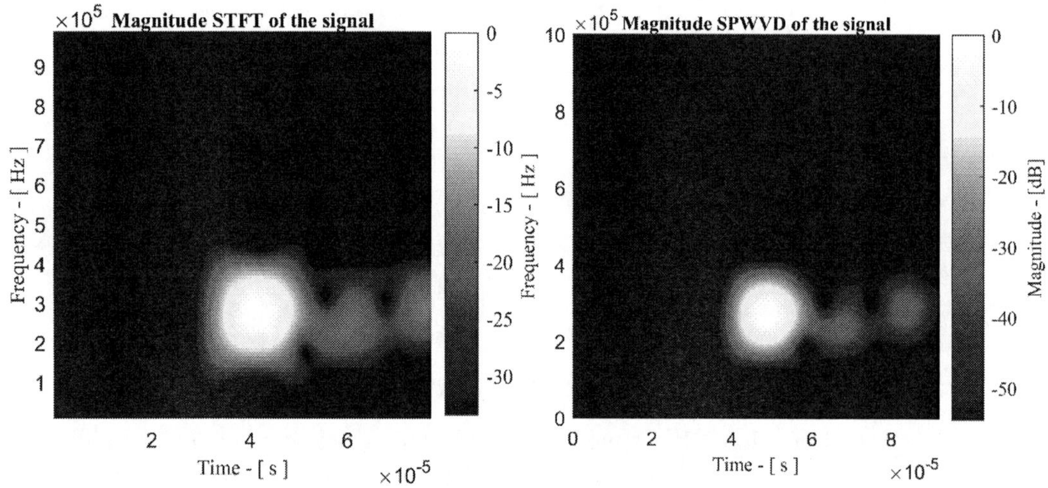

Fig 9: Results for crack depth 1.5 mm with STFT (left) and SPWVD (right). For both TFRs a window length of 151 μs was used.

4.5 Statistical methods

The following statistical methods are KPIs in which the baseline signal is compared to the signal from the damaged specimen.

4.5.1 Linear Correlation Coefficient

Two signals are compared by using a linear correlation of Pearson type defined by equation (6) which compares the two-time instances in this case being two signals x and y which return a value based on their similarity. It is normalized between 0 and 1 where 1 represents totally identical signals. In the equation, μ is the mean, Cxy the covariance, σ the standard deviation and k the total length of the signal respectively.

$$Corr=\frac{Cxy}{\sigma_x\sigma_y}=\frac{\sum_{k=1}^{k}(x_k-\mu_x)(y_k-\mu_y)}{\sqrt{\sum_{k=1}^{k}(x_k-\mu_x)^2}\sqrt{\sum_{k=1}^{k}(y_k-\mu_y)^2}} \tag{6}$$

4.5.2 *Autocorrelation Coefficient*

Correlation is a technique which is used to compare two variables or measured data signals (cross correlation) to see how similar they are to each other [19]. Autocorrelation or serial correlation is the correlation of a signal with a delayed copy of itself as a function of delay (time lag). The result in the statistical method is normalized between 0 and 1 while in the signal processing method the result is not normalized [20]. Usually the statistical method is often used because the normalized values can be used for further analysis. The autocorrelation coefficient r_k is defined as [21],

$$r_k = \frac{c_k}{c_0}\ ;\ c_k = \frac{1}{T}\sum_{T-1}^{T-k}\left(y_t - \bar{y}\right)\left(y_{t+k} - \bar{y}\right) \tag{7}$$

where, c_0 is the sample variance of the time series y_t, $\bar{y}$ is the mean of the respective data set and k is the lag of the data set. The ratio of the envelope of the autocorrelated signal of the damaged specimen to the baseline signal (Fig 10(b)) shows at which time the Lamb wave modes are interacting with the crack.

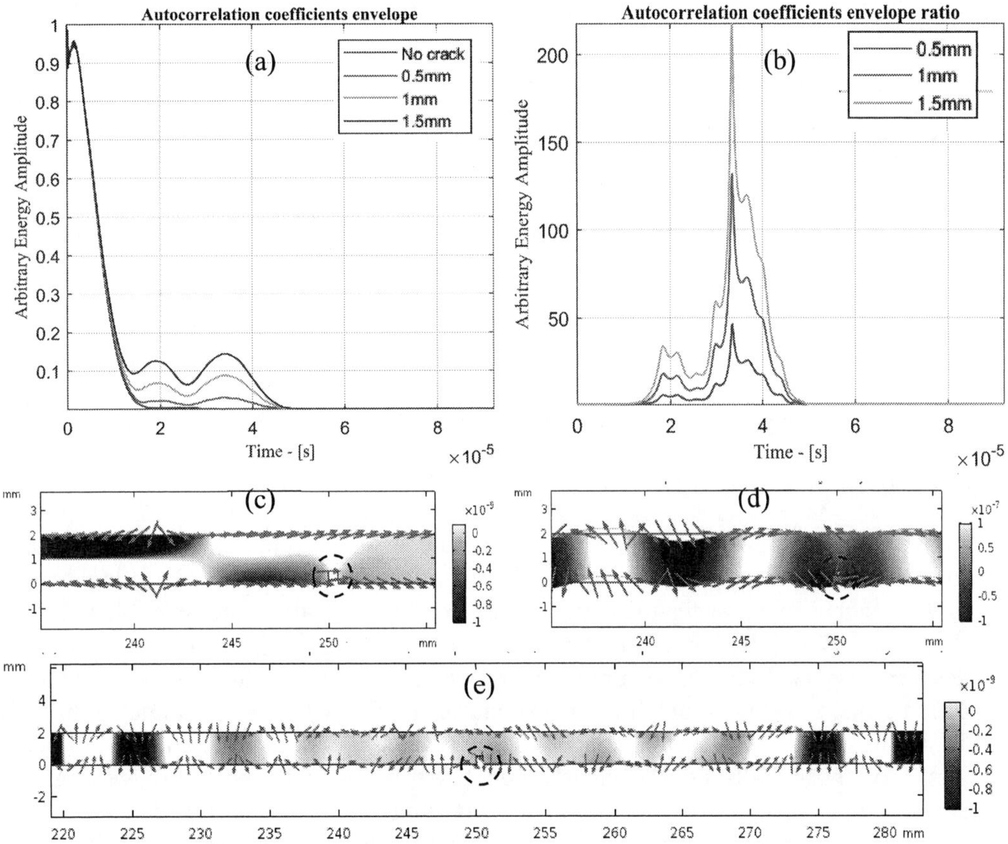

Fig 10 : (a) Envelopes of the autocorrelated signals (b) ratio of envelopes of signal from damaged specimen to the baseline signal. (c) and (d) are snapshots from simulation showing the wave interaction at 18 and 33 μs respectively (e) waves at around 49 μs having finished interaction with the damage. Dashed circles indicate the region of the artificial crack.

The snapshots from simulation are shown in Fig 10 (c), (d) and (e) and indicate that the peak at 18 μs is related to the S0 packet starting to interact with the crack, the peak at 33 μs is related to the S0 wave being mode converted to the A0 mode at highest energy and the time around 49 μs indicates that the interaction of the waves with the damage is complete respectively. The amplitudes of peaks at 33 μs and 49 μs can be used as the KPIs and this information can be used to calculate the crack location (distance) from group velocity.

4.5.3 Root-Mean Square Deviation (RMSD)

RMSD is defined as, "A measure of accuracy, to compare forecasting errors of different models for a particular dataset and not between datasets, as it is scale-dependent" [22]. The respective equation can be written as [23]

$$RMSD = \sqrt{\frac{\sum_{t=1}^{k}\left(x(t)-y(t)\right)^2}{k}} \tag{8}$$

where, x is the reference signal and y is the signal to be compared. k is the total length of the signal. The results are between 0 and 1 where 0 being the signal with no error which is identical to the reference signal.

4.5.4 Mean Absolute Error

Mean absolute error is the measure of distance between two continuous variables and is simply the ratio mean of absolute values of the difference between the data of signals and the total length of the signal. It is given by [23] as,

$$MAE = \frac{\sum_{t=1}^{k}\left|x(t)-y(t)\right|}{k} \tag{9}$$

where x and y are the baseline signal and the signal from the damaged specimen respectively. k is the total length of the signal and the values are again normalized from 0 to 1 where 0 being the signal with no error which is identical to the reference signal.

5 KPI Determination Results

The above analysis has shown for a first example of low complexity that Lamb wave simulation can serve as a means to identify KPIs. KPIs, which have been considered here as Lamb wave signal inherent features, are features either generally used or published in the open literature. The KPIs shown here do describe a clear relationship with damage size. This has been observed with the classical parameters such as amplitude and phase (Fig 7) but also the Hilbert transform (Fig 8) and the autocorrelation (Fig 10(b)). A similar conclusion can be drawn with respect to the statistical methods. The results are summarized in Table 1 below.

All of those KPIs can be clearly quantified and it may be worth getting those normalized such that a true comparison regarding their sensitivity can be shown. Autocorrelation can even determine the time when the Lamb wave interacts with the damage and it weakens the influence from boundary reflections. STFT and SPWVD are KPIs, which show a sensitivity to damage as well, however, its result is displayed in a frequency-time diagram from which a quantifiable number first needs to be derived. The next questions to be answered now are if the same 'topology' of normalized KPIs is obtained for the same type of damage when the complexity of the structure to be analyzed increases and if a combination of KPIs may enhance POD.

		(X=Baseline, Y=Damaged) (different depths of crack)		
Raw Signal	**X**	**Y(0.5mm)**	**Y(0.5mm)**	**Y(0.5mm)**
Amplitude (V)(**S0**)	0.131	0.130	0.105	0.068
Time (μs) (**S0**)	48.8	48.9	49.0	49.1
Difference of Signals	**X-X**	**X-Y(0.5mm)**	**X-Y(1mm)**	**X-Y(1.5mm)**
Amplitude (V) (**S0**)	0	0.017	0.045	0.080
Time (μs) (**S0**)	0	48.0	48.3	48.6
Hilbert Envelope	**X**	**Y(0.5mm)**	**Y(0.5mm)**	**Y(0.5mm)**
Energy (Normalized)(**S0**)	1	0.994	0.807	0.522
IF (kHz) (**S0**)	270.79	271.78	271.07	269.96
Time(μs) (**S0**)	48.6	48.7	48.8	48.8
Energy (Normalized)(**A0**)	0.0046	0.0280	0.0674	0.0802
IF (kHz) (**A0**)	254.29	257.19	254.74	252.96
Time(μs) (**A0**)	72.3	68.6	68.5	68.4
Autocorrelation Ratio	**X/ X**	**Y/X(0.5mm)**	**Y/X(1mm)**	**Y/X(1.5mm)**
Amplitude (**S0**) at **18.5 μs**	0	6.052	18.304	34.099
Amplitude (**A0**) at **33.5 μs**	0	46.532	131.953	217.448
Other Statistical Methods	**X-X**	**X-Y(0.5mm)**	**X-Y(1mm)**	**X-Y(1.5mm)**
LCC	1	0.990	0.943	0.838
RMSD	0	0.003	0.009	0.016
MAE	0	0.001	0.004	0.007

Table 1 Results for the different methods with different crack conditions

6 Conclusion and Future work

When considering the Lamb wave signals, the information about the delamination, crack and deformation in the structure is hidden in the signal and the fact that Lamb wave signals are non-linear (what the sensor is measuring is not directly related to the excitation signal of the actuator), definitely one has to perform different kinds of transformations and signal processing in order to study about the Lamb wave signal. All the above methods were studied for the single plate with an artificial surface crack so far and will be expanded to the patched repair shown in Fig 2 above. The selection of KPIs is not limited to what has been presented but is rather open to whatever may be proposed otherwise or developed in the future. This may include techniques such as principal component analysis (PCA), time reversal methods or the Gaussian mixture model, to just name a few. However, the flavour of such a comparative study and the approach in terms of KPIs can serve as a template on how to benchmark KPIs at its best. A next step is now to combine the various KPIs such that an enhanced result in terms of classification, quantification and location of the damage can be found. This is considered to be done in terms of an ANN. The different KPI results will therefore be combined to an input vector, which tells the information of the structure in different aspects. With this the ANN can learn on how to converge the different KPIs to the desired outputs such as an enhanced POD for a complex damage tolerant structure in a controlled space represented by the structural repair mentioned above. The approach to be taken in that regard is shown in Fig 11 and may also serve as an instrument on how to handle the SHM system design issue in the sense of big data handling in the longer term.

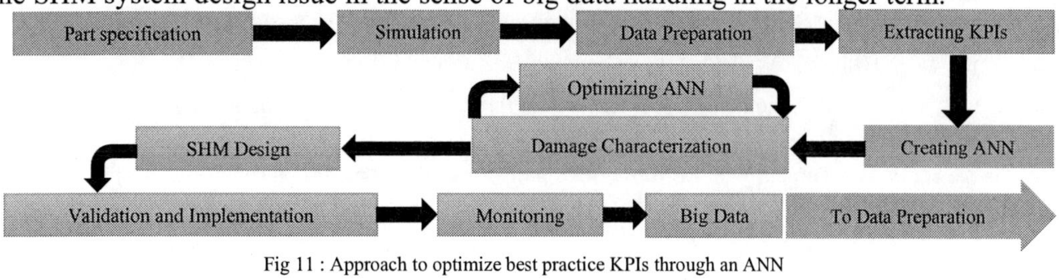

Fig 11 : Approach to optimize best practice KPIs through an ANN

7 References

[1] C. Boller, F.-K. Chang and Y. Fujino (Ed.s), "Encyclopedia of Structural Health Monitoring," in *Vol 5*, Chichester/UK, J. Wiley & Sons, 2009.

[2] S. Yashiro et al, "An NDT technique for composite structures using visualized Lamb-wave propagation," *Composites Science and Technology,* vol. 67, no. 15-16, pp. 3202-3208, December 2007.

[3] M. Castaings et al, "The generation, propagation, and detection of Lamb waves in plates using air-coupled ultrasonic transducers," *The Journal of the Acoustical Society of America 100,* vol. 3070, 1996.

[4] N. Hosoya et al, "Non-contact and non-destructive Lamb wave generation," *International Journal of Mechanical Sciences using laser-induced plasma shock wave,* vol. 140, pp. 486-492, May 2018.

[5] M. Luo et al, "Development of a Novel Guided Wave Generation System Using a Giant Magnetostrictive Actuator for Nondestructive Evaluation," February 2018.

[6] Zhiqi Guo et al, "EMAT GENERATION AND LASER DETECTION OF SINGLE LAMB WAVE MODES," 1998.

[7] Ye Lu et al, "Artificial Neural Network (ANN)-based Crack Identification in Aluminum Plates with Lamb Wave Signals," *Intelligent Material Systems and Structures,* January 2009.

[8] Z. Lasova et al, "Determination of group velocity of propagation of Lamb waves in aluminium plate using piezoelectric transducers," *Applied and Computational Mechanics,* vol. 11, pp. 23-32, 2011.

[9] P. Gómez et al, "Lamb Waves and Dispersion Curves in Plates and its Applications in NDE Experiences Using Comsol Multiphysics," in *COMSOL*, Mieres, Spain., 2011.

[10] Joseph L.Rose, Ultrasonic Guided Waves in Solid Media, Section 6.3, Cambridge University Press, 2014.

[11] J. L. Prego, "Matlab File exchange," 2010. [Online]. Available: https://www.mathworks.com/matlabcentral/fileexchange/28367-the-lamb-toolbox.

[12] A. Taltavull, *Structural Health Monitoring Solutions for Patched Metallic Aircraft Repairs Based on Guided Ultrasonic Waves,* Saarbrücken, Germany, May 2017.

[13] C. Boller et al, "Integration of Non-Destructive Evaluation-based Ultrasonic Simulation: A means for simulation in structural health monitoring," *Structural Health Monitoring,* vol. 16, no. 5, pp. 611-629, 2017.

[14] R. Sridaran Venkat, M. El Bakouri and C. Boller, "Damage Indices as a Measure for Optimal Sensor Placement in a," in *9th EWSHM*, Manchester,UK, 2018.

[15] "LABORATORY FOR ACTIVE MATERIALS AND SMART STRUCTURES," University of South Carolina, [Online]. Available: http://www.me.sc.edu/research/lamss/html/software.html.

[16] L. Cheng, "Identification of corrosion damage in submerged structures using," *Smart Materials and Structures,* November 2009.

[17] Niethammer et al, "Time-frequency representation of Lamb waves using the reassigned spectrogram," *Acoustics Research Letters Online,* March 2000.

[18] J. Li and S. Liu, "The Application of Time-Frequency Transform in Mode Identification of Lamb Waves," in *17th World Conference on Nondestructive Testing*, Shanghai, China, 2008.

[19] "Wikipedia," [Online]. Available: https://en.wikipedia.org/wiki/Cross-correlation ; https://en.wikipedia.org/wiki/Correlation_and_dependence.

[20] "Wikipedia," [Online]. Available: https://en.wikipedia.org/wiki/Autocorrelation.

[21] G. E. P. Box, "Time Series Analysis: Forecasting and Control," Englewood Cliffs, NJ, 1994.

[22] "Wikipedia," [Online]. Available: https://en.wikipedia.org/wiki/Root-mean-square_deviation.

[23] R. J. Hyndman et al, "Another look at measures of forecast accuracy," *International Journal of Forecasting,* vol. 22, no. 4, p. 679–688, 2006.

[24] Ghose et al, "Two Dimensional FEM Simulation of Ultrasonic Wave Propagation in Isotropic Solid Media using COMSOL," in *COMSOL*, India, 2010.

Guided wave monitoring of a riveted metallic patched repair using a model-based approach

Haniza MOHD NOOR [1], Christian DÜRAGER [2], Ramanan SRIDARAN VENKAT [3]
Christian BOLLER [1,3]
[1] Dresden International University (DIU), Dresden/Germany
[2] IMITec GmbH, Meilen/Switzerland
[3] Chair of NDT & Quality Assurance (LZFPQ),
Saarland University/Saarbrücken, Germany

Contact: christian.duerager@imitec.ch

Abstract.
A challenge with respect to structures based on structural health monitoring (SHM) is to extract the right features from the signals having been measured. A way on how to get this done is on the basis of a model-based damage feature extraction procedure. In that case a numerical model of the structure considered is taken as a basis to determine sensor signals such as being generated through guided ultrasonic waves. Signals are measured experimentally in parallel and are compared to the simulated results. In case of a mismatch the model is then adapted through a learning process to what is monitored in reality. The model can be of different levels of complexity starting from a baseline measurement being compared to an actual measurement and ending in a rather complex mathematical (i.e. numerical) model to be considered as the reference that may adaptively be adjusted through non-linear least mean square fits. In the paper to be presented the model-based feature extraction process mentioned above is intended to be validated experimentally on different components of increasing structural complexity starting from a flat plate only over the same type of plate with three holes and a patched metallic repair both considering an undamaged and an artificially damaged condition. The components are monitored with a wireless guided wave based SHM system. Locations of the transducers to be placed on the components were decided in accordance to numerical simulations performed in parallel.

1. Introduction

Ultrasonic testing (UT) has a well-established performance for detection of defects and will be increasingly used in the future in view of the predicted increase of composite structures to be used. However, testing of large structures with UT is slow because the testing region is limited to the area immediately surrounding the transducer.

An alternative and more elegant approach is the use of guided ultrasonic waves because they can be excited at one location on the structure and propagate over a long distance (e.g., [1-3]) provided the geometry (thickness) is kept constant. By analyzing the characteristic of the guided waves such as the returning echoes, change of dispersion relationships, etc., the presence of flaws may be detected. However, because of the non-linear propagation behavior and the multimode nature of guided ultrasonic waves, the interpretation of the measurement signals is more complicated and makes the localization of the flaw position in the structure less precise. The schematic drawing in Fig. 1 shows the physical principle when guided ultrasonic waves are used for damage detection and all the aspects, which can have an influence on the measured results. This is the reason why signal processing is a crucial aspect in any guided ultrasonic wave application. The objective is to extract information from the signals measured to decide if a flaw has propagated in the structure and if so, characterize it in terms of its location.

The process of damage feature extraction or signal processing is not a new concept; it was required from the outset of the development of structural-health monitoring (SHM) in order to identify the interesting features within a measured response of a structure. Concepts already in use include some well-known signal analysis methods, e.g., Fourier or wavelet transforms, and other methods related to the scientific field of data prediction and parameter estimation. Different methods provide different types of results. For example, results from analysis methods based on transforms provide information about the spectral representation of the signal, whereas results from prediction methods contain more physical information about the underlying processes. In Fig. 2 the different methods for damage feature extraction currently used are listed beginning with the lowest level (Level 1) of extraction possibility up to the highest level (Level 6) where even complex non-linear measured signals can be analyzed for damage features. Typical feature extraction processes for different levels of extraction possibility can be found in [4-9].

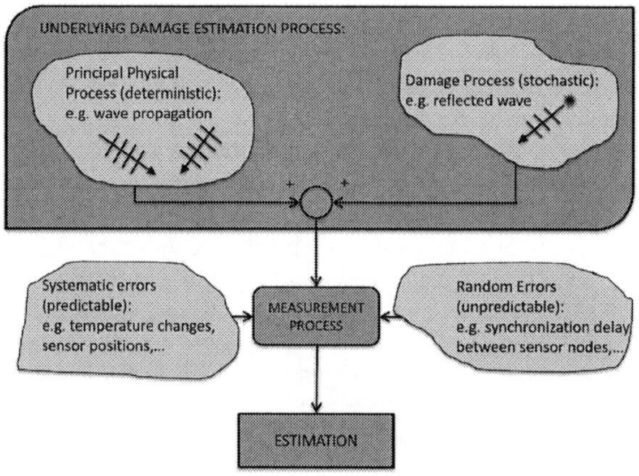

Fig. 1. Principle physcial processes for damage detection with guided ultrasonic waves and the influential effetcs on the measurement results.

In this paper a new approach for extracting the damage relevant features out of a measured signal is presented, which provides significant advantages over the damage feature extraction methods actually in use. The method developed in the framework of this work resembles the matching pursuit approach, however, with the difference that the dictionary of the waveforms has been replaced by an adaptive model of the expected guided ultrasonic wave signal. One advantage of this method is its possibility to extract the damage relevant features out of dispersive multi-mode signals even when these signals are overlapped by non-damage related features such as e.g. incident wave pulses. The ability to adapt to changing environmental conditions which may affect the measured signal is a further advantage of the present damage feature extraction process. This makes the present method very well suited for the analysis of guided ultrasonic signals and particularly for real world applications. Because of the ability of the model-based damage feature extraction process to handle also non-linear and noisy measurement data it is classified within Level 5 and/or Level 6 in Fig. 2.

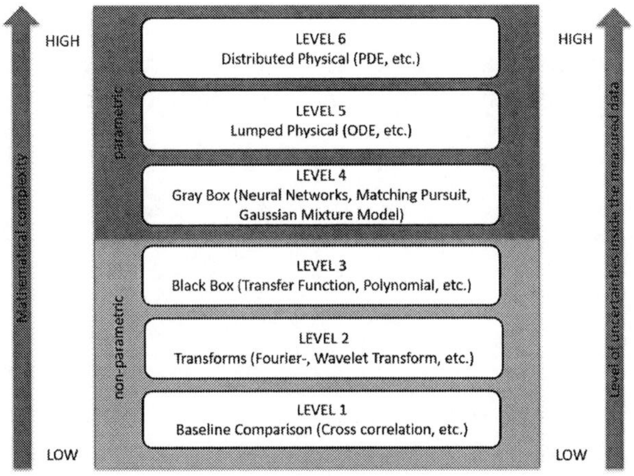

Fig. 2: Damage feature extraction analyzing methods and their possibilities to extract features out of complex signals.

2. Model-based damage feature extraction

The process for the extraction of the damage relevant features in this paper being described in more detail in [10] consists of two main parts, an approximated mathematical model of the expected measured test signal and a non-linear least mean square (NLMS) algorithm for the estimation of the damage relevant parameters. The approximated model is based on the two-dimensional wave equation whereby the dispersion and the chirping of the wave pulses have been taken additionally into account. In this section first the development of the approximated mathematical equation for the model of the expected measured signal is explained, followed by the explanation of the model-based damage feature extraction process. The proof of concept for the proposed process and the main results are described in the subsequent sections.

2.1 Approximated mathematical wave pulse equation

The derivation of the approximated mathematical equation starts with the equation for a time-harmonic two-dimensional plane wave with the angular frequency ω and propagating along the x-axis [10],

$$y(x,t) = y_0 exp^{j(kx-\omega t)}. \qquad (1)$$

Here k is the wave number and t the time of the propagating plane wave. Due to the dispersive behavior of the guided waves the wavenumber k is a function of ω, which can be approximated by a parabolic function about the angular frequency ω_0,

$$k = k_0 + a(\omega - \omega_0) + \frac{1}{2}b(\omega - \omega_0)^2 \tag{2}$$

where $a = {c_g}^{-1}$ is the invers of the group velocity $c_g = \frac{d\omega}{dk}$ at ω_0 and $b = d^2k/d\omega^2$ expresses the dispersion of the medium, i.e. the fact that the group velocity depends on ω. When we consider a wave pulse whose duration involves a fairly large number of oscillations of the field, the frequency spectrum is no longer infinitely narrow and the various frequency components travel at different speeds. However, the intensity is significant only if t/x is not too different from $a = {v_g}^{-1}$, because the frequency spectrum of the pulse is narrow.

When a wave pulse propagates from its origin $(t = x = 0)$ to a point (t, x) the shift of the wave pulse can be by expressed by,

$$S(x,t) = k(\omega)x - \omega t \tag{3}$$

If one replaces $k(\omega)$ by its expression from Equation 2 and takes ω from Equation (3), one obtains after rearranging,

$$S(x,t) = k_0 z - \omega_0 t - \frac{1}{2}(t - ax)^2/bx \tag{4}$$

The amplitude reaching the space-time point (x, t) is,

$$y(x,t) = y_0 exp[jS(x,t)] \tag{5}$$

One now assumes that the wave pulses are propagating in the shape of a Gaussian pulse and furthermore the pulses are propagating circular away from the exciting source $(x = r)$. After adding this additional information and rearranging Equation (5) one gets the final approximated mathematical equation for the wave pulse for the approximated mathematical model of the expected measured signal used,

$$y(r,t) = \underbrace{\left[\frac{{\tau_0}^4}{{\tau_0}^4 + (\kappa r)^2}\right]^{\frac{1}{4}}}_{\text{Amplitude wave pulse}} \underbrace{exp\left[-\frac{(t - k_0 r^2 {\tau_0}^2)}{2\sqrt{{\tau_0}^2 + (\kappa r)^2}}\right]}_{\text{Envelope wave pulse}} \underbrace{e^{j\left(\omega_0 t + \frac{\psi t^2}{2} - k(\omega)r\right)}}_{\text{Carrying wave}}. \tag{6}$$

Here, τ_0 is the initial width of the Gaussian pulse, κ is the dispersion coefficient and ψ is the chirping coefficient.

2.2 Damage feature extraction algorithm

In the following the principle functionality of the damage feature extraction process developed in the framework of this paper is described below sequentially and is furthermore listed in Table (2).

i. The measured response signal at the sensor position $y(r,t)$ is the input for the damage feature extraction process. In a first step, the envelope F of $y(r,t)$ is calculated by applying the Hilbert transform. At the same time, an approximated signal y_{est} of the expected measured signal is calculated by using the approximated mathematical equation for each expected wave pulse inside the measured signal. Subsequently the envelope F_{est} of the signal y_{est} is calculated.

ii. In the next step, the parameter vector a^k of the approximated model of the expected measured signal is adjusted by a non-linear least mean square (NLMS) algorithm in the way that the envelope from the calculated signal F_{est} fits best the envelope of the measured signal F. The *Levenberg-Marquard* algorithm is used for the NLMS parameter estimation. The adjustable parameters of the approximated model are listed in Table (1).

Parameter, a^k	Description
A_n	Wave pulse amplitude
r_n	Propagation distance
κ_n	Dispersion parameter
ψ_n	Chirping parameter

Table 1: Adjustable parameters of the approximated mathematical model.

iii. If the estimated squared error $S(a^k)$ between the envelope of the calculated signal F_{est} and the envelope of the measured signal F reaches a certain threshold the damage feature extraction algorithm stops and it is assumed that no damage features are present within the measured signal $y(r,t)$. In this case the estimated parameter vector a^k corresponds to the wave pulses contained in the signal without damage (e.g., incident wave pulses and/or reflected wave pulses from the edges of the structure).

iv. If the NLMS algorithm stops after a defined number of runs and there will be a remaining squared error $S(a^k)$ between F and F_{est}, then it is assumed that further wave pulses are within the measured signal $y(r,t)$. For this case the approximated model will be expanded by two additional wave pulses, the supposed S_0 and A_0 reflected wave pulse from a defect to the structure. Now, step (ii) will be carried out again and after the NLMS stops, the final estimated parameter vector a^k contains the values for the reflected wave pulses from the defect.

3. Application of the approach

The damage feature extraction process described above was tested by using a synthetic signal inspired by measurements taken from guided ultrasonic waves travelling in a 1.6 mm thick plane sheet of aluminium. This allows certain parameters of the synthetic test signals to be changed, which enables the range of application for the present algorithm to be estimated. The signals used for the test were simulated based on the basic wave Equation (6) described in the previous section. The simulated test signal was once assumed without damage features and further with damage features. Both cases were tested in the validation phase but here only the case with damage features will be explained. In Fig. (4) the upper graph shows an example of a simulated test signal $y(r,t)$ in a plane plate and its envelope used for the concept proof. The lower graph in Fig. (4) shows the different wave pulses inside the simulated signal. The parameters used for the simulation are listed in Table (3).

Input signal:	
F;	Envelope of measured input signal
Initial parameters:	
$a^0 = [a_{1n}, a_{2n}, \dots, a_{mn}]$;	Parameter vector
μ^0;	Initial step size
$\beta_0 = 0.25, \beta_1 = 0.75$;	Criteria for step size
ϵ_1;	Threshold minimum squared error
ϵ_2;	Threshold minimum change of
k_{max};	Maximal number of iterations
Step 1:	
$F_{est}(a^k) = Hilbert(y_{est}(a^k))$	Calculate envelope of model signal y_{est}
$S(a^k) = \|F - F_{est}(a^k)\|^2$	Calculate sum of squared error $S(a^k)$
$\frac{\delta S(a^k)}{\delta a^k} = \frac{\delta \|F - F_{est}(a^k)\|^2}{\delta a^k} = J(a^k)$	Calculation of the Jacobian
Step 2:	
$\left\| \begin{pmatrix} J(a^k) \\ \mu I \end{pmatrix} s^k + \begin{pmatrix} S(a^k) \\ 0 \end{pmatrix} \right\| \Rightarrow min$	Minimize the linear least mean square problem
Step 3:	
$a^{k+1} = a^k + s^k$	
$S(a^{k+1}) = S(a^k + s^k)$	
$S(a^k) + J(a^k)s^k$	
Step 4:	
$\rho_\mu = \frac{\|S(a^k)\|^2 - \|S(a^k - s^k)\|^2}{\|S(a^k)\|^2 - \|S(a^k) + J(a^k)s^k\|^2}$	Calculate criteria for new step size
if $\rho_\mu \le \beta_0$	s^k is not accepted; μ is doubled; new correction factor s^k is calculated
if $\beta_0 < \rho_\mu$	s^k is accepted; μ is retained
if $\rho_\mu \ge \beta_1$	s^k is accepted; μ is halved
Step 5:	
$\|S_{a^k}\| \le \epsilon_1$ or	
$\|a^k - a^{k-1}\| \le \epsilon_2(\|a^{k-1}\| + \epsilon_2)$ or	
$k \ge k_{max}$	
Result:	
$a^k = [a_{1n}, a_{2n}, \dots, a_{mn}]$	Estimated parameter vector

Table 2: Sequence of the damage feature extraction algorithm

A_{A0inc} [V]	r_{A0inc} [m]	κ_{A0} [-]	ψ_{A0} [-]	A_{S0inc} [V]	r_{S0inc} [m]	κ_{S0} [-]	ψ_{S0} [-]	A_{A0dam} [V]	r_{A0dam} [m]
1	0.13	1e-4	1e-8	0.5	0.05	0	0	0.1	0.18

A_{S0dam} [V]	r_{S0dam} [m]
0.03	0.08

Table 3:-Wave pulse parameters for the simulated test signal

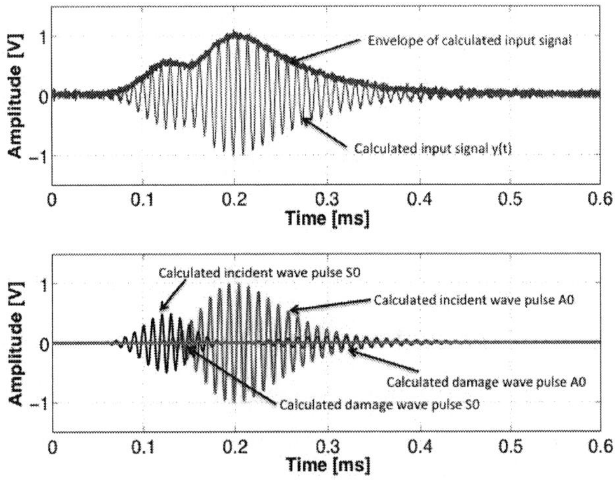

Fig. 3. Example of a simulated signal used for the proof of the damage feature extraction process (picture above) and whole simulated signal with its envelope (picture below); single wave pulses inside the simulated signal.

In order to demonstrate a more realistic test case additional noise was added to the test signal. The values for the initial parameters used for the damage feature extraction process are listed in Table (4) below.

A_{A0inc} [V]	r_{A0inc} [m]	κ_{A0} [-]	ψ_{A0} [-]	A_{S0inc} [V]	r_{S0inc} [m]	κ_{S0} [-]	ψ_{S0} [-]	A_{A0dam} [V]	r_{A0dam} [m]
0.2	0.1	0.001	0.001	0.1	0.01	0	0	0.01	0.13

A_{S0dam} [V]	r_{S0dam} [m]
0.001	0.06

Table 4. Initial wave pulse parameters

4. First Results

The result of the damage feature extraction process is the estimated parameter vector a^k. It contains the values about the respective wave pulse amplitude A_n and the propagation distance r_n for each single wave pulse contained in the simulated test signal. For a better illustration the envelope of the estimated signal, and here only the damage relevant wave pulses, is calculated based on the estimated wave pulse parameters and displayed in Fig. (5). There, the envelopes of the simulated test signal (damage relevant wave packets) are compared to the estimated envelope described above.

In order to test the stability of the proposed process the A_0 incident wave pulse was stepwise overlapped by the reflected A_0 wave packet from the damage. The overlapping was done in a ratio range between 0% (no overlapping) to 100% (full overlapping). Furthermore, the amplitude ratio between the two wave pulses was varied between 0.01 (low amplitude of the A_0 reflected wave packet from the damage) and 0.1. In Fig. (6) the results for the test with different overlapping ratios and varying wave pulse amplitudes are shown.

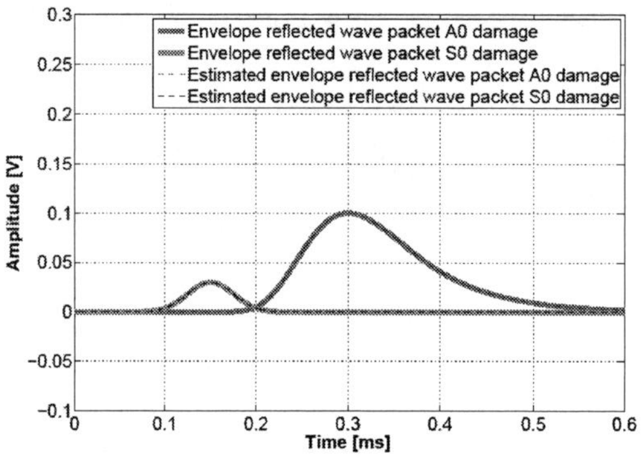

Fig. 4. Example of the resulting envelope signal after applying the damage feature process

It can be seen in Fig. (6) that with a relatively small amplitude ratio between the incident and the reflected wave packet the error δr between the estimated and simulated propagation distance of the reflected wave packet from the damage is relatively high (>200 mm).

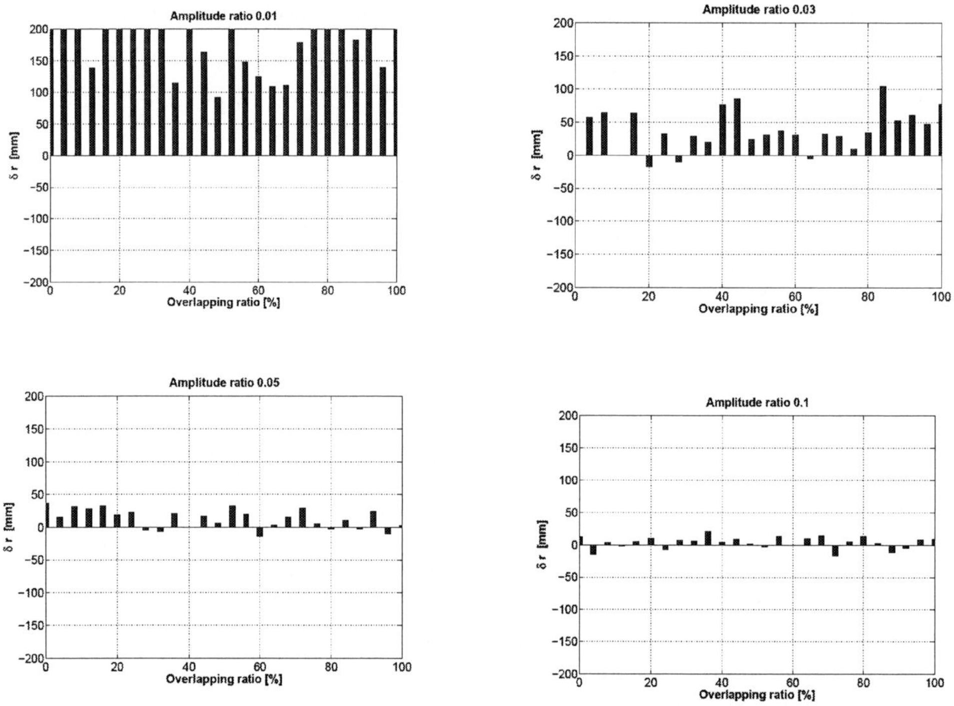

Fig. 6. Results after applying the damage feature extraction process to a simulated signal with different overlapping ratios between wave pulses and changing wave pulse amplitude

This is because of the additional noise inside the simulated signal, which makes an estimation of the damage relevant features with low amplitude not possible. With increasing amplitude of the reflected A_0 wave packet from the damage the error δr decreases and for an amplitude ratio between incident and reflected signal the average error is approximately 5 mm which is a reasonable value for the estimation of the damage position.

5. Next Evaluation Steps

In view of increasing the complexity of the structures considered the plate described above will be analyzed with an artificial crack as it is shown in Fig. 7 and a patched repair as shown in Fig. 8 below. The numerical results have been obtained for the patched repair originally from [12] and have been further elaborated along the work described in [13].

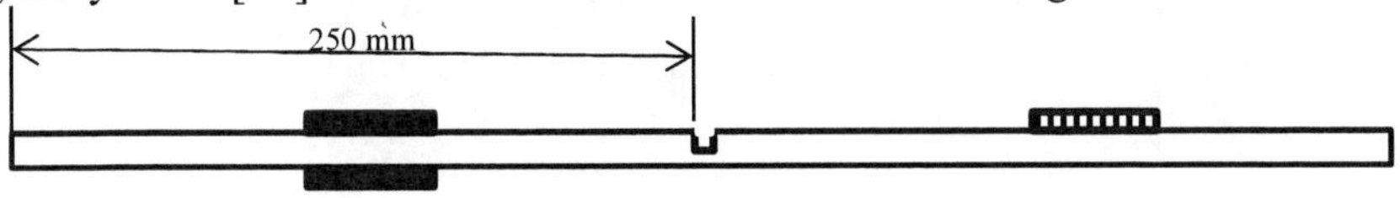

Fig. 7. Plate with a crack at 250 mm from the corner; crack width is 0.5 mm and depth is 1.5 mm

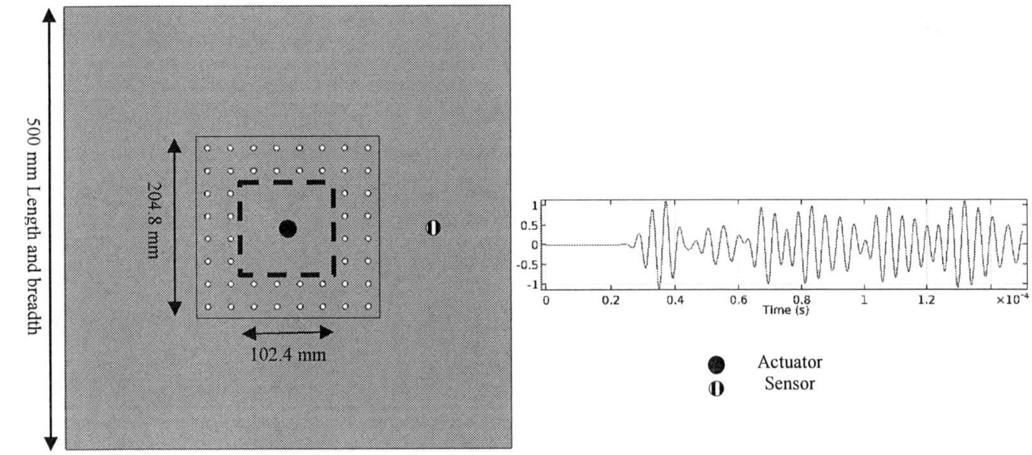

Fig. 8. Model of patch-repaired specimen according to [12] (dashed lines represent the cut-out region below the patch) and the signal recorded by the sensor

With those simulations available an experimental validation will be performed of which the equipment has been realized in [10]. The experimental setup to be applied is shown in Fig. 9. It consists of the specimen (1), the guided wave transducers (2), the wireless sensor node (3) and the central server including the graphical user interface (4). Compared to existing wireless systems for similar applications, this system used here is relatively small in size, light in weight, allows for a high amplitude of excitation voltage, has embedded anti-noise filtering and is optimized in terms of time synchronization between the central server and the wireless sensor units. Furthermore, the graphical user interface enables convenient use of the wireless system. The electric circuit of the wireless sensor node is shown in Fig. 10. The excitation signal for the piezoelectric transducer is a sine sweep with 18 cycles, modulated with a Hamming window with an amplitude of 60V peak-to-peak. The excitation frequency can be selected on the central server in a range from 30 to 300 kHz with a step size of 1 kHz. In order to measure the voltage from the piezoelectric transducers, the voltage drop over an internal resistor R_{pe} (Fig. 10) is measured. This leads to a relatively high input impedance for the wireless signal processing unit (in comparison with the internal resistor), a problem which is solved by using a non-inverting amplifier at the input

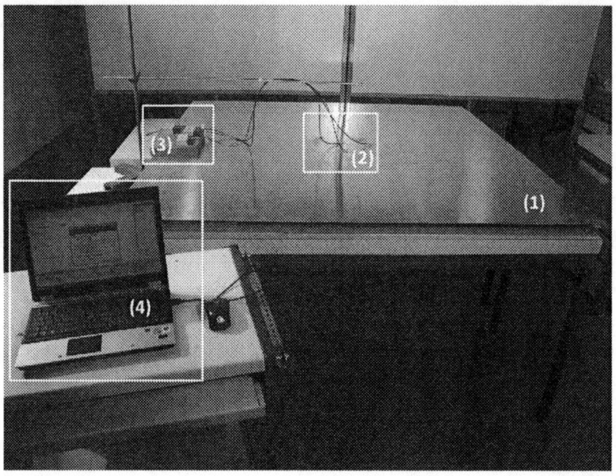

Fig. 9. Experimental setup: aluminum alloy plate (1), piezoelectric transducers bonded onto plate (2), wireless measurement devices (3) and central processing unit with wireless receiver (4)

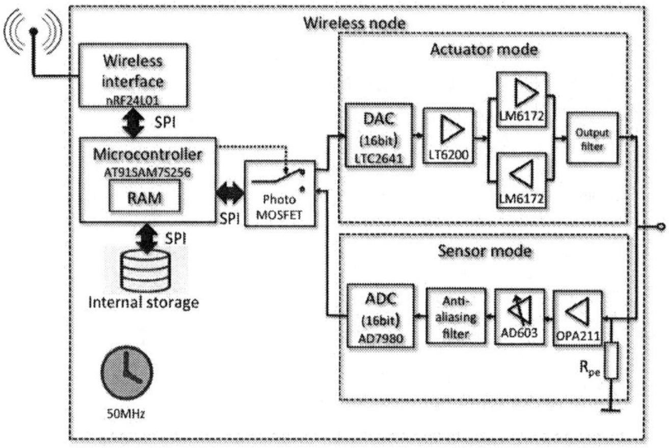

Fig. 10. Electric circuit diagram of a wireless sensor node

For a smooth signal, an anti-aliasing filter is integrated into the circuit just before the analog-to-digital converter (ADC). The ADC has a 16-bit resolution with a sampling frequency of 1 MHz. This leads to a signal-to-noise ratio (SNR) of about 87 dB. By taking the Nyquist sampling theorem into account (sampling frequency > 2 x signal frequency) a sampling frequency of about 1 MHz is (more or less) enough for a signal frequency up to 300 kHz.

6. Summary and conclusion

Monitoring of structures based on structural health monitoring (SHM) provides a variety of significant challenges. One of them is to extract the right features from the signals having been measured. A way on how to get this done is on the basis of a model-based damage feature extraction procedure as proposed in [10]. In that case a model of the structure considered is taken as a basis to generate sensor signals where the model is then adapted through a learning process to what is monitored in reality in case a mismatch between simulated and recorded signal is too big. The model can be of different levels of complexity starting from a baseline measurement being compared to an actual measurement and ending in a rather complex mathematical (i.e. numerical) model to be considered as a reference that may adaptively be adjusted through non-linear least mean square fits.

Within this paper the process of validating the model-based feature extraction process experimentally has been introduced considering a flat plate and a patched metallic repair for an undamaged and an artificially damaged condition. The numerical results will be taken from work described in [13] and will be validated with the SHM hardware described above, specifically in terms of extracting the appropriate features from the signals being measured. The results will show, how far this is possible and what additional features may be measured which the numerical model may not be able to describe in its current state. This may possibly trigger a modification in the numerical model leading to an iteration process between simulation and measurement demonstrating the need for adaptation of an SHM system, not just between the measurements the system takes but also between simulation and hardware in general.

References

[1] Y. Cho, D.D. Hongerholt, J.L. Rose, Lamb wave scattering analysis for reflector characterization. *IEEE Transactions on Ultrasonics, Ferroelectrics, and Frequency Control,* **44**, 44-52, 2002

[2] J.L. Rose, M.J. Avioli, P. Mudge, R. Sanderson, Guided wave inspection potential of defects in rail, NDT S E International, **37**, 153-161, 2004

[3] Z. Su, L. Ye, Y. Lu, Guided lamb waves for identification of damage in composite structures: A review, *Journal of Sound and Vibration,* 295, 753-780, 2006

[4] G. Beresford-Smith, I.M. Mason, A parametric approach to the compression of seismic by frequency transformation, *Geophysical Prospecting*, **28**, 551-571, 1980

[5] A.K. Booer, J. Chambers, I.M. Mason, Fast numerical algorithm for the recompression of dispersed time signals, *IEEE Electronic Letters,* **13**, 453-456, 1980

[6] H. Joeng, Y-S. Jang, Wavelet analysis of plate wave propagation in composite laminates, *Composite Structures,* **49**, 443-450, 2000

[7] B. Xu, V. Giurgiutiu, L. Yu, Lamb Wave Decomposition and Mode Identification Using Matching Pursuit Method, *Report Department of Mechanical Engineering, University of South Carolina,* page 12, 2009

[8] J-C. Hong, K.H. Sun, Y.Y. Kim, The matching pursuit approach based on the modulated Gaussian pulse for efficient guided wave inspection, *Smart Materials and Structures,* **14**, 548-560, 2005

[9] A. Raghavan, C.E.S Cesnik, Guided-wave signal processing using chirplet matching pursuits and mode correlation for structural health monitoring, *Smart Materials and Structures,* **16**, 355-366, 2007

[10] C. Dürager, Model_Based Damage Feature Extraction for Structural Health Monitoring Applications; Doctoral Thesis, Saarland University, 2018

[11] D. Royer, E. Dieulesaint, Elastic Waves in Solids I, free and guided propagation, *Springer- Verlag Berlin Heidelberg New York,* ISBN 3-540-65932-3, 1996

[12] A. Taltavull, *Structural Health Monitoring Solutions for Patched Metallic Aircraft Repairs Based on Guided Ultrasonic Waves,* Master Thesis, Saarland University, 2017

[13] A. Asokkumar, R. Sridaran Venkat and C. Boller: Simulation based key performance indicator determination in guided wave monitored metallic patched repairs; Proc. of 10th Int. Symp. on NDT in Aerospace, Dresden/Germany, 2018

Detection of delamination and impact damage in multilayered lightweight materials

Tobias GAUTZSCH [1], Andreas BODI [1], Manuel LUCAS [2], Ralf STEINHAUSEN [3], Mario KIEL [3]
[1] SONOTEC Ultraschallsensorik Halle GmbH, Halle (Saale), Germany
[2] SONOTEC US Inc., Islandia, New York, USA
[3] Forschungszentrum Ultraschall gGmbH, Halle (Saale), Germany
Contact e-mail: t.gautzsch@sonotec.de

Abstract. A variety of lightweight composite materials are used in aerospace construction. Most common are multiple layers of aluminium with adhesive interfaces of different thicknesses, GFRP sandwich structures with and without honeycomb cores and glass fibre reinforced aluminium. The material properties of all materials are highly dependent on the integrity of all adhesive and matrix structures. These can be impaired by impacts, production problems, physical stress or aging.

Air-coupled ultrasonic testing (ACUT) can be a method for non-destructive quality control for semi-finished composite materials as well as products. Additionally, it can be applied for maintenance testing of parts to find impact and stress damage.

We will present a case study on the detection, measurement and evaluation of delamination in multi-layered aluminium bonds with up to 4 sheets of aluminium in various thicknesses. The aviation-specific specimen is measured in through-transmission with ACU with multiple frequencies, filters and averages with classic piezocomposite as well as phased-array ACU transducers. Delamination and flaws can be found in every adhesive layer. Depending on the composite thickness, different frequencies have to be applied.

Introduction

Air-coupled ultrasonic testing (ACUT) is an aerospace proven technology for inspecting FRP-components. Based on continuous improvements in converter technology, transmitting and receiving electronics and advanced evaluation algorithms, we have been working intensively on the development of new fields of application for air-coupled ultrasonic systems. These efforts are aimed in particular at testing high-damping materials as well as materials with multiple boundary layers of different properties. Even inspection tasks that could previously only be covered by conventional ultrasound, such as the testing of sheet metal welds, can now be partially replaced by air-coupled ultrasonic tests. (Compare [1])

Due to the need for lightweight construction in all areas of life, composite materials are rising in popularity. In the process of construction, the lightweight potential of the individual composite parts is increasingly being exploited. For this reason, the focus in quality control even with simple materials such as insulation sheets or sandwich structures is more placed on the structural integrity. The proof of the complete adhesion of cover layers to honeycomb cores and laminar bonds can be performed in the same way as the verification of the consolidation of pressed insulation boards. Delamination, air and foreign body inclusions as well as impact damage can be detected with the imaging air-coupled ultrasound technology. [1], [2]

The hybrid design, which is prevalent in many areas with a multi-material mix of metals, reinforced and unreinforced plastics as well as ceramics, is causing a change in the established joining technologies. Areal bonding is increasingly being used in these areas, posing new tasks for materials testing. A major challenge for non-destructive testing are "Kissing Bonds", touching but non-adhesive areas in bonds. One approach to detect these flaws is the one-sided test in pitch / catch arrangement. (see [3], [4])

In this paper, we will describe the application development of the ACUT on multi-layered adhesive bonded aluminium sheet metal. The development was performed on an aerospace specific calibration specimen with build in test flaws. Additionally to the multiple bonding layers, an increase in standoff distance was required on the transmitter side.

Measurement Equipment

The measurements in this article were performed using the High-End SONOAIR ACUT System. With four default input and output channels, this enables the operation of the SONOTEC Phased-Array ACUT transducers and the simultaneous measurement with four conventional transducers. Each channel allows a free configuration of the transmitter characteristics up to a voltage of 800 V. The combination of a preamplifier at the receiver and an internal amplifier in the electronics achieves an extremely low inherent noise of <1 nV / √Hz. The possible amplification dynamics of 120 dB is necessary to generate sufficient signal-to-noise ratio (SNR) for a meaningful and reproducible measurement even with highly attenuating materials. The excitation and detection of the air-coupled ultrasound is carried out with the CF series transducers listed in Table 1.

Table 1Overview ACUT-Transducers

Transducer	Frequency	Type	Focal distance	Sensitivity
CF 075	75 kHz	planar	30 mm	-34 dB
CF 200	200 kHz	planar	15 mm	-35 dB
CF 400	400 kHz	focused	50 mm	-31 dB
CF 400 3E	400 kHz	Phased-Array	20 mm	-31 dB

Test Specimen

The test specimen is an aviation-specific step wedge consisting of multiple adhesively bonded aluminium plates of various thicknesses as seen in Fig.1. The bottom two sheets, 1.7 mm respectively 1.5 mm thick, spread out over the full area of the specimen. The rest of the steps are build up by smaller pieces of varying size and thicknesses. The value for each step can be seen in table 1. Each step contains 2 – 4 sheets of aluminium.

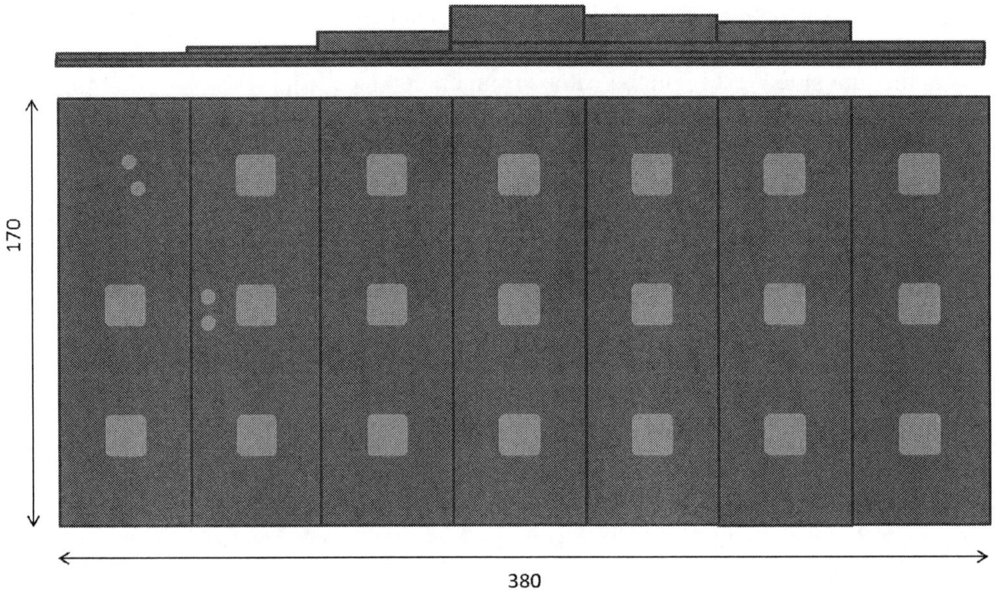

Figure 1: Test Specimen

There are two types of test flaws. The rectangular flaws are Teflon sheets in between the bottom two layers. These 20 x 20 mm areas simulate delamination or kissing bonds in the adhesive. Additionally, there are for drilled holes in the top sheet of the first two steps. These holes are filled with a composite filling material and simulate repaired damages in the aluminium sheet.

Table 2: Step sizing of the test specimen

	Step 1	**Step 2**	**Step 3**	**Step 4**	**Step 5**	**Step 6**	**Step 7**
1st Sheet				10.5 mm	8.7 mm	6.0 mm	
2nd Sheet		1.5 mm	5.7 mm	1.9 mm	1.9 mm	1.9 mm	1.9 mm
3rd Sheet	1.5 mm	1.5 mm	1.5 mm	1.5 mm	1.5 mm	1.5 mm	1.5 mm
4th Sheet	1.7 mm	1.7 mm	1.7 mm	1.7 mm	1.7 mm	1.7 mm	1.7 mm
Total	3.6 mm	5.2 mm	10.0 mm	17.2 mm	14.7 mm	12.5 mm	6.0 mm

Inspection task specific restrictions

The inspection task is to perform a contactless inspection of adhesively bonded aluminum. Because of interfering contours on the top side of the test object, a minimum standoff distance of the transmitter of about 115 mm is required during the scan. The sampling distance is set at 2 x 2 mm with a measuring speed of 600 mm/s, resulting in 300 kHz PRF. Because of the high scan speed, no averaging is possible. Data-processing is limited to online filtering of the acquired data. No additional post-processing like data-converting, interpolation or mean value filtering was allowed. The primary measuring frequency is 200 kHz.

The SONOTEC CF200 transducer is the choice for the required mean frequency of 200 kHz. However, as can be seen in table 1, the focal distance of this transducer type is 15 mm. In trough transmission measurement, the best results are obtained if the standoff distance of the transducers is the focal distance or near field length. It is well known, that this distance is a function of the frequency and the aperture size. It is possible to reduce the focal distance by applying a mechanic focus like a lens or to focus the sound beam electronically with multi-element, phased-array transducers [2]. But for this application, it was necessary to increase the focal distance. For any given frequency, this can only be done by increasing the aperture.

The method of choice to increase the aperture was a circular array of multiple CF200 piezo-ceramic transducers because of its high sensitivity combined with the long lifetime of the CF200 ceramic. The array was built with a center transducer and several others in a circle surrounding it. This circle was calculated to have the aperture creating the required 115 mm near field length. Additionally, the surrounding transducers were slightly angled, so that the sound axis of all individual sound beams crossed in the aspired focal spot. The resulting sound beam of the array can be seen in Fig. 2. Please note, that an additional 20 mm need to be added at the left side of the figure.

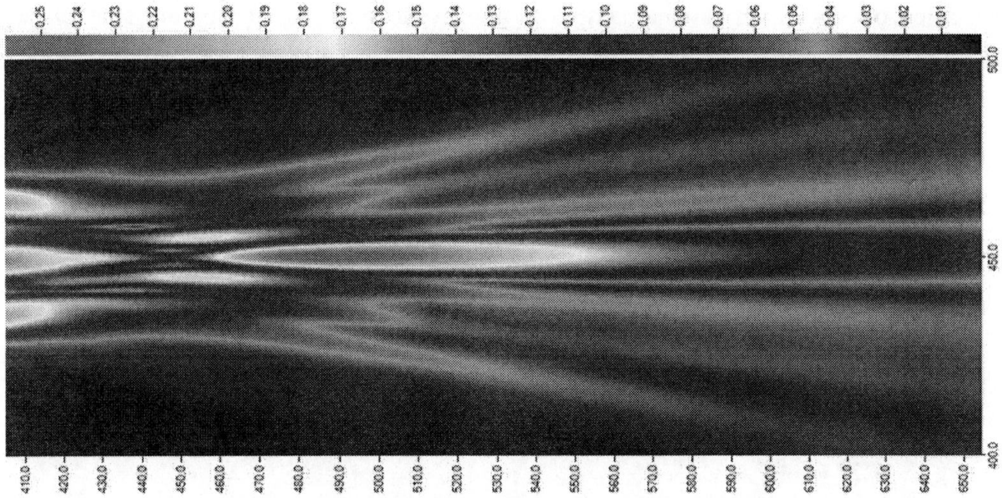

Figure 2: Sound Beam of the 200 kHz "far-standoff' Transducer

Measurement Results

The first assessment of the specimen was done with focused 400 kHz transducers at normal standoff distance and with a slow scanning speed. For a maximum defect resolution, a stepping of 1 x 1 mm was chosen. Because of the high attenuation of 400 kHz ultrasound, an average of 8 was applied to improve the SNR. The result of the scan can be seen in Fig. 3. It is apparent that all flaws could be detected. The rectangular flaws could not be transmitted. This is shown by the black color in this area. You can also notice, that the permeability of the steps varied with step 4, a four sheet step with the maximum sheet size of the test object being the one with the highest drop in amplitude. The circular flaws, marked with the circle a, did not affect the transmission signal as much. They appear as a round shade but are well detectable. So are some round shades in Step 3, which are expected to be air bubbles in the adhesive layer. At the lower edge of the specimen, marked with the circle b, an edge induced interference pattern is visible.

Figure 3: 400 kHz scan of the test specimen

The next step in the assessment of the application was a detailed look into the regions of interest of Step 1 and 2. For this measurement, the frequency was changed to 200 kHz and the standoff distance reduced to 15 mm on the transmitter as well as the receiver. Fig. 4 shows that all flaws can be detected as well as resolved at this frequency as well. Due to the lower attenuation of 200 kHz, the measurement could be performed at a satisfactory SNR level without averaging. The rectangular flaws are not detected as rectangles but have amplitude maxima at the corners of the rectangle. This effect is expected to be caused by interference of ultrasonic waves propagating around the flaw.

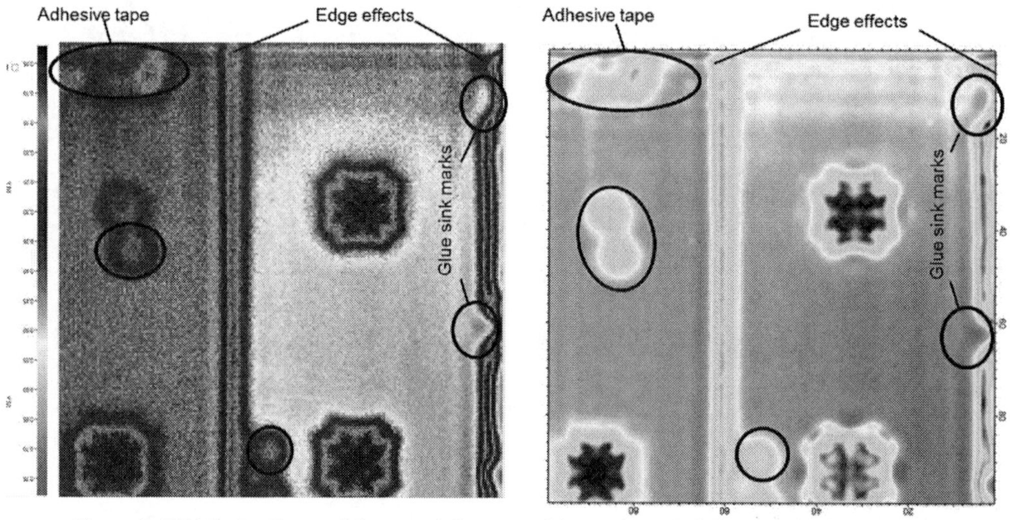

Figure 4: 200 kHz detail scan of the upper left corner of the specimen. left: linear scale right dB-scale

The third measurement was done with the customer specific boundary conditions and the newly developed transducers with a standoff distance of 115 mm. To achieve the high PRF of 300 kHz with the maximum laboratory scan speed of 100 mm/s, the stepping in scanning direction had to be reduced to 0.3 mm. The result of this scan is shown in Fig.5. Comparing this results with the detailed scans in Fig 4, it is apparent that the newly build transducer is capable of producing a focal diameter in 115 mm comparable to the focus spot of the CF200 in 15 mm. All significant spots of the detail scans can be found in this result.

Once again the ultrasound is attenuated differently by the steps, as observed in Fig. 3. However, the step 3, which is a 4 layer step, has the second highest amplitude even though it has more layers than step 1 and is thicker than step 2. The highest amplitude can be observed in step 7. This is a different behavior than with 400 kHz. This is a sign, that the transmission of ultrasound is frequency dependent.

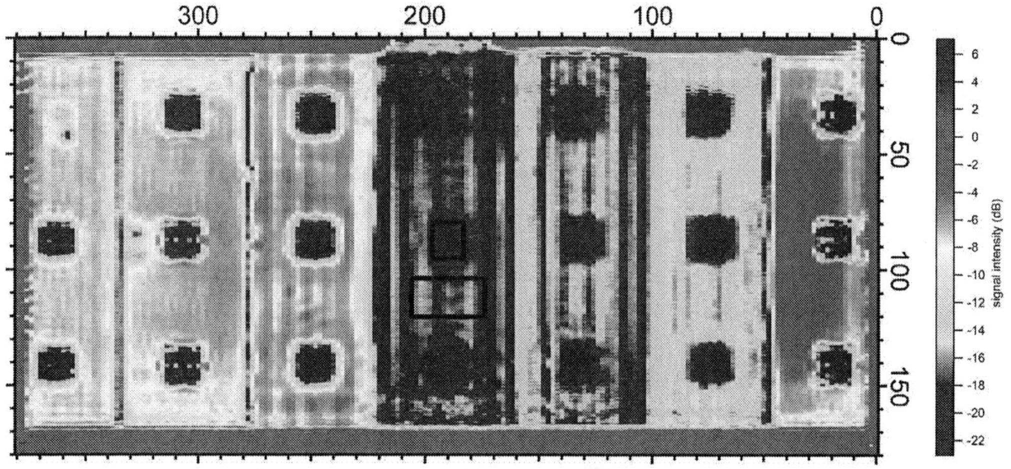

Figure 5: 200 kHz scan with customer specific scan settings of 2 x 0,3 mm resolution and 300 Hz PRF

Another indication of the influence of the frequency and wavelength on this measurement is the intensity distribution throughout each step. Especially in the low amplitude steps 4 and 5, strong interference pattern are apparent. This effect was also detected in the 400 kHz scan but with less intensity, more maxima and a smaller wavelength in the pattern.

Conclusion

The presented application shows that the detection of delamination in multilayered adhesive bonded aluminum with air-coupled ultrasound is possible. The challenge of the implemented customer specific restrictions on the inspection task was solved with a newly developed transducer. The influence of the frequency on this inspection task will be the subject of further investigations.

References

[1] Oster, R.: "Non-destructive testing methodologies on helicopter fiber composite components challenges today and in the future" - 18th World Conference on Nondestructive Testing, 16-20 April 2012, Durban, South Africa

[2] Steinhausen, R.; Kiel, M.; Pientschke, C.; Scheunemann, S.: Ortsaufgelöste Reflexionsanalyse mittels Luftultraschall - Einseitige Prüfung von CFK-Wabenstrukturen DGZfP - Seminar des FA Ultraschallprüfung 2015

[3] Gaal, M.; Döring, J.; Prager, J.; Brackrock, D.; Dohse, E.; Homann, T.; Grzeszkowski, M.: „Anwendung geführter Ultraschallwellen für die Prüfung von Klebeverbindungen", in DGZfP-Jahrestagung 2014

[4] Kiel, M, Steinhausen, R., Bodi, A, „Einseitige Luftultraschallprüfung von Blech-verbindungen - Eine Sache der Geometrie", Berichtsband DGZfP Jahrestagung 2017, Koblenz

SITAU TRITON: High Speed Ultrasound Inspection System for Complex Geometry Composites

Montserrat ACEBES [1], Ricardo GONZALEZ BUENO [2], Rafael DELGADO DE MOLINA[1], Alfredo MORALES[3], Jorge F. CRUZA [2], Iñaki GAUNA [1], Roberto GIACCHETTA [2]

[1] TECNITEST INGENIEROS, Calle Ciudad. de Frías, 1, 28021, Madrid, Spain

[2] DASEL, Avda del Cañal 44 N328500 Arganda del Rey, Madrid, Spain

3 Aciturri Aeronáutica, S.L.U.Avda. de la Industria, 19, Tres Cantos, Madrid, Spain

Contact e-mail: comercial@tecnitest.com / roberto@daselsistemas.com

Abstract. The aerospace composite industry demands the inspection of 100% of the manufactured components, which places the non-destructive-evaluation methods in the critical path of the manufacturing process. In one hand, inspection time has to be minimized to avoid the NDT process to be a bottleneck in the process, which requires increasing the inspection speed and/or incrementing the number of probes and electronic channels in the case of ultrasound inspections (UT).

We present in this work the result of 2 years of work, with the achievement of a highly modular, scalable and affordable solution for high-speed ultrasound automated inspection, called SITAU TRITON.

SITAU TRITON is the most advanced and flexible automated UT solution in the market, integrating in a single machine, three UT inspection systems:

1) SITAU TRITON GIMBAL: Able to perform like a 3 axes Cartesian system to inspect planar and smooth curvature parts, by using PA technology for high speed operation.

2) SITAU TRITON COMPLEX: With up to 896 ultrasonic channels, it is designed to inspect the 100 % of complex geometry components in one pass, which guarantees the minimum possible inspection time. A quick and automated interchangeable system allows using specific inspection heads for each component.

3) SITAU TRITON TTU: Is a through transmission system (TTU), able to work with conventional water coupled technology or with air-coupled ultrasound, and with single or dual frequency transducers. Trajectories are optimized for each component ensuring a collision-free operation.

One of the unique features of SITAU TRITON is phased-array AUTOFOCUS, which takes automatic UT inspections to the next level. Geometry changes and manufacturing tolerances along the component are automatically compensated in real-time by the UT equipment, which make the SITAU TRITON system one of the most robust and reliable on the market, even for radius and complex geometries. Furthermore, our air-coupled ultrasound sub-system based on the AIRSCOPE technology, has proven to be one of the more sensitive and robust solutions in the market, able to face inspection problems only solved with water-coupled system up to date.

In this work we present the key innovations of the SITAU TRITON technology and the results obtained with the system.

Introduction

The increasing use of composite materials in more industries and the high-quality standards, required for this type of material, implies a 100% quality control of the structural elements before these parts are assembled and subsequently used. This comprehensive control implies finding a suitable way for testing and evaluation. Ultrasound Non-destructive testing is an appropriate method to detect defects in these types of materials.

When the NDT inspections began, at the end of the last century, the need for their automation arose. This automation would increase the velocity of inspections one order of magnitude, at the same time that would allow reaching great precision and repetitiveness, making possible the generalized use of the images for evaluation.Many of the inspections of complex surfaces have been carried out for years by specialized human operators. Although, as time passed technological background regarding the use of robots in the field of non-destructive inspection appeared [1].

The automatic systems are subject to great requirements of stability, velocity and precision in the whole trajectory of the movement and, in many occasions, of size to be able to inspect large pieces. For this reason, the first designs were based on gantry system, which are still the most used in NDT.

However, in most current industrial processes, prevail the use of articulated robotic arms, that usually have six axes, unlike the typical gantry systems used in NDT inspections that are a combination of linear axes (X axes, Y, Z) and rotary ones.

In this way, some of the modern articulated arm robots have appropriate characteristics to be used in automated NDT systems [2]. Thus, in order to increase productivity, there is a growing interest in robotic solutions within the aerospace sector. However, it is necessary to look for the best mechanical design that allows the probes to adapt to the geometry of the piece in order to achieve automated inspection without loss of signal. Nevertheless, this is not enough in the case of complex and variable geometry components, and it is necessary for the ultrasonic system to dynamically adapt its operation parameters to guarantee the best possible image at all time. This is known as adaptive imaging or auto-focusing, a research hot topic among the ultrasound scientific community [3-6].

In summary, both in the automatic inspections with ultrasounds and in other NDT techniques, the images quality is totally related to the quality of the robotic system, to the signal acquisition and processing system and to the position and stability of the probes during the acquisition. These parameters become even more critical in elements that have complex geometries. In this case there is an additional challenge in the design of automatic systems and UT equipment [7].

1.Objectives

The importance of quality control in the aeronautical sector, the complex structure of new components, and the increase in production volume requires to develop new flexible, accurate and fast inspection machines.

Conventional ultrasonic testing based on single-element probes is limited by its lack of capacity to reduce the inspection time and to tackle complex geometries components. The alternative and current trend for composite materials is using phased-array (PA) systems, based on multi-element probes able to control the beam properties. Besides increasing the inspection speed by electronic scan, they allow to dynamically focusing inside the material, achieving the best possible image even in the presence of complex shape interfaces.

The objective of this work is to present a flexible and robust platform based on the PA technique, for the development a whole new range of automated inspection machines. It uses a robot and an advanced UT equipment, which inspects, not only flat parts, but also curved geometries following the contour of the part without loss of signal, removing the necessity to delimit different zones of inspection, optimizing the representation of results and reducing the time of evaluation and errors in interpretation.

This system has robotized trajectory tracking, applying Pulse-Echo, and Double Transmission techniques with Phased Array auto-focusing technology, as well as TT transmission, with water coupling, so that it is possible to inspect one hundred percent of monolithic structures of solid laminates and sandwich structures.

2. Development

In this way, Tecnitest and Dasel have worked on the development of a robotic inspection system that allows the inspection of the pieces with a complex geometry.

Technical specifications of the system were defined by Caetano Aeronautic in order to develop a multi-technical installation able to follow the outline and inspect in an only pass all Airbus Defence Space products, type Omega. All this, based on the specification defined 24 months earlier by Aciturri Aeronautics to inspect all profile type products with pulse-echo technique using Phased Array

The SITAUT TRITONsystem is a device designed to perform inspections, using ultrasound, in pieces made of composite materials.

2.1 Parts of the system

To be able to track this type of trajectories in a robotized way, the system consists of:

- A modular immersion tank.
- A flexible ultrasound system DASEL SITAU ST 187 designed to operate in automatic systems which require multiple phased-array probes to cover 100% of the area to be inspected. The connection of up to 7 phased-array probes with 128 elements each, gives the necessary flexibility to set the most appropriate configuration depending on the geometry of the component to be inspected.
- Advanced AutoFocusalgorithm for surface detection and real-time Dynamic Depth Focusing, ensuring the best possible image along the whole inspection.
- An articulated industrial robot with programmable manipulator in its six axes, automatically controlled and programmable to work in a linked and independent way in different areas of movement simultaneously, applying the Pulse-Eco techniques with Phased Array, double transmission with plate reflective, and transmission for sandwich structures, so that the inspection of one hundred percent of the surface of the piece is possible. A fast and automatic tool exchange mechanism allows easy and reliable switching between operation modes.
- A robot guidance system with linear axes with a programmable length.
- A Security system, composed by: physical barriers, optical barriers, security stop buttons, control cabinet.
- VisualScansoftware comprises of user-fiendly interface modules for motion programming and control, data acquisition in all operation modes, signal representation in single or multiple window frames and evaluation.A, B, C and D-scan representation are available with complete A-Scan data recording and reporting functions.

Fig. 1. Immersion tank with robotized trajectory tracking.
a) Prototype installed in Tecnitest; b) System installed in a company of the aeronautical sector

3. Operation modes

Currently this system allows working in three operating modes

3.1Operation mode 1(SITAU-TRITON Gimbal)

The immersion tank and the robot are used together and the displacement is done following a linear track (Fig. 2). The use of encoders and a gimbal allows the inspection of surfaces with certain curvature and a trajectory tracking according to a plane that is not necessarily horizontal.

The available inspection techniques are: PE and DT, with PA technology, for inspection of solid laminates with a flat or slightly curved shape. In this mode of operation, the pieces are placed in immersion in the bottom of the tank and the scan is performed according to a plane that can be vertical or horizontal.

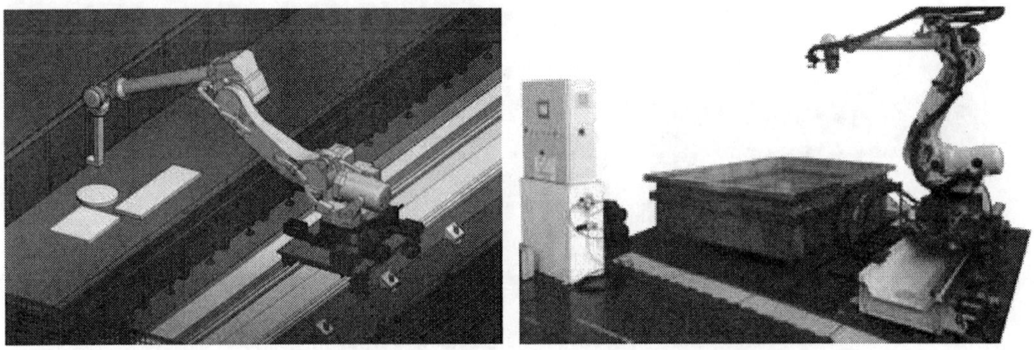

Fig. 2. Process of inspection with PA in immersion for flat and slightly curved pieces
a) Simulation; b) System working in operation mode 1 installed in Tecnitest

3.2Operation mode 2(SITAU-TRITON Complex)

The tank is used and the pieces with complex geometry (e.g. omegas type) are placed on an ad-hoc tool (previously designed for this use) immersed in the tank (Fig. 3). The robot, with the previous information corresponding to the CAD design, is responsible for carrying out the inspection completely automatically using a custom-made probe head that allows absorbing deviations in the geometry. Final image improvement is achieved by the AutoFocus capacity of the SITAU ultrasound equipment.

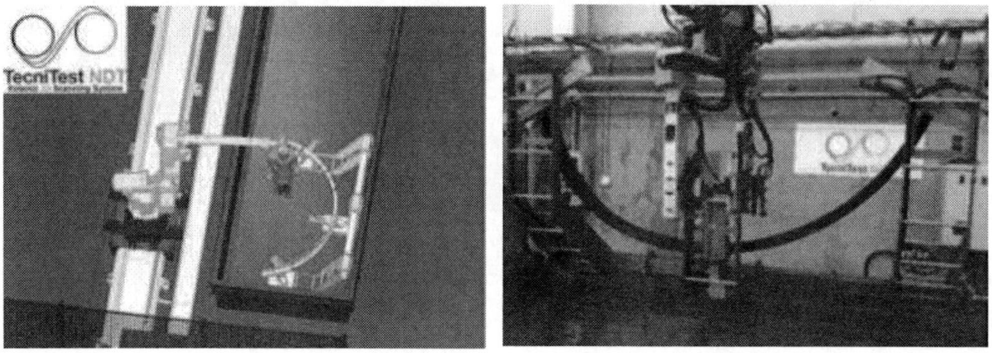

Fig. 3. Process of inspection with PA for elements with complex geometry

The inspection technique available is PE, with PA probes (Fig. 4). Inspection trajectories are programmed, based on the information of the coordinates of the piece to be inspected, so that the robot together with the head can follow the defined geometry.In this mode of operation, omega-type pieces with complex geometry, height and variable thickness, have been used. Despite the deviations of the piece, it is possible to perform the inspection in a single scan. In order to increase the productivity, the robot has been equipped with an automatic tool change system, which avoids manual intervention and reduces the time in the test. This element allows, in turn, a tool change between modes of operation. The head is designed to inspect 100% of the surface of the piece in one step, using 7 phased array probes, up to 896 Ultrasonic Channelsin less than 15 minutes

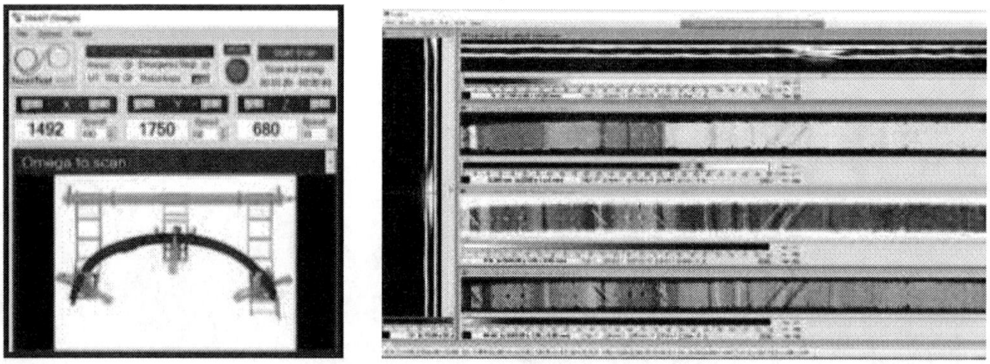

Fig. 4. Presentation of results for a defective type omega piece

3.3Operation mode 3

Based on the use of TT inspection mode using a single-transducer dual-frequency with water or Air-Coupled ultrasound with Airscope technology. It allows inspecting sandwich components, both flat and with complex geometry using a transmission tool (Fig. 5).

The pieces are placed following an axis, the inspections can be done by means of a single scan, when the geometry is slightly curved, or when more complex geometries may be necessary to program a trajectory for each piece. So far two dual frequency ultrasound probes have been used for transmission in water, but the development can be extended to an in-air inspection system based on DASEL AirScope.

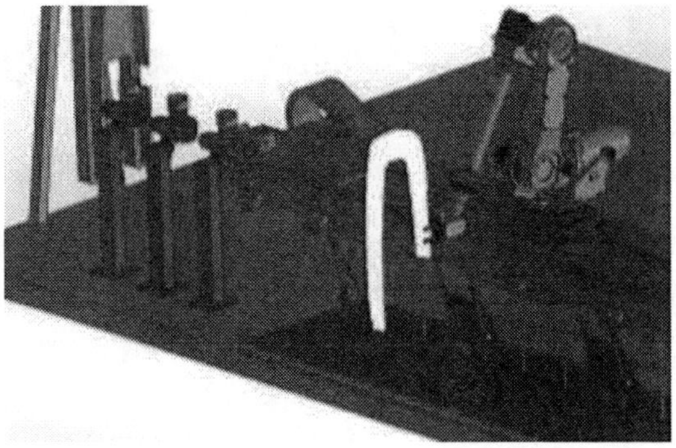

Fig. 5..Process of inspection using TT inspection modeto inspect sandwich components

4. SITAU Phased-Array System

Ultrasound phased-array systems are based on individually controlling, in emission and reception, the elements of an array. This usually requires a large number of ultrasound channels (typically between 32 and 256), increasing the hardware and software complexity of this kind of systems. Furthermore, the image formation process requires to precisely know the time of flight from each array element to each sample in the image (focal law), which is not a simple task for complex shape components in water immersion.

These two aspects have limited the development of automated inspection machines for large and complex shape components in the aerospace industry. In one hand, there is a strong demand for reducing the inspection time, which requires increasing the number of arrays and UT channels in order to inspect the components in only one pass. On the hand, the geometry variations of the component along the scan require continuously changing the focal laws to achieve normal incidence, a necessary condition to inspect highly anisotropic materials like CFRP.

When the geometry is exactly known by design, it would be possible to pre-calculate the focal laws for the whole component and to apply them during the inspection in a second step. Nevertheless, this approach is useless in practice, due to the manufacturing geometry tolerances and positioning uncertainties of real components and machines. The alternative is to develop systems able to automatically calculate the focal laws with no previous knowledge of the component geometry, which is known as *auto-focusing*, a research hot topic among the ultrasound scientific community.

To face these challenges, DASEL has developed the SITAU phased-array technology, one of the most advanced UT platforms in the market. *Its modular design allows building systems with up o 2048 channels*, which enables to design complex scanning heads based on large arrays count. Furthermore, *it implements a novel auto-focusing algorithm able to guarantee normal incidence regardless the geometry variations and positioning errors* along the scan, a unique feature among current state-of-the-art ultrasound equipments.

In particular, the SITAU-TRITON system counts with an 896 channels electronics (Fig.6 left), able to simultaneously control up to 7 phased-array probes. With this configuration omega type components can be inspected in only one pass, considerably reducing the inspection time (Fig.6 right).

Fig. 6. (Left) SITAU phased-array system with 896 channels (Right) example of PA probes distribution for inspecting an omega stiffener with SITAU-TRITON.

The common UT practice in aerospace NDT is to perform linear scans with 0º incident beams. The main reason for this is that propagation inside CFRP is highly anisotropic, which means that a preference detection direction exists: Normal to the carbon fiber layers. While in a plane component with ideal positioning a simple linear scan can be used for this task, for curved components or in presence of positioning errors, it is necessary to adapt the beams direction to achieve normal incidence. Some alternatives based on iterative algorithms have been presented [3-4], but can be only applied to small thickness components as no focusing exists. These methods are based on generating a plane wave that adapts to the surface shape ensuring normal incidence at all points, but no focusing is produced into the component.

By contrary, the AutoFocus algorithm developed by DASEL, estimates the component surface shape from a set of single element acquisitions, and then, *it calculates the focal laws for dynamic depth focusing at all depths*. This way, the best possible image is obtained regardless the component geometry, even if it changes along the scan. The AutoFocus algorithm operates in strict real time, not reducing the inspection speed and with no previous knowledge of the component shape. It is based on the Virtual Array concept [9], which allows simplifying the focal law calculation in two media scenarios (i.e. water coupling and solid component).

AutoFocus is necessary for reliably inspecting radii and corners in T and omega type stiffeners. Current practice is to use a circular PA probe with its curvature center coinciding with the radii center (Fig.7 left). This way, a linear scan with zero-delay focal laws can be used, which guarantees normal incidence. However, if the curvature center does not match the curvature center of the radii (Figure.7 right), the normal incidence condition is lost, and the zero-delay approach is not valid. This kind of misalignments is common in practice and can have multiple sources: Component positioning errors, manufacturing tolerances, variations in the radius along the component, robot trajectory errors or mechanical clearances. Because of the large refraction index between water and CFRP, slight probe misalignments severely reduce the amplitude of the back-wall echo and the probability of delaminations detection.

Although probe misalignment can be minimized by carefully designed mechanics like that of the SITAU-TRITON scanning head, normal incidence cannot be guaranteed if the component shape is not perfectly circular. This is a quite common situation in aerospace CFRP components, which are manufactured by layer stacking without using a rigid mold in the upper face.

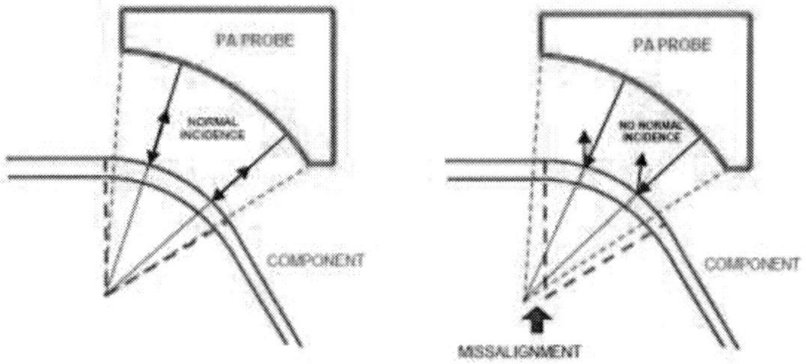

Fig. 7. Typical configuration for radii inspection with curved PA probe (left) perfect positioning (right) probe misalignment.

Fig.8 top shows a B-Scan of the upper radii of an omega structure with the probe configuration of Fig.7. Although the curvature center of the radii and the probe coincide, the radii is not perfectly circular, which severely distorts the B-Scan. In one hand, the amplitude and position of the interface echo changes. But more important, the amplitude of the back-wall echo is reduced in some regions, which will be misinterpreted as porosity.
Fig.8 bottom shows the B-Scan image obtained with the same set-up and the SITAU AutoFocus algorithm. By automatically detecting the interface echo and calculating, in real time, the correct focal laws, homogeneous interface and back-wall echoes are obtained, greatly improving inspection quality and reliability. This is performed without a-priori knowledge of the surface geometry, which makes the ultrasound inspection completely independent of the robot movement and/or the component shape.

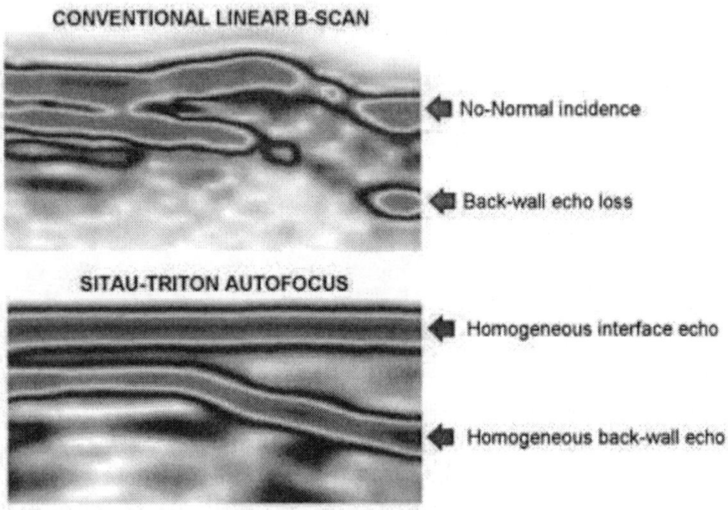

Fig.8. B-Scan image of an omega radii with the array set-up of Figure 2 (top) conventional B-Scan (bottom) SITAU AutoFocus algorithm.

Different from other approaches that use non focused emission with normal incidence only, the SITAU AutoFocus algorithm is able to perform arbitrary angle scans with emission focus and dynamic-depth-focusing on reception (Patented ProFoctechinque [8]), which allows inspecting not only small thickness components but also larger complex structures. Fig. 9 shows the S-Scan obtained from a 80 mm thick aluminum component, where the details of the bottom surface can be clearly identified regardless the complex shape of the water-aluminum interface.

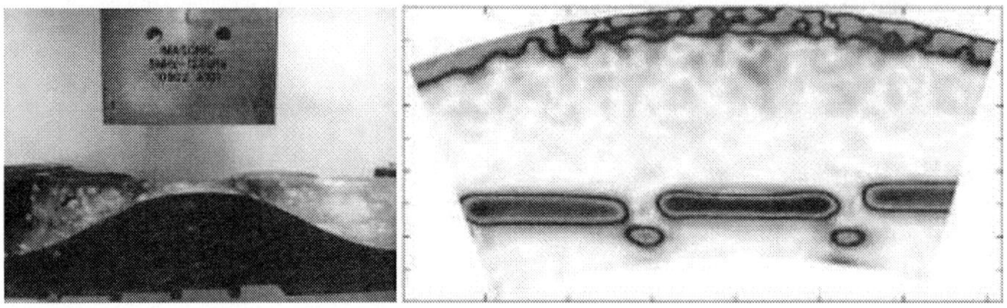

Fig.9. Inspection of a complex geometry and large thickness component (left) picture (right) AutoFocus S-Scan image.

In summary, SITAU AutoFocus technology represents a new paradigm for the automated inspection of large and complex components in the aerospace industry: No matter the surface profile, *the best possible image is guaranteed by the ultrasound system at all time*, greatly simplifying the inspection set-up and increasing the machine reliability. Its real-time operation, along with the large channel count available with SITAU architecture and a carefully designed scanning head and robotic system, allow reducing the inspection time of large and complex shape aeronautic components to unprecedented levels.

5. Conclusions

An automatic system using a robot to inspect complex materials has been presented. Its flexibility, both in the mechanical and the UT equipment configuration turn it in a flexible inspection platform for multiples uses. Up to now, three operation modes are implemented :

- *Operation mode 1*: the immersion tank and the robot are used together and the displacement is done following a linear track. It is able to perform like a 3 axes Cartesian system to inspect planar and smooth curvature parts, by using PA technology for high speed operation.
- *Operation mode 2*: The pieces with complex geometry (e.g. omegas type) are placed on an ad-hoc tool (previously designed for this use) immersed in the tank. With an UT module of up to 896 ultrasonic channels and 7 PA probes, the scanning head is designed to inspect the 100 % of complex geometry components in one pass, which guarantees the minimum possible inspection time. Furthermore, image quality is guaranteed by a innovative AutoFocus algorithm.
- *Operation mode 3:* Based on the use of TT inspection mode with water or air-coupled ultrasound, and with single or dual frequency transducers. Trajectories are optimized for each component ensuring a collision-free operation.

Operation modes 1 and 2, have already been certified by Airbus Defence Space following Caetano Aeronautic specifications. Mode 3 is expected to be certified during 2019.

6. Acknowledgement

This work has been partially founded by Torres Quevedo Grant PTQ-17-09089.

References

[1] R. Bogue. The role of robotics in non - destructive testing. Industrial Robot: An International Journal, 37 (5),421-426, 2010.
[2] X. Zhen, et al. Profile tracking with ultrasonic alignment for automatic Non-destructive testing of complex structures. Robotics and Computer–Integrated Manufacturing 49, 134–142, 2018
[3] S. Robert, O. Casula, O. Roy and G. Neau, “Real-time Inspection of Complex Composite Structures with a Self-Adaptive Ultrasonic Technique”, 18th World Conf. on Non Destructive Testing, Durban (South Africa), 2012.
[4] Evaluation of conventional ultrasonic inspection facilities, equipment and probes. AITM6-0013 Issue 6Page 1 of 76, October 2014
[5] Richard D., Monette M., Maes G., “Iterative Time Reversal Technique for Inspection of Composite Structures”, WCNDT, Munich Germany, 2016.
[6] A. Hunter, B. Drinkwater and P. Wilcox, “Least-Squares Estimation of Imaging Parameters for an Ultrasonic Array Using Known Geometric Image Features”, *IEEE Trans. on UFFC*, 58, 2, pp. 414-426, 2011.
[7]I. Cooper, I. Nicholson, D. Yan, B. Wright, D. Liaptsis, C. Mineo. Development of a fast inspection system for complex composite structure - the Intacom project. 5th International Symposium on NDT in Aerospace, 13-15th November 2013, Singapore.
[8] Patent ES2525600 "Method for real-time control of the dynamic approach in ultrasonic imaging systems and associated sampling advance calculating device", 2014.
[9] Patent WO 2005073957 "Coherent composition of signals by means progressive focal correction", 2005

Coherent Adaptive Focusing Technology for the Inspection of Variable Geometry Composite Material

André LAMARRE [1], Etienne GRONDIN [2]
[1] Olympus Scientific Solutions Americas, Quebec City, Canada
[2] Olympus Scientific Solutions of the Americas, Quebec City, Canada
Contact e-mail: andre.lamarre@olympus-ossa.com

Abstract

The aviation industry has seen above normal growth in recent years, owing in part to lower oil prices contributing to millions of dollars in savings for aircraft operators. As a result of this growth, production rates for new airplanes have increased, and new aircraft programs are being launched. Consequently, aviation component manufacturers are facing new challenges, including a rise in production rates, higher probability of detection (POD) requirements due to the critical nature of the parts being manufactured, a lack of skilled operators, and parts with increasingly complex geometry. To respond accordingly, ultrasonic phased array (PA) instruments have evolved, enabling the implementation of advanced acquisition strategies, such as adaptive focusing. Coherent adaptive focusing simplifies the inspection of variable radiuses, variable opening angles, and twisted components, and it also compensates for probe misalignment through innovative signal-processing algorithms. This paper presents an overview of coherent adaptive focusing technology with the goal of helping NDT integrators and composite material manufacturers address system performance, production output, and quality control issues.

OLYMPUS

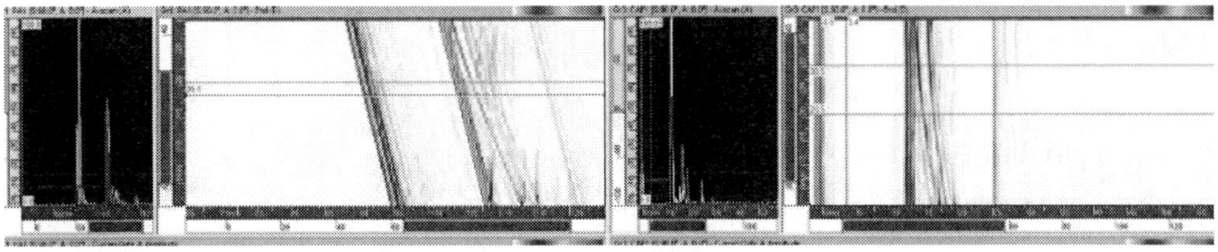

Coherent Adaptive Focusing Technology for the Inspection of Complex Geometry

Olympus Scientific Solutions Americas | Andre LAMARRE

NDT in Aerospace 10th International Symposium on NDT in Aerospace DGZfP

Introduction

OLYMPUS

Ultrasonic Phased-array benefits

- Ultrasonic Phased-array technology benefits for composite inspection
  - Faster inspection
  - High-resolution
  - Improved coverage
  - Different geometries
    - Flat and curved parts
  - Scalable instruments allows parallel firing of phased-array probes for high productivity

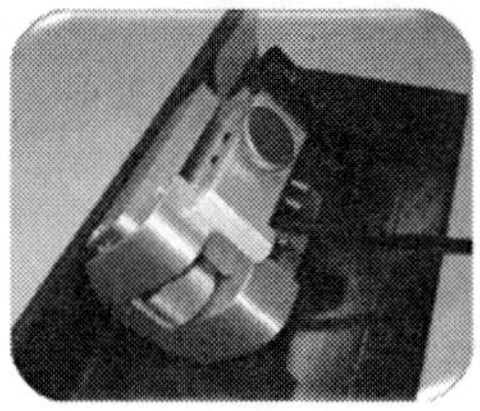

OLYMPUS

Ultrasonic phased-array limitations

- Beam forming is produced with pre-determined focals laws(time delays)
  - Inspection configuration must fit the model used by the calculator (deterministic model)
- Linear array probe must be parallel to the surface of the part to work properly
  - Probes could be badly positioned (angled)
  - Flat part might not be perfectly flat (pseudo-flat)
- Curved array probe must be concentric with the radius of the part to work properly
  - Probes could be badly positioned
  - The radius itself could be different from its nominal value
  - The radius could be a non-perfect radius
- A way to adapt the focal laws(beams) to the real inspection configuration is required:

COHERENT ADAPTIVE FOCUSING

OLYMPUS

Coherent Adaptive Focusing (CAF) Objectives

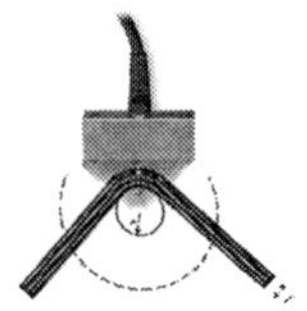

- Adapt the ultrasonic beams(focal laws) to the real inspection scenario
- Improve & simplify the inspection of complex geometry components
  - Curved and twisted parts
  - Variable and non perfect radius

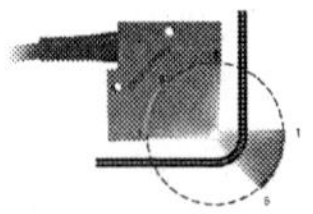

- Compatible with
  - Linear PA Probe on Pseudo Flat Surface
  - Linear & Curved PA Probe on Radius (inside and outside)
- Performance equivalent to phased array
  - Inspection speed has to be similar to standard phased array

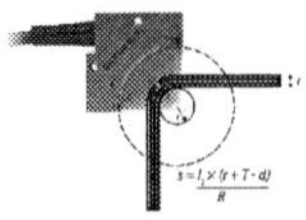

- Reliable & Robust
  - No data loss or convergence issues

Inspection of Inner radius

OLYMPUS

Possible issues with radius inspection

- Uncorrect probe positioning:

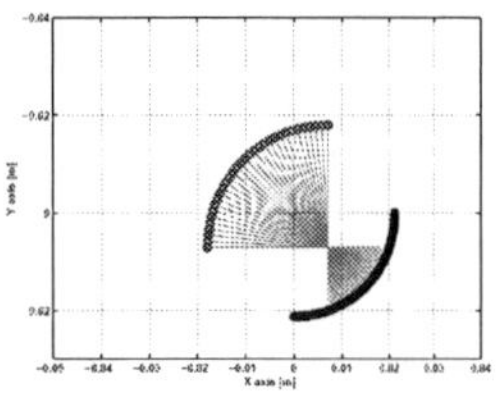

Inner Radius – Out Of Conc. (-)

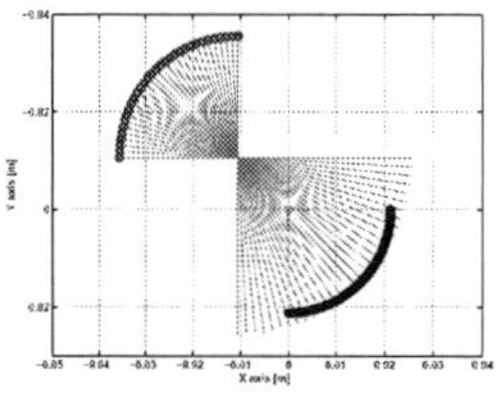

Inner Radius – Out Of Conc. (+)

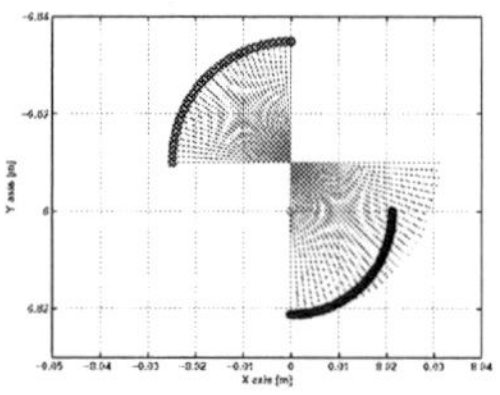

Misalignment

- Others:
  - The radius itself could be different from the nominal value
  - The radius could be a non perfect radius

OLYMPUS

Phased-array inspection of a radius (perfect concentricity)

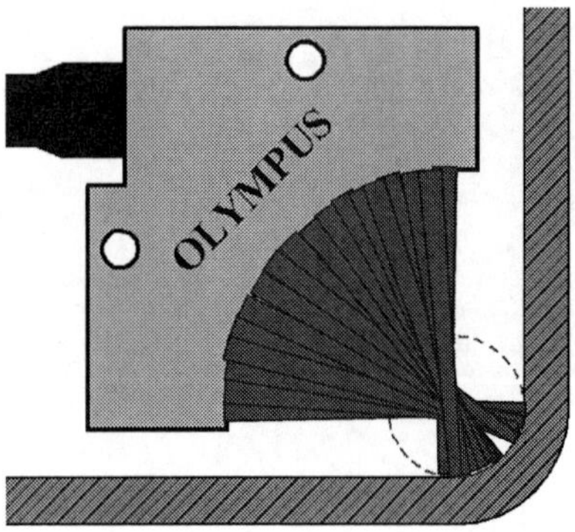

The distance from the probe to the part must be adjusted so as to coincide the center of radius of the probe to the center of the inner radius of the corner.
The entry beams are normal to the radius surface
Linear scan is performed covering the whole radius

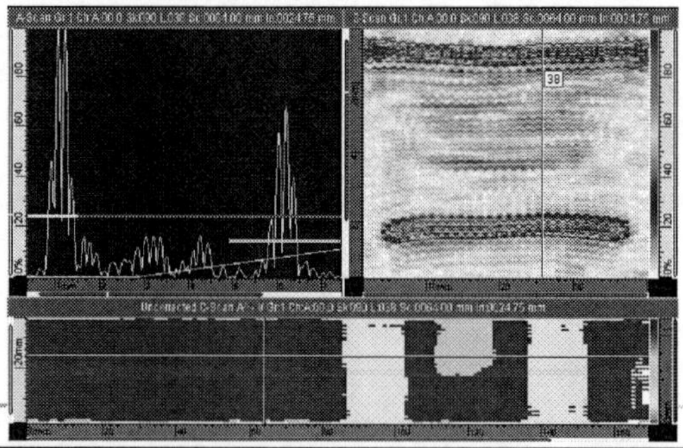

OLYMPUS

Phased-array inspection of a radius (out of concentricity)

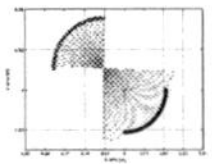

- **Part & Scan Parameters**
  - **Geometry**: L shape
  - **Radius** : 5mm
  - **Out of concentricity** : +15mm
  - **Scan** : Internal face
  - **Probe** : 3.5CC25-32R4
  - **Wedge** : SR4-IE90

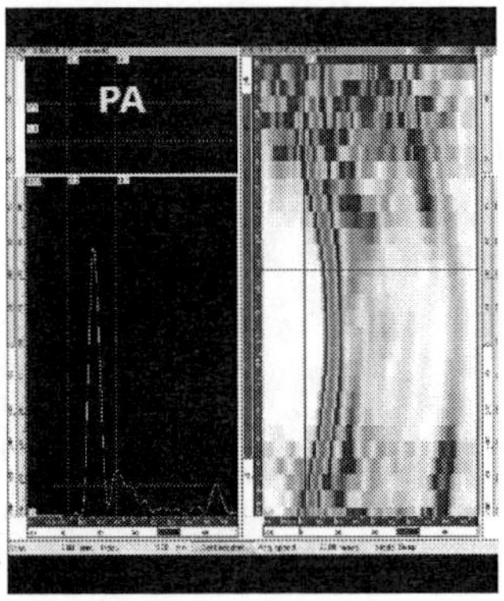

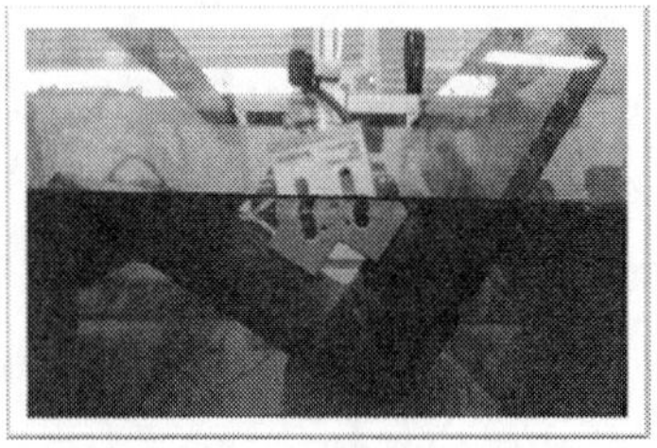

OLYMPUS

Coherent Adaptive Focusing of a radius (out of concentricity) Results

- **Out of concentricity** : +15mm

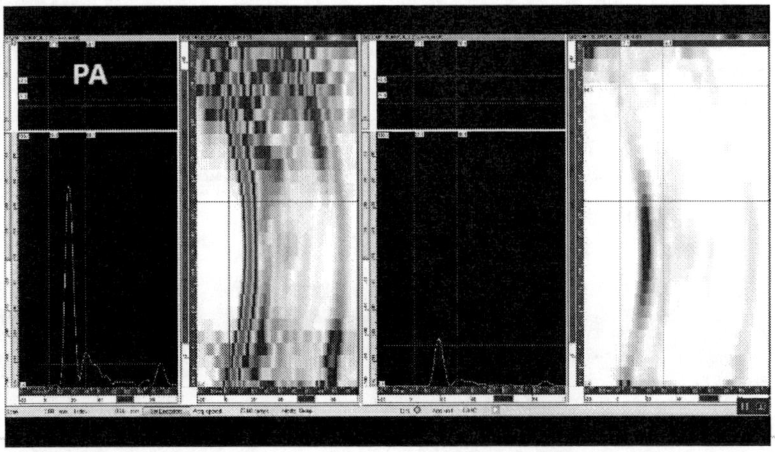

Coherent Adaptive Focusing
How does it works?

OLYMPUS

Coherent Adaptive Focusing

- Adaptive Focusing is an iterative process that allows the transmission of a wave-front parallel to the part
- When achieved, a synthetic linear scan is played off-line to examine the part

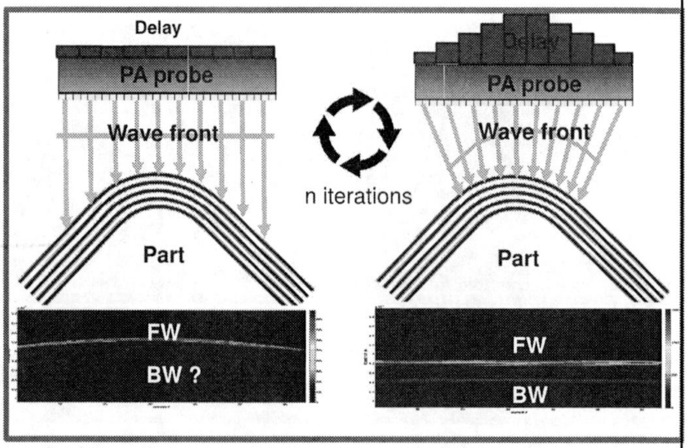

OLYMPUS

Coherent Adaptive Focusing – How it works

- **Step 1 :** Acquisition with known delays in Tx and Rx (0 or pre-defined)

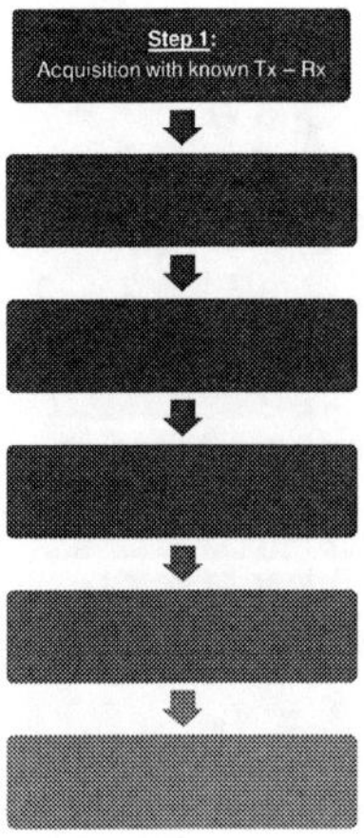

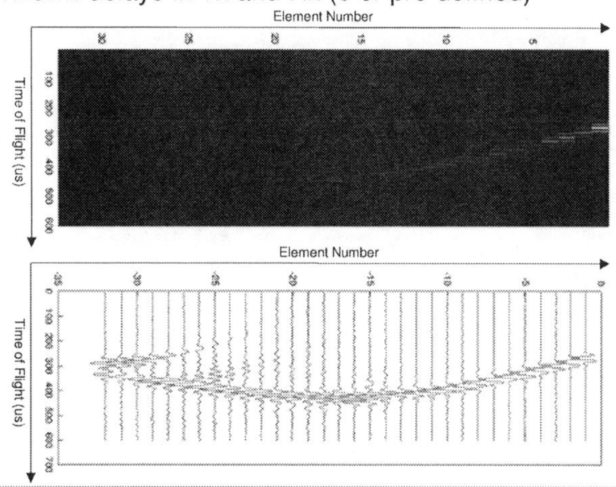

OLYMPUS

Coherent Adaptive Focusing – How it works

- **Step 2 :** Identification of potential Front Wall Echo location

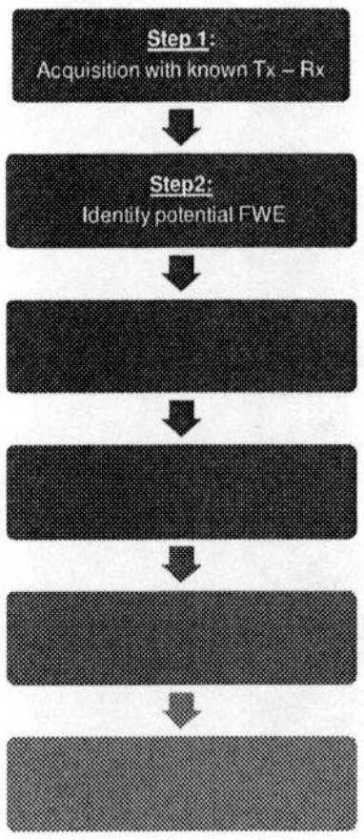

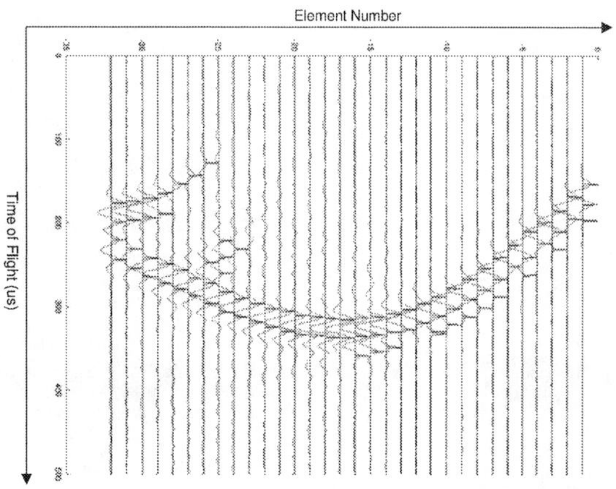

OLYMPUS

Coherent Adaptive Focusing – How it works

- **Step 3 :** Distinguish Artefacts from front wall echo location
- s

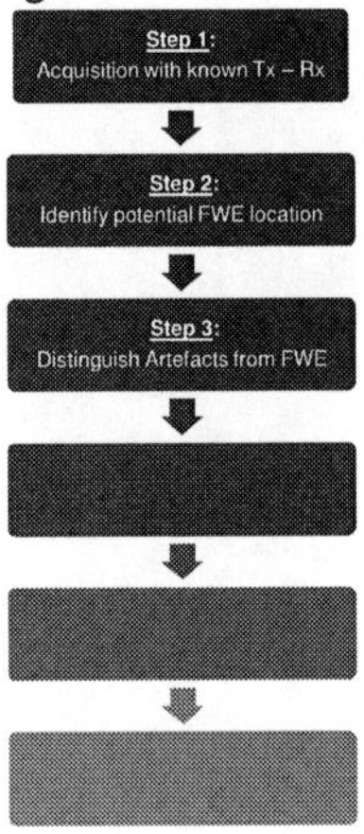

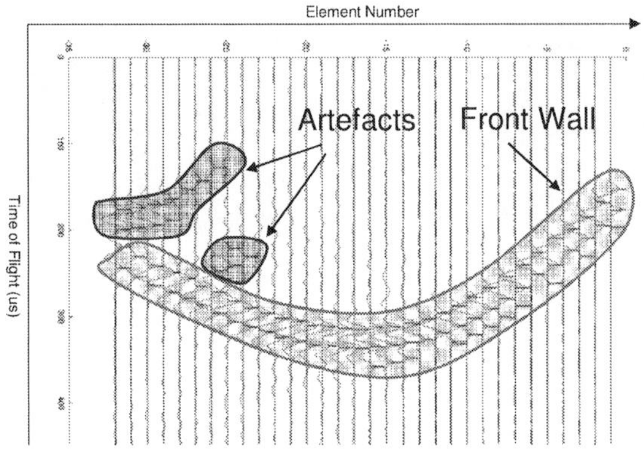

4. *Physics, Optics.* of or relating to waves that maintain a fixed phase relationship, as in coherent light, or light in which the electromagnetic waves maintain a fixed and predictable phase relationship with each other over a period of time.: See also laser.

OLYMPUS

Coherent Adaptive Focusing – How it works

- **Step 4 :** Discard artifacts

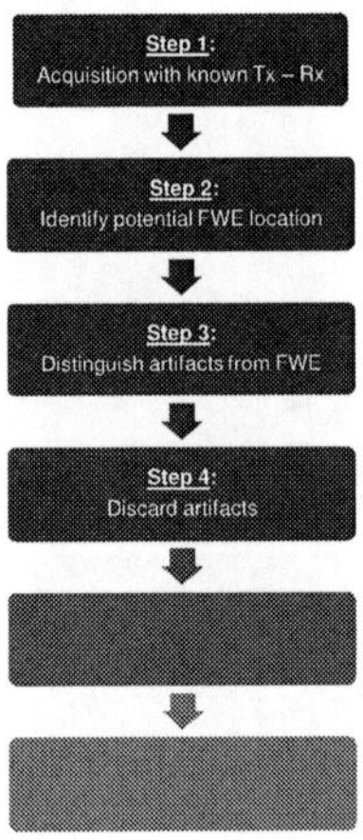

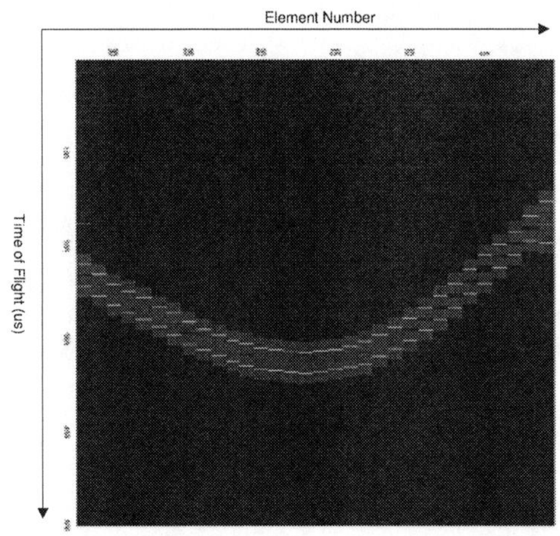

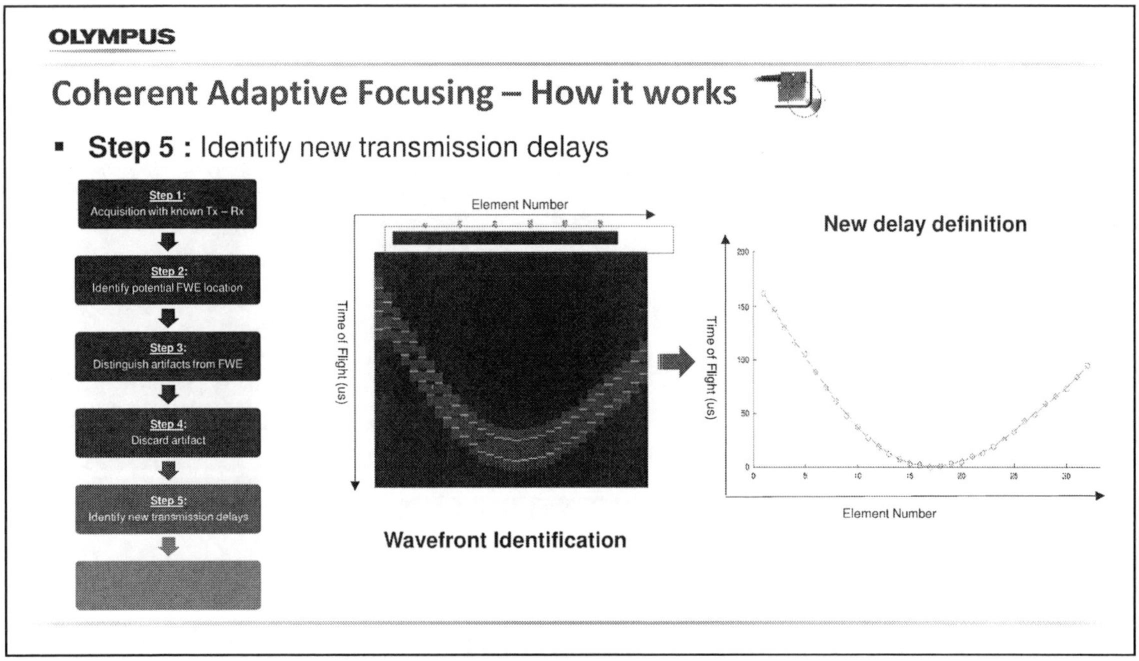
OLYMPUS
Coherent Adaptive Focusing – How it works
Step 5 : Identify new transmission delays
Step 1: Acquisition with known Tx – Rx
Step 2: Identify potential FWE location
Step 3: Distinguish artifacts from FWE
Step 4: Discard artifact
Step 5: Identify new transmission delays
Element Number
Time of Flight (us)
Wavefront Identification
New delay definition
Time of Flight (us)
Element Number

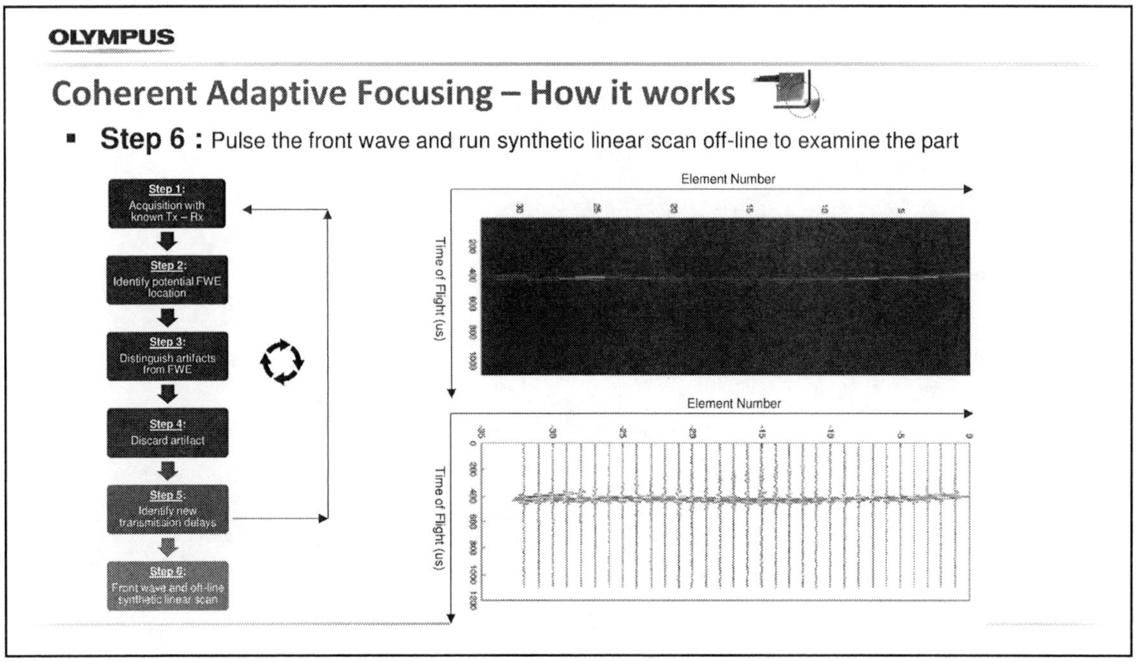
OLYMPUS
Coherent Adaptive Focusing – How it works
Step 6 : Pulse the front wave and run synthetic linear scan off-line to examine the part
Step 1: Acquisition with known Tx – Rx
Step 2: Identify potential FWE location
Step 3: Distinguish artifacts from FWE
Step 4: Discard artifact
Step 5: Identify new transmission delays
Step 6: Front wave and off-line synthetic linear scan
Element Number
Time of Flight (us)
Element Number
Time of Flight (us)

Other configurations results

OLYMPUS

Coherent Adaptive Focusing ID inspection with varying radius

- **Part & Scan Parameters**
  - **Geometry**: L shape
  - **Radius** : 0.35 in to 0.5 in.
  - **Scan** : Internal face (ID)
  - **Probe** : 5CC25-32R4
  - **Wedge** : SR4-IE90
  - Focus PX hardware
  - Focus PC software

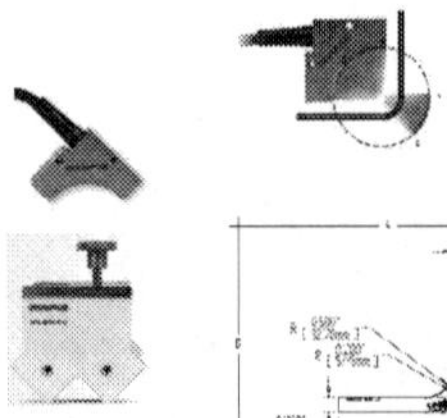

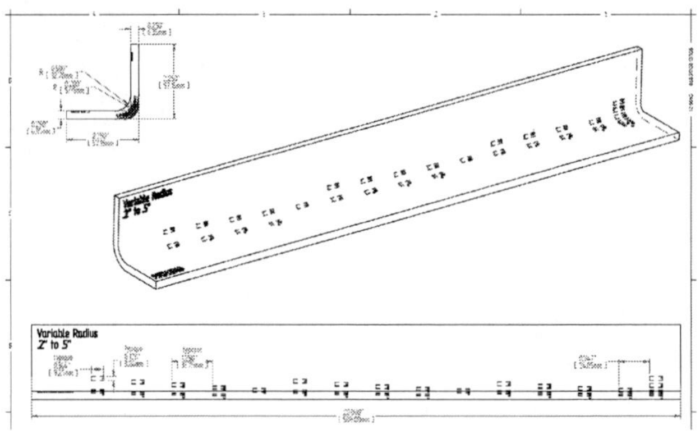

- Probe is concentric at the beginning of the scan

OLYMPUS

PA - CAF on ID variable radius 0.35 to 0.60 in.

Phased-array CAF

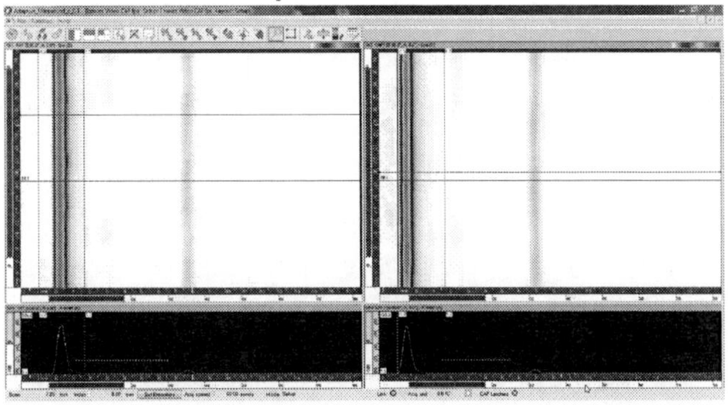

Probe is concentric at the beginning of the scan

OLYMPUS

Coherent Adaptive Focusing OD inspection with varying radius

- **Part & Scan Parameters**
  - **Geometry**: L shape
  - **Radius** : 0.35 in to 0.5 in.
  - **Scan** :External face (OD)
  - **Probe** : 5CC25-32R4
  - **Wedge** : SR4-IE90
  - Focus PX hardware
  - Focus PC software
- Probe is concentric at the beginning of the scan

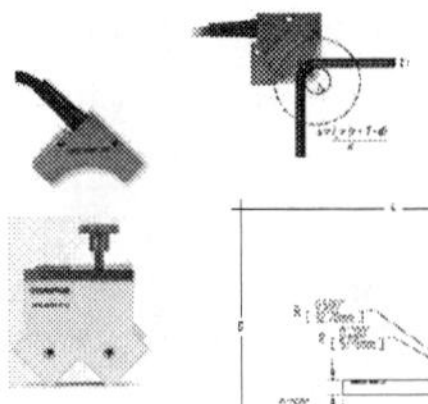

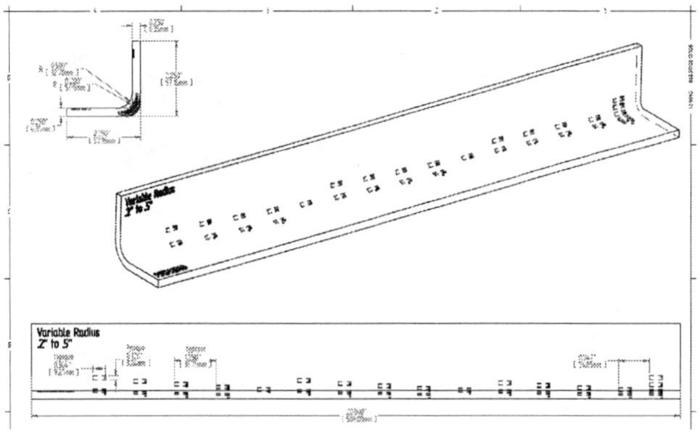

OLYMPUS

PA - CAF on ID variable radius 0.35 to 0.60 in.

Phased-array CAF

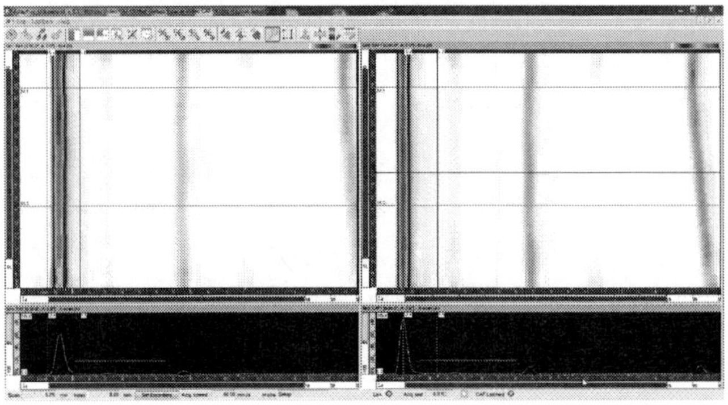

Probe is concentric at the beginning of the scan

OLYMPUS

Coherent Adaptive Focusing on flat surface with angled probe

- **Part & Scan Parameters**
  - **Geometry**: flat part
  - **Water column:**
  - **Probe : 3.5L64-NWI**
  - Focus PX hardware
  - Focus PC software
- Probe is angled 7 degrees

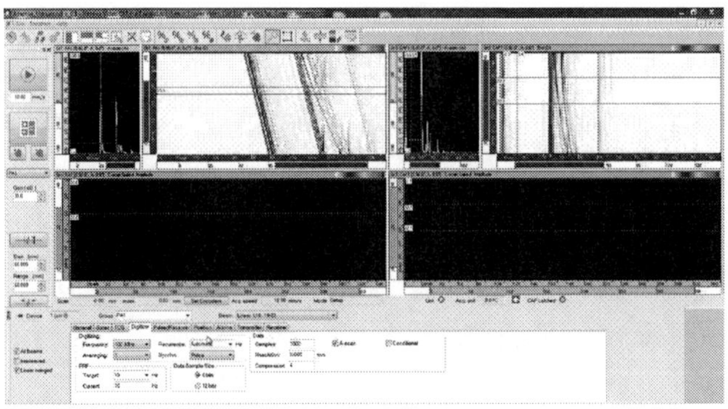

Coherent Adaptive Focusing capabilities

OLYMPUS

Coherent Adaptive Focusing Solution Results

- **Boundaries**

Case 1 : Inner Radius Inspection

Name	Results
Maximum Out Of Concentricity	(+) 30.0 mm (-) 2.0 mm
Maximum Misalignment Horizontal	(±) 6.0 mm
Maximum Misalignment Vertical	(±) 6.0 mm
Coverage	90.0°
Detection (Reliability & Stability)	All flaws detected & no data loss

Case 2 : Outer Radius Inspection

Name	Results
Maximum Out Of Concentricity	(+) 9.5mm (-) 10.0mm*
Maximum Misalignment Horizontal	(±) 6mm
Maximum Misalignment Vertical	(±) 6mm
Coverage	90.0°
Detection (Reliability & Stability)	All flaws detected & no data loss

* Until the probe touches the part

Case 3 : Tilted Probe/Part

Name	Results
Maximum tilt	25.0 °
Coverage @Max. tilt	**d = L*cos(alpha)** d: coverage L: probe length alpha: tilt angle
Detection (Reliability & Stability)	All flaws detected & no data loss

OLYMPUS

Coherent Adaptive Focusing Solution Results

- **Productivity**

Parameters	
Nb. Of Elements	32
Aperture Size	4
Aperture Steps	1
Water path	25 mm
Material Thickness	8.5 mm
Material Velocity	2700 m\s
Nb. Of iterations	5

PRF with CAF	PRF without CAF
300 Hz*	517 Hz

FPGA implementation should result in a PRF improvement of up to 450 Hz

Conclusions

OLYMPUS

Coherent Adaptive Focusing

- Allows to adapt the beams(focal laws) to the real inspection configurations
- Relaxes the need on mechanics to be perfectly aligned
- Is compatible with linear and curved phased-array probes
- Ease the inspection of components with changing configurations (varying radius, varying aperture, pseudo-flat...)
- Has detection and speed performances comparable to phased-array
- Easy to use by operator

OLYMPUS®

Quantitative evaluation of delamination depth in CFRP based on pulse compression eddy current pulse thermography

Q. Yi1, G.Y. Tian1, J. Zhu1, S. Laureti2, H. Malekmohammadi2, M. Ricci3
1Department of Electrical and Electronic Engineering, Newcastle University, Merz Court, NE1 7RU, Newcastle upon Tyne, UK
2Dept. Of Engineering, University of Perugia, Polo Scientifico didattico di Terni, Strada di Pentima 4, 05100 Terni, ITALY
3Department of Informatics, Modeling, Electronics and System Engineering, University of Calabria, Rende, 87036, ITALY
Contact e-mail: qiuji.yi@newcastle.ac.uk

Abstract. The growing application of composites material in aerospace leads to the urgent need of non-destructive testing and evaluation (NDT&E) of fatal defects such as impact damage and delamination in these materials. Eddy Current Pulsed Thermography (ECPT) is an emerging NDT technique capable of detecting such defects (e.g. impact damage and delamination). However, the characterization of delamination within a CFRP specimen is difficult to be accomplished by a single pulse excitation due to the challenging extraction of thermal diffusion in multi-layered structures. To cope with this problem, this paper combines the use of Eddy Current(EC) stimulus and barker coded modulated pulse compression technique for enhancing the delamination characterization ability for CFRP. Based on the estimated impulse response obtained by exploiting the PuC algorithm, delamination area is located by thermal pattern enhanced method based on K-PCA(kernel principal component analysis) techniques. Crossing point feature is extracted between defective area and non-defective area impulse response. Results show that delamination can be detected with depths from 0.46mm to 2.30mm and the proposed feature crossing point has a monotonic relationship with delamination depth.
Keywords: Pulse Compression, Eddy Current Pulsed Thermography, Quantitative evaluation, CFRP, Delamination depth

1. Introduction

The use of CFRP materials have become increasingly popular among conventional engineered materials due to their extraordinary mechanical and thermal properties, such as high strength-to-weight ratio, corrosion resistance, improved fatigue performance and low coefficient of thermal expansion [1]. Delamination is one of the most common defects for composite materials, which might arise either during the manufacturing process or during their onin-service life, e.g. during the standard life-time of aircraft components.
A promising NDT technique to detect delamination in CFRP is the Active Thermography (AT) [2, 3]. Being a non-contact method with the capability of inspecting a large area of the Sample Under Test (SUT). In AT, the desired thermal contrast is achieved by applying an external heating stimulus over the SUT [4] by mans of various illumination sources [5]. SUT's illuminated surface heats up and then diffuses into the sample. Among the AT techniques, one of the extensively applied AT technique is ECPT. Based on the nature of EC

technique, and since CFRP exhibits low electrical conductivity, in ECPT the stimulation is volumetric [6, 7]. Hence, ECPT is not influenced by the surface conditions of the SUT [8]. However, despite the progresses made on both evaluating the so-called barely invisible impact damage (BVID) [9] and visualizing delamination [6] by means of ECPT, the quantitative evaluation of delamination depth is still challenging. In fact, the complex structure of CFRP i.e. layered structure, anisotropic conductivity and different fiber direction, results in non-uniform thermal distribution, thus hampering the faithful quantitative evaluation of delamination depth from thermal response.

The PuC algorithm outputs a good estimate of impulse response that is close in both characteristics and quality to that obtainable from PT. In PuCT, the heat source emission is commonly modulated either by a frequency-modulated "Chirp" signal [10], or by a phase-modulated signal [11], e.g. Barker code. These coded excitations have a unique feature: their bandwidth B and their time duration T are uncorrelated. As a consequence, the frequency content of the coded signal can be tailored to suit the investigation of a given sample, while T can be increased almost arbitrarily to achieve the desired SNR [12]. Although PuCT is now a quite established technique that shows premises promises for maximising the SNR even when low power heat sources are employed, very few attempts are reported in literature to extend this approach to ECPT [13, 14].

Fig.1 shows the block diagram of the processing steps accomplished on the raw data. A man-made CFRP delamination sample was excited by a Barker-modulated induction heating stimulus and the raw thermal data are acquired. Then, a denoising algorithm is applied pixelwise to the raw data, as well as a non-linear fitting algorithm to remove the step-heating contribution [11]. PuC is then performed by convolving the detrended data with the so-called 'matched filter', so as to retrieve the impulse response h(t) pixelwise. A thermal pattern enhancement method is then applied over the h(t)'s to visualize and locate each delamination. A new feature extraction algorithm based on analysing the time delay of h(t) resulting from pixels area onto delamination buried at different depths into the sample is also proposed to conduct a quantitative study of delamination depth.

This paper is organized as follows: Section 2 presents the theoretical background of ECPT; the basic theory of PuC and the methodology of impulse response based thermal enhanced pattern technique. Section 3 provides details about the Barker code, the ECPuCT experimental setup and the benchmark delaminated sample. Section 4 shows how to perform PuC correctly and highlights the analysis of various extracted thermal pattern by the proposed technique. A Quantitative study of depth information of delamination by the proposed feature is also included (Depth QNDE). Conclusion will be made in Section 5.

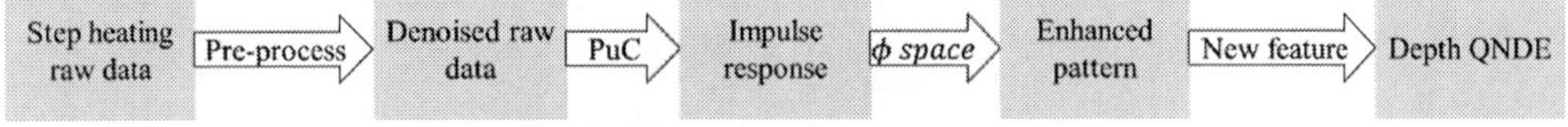

Fig 1 Block diagram of the processing steps

2. Theoretical background of eddy current pulse compression thermography

2.1 Theory of eddy current simulated volumetric heating in CFRP

The theory of ECPT on ferromagnetic material has been addressed in previous researches [15-17]. The principle of ECPT for CFRP, is different from that of ferromagnetic materials, concerning both the heating mode and the thermal wave diffusion pattern. According to electromagnetic wave penetration depth equation (Eq.(1)), the penetration depth δ has a monotonic decreasing relationship with increasing current frequency:

$$\delta = \frac{1}{\sqrt{\pi\mu\sigma f_{carrier}}} \quad (1)$$

where $f_{carrier}$ is the frequency of the excitation current [Hz], σ is electrical conductivity [S·m^{-1}], and μ is magnetic permeability [H·m^{-1}]. In general, for CFRP the bulk conductivity σ value is in the order of 15000 S/m and it is non-magnetic [18]. Considering $f_{carrier} = 300$ kHz, the corresponding penetration depth is 7.50 mm while for the same value of $f_{carrier}$, the penetration depth in steel ($\sigma = 9.93 \times 10^6$ S/m, $\mu = 3.77 \times 10^{-4}$ H/m) is equal to 29.08 µm, which is significantly smaller than one obtained for CFRP. Thus, since the thickness of the CFRP sample is lower than corresponding value of δ, the heating mode is volumetric.The unique advantage of volumetric heating compared to the surface heating scenarios (e.g. flash thermography) comes from the direct interaction between the buried defect and the induced eddy current field. Overall, the heating pattern of EC excitation in transmission arrangement can be considered as follows: firstly, the whole specimen is heated-up by induced eddy current by Joule's effect . Then, the defected area is heated more than the sound areas at the same depth.

2.2 Pulse compression basic theory

The working principle of the PuC technique is sketched in Fig.3. Given a coded excitation $s(t)$ of duration T and bandwidth B, and another signal $\Psi(t)$, which is the so-called matched filter, such that their convolution "$*$" approximates the Dirac's Delta function $\delta(t)$ as:

$$s(t) * \Psi(t) = \tilde{\delta}(t) \approx \delta(t) \quad (4)$$

If Eq.(4) holds, then an estimate $\tilde{h}(t)$ of the $h(t)$ is obtained by convolving the recorded output signal $y(t)$ with the matched filter $\Psi(t)$. The process is mathematically showed below for a single pixel of the acquired thermograms, in the presence of an Additive-White-Gaussian-Noise $e(t)$, which is uncorrelated with $\Psi(t)$. By convolving the output signal $y(t)$ with the matched filter $\Psi(t)$,the impulse response can be obtained as:

$$\tilde{h}(t) = y(t) * \Psi(t) = h(t) * \underbrace{s(t) * \Psi(t)}_{=\tilde{\delta}(t)} + e(t) * \Psi(t) = h(t) * \tilde{\delta}(t) + \tilde{e}(t) \approx h(t) + \tilde{e}(t) \quad (5)$$

The main advantages of PuC over pulsed excitation is that an estimate of the impulse response can be achieved at end of the procedure, while delivering energy to the system over an extended time. In this way, it is possible to provide more energy, and hence to increase the SNR and detectability of eddy current thermography system.

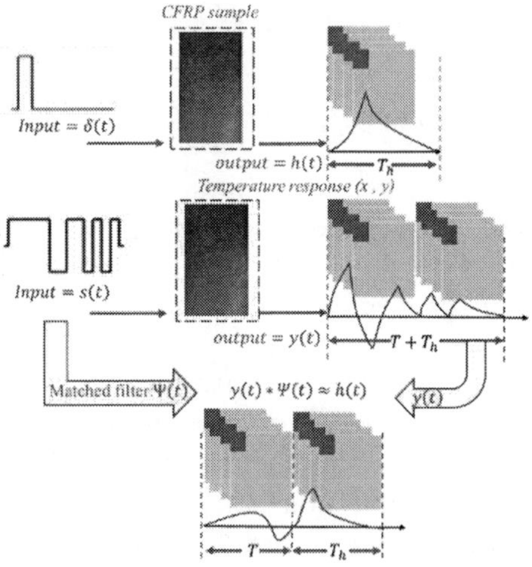

Fig 3 Comparison between pulse exaction and coded excitation

2.3 Thermal pattern enhancement techniques based on impulse response and feature extraction

To gain insight on how K-PCA has been applied here for enhancing ECPuCT data, the implementation of the enhancement method is schematically depicted in Fig.4 and mathematically introduced here below. Considering the calculated impulse response $\tilde{h}$'s retrieved pixelwise by exploiting the procedure described in Eq.(5) being reshaped as:

$$\left[\tilde{h}_1, \tilde{h}_2, \dots, \tilde{h}_{N_x N_{y-1}}, \tilde{h}_{N_x N_y}\right] \tag{6}$$

where, $N_x$and N_y are the number of x and y pixel of the acquired IR thermograms. Thus, the reshaped data can be recognized as a matrix having dimension of $Q \times N_x N_y$, where Q denotes the number of frames recorded by the IR camera, *i.e.* $T_h \times FPS$, where FPS stands for the acquired Frames Per Second. By using the kernel method, the impulse response is projected to kernel space ϕ, thus obtaining the kernel matrix $K(i,j)$ as:

$$K(i,j) = \frac{1}{M}\sum_{i=1}^{M}\left(\phi(\tilde{h}_i) - \frac{1}{M}\sum_{j=1}^{M}\phi(\tilde{h}_j)\right)\left(\phi(\tilde{h}_i) - \frac{1}{M}\sum_{j=1}^{M}\phi(\tilde{h}_i)\right)^{\mathrm{T}} \tag{7}$$

where M is equal to $N_x N_y$ and ϕ is Gaussian kernel function, defined as Eq.(8):

$$\phi(\tilde{h}_i^1, \tilde{h}_i^2) = \exp\left(\frac{\|\tilde{h}_i^1 - \tilde{h}_i^2\|_2^2}{2\sigma^2}\right) \tag{8}$$

The kernel matrix $K(i,j)$ of Eq.(7), can be simply named as K. The eigenvector α of K can be obtained as:

$$\lambda_i \alpha_i = K\alpha_i \tag{9}$$

Based on the obtained eigenvector α_i, the enhanced thermal pattern can be projected as:

$$H_{enhanced} = [\alpha_1, \dots, \alpha_T]H_{original}^{\mathrm{T}} \tag{10}$$

where $H_{enhanced}$ contains different extracted thermal patterns.

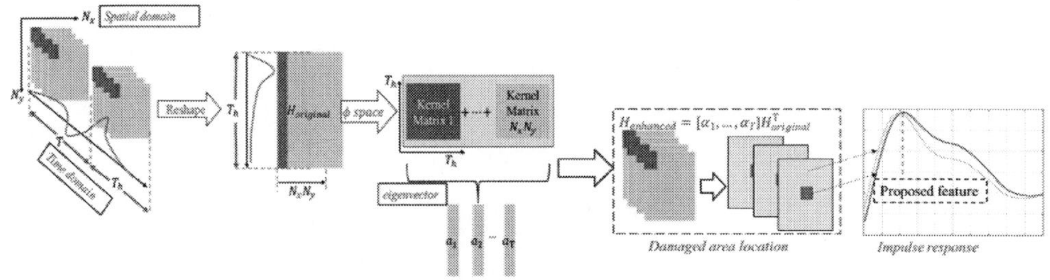

Fig 4 Diagram of feature extraction

3. Experiment setup

3.1 Eddy current pulse compression thermography setup

The ECPuCT system diagram is illustrated in Fig.5. Signal generator is used to send both the BC modulating signal to the induction heating coil and a reference clock trigger to the IR camera to acquire thermograms at 50 FPS. A Cheltenham Easyheat 224 Induction heating unit is used for coil excitation with a maximum excitation power and current values of 2.4 kW and 400 A respectively with tuneable $f_{carrier}$ from 150 to 400 kHz. For the reported experimental results, values of excitation current I equal to 40 A and $f_{carrier}$ of 240 kHz were selected to avoid eventual damage on the CFRP sample due to the long excitation time using BC signal. IR camera was the FLIR SC655, equipped with an un-cooled microbolometer detector array with the resolution of 640 × 480 pixels, the spectral range of 7.5 - 14.0 μm and NETD < 30 mK. The IR camera records the surface temperature distribution of the 13 s BC as well as additional 30 seconds cooling period . Finally, captured thermal videos were transmitted to a PC for visualization and postprocessing, including signal pre-process, pulse compression, thermal pattern enhancement and quantitative depth evaluation.

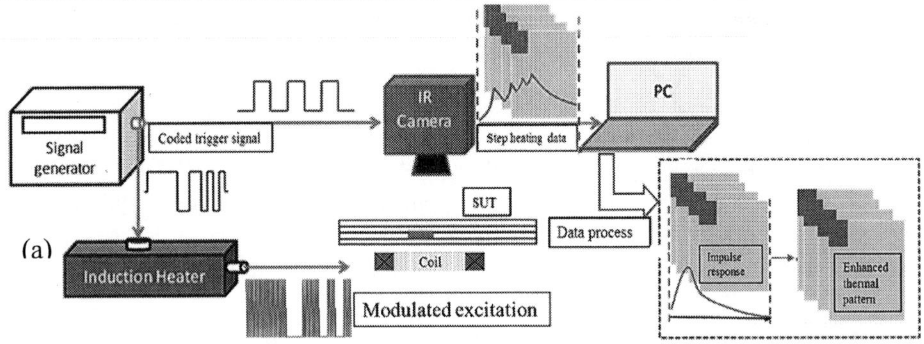

Fig.5 Sketch of the experimental setup

3.2 Sample under test

The CFRP sample was realized by MDP company (Terni, Italy) [11].The CFRP laminate sample contained twelve plies of carbon fibre fabric with an areal density of 0.2 g/m^3. Its lateral dimensions were 240 mm×200 mm for an overall thickness of ~2.80 mm which is shown in Fig 6. The fibres orientations were 0° and 90° and the matrix was an Epoxy Resin RIM 935. The artificial delamination were realized by inserting thin square pieces of Teflon tape having lateral dimensions of 20 mm×20 mm and thickness equal to 75 μm between the plies. Nine artificial defects were inserted at increasing depths: the shallower was placed under the 2nd ply at a depth *d*~0.46 mm (defect #1), the deepest (defect #9) under the 10th ply at a depth of *d*~2.3 mm.

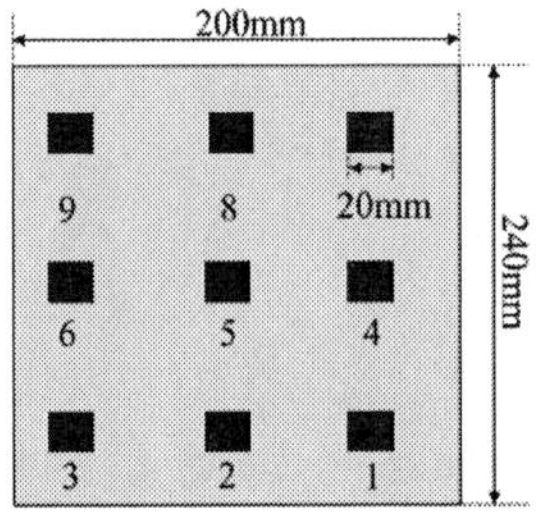

Fig 6 Schematic diagram of SUT

4. Results and analysis

4.1 Defect location based on pattern enhancement method

The impulse response $\tilde{h}(t)$ is a sequence of 1500 frames. However, when it comes to visualize the delamination, most of the frames contribute little to the defect location. To solve this problem, the kernel method described in Sec.2.3 is applied to find and learn the mutual relation of impulse response in individual pixels. Thus, the pixelwise-retrieved $\tilde{h}(t)$'s are embedded into a suitable high-dimensional feature space through Gaussian kernel function. In addition, the kernel method is also an efficient way to imply the nonlinear characteristics of the raw data.Fig.7 shows the extracted thermal pattern from BC raw data (threshold). It is noted that first thermal pattern summarizes the overall heating phenomena during experiment because the heating is conducted on one side of rectangular coil, which generates linear heating. The sensitivity of pattern 1 for visualizing defect is heavily reduced for increased delamination depth. This is because pattern 1 contains most of information about heating pattern, which might cover up the faithful defect information. It is observed that patterns 2 and 3 obtain opposite profile for delamination. Pattern 2 generally presents the delamination in lower contrast while pattern 3 shows higher contrast compared to the surrounding area. This is because the eigenvectors extracted were supposed to be orthogonal with each other in kernel space ϕ. The indicated relationship between pattern 2 and pattern 3 can enhance the detectability in delamination area through logic operation. In fig 8, it is shown that defect SNR is reduced after D5, where the square shape of delamination profile cannot be identified. However, the delaminated area can still be characterized by fibre structure in 0° and 90° from D6~D9 since induced eddy currents are parallel to 0° and 90°, which have the largest electrical conductivity and thermal conductivity in SUT thus generating abnormal thermal wave with diffusion length μ longer than the delamination depth d.

To conclude, artificial delamination in CFRP has been demonstrated to exhibit different patterns by kernel method. In the following section, the impulse response retrieved at the most abnormal pixels area are investigated for quantitative study

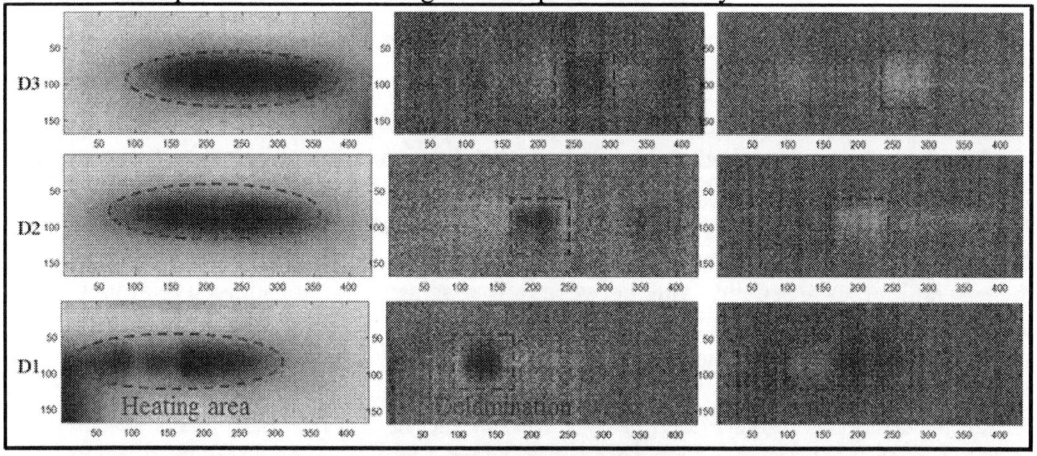

Fig 7 Extracted thermal pattern (D1-D3)

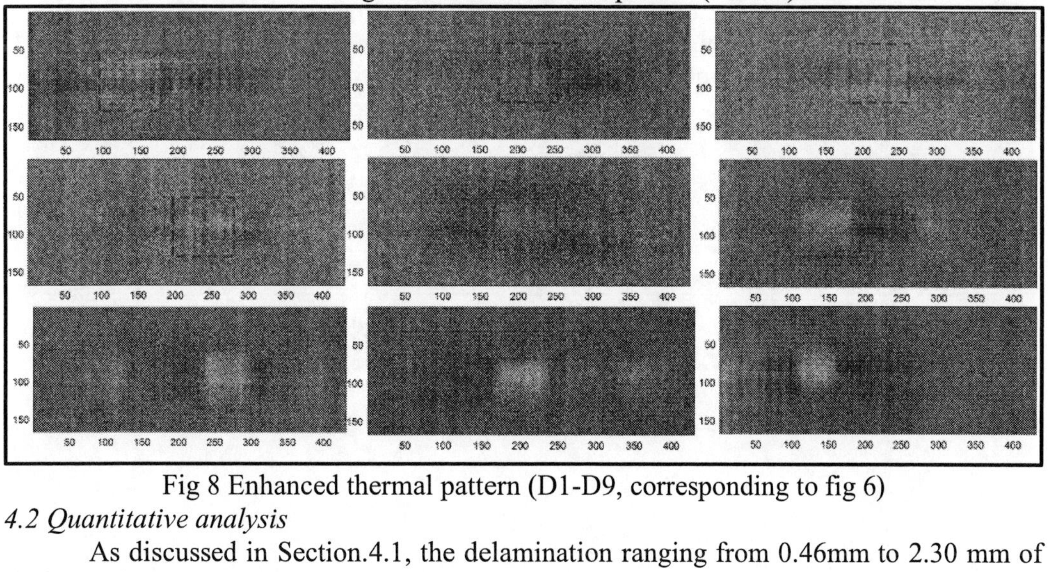

Fig 8 Enhanced thermal pattern (D1-D9, corresponding to fig 6)

4.2 Quantitative analysis

As discussed in Section.4.1, the delamination ranging from 0.46mm to 2.30 mm of depth can be visualized by thermal pattern enhanced method based on the impulse response. Moreover, to determine the delamination depth, the phase contrast between defected and non-defected areas can be exploited, by comparison of the obtained impulse responses based on

PT theory [19].Each defected area and non-defected area is formed by five lines of data in the centre of higher contrast and lower contrast based on enhanced pattern.

It can be noted the defective area is cooling slower due to the thermal diffusivity, which means the material ability to exchange heat with its surroundings has become worse with the increase in depth. The deeper the delamination, the worse is the ability to exchange heat with non-defective area. Results showed in Fig.9 demonstrates that the impulse response amplitude is less sensitive than the phase change of crossing point of defective and non-defective area. Thus, it is observed that the position of the first crossing point of impulse responses of defected and non-defected areas has a monotonic relationship with delamination depth, as shown in Fig.9 and Fig.10. This can be served as an excellent feature to quantify the delamination depth. From Fig.10, it can be noted that the feature based on the crossing point value follows a linear trend if three defects in a row over the SUT are considered, *e.g.* D1-D3 or D6-D9, which is due to the actual linear coil used in this experiment. In fact, the experimental condition can be considered the same if defects are on the same line. Better results could be possibly obtained by achieving a more uniform heating through optimizing coil shape.

However, after defect 6, the crossing pattern of defected area and non-defected area has been disturbed by noise, (see Fig.9(c)). Moreover, extra crossing points observed in D7, continuing equal value of defected and non-defected areas in D8 and initial disturbed crossing point in D9 are caused by the low SNR of original signal. Thus, despite the effectiveness of the denoised approach applied in this work, the clear information of defect is still challenging to be extracted in D7~D9.

Fig.11 shows the fitted curve for the depth versus time instant of the first crossing point of defective and non-defective areas acquired on 5 different lines. Based on the fitting curve, there is still a crossing point value between defective area and non-defective area when the depth is zero, *i.e* surface defect. Considering the crossing point feature is influenced by defect's depth and the electrical and thermal properties, when depth is zero, this feature can be only related to defect's electrical and thermal properties, which can be useful to derive these properties. Moreover, judging by error bar in Fig. 11(a) and from the selected data distribution in Fig.11(b), the deviation generated by the selected position is acceptable.

To conclude, monotonic relationship between crossing points of retrieved impulse responses illustrates that depth evaluation of delamination in CFRP can be conducted by calculating the phase change of defected and non-defected areas.

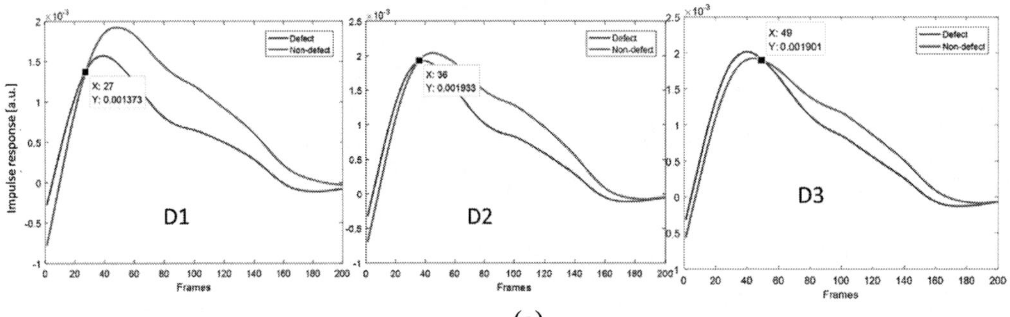

(a)

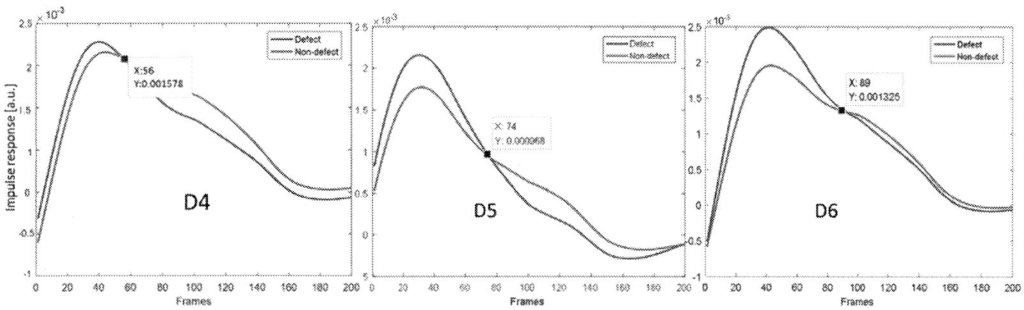

(b)

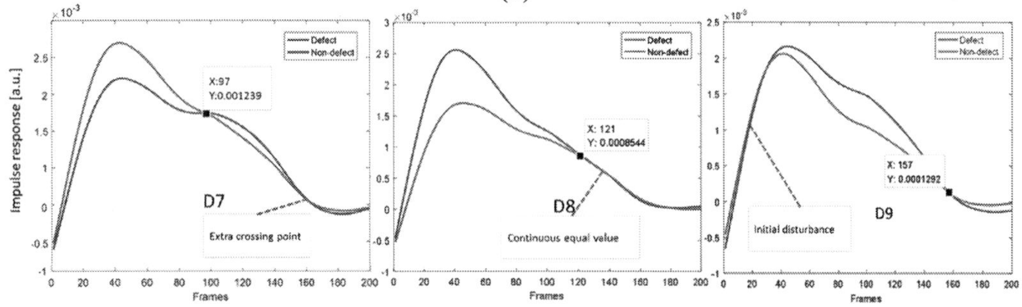

(c)

Fig 9 mean value of selected impulse response at:(a) D1/D2/D3, (b) D3/D4/D5 (c)D7/D8/D9

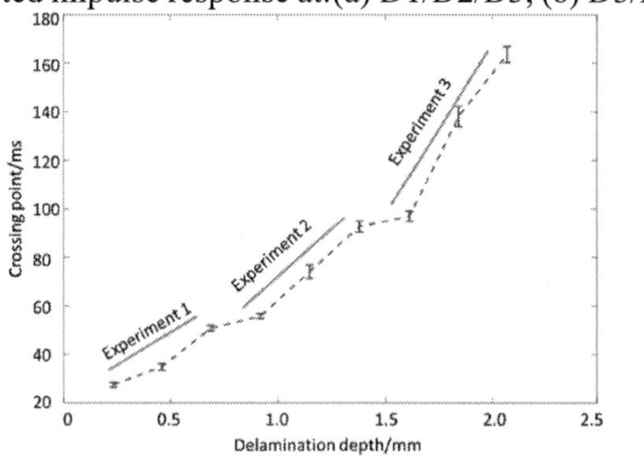

Fig 10 Monotonic relationship of delamination depth(D1-D9) and crossing point

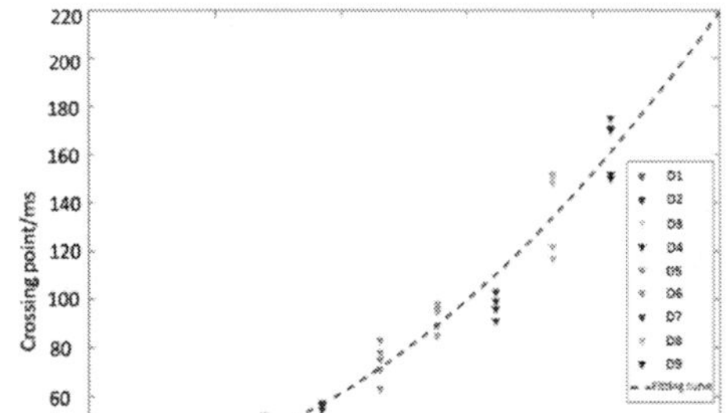

(a)

Data(ms) / Defect No	Line 1	Line 2	Line 3	Line 4	Line 5
D1	26	28	28	28	28
D2	33	35	34	38	35
D3	50	49	52	51	52
D4	57	56	56	56	54
D5	78	75	72	83	63
D6	85	96	89	98	95
D7	99	96	103	91	96
D8	151	148	153	122	117
D9	174	150	153	170	165

(b)

Fig 11 Acquired crossing points of 5 lines of data versus delamination depth: (a) fitting curve, (b) specific data for D1~D9

5. Conclusion

In this paper, proposed feature based on ECPuCT for CFRP delamination evaluation is investigated. The main contribution of this work includes:
(1) The feasibility of Barker code modulated eddy current pulse stimulation on CFRP is experimentally investigated. Results show that delamination can be detected with depth from 0.46 mm to 2.30 mm using ECPuCT approach.
(2) Quantitative evaluation of depth information of different delamination is made based on proposed feature which is the crossing point of the impulse response acquired from selected abnormal pixel. The proposed feature has monotonic relationship with delamination depth.

Reference

[1] Zhou, A., et al., *Effect of moisture on the mechanical properties of CFRP–wood composite: an experimental and atomistic investigation.* Composites Part B: Engineering, 2015. **71**: p. 63-73.
[2] Yang, R. and Y. He, *Optically and non-optically excited thermography for composites: A review.* Infrared Physics & Technology, 2016. **75**: p. 26-50.
[3] Milovanović, B. and I. Banjad Pečur, *Review of active IR thermography for detection and characterization of defects in reinforced concrete.* Journal of Imaging, 2016. **2**(2): p. 11.
[4] Maldague, X., *Theory and practice of infrared technology for nondestructive testing.* 2001.
[5] Schlichting, J., C. Maierhofer, and M. Kreutzbruck, *Crack sizing by laser excited thermography.* NDT & E International, 2012. **45**(1): p. 133-140.
[6] Yang, R. and Y. He, *Eddy current pulsed phase thermography considering volumetric induction heating for delamination evaluation in carbon fiber reinforced polymers.* Applied Physics Letters, 2015. **106**(23): p. 234103.
[7] He, Y. and R. Yang, *Eddy current volume heating thermography and phase analysis for imaging characterization of interface delamination in CFRP.* IEEE Transactions on Industrial Informatics, 2015. **11**(6): p. 1287-1297.
[8] Cheng, L. and G.Y. Tian, *Transient thermal behavior of eddy-current pulsed thermography for nondestructive evaluation of composites.* IEEE Transactions on Instrumentation and Measurement, 2013. **62**(5): p. 1215-1222.
[9] Ren, W., et al., *Quantitative non-destructive evaluation method for impact damage using eddy current pulsed thermography.* Composites Part B: Engineering, 2013. **54**: p. 169-179.
[10] Laureti, S., et al., *The use of pulse-compression thermography for detecting defects in paintings.* NDT & E International, 2018. **98**: p. 147-154.
[11] Silipigni, G., et al., *Optimization of the pulse-compression technique applied to the infrared thermography nondestructive evaluation.* NDT & E International, 2017. **87**: p. 100-110.
[12] Laureti, S., et al., *Comparative study between linear and non-linear frequency-modulated pulse-compression thermography.* Applied Optics, 2018. **57**(18): p. D32-D39.
[13] Laureti, S., et al., *Detection of rebars in concrete using advanced ultrasonic pulse compression techniques.* Ultrasonics, 2018. **85**: p. 31-38.
[14] Yang, R. and Y. He, *Pulsed inductive thermal wave radar (PI-TWR) using cross correlation matched filtering in eddy current thermography.* Infrared Physics & Technology, 2015. **71**: p. 469-474.
[15] Tian, G.Y., et al., *Eddy current pulsed thermography with different excitation configurations for metallic material and defect characterization.* Sensors, 2016. **16**(6): p. 843.
[16] Peng, J., et al., *Investigation into eddy current pulsed thermography for rolling contact fatigue detection and characterization.* NDT & E International, 2015. **74**: p. 72-80.
[17] Gao, Y., et al., *Electromagnetic pulsed thermography for natural cracks inspection.* Scientific reports, 2017. **7**: p. 42073.
[18] Li, Z. and Z. Meng, *A review of the radio frequency non-destructive testing for carbon-fibre composites.* Measurement Science Review, 2016. **16**(2): p. 68-76.
[19] Zeng, Z., et al., *Depth prediction of non-air interface defect using pulsed thermography.* NDT & E International, 2012. **48**: p. 39-45.

Improved shear horizontal wave piezoelectric fiber patch (SH-PFP) for structural health monitoring applications

Yongtak KIM [1], Bernd KÖHLER [1]
[1] Fraunhofer-Institut für Keramische Technologien und Systeme IKTS, Dresden, Germany
Contact e-mail: yongtak.kim@ikts.fraunhofer.de

Abstract. Controlling the wave directivity of transducers is an important aspect of guided wave based structural health monitoring (SHM), because a strong directivity is not only energy efficient but also has an advantage of simple signal analysis. The fundamental shear horizontal (SH0) wave is gaining its popularity due to its non-dispersive characteristics. Piezoelectric fiber patches (PFP) are lightweight, thin, and flexible and therefore adaptable to curved surfaces. PFP can be designed to preferentially generate and receive either Lamb or shear horizontal guided waves. Building on previous work for a SH-PFP transducer [1] its geometry is modified to optimize the mode purity and the directivity. Finally, a new version of SH0 wave PFP with improved directivity is proposed. The new transducer concept and its working principles are explained, and the numerical simulation is conducted to validate its performance. Results show that the new transducer can generate almost pure SH0 waves along the desired direction with high directivity.

Introduction

Non-destructive testing (NDT) denotes a procedure to verify, whether a material or object has still the properties relevant for its intended use without affecting the object's ability [2]. Therefore it is normal that NDT technique is accompanied by technicians who can use test equipment. In contrast to that, structural health monitoring (SHM) denotes either the permanent monitoring of a structure with sensors or the periodic reading out of such permanently installed sensors [3]. Therefore, the importance of reliable SHM techniques in aerospace industry increases because of their lightness and non-human dependable nature.

Guided wave (GW) is one of the most promising technique in SHM implementation. GW uses wave propagation in a structure and investigates signal changes such as reflection and mode conversion, which could be considered as indicators of damage in the structure. There are various types of modes in GW, and most of them are dispersive. Dispersive characteristic means that the speed of the wave is frequency dependent. This makes the measured signal more complex and more difficult to analyze to detect damage.

Compared with the dispersive wave modes, the non-dispersive fundamental shear horizontal (SH0) wave in plate-like structures or the fundamental torsional [T(0,1)] wave in pipe-like structures is more promising in practical application [4]. In the past decades, SH0 wave considered to be limited in NDT techniques such as EMAT and used for metallic structures

[5, 6]. Obviously, such techniques are not suitable for SHM in aerospace due to their relative large size and low energy conversion efficiency [7].

Wilcox et al. employed a thickness-shear (d15) mode piezoelectric transducer to generate SH0 wave [8]. However the signal-to-noise ratio (SNR) of this transducer is rather small without any backing forces or seismic mass. Miao et al. proposed face-shear d24 PZT transducer [7]. Since the face-shear transducers don't need a backing force, they show superior to thickness-shear ones. Miao used these face-shear transducers to introduce an omnidirectional SH0 wave ring array in their further research. However, the array has limitations in complex geometry and brittle material problems. Koehler et al. proposed a new type of SH0 transducer using piezoelectric fiber path (SH-PFP) based on flexible piezoelectric fibers using double layer crossed fiber arrangement [1]. Even though SH-PFP is already very fascinating SH0 wave transducer, further work is needed to utilize the full potential of the transducer, especially with the mode purity and the directivity.

In this work, we propose a new version of SH0 wave PFP with improved directivity. The new version of SH-PFP, so called Dual SH-PFP is made out of two original SH-PFP. We will show that Dual SH-PFP (DSH-PFP) can generate SH0 wave in only two opposite directions with rather high purity. The configuration and the working principles of the suggested new transducer is explained, and the numerical simulation is conducted to validate its performance.

1. Configuration and working principle

1.1 Original SH-PFP

Two rectangular shaped PFPs with fiber orientations tilted at ±45° relative to the patch length orientation are used to make the original SH-PFP [1]. As shown in Figure.1, when two different PFPs with same electrodes are overlapped, SH0 waves are generated from 4 of each side of the SH-PFP.

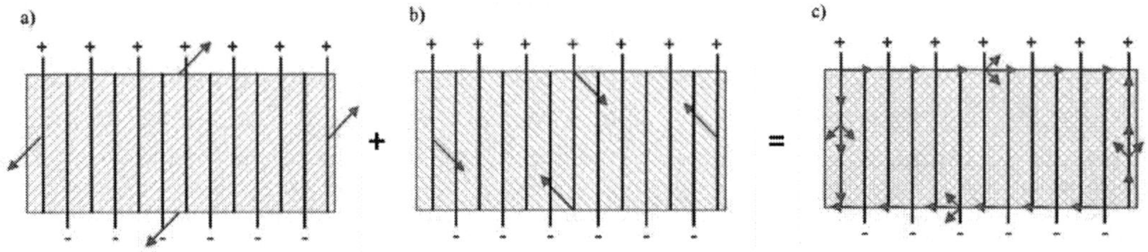

Figure 1. Configuration of the original SH-PFP [1]: The yellow lines show the piezo fiber directions, and the red lines show the directions of surface traction when a voltage is applied

Since the shape of a PFP is rectangular, there are two main directions and minor directions of wave propagation. To increase the directivity, the effect of two minor directions should be eliminated, while the wave propagation from the main directions should be maintained. The general method is to choose the length of the PFP as a multiple of the wavelength (λ) of target SH0 wave, while the width as an odd multiple of the half wavelength. However, since input signals are normally modulated by a window, even though the width of the PFP is matched as an odd multiple of the central frequency half wavelength, near-central frequencies will not be completely eliminated.

1.2 New version of SH0 wave PFP: Dual SH-PFP

To eliminate the effects of minor directions and to increase the directivity, a new version of SH0 wave PFP is proposed. As shown in Figure 2, two SH-PFPs with different piezo fiber directions are used. For example, the first SH-PFP (Left, from Figure 2) has a bottom PFP with +45° tilted fiber direction, and a top PFP with -45° tilted fiber direction. The second SH-PFP (Right, from Figure 2) has -45° tilted fibers for bottom PFP, and +45° tilted fibers for top PFP. Green arrows show the surface tractions driven by bottom PFPs, and red arrows show the surface tractions driven by top PFPs.

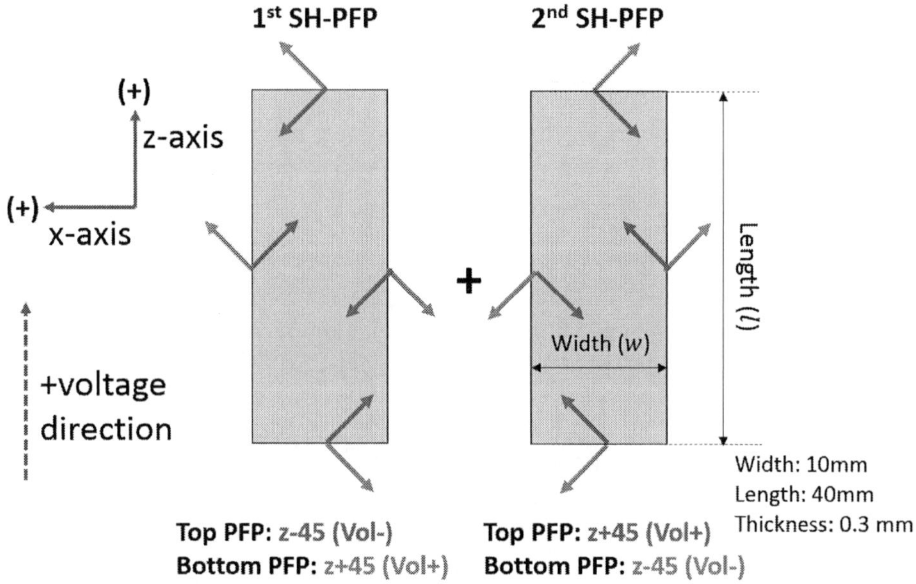

Figure 2. Configuration of the Dual SH-PFP. The red and green arrows show the surface tractions driven by the top and the bottom PFPs respectively.

The driven surface tractions are symmetric with respect to the z-axis. Therefore, when the two different SH-PFPs are put side by side, the surface tractions along the width of each PFP are of opposite sign. The sum of the tractions is zero and thus the effect of the guided wave will cancel out to a large extend. This will result in bi-directional SH0 wave propagation as shown in Figure 3.

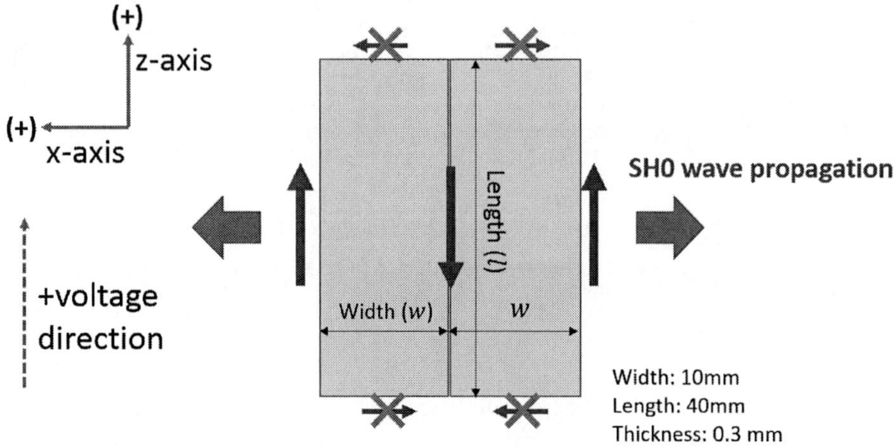

Figure 3. Dual SH-PFP wave propagation. The surface tractions along the width of each PFP are marked with red crosses, because they nearly cancel out in their effect on the generated wavefield.

The purple arrows show the summation of the driven surface tractions, and the blue arrows show the SH0 wave propagating directions. No matter what frequency is used for driving signal, the minor side surface tractions always cancel each other out.

2. Numerical simulations and results

Commercial finite element method (FEM) simulation software ANSYS was used to compare the performance of the new Dual SH-PFP with that of the original SH-PFP. Each PFP has an active area of 40 * 10 mm^2 and 0.3 mm of thickness. The actual configuration of PFP is too complicated to be modelled in ANSYS, so a homogenized PFP design was used as described in [9]. A steel plate of 2mm thickness was used as a target structure. The theoretical SH0 wave speed on the structure was 3250 m/s. The transducer was driven by a 3-cycle Hanning windowed sine signal and the central frequency was varied from 40 to 225 kHz.

2.1 General comparison

The Figure 4 shows the calculated tangential velocity component of the wavefield snapshots of the original SH-PFP (Figure 4 (a)) and the Dual SH-PFP (Figure 4 (b)) in a cylindrical coordinate system.
As shown in Figure 4 (a), the original SH-PFP generates SH0 waves along 4 orthogonal directions; two main directions (0° and 180°) and two minor directions (90° and 270°). In comparison, for the Dual SH-PFP only the main wave packets are visible. Other wave packets perpendicular to that might be present, but they are too weak to be visible in the image.

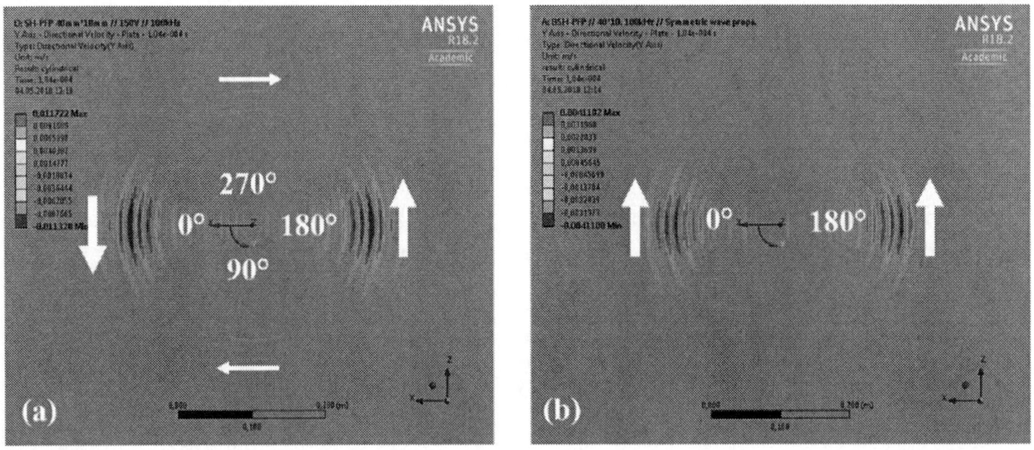

Figure 4. Tangential velocity wave fields of (a) the original SH-PFP and (b) the Dual SH-PFP

2.2 Directivity comparison

High directivity is preferred because it simplifies the signal analysis. The length of a PFP is 40 mm, which is comparable with the half wavelength of the SH0 wave at 40 kHz (40.6 mm). In this case, strong wave propagations are expected not only from the main directions but also from the minor directions of the original SH-PFP. 40 kHz central frequency signal was driven to both SH-PFP transducers to compare their directivity. Figure 5 shows the result.

The x-axis indicates one of the main wave propagating direction (0°) and the y-axis indicates one of the minor direction (90°). Due to symmetry of the transducers, we restricted the plot of the directivity to the first quadrant. The amplitudes were normalized by the maximum amplitude value of the Dual SH-PFP.

As shown in the graph, the Dual SH-PFP generates SH0 wave mainly along the main direction, while original SH-PFP generates SH0 waves along both main and minor directions.

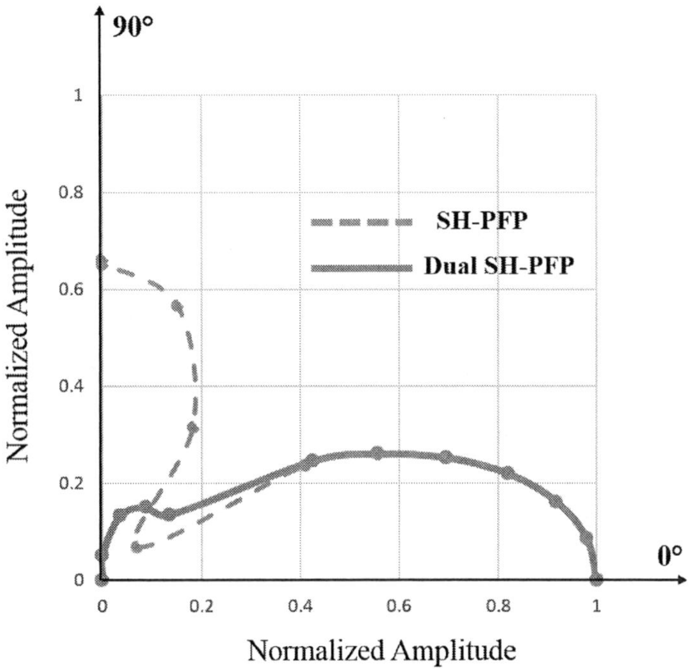

Figure 5. Directivity comparison of SH0 wave between the SH-PFP (Orange dotted line) and the Dual SH-PFP (Blue line) from 0° to 90°. The amplitudes are normalized by the maximum amplitude of the Dual SH-PFP. The central frequency is 40 kHz

2.3 Purity comparison

At the corners of a SH-PFP, two surface tractions from both adjacent sides add up under 90° leading to a resulting traction acting under 45°. This can be considered as the source of the S0 lamb wave [1]. Since the surface tractions along the minor sides of the Dual SH-PFP are working against each other and cancelling each other out partially, the amplitude of the generated S0 lamb wave is expected to be smaller. So the purity ratio defined as a ratio between the SH0 and the unwanted S0 signal amplitudes is expected to be improved.

To confirm this expectation, we compare SH0 and S0 signal amplitudes for both transducer types. Figure 6 shows graphs of the tangential velocity wave fields (SH0 wave, blue) and the radial velocity wave fields (S0 lamb wave, orange) at 50 and 100 kHz central frequency signals from 0° (x-axis) to 90° (y-axis). The amplitudes are normalized by the maximum amplitude value of each transducer at each central frequency.

As it is shown in the graphs, not only the directivity but also the purity is improved in the Dual SH-PFP. To quantitatively evaluate the purity improvement, the purity ratio is calculated for each case. The maximum purity ratio of the original SH-PFP was 5.52 at 150 kHz central frequency, while the Dual SH-PFP shows the maximum purity ratio of 9.01 at 100 kHz central frequency (Table 1).

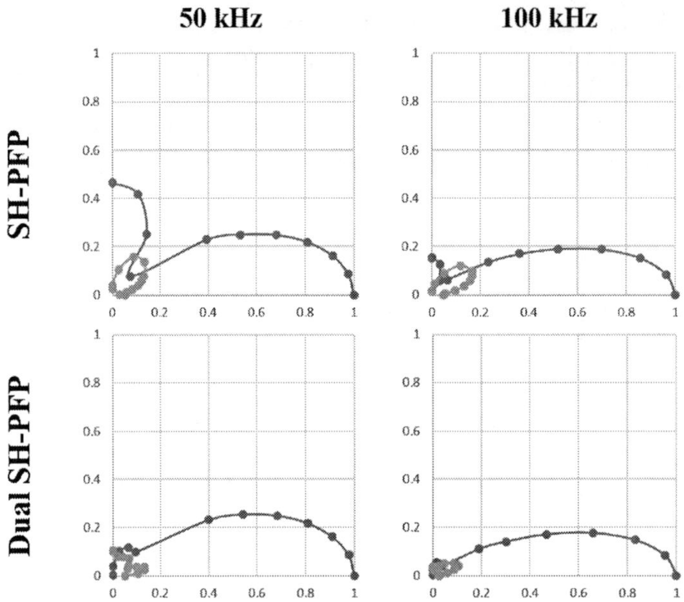

Figure 6. Purity comparison of the SH-PFP (Upper) and the Dual SH-PFP (Lower): SH0 wave amplitudes (Blue) and S0 lamb wave amplitudes (Orange) are normalized by the maximum amplitude of each transducer at each central frequency; at 50 and 100 kHz

Table 1. The purity ratio comparison

Frequency	40kHz	50kHz	100kHz	150kHz	200kHz
Dual SH-PFP purity ratio	6.08	7.25	9.01	5.92	6.21
Original SH-PFP purity ratio	3.33	5.21	5.15	5.52	5.05
Purity ratio increment	183%	139%	175%	107%	123%

2.4 Amplitude comparison

The new design also affects the maximum amplitudes. Figure 7 shows the amplitude comparison at the main direction (0°) between the original SH-PFP (Orange) and the Dual SH-PFP (Blue) from 50 to 225 kHz of central frequencies. The amplitudes are normalized by the maximum value of the Dual SH-PFP. The amplitudes of the Dual SH-PFP are indicated in parentheses as a percentage of the original SH-PFP at each central frequency. It is shown that the Dual SH-PFP can generate stronger SH0 wave, while maintaining the superiority of the directivity and the purity. Both transducers will have their maximum amplitudes when the PFP width equals to the half wavelength ($\lambda/2$) of the signal, but the increment of the Dual SH-PFP is much larger.

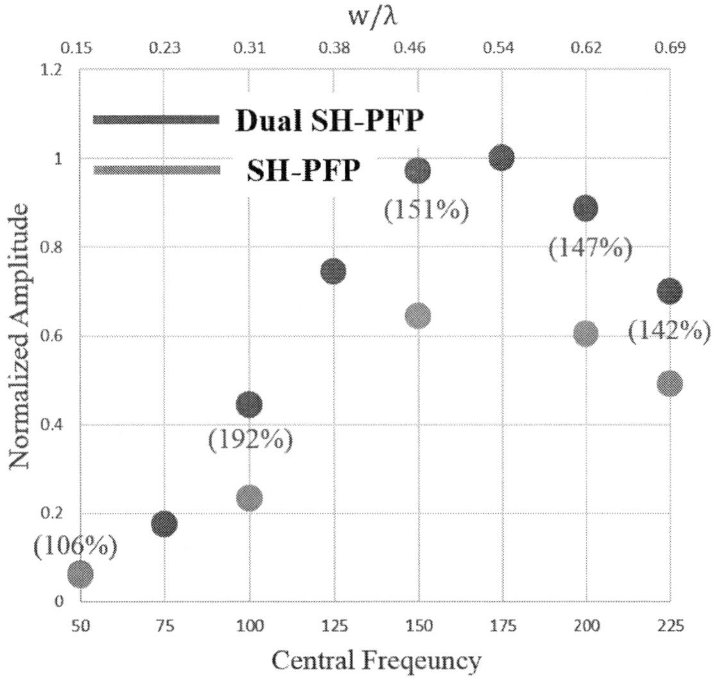

Figure 7. Amplitude comparison of the Dual SH-PFP (Blue) and the original SH-PFP (Orange). Normalized by the maximum amplitude of the Dual SH-PFP. The horizontal axis at the top of the graph is the ratio between the patch width (w) to the wavelength (λ).

3. Summary and conclusions

In this study, a new version of SH-PFP, so called Dual SH-PFP for the high directivity SH0 wave propagation is proposed. The configurations and the working principles of the original SH-PFP and the Dual SH-PFP were shown. The performance of the Dual SH-PFP was investigated by comparing with the original SH-PFP. Results show that the new transducer can generate almost pure SH0 waves along the desired direction with high amplitude. The high directivity and the high purity of the proposed new transducer suggest that this new version of SH-PFP could be widely used for SHM techniques in aerospace industry.

Acknowledgement

This research was supported by the NDTonAIR project (Training Network in Non-Destructive Testing and Structural Health Monitoring of Aircraft structures) funded under the action: H2020-MSCA-ITN-2016- GRANT 722134

References

[1] B. Koehler, T. Gaul, U. Lieske, and F. Schubert, Shear horizontal piezoelectric fiber patch transducers (SH-PFP) for guided elastic wave applications, *NDT&E Int.*, vol. 82, (2016), 1–12

[2] D.S. Forsyth, 5-Nondstructive testing of corrosion in the aerospace industry, *Corrosion Control in the Aerospace Industry*, (2009), 111-130

[3] V. Giurgiutiu, 16-Structural health monitoring (SHM) of aerospace composites, *Polymer Composites in the Aerospace Industry*, (2015), 449-507

[4] R. Ribichini, F. Cegla, P.B. Nagy, P. Cawley, Study and comparison of different EMAT configurations for SH wave inspection, *IEEE Trans. Ultrason. Ferroelectr. Freq. Control*, vol. 58, (2011), 2571–2581.

[5] RB. Thompson, Physical principles of measurements with EMAT transducers, *Phys. Acoust,* vol. 19, (1990), 157–200.

[6] H.M. Seung, C. Park, Y.Y. Kim, An omnidirectional shear-horizontal guided wave EMAT for a metallic plate, *Ultrasonics*, vol. 69, (2016), 58–66.

[7] H.C. Miao, Q. Huan, F.X. Li, Excitation and reception of pure shear horizontal waves by using face-shear d24 mode piezoelectric wafers, *Smart Mater. Struct.*, vol. 25, (2016), 11LT01.

[8] P.D. Wilcox, M. Lowe, P. Cawley, Omnidirectional guided wave inspection of large metallic plate structures using an EMAT array, *IEEE Trans. Ultrason. Ferroelectr. Freq. Control*, vol. 52, (2005), 653–665.

[9] C.R. Bowen, P.F. Giddings, A.I.T. Salo, and H.A. Kim, Modeling and characterization of piezoelectrically actuated bistable composites, *IEEE Trans. Ultrason. Ferroelectr. Freq. Control*, 58, (2011)

Applying features of nonlinear ultrasonic modulation for defect detection in vibrating structures

Sevilia SUNETCHIIEVA [1], Helge PFEIFFER [1], Sebastiaan CRETEN [1], Christ GLORIEUX [2], Martine WEVERS [1]
1 KU Leuven (Catholic University of Leuven), Department of Materials Engineering, Belgium
2 KU Leuven (Catholic University of Leuven), Department of Physics and Astronomy, Belgium
Contact e-mail: sevilia.sunetchiieva@kuleuven.be

Abstract. Aircraft structures are subjected to various external factors that influence their lifetime. Due to the high need for structural health monitoring in aerospace applications, numerous, quite mature linear ultrasonic NDT techniques have been developed for the detection of defects. Typically guided waves are used, which are based on mode conversion and reflection of probe waves by a defect, provided the defect is open, resulting in an acoustic impedance mismatch. However, in practical applications defects are often 'closed' when not under substantial stress.

Moreover, in most nonlinear ultrasonic NDT techniques the lack of differentiation between sources of nonlinearity makes defects indistinguishable from e.g. nonlinearity induced by mechanical contacts.

Here, we aim to detect damage, addressing the practical difficulties of monitoring vibrating structures. Typical defects were created by fatigue facilities and detected in aluminium plate-like samples using PZT transducers to generate and detect probe waves. The presented diagnostic algorithms compare features of the nonlinear relation between the amplitude of the transmission probe wave and the load on the sample with a threshold value, in order to assess the state of the sample. The applications are robust to environmental changes, are based on durable components, while being sensitive to vibrating defects.

Part of the research leading to these results has received funding from the European Community's Seventh Framework Programme [FP7/2007-2013] under grant agreement n°212912 and the "NDTonAIR" project (Training Network in Non-Destructive Testing and Structural Health Monitoring of Aircraft structures) under the action: H2020-MSCA-ITN-2016- GRANT 722134.

Introduction

Due to the insreased usage of lightweight materials for aircraft industry there is strong demand in the development of non-destructive evaluation technologies having monitoring capabilities. If the crack initiation leading to the major or even fatal failure can be detected, action can be taken to prevent or avoid the destruction of the structure, and appropriate repairs can be carried out. Structural Health Monitoring (SHM) offers interesting options to assess the state of components, thereby reducing the costs spent on scheduled maintenance and the possibility to prolong the operational lifetime of these components than it was originally foreseen. Most SHM systems make use of one or a combination of more non-destructive testing (NDT) techniques. Some of them, for example acoustic emission and resonance analysis, are not appropriate for use in a complex noisy environment with restricted implementation areas for the sensors. The development of a successful SHM technologies relays on boundary conditions set by different factors. Taking into account the possible environmental and boundary conditions, the ultrasonic method is one of the most promising choices.

1. Methodology and experimental details

Several linear and nonlinear ultrasonic NDT techniques are available, nevertheless their practical implementation is still limited. This promote the development of 'online' in-flight monitoring [1]. Linear ultrasonic NDT is based on the detection of mode conversion, attenuation and reflection of probe waves by a defect. Provided that the defect is open, it results in an acoustic impedance mismatch. However, in practical applications, in particular in unloaded conditions, defects can 'close' and return to their quasi-homogeneous condition. In such cases, effects on the probing acoustic waves are absent. Technologies, based on nonlinear ultrasonic, have been developed to resolve the problem of 'closed' cracks. Acoustic nonlinearity arises when acoustic waves propagate through regions with wave dependent material parameters, such as nonlinear elastic properties or nonlinear effects at a defect.

Consider a crack in a sample that is closed when the sample is not loaded and that opens when the sample is under load. Depending on the applied cyclic load on the sample the crack opens and closes accordingly. In the proposed method, throughout the cyclic load, a transmission measurement is performed using a high frequency (HF) probe wave. If the crack is closed, the probe wave does not encounter a region of acoustic mismatch and achieves normal transmission through the crack zone. When the load increases, the crack starts to open resulting in mode conversion of the probe waves at the open part of the crack. The transmission amplitude cyclically changes from high transmission at low applied load to low transmission at high loads, and thus the probe wave transmission amplitude undergoes nonlinear modulation and more mode conversion [2]. On the other hand, cyclic loading of an intact sample does not change the propagation properties in the monitored region. Hence, no modulation of the probe wave transmission is expected. Therefore, monitoring the properties of the modulation can be used to differentiate between an intact and a damaged sample [3].

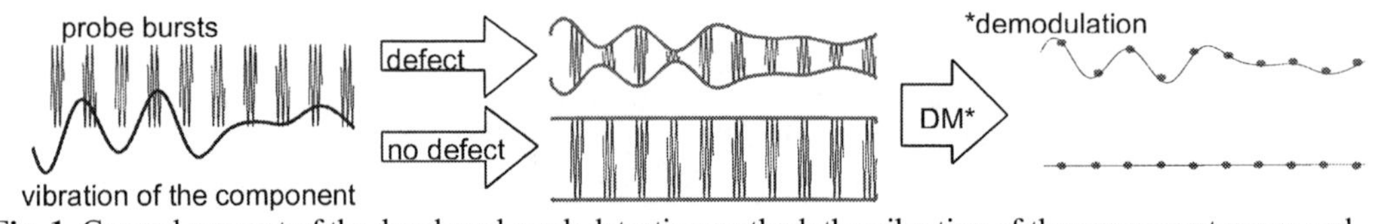

Fig. 1. General concept of the developed crack detection method: the vibration of the component opens and closes a crack and thereby modulates the probe waves. In the absence of a crack no modulation is detected[3]

In order to validate the proposed approach, a fatigue crack was initiated and grown in a small plate using a MTS-810 load bench. The sample was an aluminium (Al 2024-T3) plate of 1 mm thick, 300 mm length and 80 mm width. In order to accelerate the formation of a crack and to force its position, a 3.5 mm diameter hole was drilled in the middle of the crack zone. The outer edges of the plate were clamped by the grips of the MTS-810 loading bench, which was programmed to apply a sinusoidally varying (10 Hz) tension loads between 4000 N and 12000 N along the longitudinal direction of the sample. The cyclic sinusoidal load served two purposes. On one hand, the pulling mimicked, in an accelerated fashion, the fatigue inducing loads that the component could undergo under operation. On the other hand, the load cycle produced periodical modulations of the elastic response of a slowly growing crack, which in turn were detected via modulations of the probe waves propagating through the crack region.

The actuating-sensing system consisted of two PZT transducers and the probe wave was excited by a 240 kHz burst of 5 periods. The mechanical load on the sample was measured by the controller hardware of the MTS-810. The transmission measurements were grouped in sets of 20 measurements for one fatigue cycle which were measured around every 15 minutes. When the crack initiation became noticeable to the eye, the measurement rate was increased to 1 per minute.

3. Results and discussion

The transmitted amplitude was quantified by the amplitude A_i^j of the peak frequency in the Fourier spectrum of the gated wave PZT response for measurement i of set j:

$$A_i^j = \max(abs(fft(gated\_PZT\_response_i^j))) \tag{1}$$

Since for every transmission measurement a corresponding load measurement was available, so that transmission at minimum (±4000 N) load $A_{4000}{}^j_{mean}$ N and the transmission amplitude at maximum (±12000 N) load $A_{12000}{}^j_{mean}$ N could be evaluated. Then two diagnostic algorithms were proposed. They are both based on the nonlinear modulation of the transmission amplitude, figure 2. The horizontal line represents the mean value of the transmitted amplitude of the intact sample.

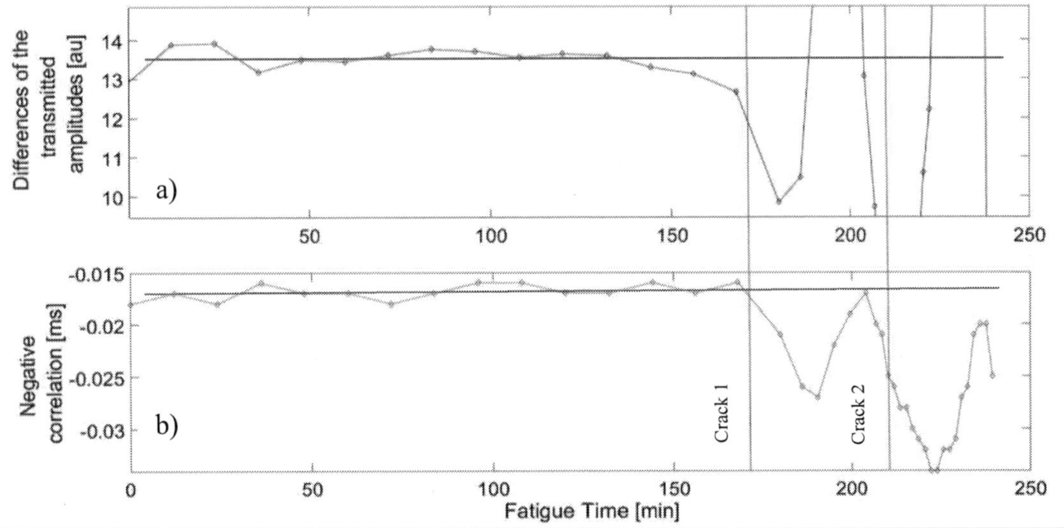

Fig. 2. Illustration of defect detection thresholds based on two tentative diagnostic algorithms: a) difference between transmitted amplitudes at high/ low - load conditions versus fatigue time b) negative correlation analysis versus fatigue time.[3]

The first algorithm, shown in figure 2 (a), is based on differences of the amplitude transmission between the high-load and low-load states during fatigue cycles. In the initial – undamaged – health state of the sample the variation of the amplitude differences is very small, whereas well before crack visibility by eye, presumably 10 minutes before (around (150 min) 90360 cycles of fatigue test) it starts to decrease dramatically. Thus, it could be beneficial for early-stage crack detection.

In the second approach, shown in figure 2 (b), the negative cross-correlation between the two signals (low-load and high-load conditions) is depicted. It provides the numerical time difference for which the high and low load signals are best aligned in time and most similar to each other in shape. In other words, the goal is to find the optimal time difference by shifting the high-load signal with respect to the low-load signal where these signals are most similar to each at corresponding fatigue time [4].

Both plots are showing very similar trends. Before the visually observable crack initiation, the negative cross - correlation approach does not show any change. In the case of the differences between two load conditions (figure 2(a)), however, there might be already an earlier indicator emerging before visual crack observation.

4. Conclusions

A laboratory setup, which allowed to acquire experimental data on initiation and development of a fatigue crack in a component under dynamic loading conditions, was designed and implemented. Preliminary results of two proposed diagnostic algorithms are presented. Monitoring of the US system response allows to detect and estimate the early-stage initiation of the cracks under loading conditions. Further cross-validation is envisaged by placing a high-speed camera and providing digital image correlation analysis of crack initiation process.

References

[1] K.-Y. Jhang, "Nonlinear ultrasonic techniques for nondestructive assessment of micro damage in material: A review", International Journal of Precision Engineering and Manufacturing 10, 123–135 (2009).

[2] I. Y. Solodov, "Ultrasonics of non-linear contacts: propagation, reflection and NDE-applications", Ultrasonics 36, 383 – 390 (1998).

[3] S. Sunetchiieva, H. Pfeiffer, S. Creten, M. Wevers, C. Glorieux, "Defect detection in vibrating structures using features of nonlinear ultrasonic modulation", conference proceedings 12th European conference on Non-Destructive Testing (ECNDT), 11-16 June 2018

[4] URL https://nl.mathworks.com/help/signal/ref/xcorr.html

Airborne testing of molded polymer compounds

Daniel KOTSCHATE[1], Saskia WENDLAND[1,2] and Mate GAAL[1]

[1]Bundesanstalt für Materialforschung und -prüfung (BAM), Germany
[2]Technical University of Ilmenau, Germany

Contact e-mail: daniel.kotschate@bam.de

Abstract: Modern and energy-efficient materials are essential for innovative designs for aerospace and automotive industries. Current technologies for rapid manufacturing such as additive manufacturing and liquid composite moulding by polymer extrusion allow innovative ways of creating robust and lightweight constructions. Commercially available printing devices often use polylactide (PLA) or acrylonitrile butadiene styrene (ABS) as raw material. Therefore, parameters like the infill ratio, influencing the ability to resist mechanical stress, may have a beneficial impact on the lifetime of components. These manufacturing technologies require a good knowledge about materials and even adapted non-destructive testing technologies and methods. Airborne ultrasonic testing has beneficial advantages for testing those lightweight constructions. It is a contact-free testing method, which does not require a liquid couplant. Therefore, it allows fast test cycles without any unwanted alternations of the material properties due to interactions with any coupling liquid. This contribution deals with the characterisation of printed specimens based on PLA by using airborne ultrasound and presents the current edge of non-destructive testing and evaluation using airborne ultrasonic transducers. The specimens, manufactured by polymer extrusion, are printed as thin plates. The infill ratio, as well as the material thickness, were varied to model density imperfections with different geometric shapes and properties. For better understanding of the limits of airborne ultrasonic testing in transmission, we compared own-developed transducers based on different physical principles: on ferroelectrets, on the thermoacoustic effect, as well as a new type of transducers based on gas discharges.

1 Introduction and background

New manufacturing techniques like additive manufacturing and the usage of lightweight and composite materials have received much attention in recent years. They enable lightweight constructions with equivalent durability and offer significant economic benefit. One common and basic additive manufacturing technique is three-dimensional printing technology with molded polymers.It has many possible use cases like biomedicine [1] or as an engineering material. However, the anisotropic structure and the inclusion of different materials (e.g. ambient air or water) during the printing process, make it difficult to test using conventional non-destructive testing (NDT) methods. In this paper, we benchmarked different air-coupled ultrasonic setups in regard to their feasibility for this testing purpose. For reliable results we varied the internal structure of the fabricated specimen to manipulate their acoustic properties. We present a set of criteria for selecting an appropriate testing arrangement and show the current edge technologies for air-coupled testing of highly structured materials.

This contribution deals with the competitive transducer technologies: polarised polypropylene (cPP) transducers [2], thermoacoustic [3] and plasmaacoustic emitters [4, 5].

2 Theory

Modern artificial materials achieve their beneficial properties by bonding different compounds and their properties or structures to connect their advantages. In lightweight constructions, higher stability are often reached by using structural elements. The most common structure, the hexagonal or honeycomb structure, achieves the elastic properties by its geometrical properties and dependencies.

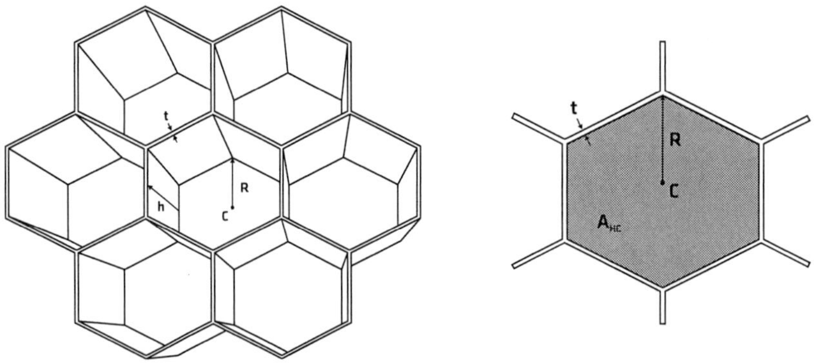

Figure 1. Regular hexagon/honeycomb structure. The elastic properties depend on the A_{HC} and the thickness t of the supports [6].

Figure 1 illustrates a schematic representation and the corresponding geometric properties of this structure. The desired weight and material reduction is reached due varying the geometric parameters of the infill structure. The volume V_{HC} of a regular hexagon structure depends mainly on the central distance R and is calculated using

$$V_{HC} = h \times A_{HC} = h \times \frac{3}{2}\sqrt{3}R^2, \tag{1}$$

where h describes height of the hexagonal infill pattern. For achieving a desired printing result in less time, the size of the infill pattern is the most important parameter to build up those structures. Assuming a constant height h (set by the final geometry) and a constant maximal central distance R_{max}, a parameter - the infill ratio - is introduced by modifying eq. (1) to

$$V_{HC}(\tau) = h \times \frac{3}{2}\sqrt{3}\left(R_{max}(1-\tau)\right)^2, \tag{2}$$

where τ is the infill ratio and varies the central distance R in a range $0 < \tau \leq 1$. Varying this parameter, also varies the acoustic parameters of the specimen.

3 Experiment

For a better comparision between the different transducer technologies, we printed a simple test piece which is shown by fig. 2 and varied the infill pattern. Therefore the fabricated specimens are equal in their dimensions, but differ in the structure size and the ratio between infill structure and raw printing material.

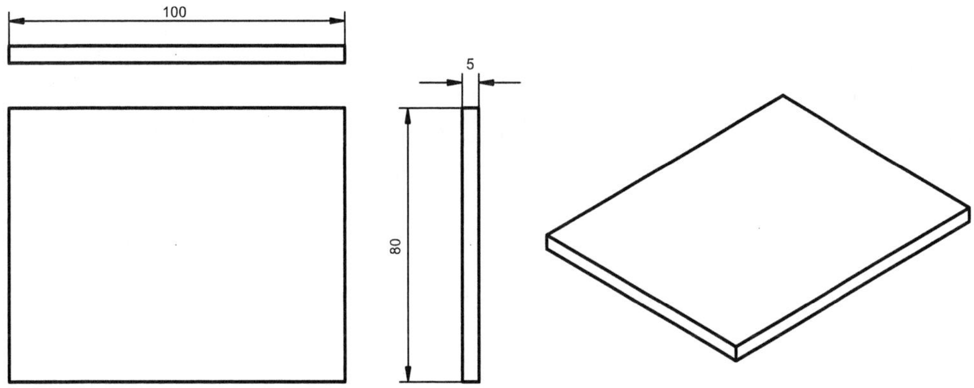

Figure 2. Technical drawing of the printed and tested specimen.

3.1 Raw material characterisation

Due to the large amount of different polymer mixtures, the density was obtained by printing a solid specimen ($\tau = 1$) to specify the density and the speed of sound of the raw material. Despite the hydrophobic surface of biodegradable polymers [7], they are able to absorb water and therefore the mixture contains two components: PLA and water. The resulting effective density of the raw material is calculated using the rule of mixtures

$$\rho_{eff}(\xi) = \xi \times \rho_1 + (1 - \xi) \times \rho_2, \tag{3}$$

where ξ is the fractional volume of each component (water or PLA), ρ_1 is the density of PLA ($\rho_1 \approx 1.24\,\mathrm{g\,cm^{-3}}$ [7]) and ρ_2 describes the density of the absorbed water ($\rho_2 \approx 1.0\,\mathrm{g\,cm^{-3}}$). Due to the absorbed water we determined the density of the raw material with $0.93\,\mathrm{g\,cm^{-3}}$. The speed of sound was quantified using a wall thickness measurement device. The resulting speed of sound is $c_L \approx 2040\,\mathrm{m\,s^{-1}}$. The calculated acoustic impedance is calculated using

$$Z_{PLA} = \rho(\xi) \times c_L \approx 2 \times 10^6\,\mathrm{m}. \tag{4}$$

Decreasing the infill ratio would increase the acoustic impedance due to the honeycomb structure inside the specimen. Therefore the transmission loss is modified by varying the size of the infill pattern. We used the printed specimen with 10% infill as reference for transdcuer characterisation. The signal amplitude and the spatial resolution of the resulting c-scan were evaluated.

3.2 Transducers

We tested this specimen (infill 10%) with three different kinds of acoustic sources and sensors: 1. Ferroelectret transducers, 2. thermoacoustic and 3. plasmaacoustic emitters. Ferroelectret transducers are based on coated and polarised polypropylene and show a thickness-mode resonance behaviour. Due to the polarisation by a significant amount of electrical field strength, electric dipoles are generated inside the existing cavities. These transducers provide an excellent signal-to-noise ratio (SNR) and, due to their working principle, a nearly perfect acoustic impedance matching. Thermoacoustic emitters using rapid gas heating to form the formation of an acoustic wave. They consist of a substrate coated with a conducting layer. Similar to the thermoacoustic emitters, plasmaacoustic sources also use the rapid gas heating of the ambient fluidic volume. In contrast to the thermoacoustic ones, the gas heating is achieved by forming a conductive channel and using the energy conversion from chemical energy to thermal energy. Due to the lack of any moving mass, the thermo- and plasmaacoustic emitters are broadband sources without any mechanical resonance behaviour.

Table 1. Investigated combinations of different technologies of air-coupled ultrasonic transducers. The abbreviations denote the used transducer technology: ferroelectric (FE), thermoacoustic (TA) and plasmaacoustic (PA).

	Combination 1	Combination 2	Combination 3
Technology	FE / FE	TA / FE	PA / FE
Center frequency [kHz]	282 / 283	wideband / 86	wideband / 86
Aperture [mm]	11 / 11	55 / 27	0.5 / 27
Pulse width [μs]	1.78	2	2
Gain [dB]	64	81	0

3.3 Experimental arrangement

Figure 3 illustrates the classic transmission setup as the experimental arrangement for the comparison of transducers. All measurements were recorded using the ultrasonic device for air-coupled ultrasonic testing USPC 4000 Airtech and a scanning and positioning system FlatScan, both by company Ingenieurbüro Dr. Hillger.

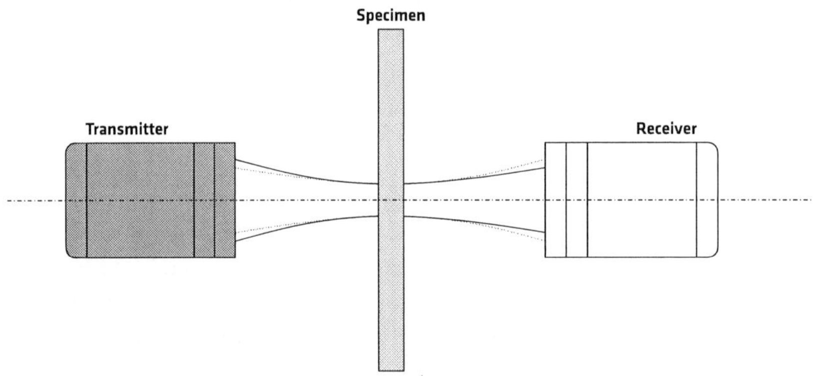

Figure 3. Setup arrangement for testing specimen in transmission using airborne ultrasound.

Booth transducers are positioned in respect to their nearfield or focal length to provide a proper spatial and amplitude resolution for best testing results as possible. Since some of the transducers are spherically focusing, the focal length was calculated using the approximation formula given by [8].

4 Results and conclusion

Figure 4 presents the achieved results and illustrates the differences between the performances of the applied transducers. For better comparison, the pre-amplification of each dataset was considered. The left picture (fig. 4 a)) shows the result obtained by a matched pair of ferroelectret transducers. These transducers have a center frequency of 280 kHz and an aperture of 11 mm. Therefore they reach a maximum spatial resolution of 2.75 mm. Figure 4 b) shows the result obtained by using a focused thermoacoustic transducer as transmitter and a ferroelectric transducer (center frequency of 86 kHz and an aperture of 27 mm) as receiver.

Due to the broadband acoustic excitation, the sensitivity is lower compared to the results achieved by the ferroelectric arrangement. The plasmaacoustic result is presented in fig. 4 c). For comparable results, the the same ferroelectric transducer as for the thermoacoustic setup was used as receiver. The plasmaacoustic source generates a much higher acoustic excitation compared to the ferroelectric and the thermoacoustic sources. Unfortunately, due to the small aperture of 500 µm the device acts like a spherical acoustic point source with wide opened directional pattern. Due to this directivity pattern no comparable resolution is achieved, although the reached sensitivity is at least 60 dB higher and may offer new possibilities. Additionally, the aperture of the plasmaacoustic device is over twenty times smaller compared to the smallest transducer combination and offers opportunities for assembling arbitrary shaped transducers or embedding highly integrated structures as transducer arrays. The presented results also illustrate the challenging task of the broadband investigation of highly anisotropic structured specimen. The specimen thickness and the associated acoustic damping is the most important point to tackle. Reducing the thickness and the usage of modern signal processing methods (e.g. denoising) might increase the signal to noise ratio [9]. However, broadband devices also offer great benefits for material characterisation methods like ultrasound spectroscopy, transducer characterisation and they reduce the need to swap the transducer on source side or changing the whole measurement setup.

5 Acknowledgment

The presented work and the corresponding results were produced during the preparation of Mrs. Wendlands bachelor thesis. The authors would like to thank Dr.-Ing. G. Lange (Technical University of Ilmenau) for this opportunity.

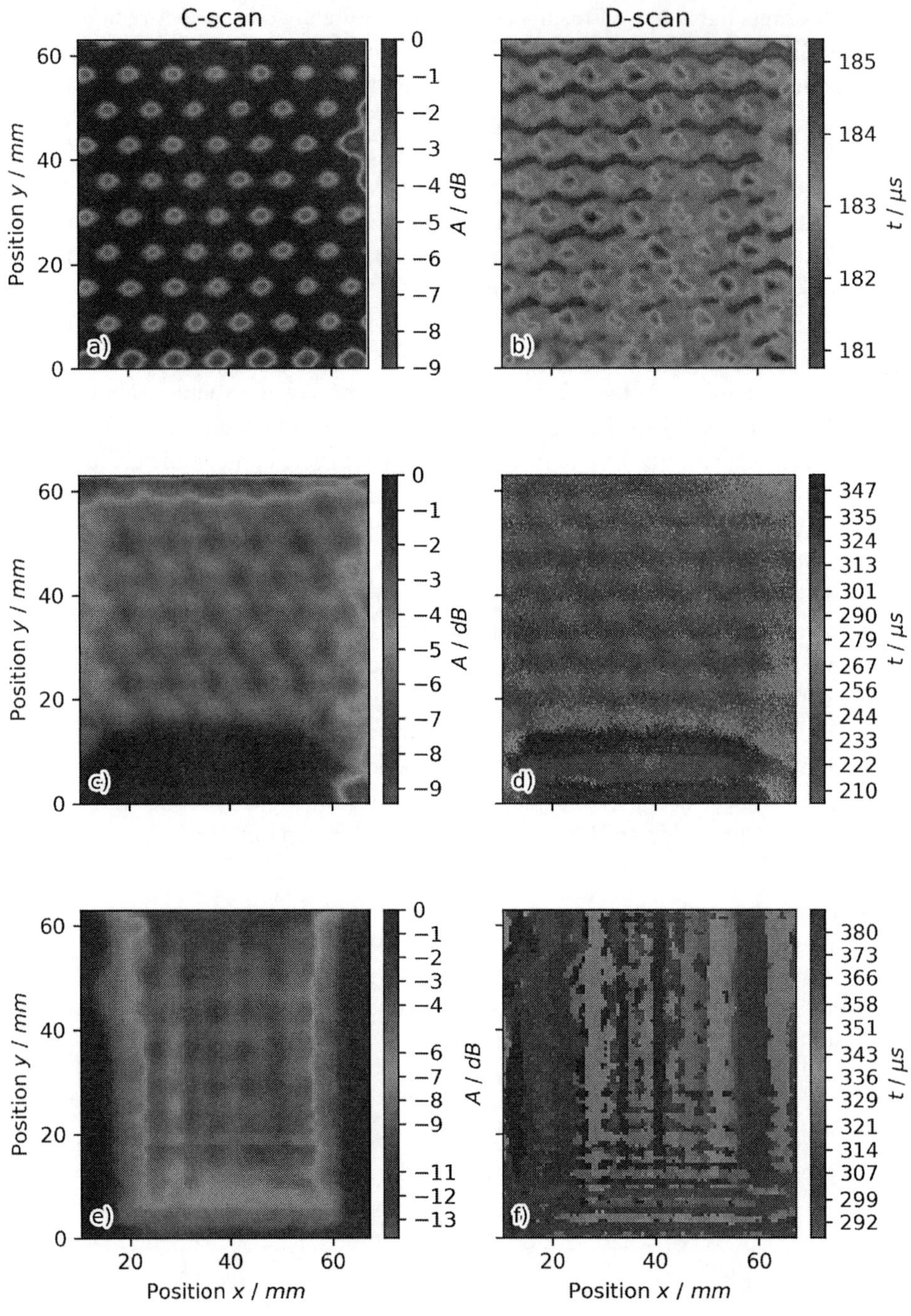

Figure 4. Resulting c-scan and d-scan images examined with three different transducer setups. a+b) cPP transducer combination c+d) thermoacousitc / cPP combination e+f) plasmaacoustic / cPP transducer combination. The images c-d) include reconstruction artefacts due to the clamping of the specimen.

References

[1] Marius Murariu and Philippe Dubois. PLA composites: From production to properties. *Adv. Drug Deliv. Rev.*, 107:17–46, 2016.

[2] Mate Gaal, Viktor Bovtun, Wolfgang Stark, Anton Erhard, Yuriy Yakymenko, and Marc Kreutzbruck. Viscoelastic properties of cellular polypropylene ferroelectrets. *J. Appl. Phys.*, 119(12):125101, mar 2016.

[3] M. Daschewski, R. Boehm, J. Prager, M. Kreutzbruck, and A. Harrer. Physics of thermo-acoustic sound generation. *J. Appl. Phys.*, 114(11):114903, sep 2013.

[4] Mate Gaal and Daniel Kotschate. New technologies for air - coupled ultrasonic transducers. In *Proceeding 12th Eur. Conf. Non-desstructive Test.*, number 1, pages 1–9, 2018.

[5] Daniel Kotschate, Mate Gaal, and Holger Kersten. Acoustic emission by self-organising effects of micro-hollow cathode discharges. *Appl. Phys. Lett.*, 112(15):154102, apr 2018.

[6] D. H. Chen and S. Ozaki. Analysis of in-plane elastic modulus for a hexagonal honeycomb core: Effect of core height and proposed analytical method. *Compos. Struct.*, 88(1):17–25, 2009.

[7] Shady Farah, Daniel G. Anderson, and Robert Langer. Physical and mechanical properties of PLA, and their functions in widespread applications — A comprehensive review. *Adv. Drug Deliv. Rev.*, 107:367–392, 2016.

[8] Hermann Wüstenberg and Jürgen Kutzner. Empfindlichkeitseinstellung beim Einsatz fokussierender Prüfköpfe in der Ultraschallprüfung an ebenen und gekrümmten Bauteilen. *Mater. = Mater. Test.*, 19(10):441–444, 1977.

[9] Klaas Bente, Daniel Kotschate, Saskia Wendland, and Mate Gaal. The thermoacoustic effect and its application in air-coupled testing of composite structures. *10th Int. Symp. NDT Aerosp.*, 2018.

New Developments and applications for Air Coupled Ultrasonic Imaging Systems

Artur SZEWIECZEK [1], Wolfgang HILLGER [1], Lutz BÜHLING [1], Detlef ILSE [1]
[1] Ingenieurbüro Dr. Hillger, Braunschweig, Germany
Contact e-mail: info@dr-hillger.de

Abstract. The ultrasonic testing of complex aerospace components is typically carried out with squirter technique. However, water coupling delivers disadvantages like pressure variations, air-bubbles, lime scales, algae and corrosion of the mechanics. Therefore, an air-coupled technique is preferable to avoid these disadvantages.

Problems caused by the large acoustic mismatch between solids and air are solved with special transducers, a powerful excitation as well as a hard- and software signal processing.

The testing is usually carried out in transmission technique with separate transducers on opposite sides of the component. The applications are mostly located in area of aerospace components, such as sandwich components. Also applications for monolithic CFRP components and metal structures are possible.

After 15 years of more or less laboratory applications the air-coupled ultrasonic technique (ACU) is used for large aerospace construction parts like the EC 145 tail boom and payload fairings. Because of the complex curved components robot scanners with 10 axes are necessary for the manipulation and the alignment of the transducers. The large size of such component requires travelling lengths of 20 meters and more.

The new developments are focussed to eight channel systems with sender and receiver arrays. This reduces the time for scanning to below 20% of the time for a one channel system. An ultra-low noise amplifier (ULNA) provides a 4 dB lower RMS value of noise.

A further area of development is the ACU with one sided access which is the research task of a LuFo project. Typical advantages of this technique is the reduction of scanning system complexity and the possibility to test hard accessible build-in components.

Introduction

Non-destructive testing with ultrasonic is mostly carried out with a coupling liquid or coupling paste which transmits the sound from the transducer to the test component. Here, a constant coupling can be very complex and cost intensive, especially for the testing of large components. The usage of water can penetrate the component and may damage it. Further disadvantages like pressure variations, air-bubbles, lime scales, algae and corrosion of the mechanics result in the need of a non-contact method like air-coupled ultrasonic testing (ACU). However, a large acoustical mismatch (more than 160 dB in a through-transmission arrangement) between the solids (transducers, test component) and the air (coupling) must be overcome [1, 2]. Furthermore, the attenuation in air increases exponentially with the frequency. Therefore, special ACU systems are required. We are involved in ACU since 1998 [3] and have developed special transducers and a special imaging system [4].

1. Air-coupled ultrasonic technique (ACU)

1.1 Testing with ACU

ACU demands a high acoustic power of the transmitter and a low-noise receiver. Echo technique applications with a single probe are not possible with ACU due to long pulse responds of the low-damped probes, a large transmission voltage and a very small reception voltage. In principle, two arrangements of transmitter and receiver are possible: pitch and catch with one sided access or transmission technique with both sided access. In pitch and catch technique, the angle beam of the transmitter probe generates transversal- or Lamb-waves into the component. After reflection at the back side the wave is reconverted into a longitudinal wave which is received with the second probe.

1.2 ACU-System

For ACU inspections the USPC 4000 AirTech has been developed [5]. It offers special adapted transducers, an ultrasonic pulser-receiver system based on PC-boards for a powerful excitation and a low-noise receiver amplification. The boards are integrated in an industrial computer. Our mechanical scanning systems can be adapted to the design of most test components. Our modular Hillgus software enables system control, data acquisition and imaging. For result evaluation our Oculus software allows detailed analysis of C-, D- or F-scans. A complete system USPC 4000 AirTech with FlatScan mechanics [6] is shown in Fig. 3. The USPC 4000 AirTech system and a CNC-controller are contained in a 19"-rack. The FlatScan system is a robust scanner for industrial applications. It allows inspections of flat components up to a thickness of 9 cm in through-transmission technique as well as in pitch and catch. This mechanic consists of two synchronously moving scanning axes on opposite sides of the component and one index axis in x-direction. It allows scanning resolutions of down to 0.15 mm and a maximum scanning speed of 500 mm/s. The scanning mechanics can be offered with different scanning areas of up to 3m x 3m. Fixtures for the transducers enable an easy adjustment of the distances between the transducers and the surface of the test component.

1.3 Transducers

Tab. 1 lists our air-coupled transducers. The average value of the transducer sensitivity up to 200 kHz is -31 dB. Using a transmitter voltage of 200V in through-transmission set-up without a specimen the receiver voltage will be about 5.5V. An additional test component between transducer and receiver typically reduces the receiver voltage of about 80 dB, additionally a defect which has to be detected causes an amplitude drop of 20 dB which leads to a receiver voltage of about 55 µV. Therefore, it is important to use transducers with a high effectivity. Otherwise, a too low transducer sensitivity could reduce defect indications below the noise level.

Our transducers are very robust and reliable. Example results of a life test with 1,600 hours (ca. two months) is presented in Figure 1. A pair of our AirTech 300-T/ 300 transducers has been tested with a pulse repetition frequency of 200 Hz and an excitation voltage of 190 Volts. Every minute an A-scan has been recorded together with the amplitude and time of flight. After 1600 hours no amplitude loss has been detected, the small amplitude changes of

1 dB are caused by air pressure variations. We have also tested our AirTech 120-T/120 and AirTech 200-T/200 transducers over a period of over a year with the same result. No signs of aging were detectable after the life tests.

Table 1. Transducers

	AirTech 50	AirTech 75	AirTech 120	AirTech 200	AirTech 300
Frequency	50 kHz	75 kHz	125 kHz	200 kHz	300 kHz
Active-ø	44.5 mm	30.0 mm	19.0 mm	11.1 mm	7.1 mm
Near field length	73 mm	50 mm	32 mm	18 mm	12 mm
Beam-ø	13 mm	8 mm	5 mm	3 mm	2 mm
Wavelength in air	6.8 mm	4.5 mm	2.8 mm	1.7 mm	1.1 mm
Sensitivity	-33dB	-31dB	-32dB	-33dB	-52dB

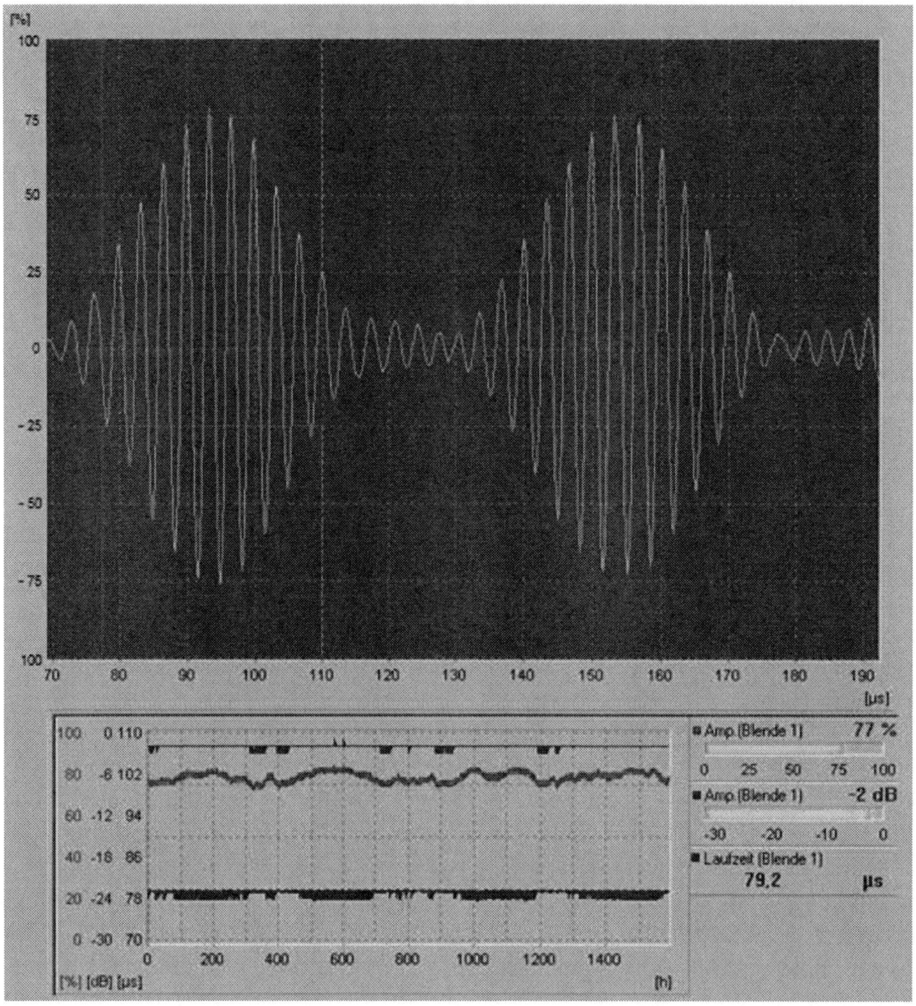

Figure 1. Results of the AirTech 300-T/AirTech 300 live test scan

Special ultra-low noise preamplifiers and filters are integrated in the receiver transducers in order to prevent electromagnetic noise. For lower inspection cycles, linear arrays with eight elements and a parallel data recording are available. Fig. 2 shows arrays from 120 to 300 kHz and a special broadband version.

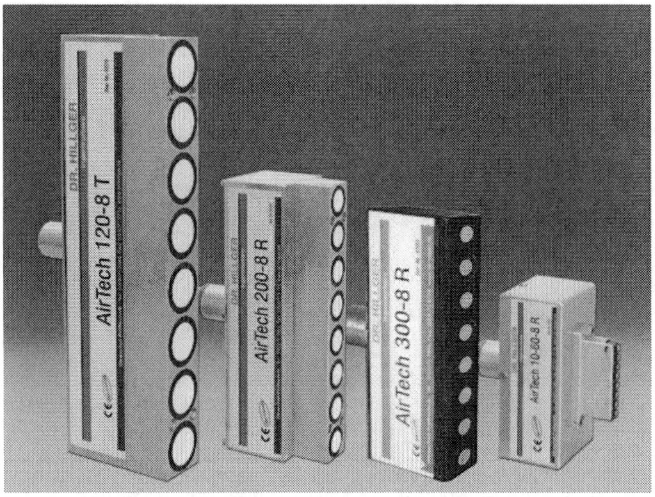

Figure 2. Arrays for ACU from 20 kHz to 300 kHz

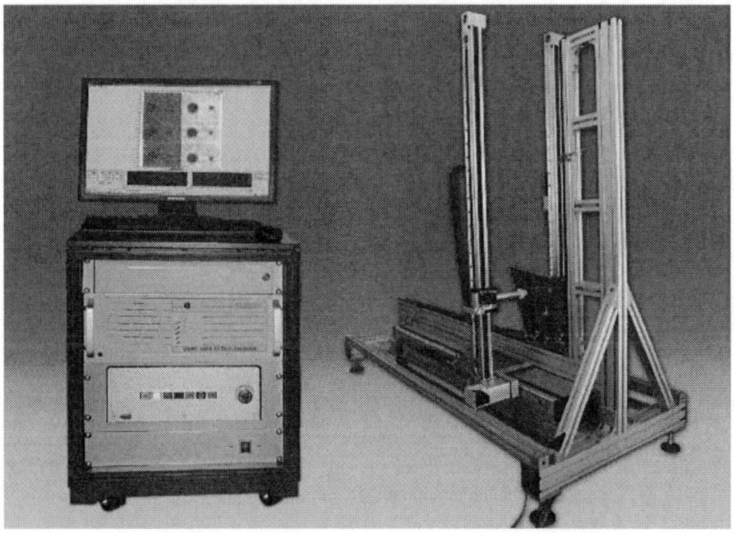

Figure 3. USPC 4000 AirTech with FlatScan

1.4 Applications

Fig. 4 and 5 present comparisons of C-scans and amplitude dynamic curves (along red lines) of air and water coupled inspections. The component in Fig. 4 is a 5.3 mm thick CFRP panel [7]. The component in Fig. 5 is a 5.3 mm thick GFRP panel. In both panels flat bottom holes in a depth of 2.8 mm and diameters between 2 and 40 mm were introduced. The ACU inspections were applied with 300kHz (AirTech 300-T and AirTech 300-R) and the water split coupling (WSC) with a 10 MHz probe.

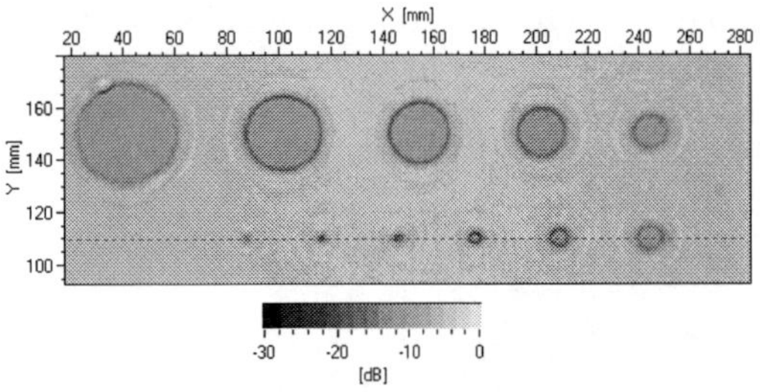

(a) C-Scan (through-transmission technique)

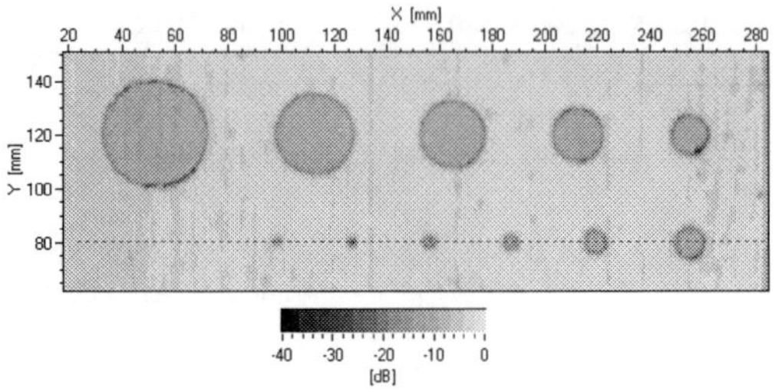

(b) C-scan (backwall echo)

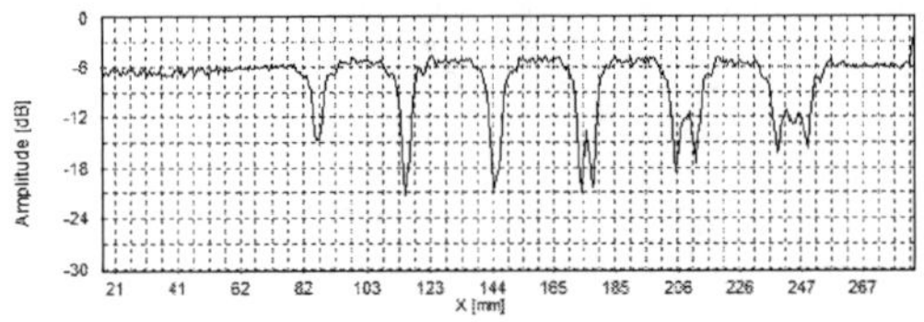

(c) Amplitude dynamic curve

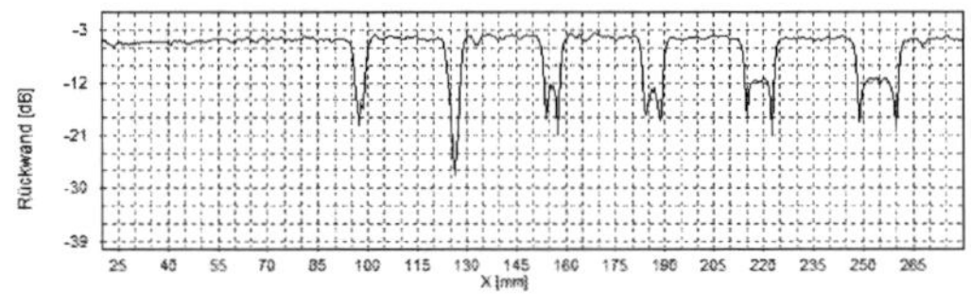

(d) Amplitude dynamic curve

Figure 4. Test results of a 5.3 mm thick CFPP component with flat bottom holes, comparison between air-couple technique (ACU) and water split-coupling (WSC)

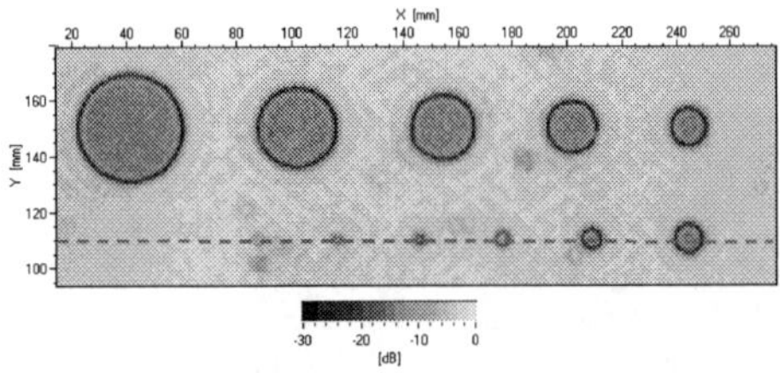

(a) C-Scan (through-transmission technique)

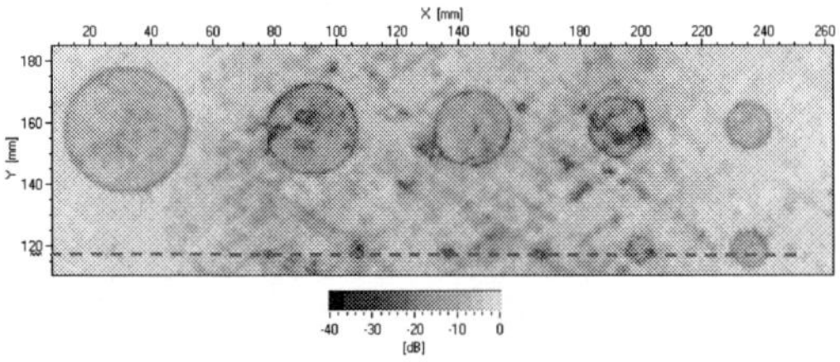

(b) C-scan (backwall echo)

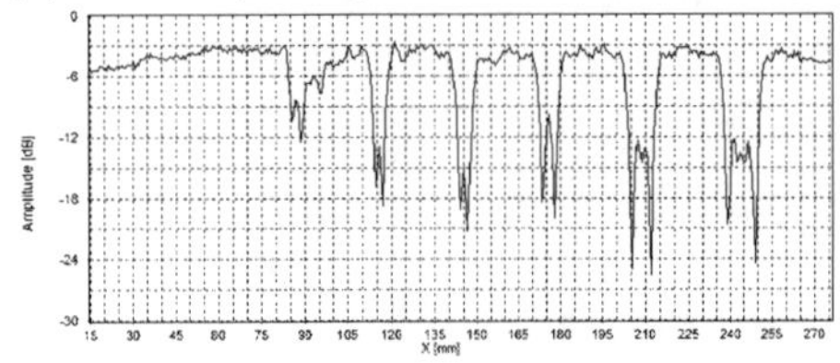

(c) Amplitude dynamic curve

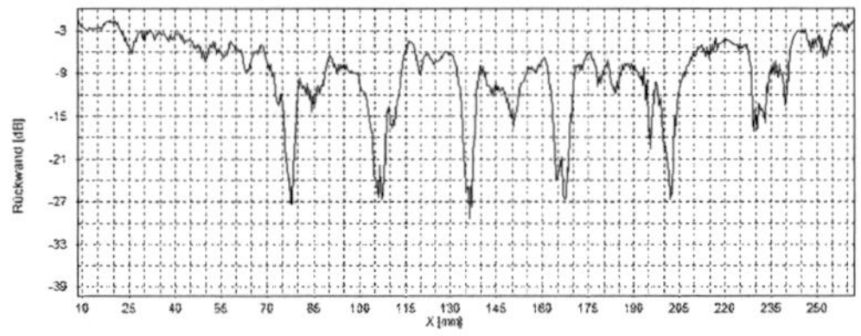

(d) Amplitude dynamic curve

Figure 5. Test results of a 5.3 mm thick GFPP component with flat bottom holes, comparison between air-couple technique (ACU) and water split-coupling (WSC),

All holes down to 3 mm are very good indicated with both techniques. At the 3 mm hole the amplitude drop is 9 dB with ACU and about 12 dB with WSC in both materials. The 2 mm flat bottom hole is situated at the right between the two hole lines and only causes a 3 dB drop in ACU and 11 dB with WSC. The total dynamic range reaches 30 dB using ACU and 40 dB for WSC.

Pitch and catch is a well-known method for ACU with one sided access (Fig. 6). Fig. 6b shows an inspection C-scan of a 2 mm thick CFRP-component. An artificial defect was positioned in the centre of the component.

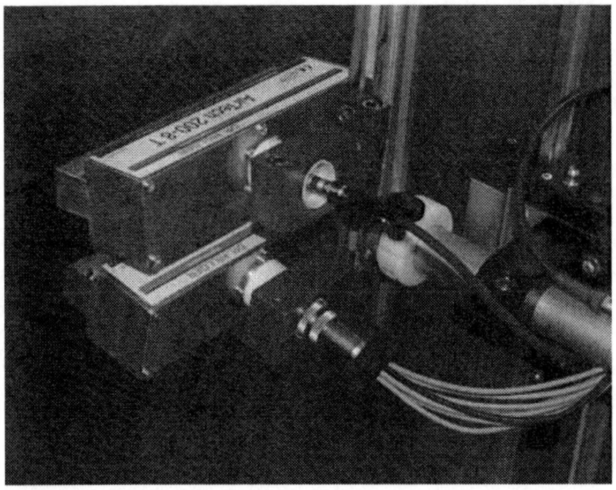

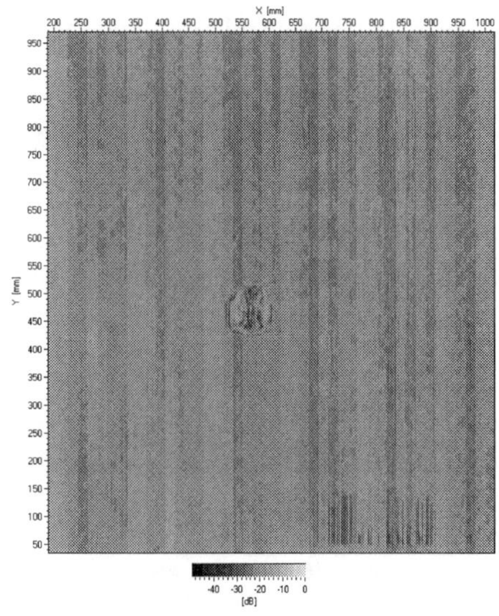

(a) Array-arrangement (b) C-scan

Figure 6. Pitch and catch arrangement and C-scan of a CFRP-component.

1.5. ACU Robot Systems

The ACU inspection has been qualified for the inspection of the EC 145 T2 helicopter tail boom. The USPC 4000 AirTech ultrasonic system with through-transmission probe arrangement has been applied for a high reliability. It has been combined with a ten axis mechanics. The detection of all relevant defects in the sandwich part of the tail boom has been proved. This challenge was only possible by a co-operation between the companies Airbus Helicopters, Innovation Works, Robo- Technology, Ostertag and Ingenieurbüro Dr. Hillger [8].

The system is shown in Fig. 7 (A). It is working since the end of 2011 and fulfills all requirements. The size of the system is about 5,3 m x 4,9 m with a height of 10,6 m. The maximum cylindrical inspection volume is given by a length of 3000 mm and a diameter range from 300 to 1100 mm. The tail boom is inspected in vertical position. The mechanic consists of two CFRP-beams which are rotatable and adjustable in height each with a three axis pivot arm. The maximum inspection speed is 500mm/s. The maximum x-offset of the system is ±1 mm. The programming of the track is carried out of a CATIA 3D model of the component.

One of the largest air-coupled systems in the world is called ANDI and has been installed in Emmen, Switzerland. The curved component surface is investigated by two AirTech transducers in through-transmission. The system is shown in Fig. 7 (b). It consists of two robots, one inside the component and one outside [8]. The outer scanning system consists of a FEM-optimized cantilever with a special CFFR robot. It has a half cylinder inspection range with a length of 21.7 m and a width of 5.4 m. The maximum velocity is 1 m/s. The inspection time depends on the scanning grid and takes about 36-72h. The accuracy reaches less than 2.5 mm (for a component length of 21m!), the automatic distance control between the component and the probes provide a tolerance of ±1 mm. In spite of a gain of more than 70 dB and a transducer cable length of 50 m no noise of the 20 powerful motors of the mechanics can be indicated in the A-scans. A highlight is the go-to function triggered by a mouse click in the C-scan. This function is very useful after scanning the component.

The C-scan can be loaded as a reference scan in the manipulator area of the user interface. In the case of a defect indication the user can click into the C-scan, the manipulation system moves to the selected point and the A-scan at this position is indicated.

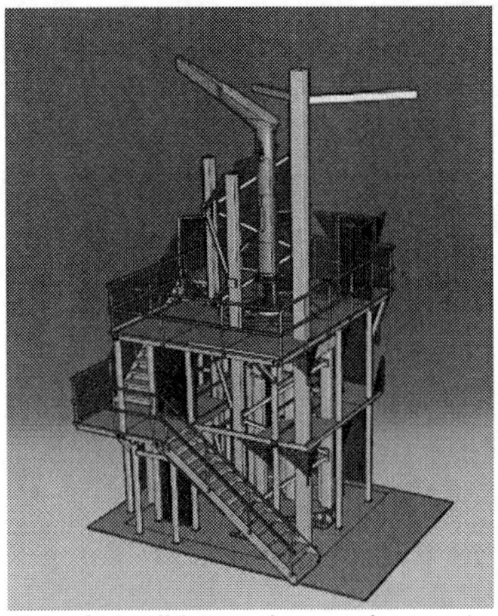

(a) ACU tail boom inspection system (Airbus Helicopters)

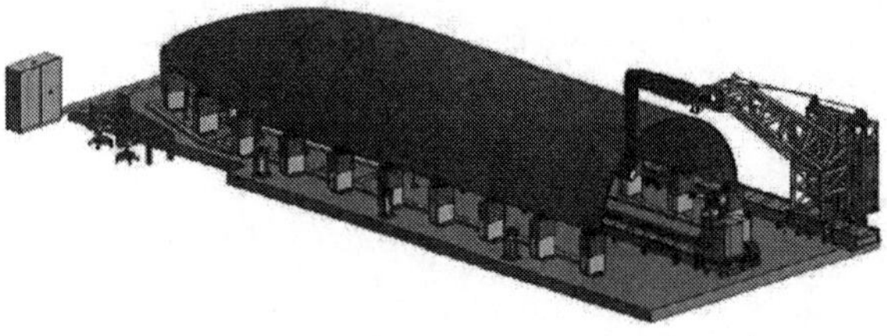

(b) ACU-equipment ANDI for space components with 21 m scanning length

Figure 7. ACU Robot systems.

2. Conclusions

Air-coupled ultrasonic technique (ACU) provides a constant coupling without any coupling liquids and pastes. It prevents all disadvantages of water coupling like corrosion and air-bubbles in the coupling path. In spite of the large acoustical mismatch inspections of composites deliver excellent results thanks to the optimized ultrasonic equipment like USPC 4000 AirTech.

However, the application of echo-technique is not possible. The inspections are carried out in through-transmission techniques with separate transducers on opposite sides of the component. Alternatively, a pitch and catch arrangement of the transducers enables a one-sided access. Components with a high degree of sound attenuation like sandwich components with honeycomb cores and even with foam cores which can be penetrated only by low frequencies can be investigated. This technique delivers a better resolution than the water jet technique. ACU can successfully be used for the defect detection in monolithic components of CFRP or even GFRP down to defects with 3 mm in diameter.

For complex aerospace components two synchronized robots with 10 axes are necessary. Examples are the tail boom inspection system (Airbus Helicopters) and two equipments for RUAG Space in Switzerland with scanning ranges of up to 21 meters.

New developments are focused on linear array-technique and to a one sided access of the component. The new eight channel system AirTech 4008 tests 1 m^2 in 4 minutes. Even a one-sided array-testing is possible in pitch and catch arrangement.

References

[1] W. Hillger, R. Meier, R. Henrich: Inspection of CFRP components by ultrasonic imaging with air coupling, Insight Vol. 46 No3 March 2004, pp 147-150.

[2] W. Hillger: Ultrasonic Testing of Composites - From laboratory Research to In-field Inspections, WCNDT Rom 2000, 15. -21.10. 2000, Conf. Proc. on CD

[3] W. Gebhardt, W. Hillger, P. Kreier: Airborne Ultrasonic Probes: Design, Fabrication, Application, 7th European Conference on Non-Destructive Testing, Copenhagen, 26-29 May 1998, Conf. Proc. pp. 3098-3105.

[4] W. Hillger, L. Bühling, D. Ilse: Air-coupled Ultrasonic Testing-Method, System and practical Applications, 11th European Conference on Non-Destructive Testing (ECNDT 2014), October 6-10, 2014, Prague, Czech Republic, Conf. Proc., see www.ndt.net.

[5] W. Hillger, L. Bühling, D. Ilse: Industrial Applications of Air-Coupled Ultrasonic Technique, 7th International Symposium on NDT in Aerospace 2015, 16 – 18 November 2015 in Bremen, Germany, DGZfP-Proceedings BB 156, ISBN 978-3-940283-76-4.

[6] W. Hillger, L. Bühling, D. Ilse: Scanners for Ultrasonic Imaging Systems, 11th European Conference on Non-destructive Testing, October 6-10, 2014, Prague, Czech Republic, Conf. Proc. on NDT.net.

[7] F. Schadowa, D. Brackrocka, M. Gaala, T. Heckel, Ultrasonic Inspection and Data Analysis of Glass- and Carbon-Fibre-Reinforced Plastics Structural Integrity Procedia 00 (2017) 000–000, www.sciencedirect.com.

[8] Presentation and live demonstration of the Air-coupled Ultrasonic Equipment ANDI in Laichingen, Eugen Ostertag, GmbH & Co KG, Germany, May, 18th, 2015.

Air-coupled ultrasonic ferroelectret transducers with additional bias voltage for testing of composite structures

Mate GAAL [1], Florian SCHADOW [1], Dustin NIELOW [1], Volker TRAPPE [1]
[1] Bundesanstalt für Materialforschung und -prüfung (BAM),
Unter den Eichen 87, 12205 Berlin, Germany

Contact e-mail: mate.gaal@bam.de

Abstract. Common air-coupled transducers for non-destructive testing consist of a piezocomposite material and several matching layers. Better acoustical matching to air is achieved by transducers based on charged cellular polypropylene (PP). This material has about hundred times lower acoustic impedance than any piezocomposite, having about the same piezoelectric coefficient. The piezoelectric properties of cellular PP are caused by the polarization of air cells. Alternatively, a ferroelectret receiver can be understood as a capacitive microphone with internal polarization creating permanent internal voltage. The sensitivity of the receiver can be increased by applying additional bias voltage. We present an ultrasonic receiver based on cellular PP including a high-voltage module providing bias voltage up to 2 kV. The application of bias voltage increased the signal by 12 to 15 dB with only 1 dB increase of the noise.

This receiver was combined with a cellular PP transmitter in through transmission to inspect several test specimens consisting of glass-fiber-reinforced polymer face sheets and a porous closed-cell PVC core. These test specimens were inspected before and after load. Fatigue cracks in the porous PVC core and some fatigue damage in the face sheets were detected. These test specimens were originally developed to emulate a rotor blade segment of a wind power plant. Similar composite materials are used in lightweight aircrafts for the general aviation. The other inspected test specimen was a composite consisted of glass-fiber-reinforced polymer face sheets and a wooden core. The structure of the wooden core could be detected only with cellular PP transducers, while commercial air-coupled transducers lacked the necessary sensitivity. Measured on a 4-mm thick carbon-fiber-reinforced polymer plate, cellular PP transducers with additional bias voltage achieved a 32 dB higher signal-to-noise ratio than commercial air-coupled transducers.

Introduction

Common air-coupled transducers are constructed with several matching layers to reduce the impedance mismatch to air. Ferroelectret materials were proposed for air-coupled ultrasonics to avoid the application of matching layers [1][2][3]. Ferroelectrets (sometimes called piezoelectrets) are defined as charged cellular polymers, uniting the properties of electrets and of ferroelectric and piezoelectric materials. Cellular polypropylene is the only ferroelectret so far applied to air-coupled transducers. This material is available as a flexible 100 μm thick film, which is suitable for cylindrical and even spherical focusing [4]. Recently the sensitivity of ferroelectret receivers was increased by applying additional bias voltage. As a source of high voltage, a separate high-voltage module was developed and built into the inspection system [5], [6].

This type of transducer was applied to testing of wind turbine rotor blades. The continuous increase of the typical size of a wind power plant causes an increase of maintenance and repair costs. Thus, nondestructive testing (NDT) methods are getting more and more accepted and they are increasingly applied. Imaging NDT techniques enable detection and localization of defects in a rotor blade, which leads to preventive action before component failure. Rotor blades are typically inspected after production within quality control, recurrently in-situ every two to four years, and after necessary repairs [7]. Mostly visual testing and acoustic emission are applied. However, these are manual methods and strongly affected by the subjectivity of the inspector. Occasionally, mechanized ultrasonic pulse-echo technique with liquid couplants is applied during production, but the liquid couplant significantly contributes to maintenance costs and it can damage sensitive composite structures. Therefore, the application of air-coupled ultrasonic technique is being investigated as a possible alternative.

Rotor blades mostly consist of many different materials. Glass-fibre-reinforced polymers (GFRP) are combined with balsa wood, cellular materials, carbon-fibre-reinforced polymers (CFRP) or honeycomb structures [8]. Two most frequent types of flaws which occur both during the production and in operation are delaminations between single GFRP layers and disbonds between the beam and the blade shell [7]. Other possible flaws are porosities, weak adhesion, insertions and wrinkles [8]. Particularly, imperfections occurring in the production may induce cracks in the face sheets, which may grow into the depth and cause failure of the shell. Here we present the inspection of undamaged and repaired rotor blade shells using newly developed air-coupled transducers based of ferroelectrets.

1. Inspection system and test specimens

1.1 Air-coupled transducers based on ferroelectrets

The air-coupled inspection in through transmission was performed using focussing ferroelectret transducers with an aperture 19 mm, curvature 50 mm and a resonance frequency of 250 kHz [4]. Bias voltage of 1.6 kV was applied to the receiver, which increased its sensitivity by about 10 dB, whereas the noise increased only by 1 dB. This means that the signal to noise ratio of this air-coupled transmission is at least 30 dB higher than with commercially available air-coupled transducers with similar parameters [5], [6].

1.2 Test specimens

A sandwich shell specimen was developed with structural mechanical properties and designed to be representative of the sandwich shell structure of large rotor blades. Figures 1 (a) and (b) shows this test specimen before damage. (All figures in this paper are based on our previous publications [9]–[11].) The dimensions were 810 mm × 510 mm × 12.7 mm. The thickness of the GFRP face sheet was 1.3 mm (consisting of two biaxially reinforced +45/-45 mats) and the foam core thickness 10 mm. The semi-finished products (mats, epoxy resin, supports) and the layer structure as well as the curvature radius were chosen to represent an actual rotor blade shell of a 27 m long rotor blade. Closed-cell foam cores were applied to prevent soaking of the epoxy resin and consequently the increase of weight. The supports (closed-cell PVC foam, AIREX C70.55) were covered with two biaxial layers (areal density 810 g/m², fibre orientation +45/-45°) on the top and bottom side of the test specimen. The brittle supporting foam was slit open with 1.5 to 2 mm bright slits, so that it could be bended to the required curvature. The applied infusion process required a low viscosity infusion epoxy resin (Hexion RIMR135/RIMH137) as a matrix system. The core is provided with regularly distributed through bores, enabling impregnation of lower layers with epoxy resin. The end faces of the test specimen were carried out as monolithic GFRP laminates and were provided with larger through bores, which were drilled to clamp the test specimen for the mechanical testing [10].

Two shell test specimens of this type were produced. The first one named BAM_SI_4 without nominal damage or imperfections was vibrationally loaded with a tension-compression load (R = −1) of 3 ‰ until failure. As a comparison, a GFRP rotor blade at rated wind has a flexural strain of ca. 2 ‰, so that 3 ‰ load is an unusually high load leading to a low cycle fatigue life. The tested load of 3 ‰ causes cracks in the laminate, but there was no typical fatigue damage (matrix cracks or cracks in transverse direction).

The second test specimen named BAM_SI_3 was identically constructed as BAM_SI_4, but had a local repair as shown in Figures 2 and 3(a). The repair was done on a nominally undamaged test specimen without any imperfections. The repair was performed manually according to common practice as it is done on rotor blades in the field, assuming the worst possible damage. The face sheet, the foam core and the „damaged“ area at the bottom face sheet were exposed and prepared for the repair. Then the laminate and the core were reconstructed. The same biaxial non-crimp fabric was used (fibre orientation +45°/-45°) as in the original part, but with the half areal density (ca. 400 g/m²) and the half of the thickness of the original material. Thinner mats are used because they are easier to handle, so that two such layers were applied to achieve the original thickness.

The repaired test specimen was subjected to a two-level oscillation fatigue test. The first load level corresponds to rated wind of 2 ‰ up to 1.5 million load cycles. The second level was a 3 ‰ load which lead to failure after 1.7 million load changes. Because of the high number of load cycles at the second load level, the test specimen shows high fatigue damage with a high level of inter-fibre failure cracks and whitening of the laminate, as shown in Figure 3. The repair area causes a significant stiffening of the test specimen and thus a local redistribution of load, which leads to massive fatigue damage in the surroundings of the repair area.

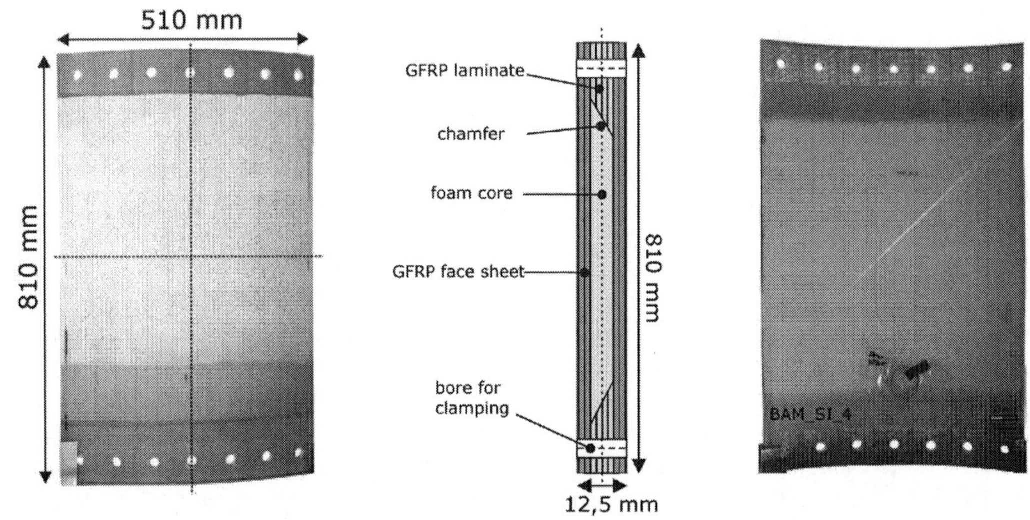

(a) (b) (c)

Figure 1. The rotor blade shell with a foam core, (a) its measures, (b) its profile and (c) the same test specimen after load.

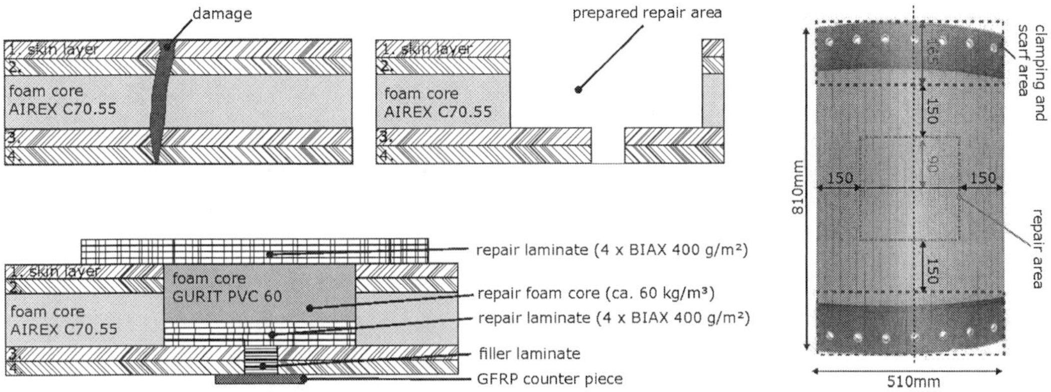

(a) (b)

Figure 2. Repair of a rotor blade shell with a foam core, (a) a draft and (b) its measures.

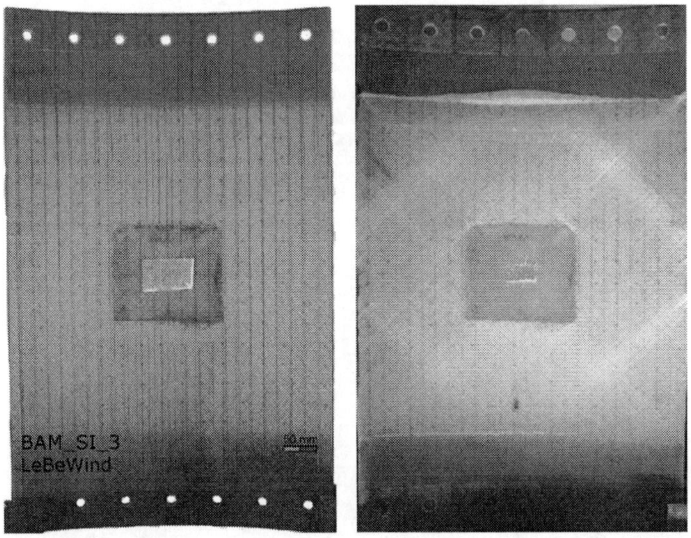

(a) (b)

Figure 3. A repaired rotor blade shell with a foam core, (a) before load and (b) after the load.

2. Inspection results

2.1 Rotor blade shell after cyclic load

The rotor blade test specimen was tested before and after the cyclic load. This load induced cracks along the fibre orientation of the surface layers (+45/-45°), which could be successfully localized by the applied air-coupled ultrasonic testing system (Figure 4). A laminographic examination has shown that the crack stretches through the surface layers as well as through the foam core. The alteration of the ultrasonic signal is particularly noticeable in the logarithmic scale. The dark vertical lines and periodical circular indications represent channels filled with epoxy resin.

2.2 Repaired rotor blade shell after cyclic load

The structure of the test specimen was locally altered through the repair. The changed material properties can be visualized by air-coupled testing in through transmission. Figure 5 shows that the time-of-flight representation (D-Scan) is more suitable to map the repair area than the more usual signal height representation (C-Scan).

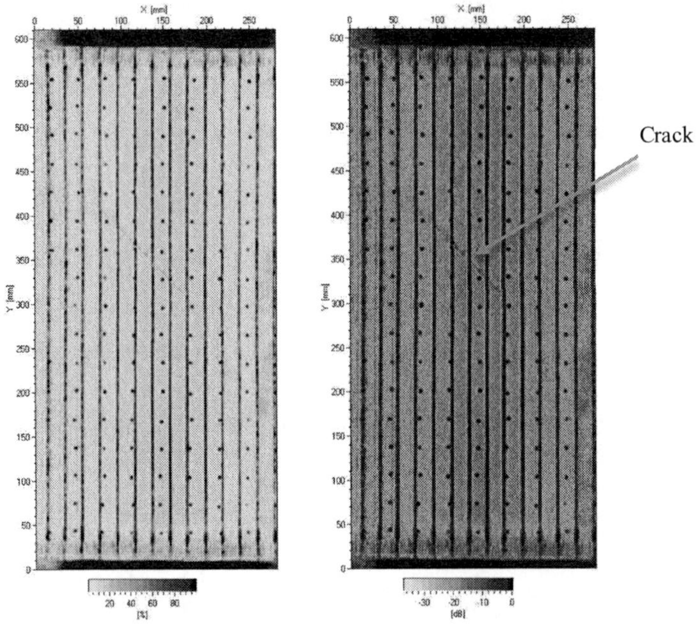

(a) (b)

Figure 4. C-Scan of a rotor blade shell inspected with focussing ferroelectret air-coupled ultrasonic transducers in through transmission, (a) linearly scaled and (b) logarithmically scaled.

(a) (b) (c)

Figure 5. (a) C-Scan and (b) D-Scan of a repaired rotor blade shell with a foam core inspected with focussing ferroelectret air-coupled ultrasonic transducers in through transmission and (c) C-Scan of the same test specimen after cyclic load.

3. Discussion and conclusions

We inspected 12.6 mm thick mockups of rotor blade shells using air-coupled ferroelectret ultrasonic transducers. In the test specimen without repair, cracks in the face sheet parallel with the fibre orientation were detected, both with visual testing and with air-coupled ultrasonic testing, with a high level of correspondence (Figures 1(c) and 4). The results of the inspection of the repaired test specimen using visual testing and air-coupled ultrasonic testing differ depending on the applied method. A comparison of Figures 3(b) and 5(c) shows that air-coupled testing and visual testing uncover different flaws, whereas some of the cracks parallel with the fibre orientation remain undetected. The areas with a reduced signal in Figure 5(c) may be larger delaminations, which are not visible, yet relevant to safety. The channels filled with epoxy resin with a thickness between 1 and 2 mm occurring during the production can be clearly detected. The corresponding locations are characterized by an elevated ultrasonic signal because the damping of the surrounding foam core is significantly higher. This high lateral resolution is the consequence of spherical focussing of the ferroelectret transducers. Ferroelectret transducers exhibit a signal-to-noise ratio 30 dB higher than commercially available air-coupled transducers, which enables testing of thicker components and components with higher damping [5].

4. Acknowledgements

We would like to thank Uwe Beck und Matthias Weise (BAM) for the electron beam evaporation on the ferroelectret transducers.

References

[1] G. M. Sessler and J. Hillenbrand, "Electromechanical response of cellular electret films," *Appl. Phys. Lett.*, vol. 75, no. 21, pp. 3405–3407, 1999.

[2] M. Paajanen, J. Lekkala, and K. Kirjavainen, "ElectroMechanical Film (EMFi) - a new multipurpose electret material," *Sensors Actuators, A Phys.*, vol. 84, no. 1, pp. 95–102, 2000.

[3] J. Döring, J. Bartusch, M. Gaal, M. Kreutzbruck, A. Erhard, and V. Bovtun, "Ferroelectret probes for non-destructive testing with airborne ultrasound," *Mater. Test.*, vol. 55, no. 2, 2013.

[4] M. Gaal, J. Bartusch, E. Dohse, F. Schadow, and E. Köppe, "Focusing of ferroelectret air-coupled ultrasound transducers," in *AIP Conference Proceedings*, 2016, vol. 1706.

[5] M. Gaal, R. Caldeira, J. Bartusch, and M. Kupnik, "Air-Coupled Ultrasonic Ferroelectret Receiver with Additional DC Voltage," in *Eurosensors 2017*, 2017, vol. 1, no. 4, p. 362.

[6] M. Gaal, R. Caldeira, J. Bartusch, and M. Kupnik, "Air-Coupled Ultrasonic Ferroelectret Receiver with Additional Bias Voltage," *Submitt. to IEEE Trans. Ultrason. Ferroelectr. Freq. Control*, vol., no., p. , 2018.

[7] T. Bruch and Y. Oberdörfer, "Prüfung von GFK Windkraft Rotorblättern mittels 500kHz Phased-Array Technologie," in *DACH-Jahrestagung*, 2015, pp. 1–8.

[8] A. Lamarre, "Improved Inspection of Composite Wind Turbine Blades with Accessible Advanced Ultrasonic Phased Array Technology," in *15th Asia Pacific Conference for Non-Destructive Testing (APCNDT2017), Singapore*, 2017.

[9] M. Gaal, F. Schadow, D. Nielow, and V. Trappe, "Prüfung von Rotorblattschalen mit Luftultraschall," in *DGZfP-Jahrestagung*, 2018.

[10] V. Trappe *et al.*, "Sandwich-Schalenstrukturen in situ charakterisiert mit Thermographie und Felddehnungsmessungen," in *DGZfP-Jahrestagung*, 2018.

[11] V. Trappe and D. Nielow, "Fatigue loading of sandwich shell test specimens with simulated production imperfections and in-situ NDT," in *ICFC 7 - The 7th International Conference on Fatigue of Composites*, 2018.

NUTHIC: Non-Contact Ultrasound Inspection Machine of Highly Integrated Composite Parts

Roberto GIACCHETTA [1], Jorge F. CRUZA [1], Rajesh TANEJA [2]
[1] DASEL SL Avda. del Cañal 44 N3 28500 Arganda del Rey, Madrid, Spain
[2] Innotecuk, Cambridge , UK

Contact e-mail: roberto@daselsistemas.com

Abstract. Aerospace industry is moving towards more integrated composite structures in order to reduce autoclave cycles and assembly operations, which leads to considerable cost, time and energy savings. However, the main drawback of this approach is that no inspection solution at a reasonable cost is given by ultrasound equipment in the market, because of the geometry complexity and access restrictions to those parts.

NUTHIC E! 8929 project (partial financed by Eurostar Program H2020) develop an adaptive and fully automated air-coupled ultrasound inspection system for highly integrated composite components, whose complex geometries and difficult access prevent using state of the art equipment.

The goal of this project is to provide the aerospace manufacturing industry with an adaptive and automated non-contact ultrasound inspection solution for highly integrated composite parts

This paper presents the final results of the project on multi spar structures, all of these measures with non-contact ultrasound technique, avoiding water coupling and reducing inspection time by integrating the inspection machine into the clean environment of the production line.

1. Introduction

Manufacturing composite parts with a high integration level introduces new challenges for ultrasound inspection systems. On one hand, as integration lowers manufacturing time, inspection time should be reduced accordingly, or at least not increased, to achieve an overall reduction of production costs. On the other hand, quality requirements cannot be relaxed, and the inspection solution should provide, at least, the same performance as current systems. Furthermore, inspection of 100% of the part must be warranted, regardless of the access restrictions that these new components impose.

In the aerospace industry, practically all automated inspection systems for composites use water as the coupling medium between the transducer and the part. This procedure usually requires moving the part from the clean zone of the manufacturing line to an inspection zone where water can be used, a time consuming process especially for large components. Furthermore, if a defect is found and the component has to return to the manufacturing zone, drying and cleaning procedures slow down the process even more.

NUTHIC E! 8929 project (partially founded by Eurostar Program H2020) is aimed to develop an adaptive and fully automated air-coupled ultrasound inspection system for highly integrated composite components, whose complex geometries and difficult access prevent using state of the art equipment. The goal of this project is to provide the aerospace manufacturing industry with an adaptive and automated non-contact ultrasound inspection solution for highly integrated composite parts.

2. NUTHIC organization, main challenges and methodology

2.1 Consortium

NUTHIC project started in the early 2015, with a duration of 36 months and a global budget around 1.6 M €. The consortium is formed by two countries (Spain and UK) and 3 companies (DASEL, ASG and INNOTECUK), with the assistance of two subcontracted research organizations (CSIC and FIDAMC), Figure 1.

SPAIN	SPAIN	SPAIN	SPAIN	UK
Dasel sistemas	CONSEJO SUPERIOR DE INVESTIGACIONES CIENTIFICAS CSIC	FIDAMC	AUTOMATION SYSTEMS GROUP	InnoTecUK
SME	RTD	RTD	SME	SME

Fig. 1. Consortium resume: DASEL, ASG and INNOTECUK partners; CSIC and FIDMAC RTD.

2.2 Main Challenges of NUTHIC project

The challenges of NUTHIC focus on three main areas:

- Develop a novel air-coupled ultrasound array technology, with large bandwidth and low insertion losses. Such a technology is not commercially available and must be developed by the consortium. The main challenge is to adapt the high acoustic impedance of the transducer to the low impedance of air to minimize insertion losses and get an acceptable sensitivity while achieving a wideband response of the transducer and a high spatial resolution [1].

- Obtain air-coupled inspection techniques with resolution and sensitivity comparable to the water coupled solutions actual state of the art. Although arrays themselves improve inspection speed when compared to single-element transducers, that is not enough for air-coupled applications. For example, slight misalignments or subtle changes in the component geometry, can lead to a strong modification of the ultrasound path because of the large refractive index between air and composite materials, probably producing severe signal losses. This is an important challenge for NUTHIC.

- Development of an adaptive inspection device able to move inside closed structures of changing shape. Such device will be placed inside the closed structure to be inspected and will be guided by its inner walls that act as a reference. No elbows or big change of curvatures are expected. A motorized solution will be provided to control the inspection device position inside the closed structure and a set of sensors will be used to monitor its real position in relation to the starting point. The mechanical structure will be designed to be easily adapted to different sizes of internal boxes. The second inspection head will track the inner device to perform through-transmission inspection. Coordination between both devices is essential, being another of the project challenges.

2.3 Consortium members and main tasks

DASEL is in charge of the ultrasound system development, due to its experience in the design and manufacture of inspection systems (conventional and phased array). For this task, DASEL subcontracted also the Spanish National Research Council (CSIC), who has developed an innovative process to manufacture transducers and array probes adapted for emission and reception in the air. Preliminary trials have shown the potential of this technology for the inspection of composite material, with improved resolution and contrast because of their focusing and linear scan capabilities, as it is shown in this work.

INNOTECUK and ASG work together to develop a highly modular and adaptive robotics. The solution devised by the consortium is to develop a compliant robot able to travel inside the parts, while adapting to the changes in geometry, coordinated with a second inspection head that will cover the part from the outside, allowing to apply trough-transmission techniques. An objective of this project is to obtain a flexible solution that allows an easy reconfiguration for different structures.ASG receive the support of FIDAMC, a composite research centre part of EADS group that has been subcontracted.

The role of FIDAMC is to define the set of components representative of actual configurations in the aerospace manufacturing industry, to which the development of the scanning head will be first targeted in the project. This is essential to get, by the end of the project, an inspection solution ready for the composite market. FIDAMC will manufacture mock-ups with artificial defects to validate the proposed solution.

3. Description of the NUTHIC technology demonstrator component

In order to demonstrate the capability of the NUTHIC technology developed by the consortium, different prototype components were evaluated, a component was selected that met the initial objectives of the project. This had to meet: low or no accessibility for inspection from the outside, geometry of the variable component.

Finally, the Multispar component was chosen, which has been manufactured in laminar carbon fiber, made up of 3 longitudinal drawers where the drawer section varies from 126.5 mm x 114 mm (minimum section) to 318x 357 mm (maximum section) These drawers extend over a maximum length of 6800 mm, with a total width that varies between 925 mm and 324 mm.

Figure 2 shows the variation in section of the component; inside these drawers is where the inspection robot must perform the sweeping of the side walls.

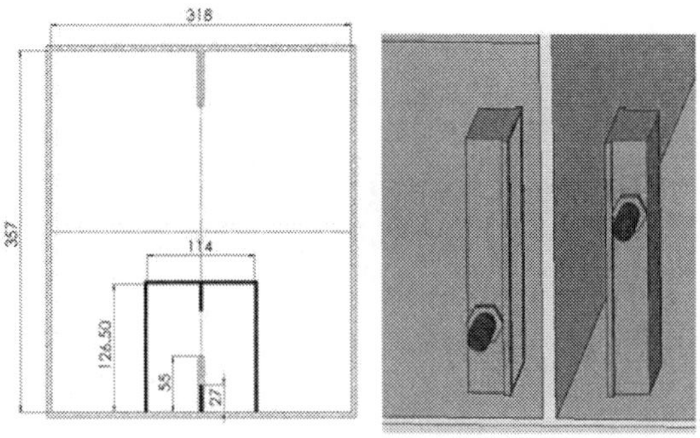

Fig. 2. a) Variation between ends of the drawer section of the Multispar component. b) Transducer array configuration.

The NUTHIC system focused on ensuring the inspection of non-accessible areas (inside the caissons), therefore the lower and upper areas are beyond the scope of this project.

4. Robot explorer

In Figure 3, you can see the robot developed by INNOTECUK which can be adapted to both the variations in height of the component and the width of the seism.

This robot has the ability to explore the entire component and adjust to variations in sections, and has a system for positioning the transducers at different heights of the component.

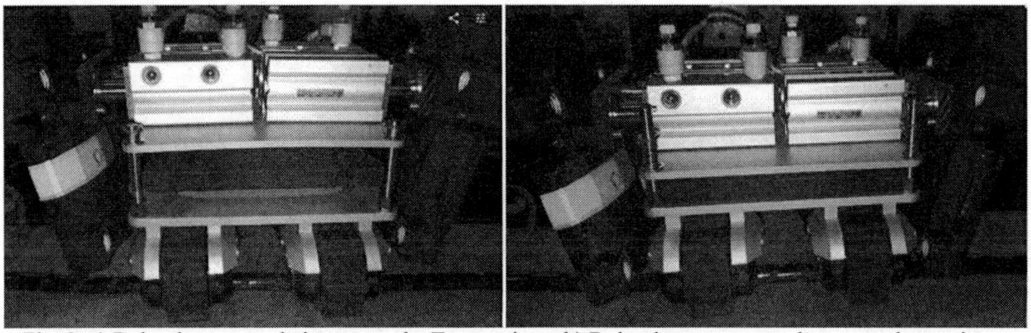

Fig. 3. a) Robot bag expanded to cover the Top sections. b) Robot bag compressed to cover the sections Botton.

4.1 Transducer alignment and guidance system

The parallelism of the transducer array is guaranteed thanks to pneumatic pistons that keep the array parallel to the wall.

In order to simplify the mechanical exploration, it was chosen to use the magnetic drive system, providing the probe holder with a set of neodymium magnets.

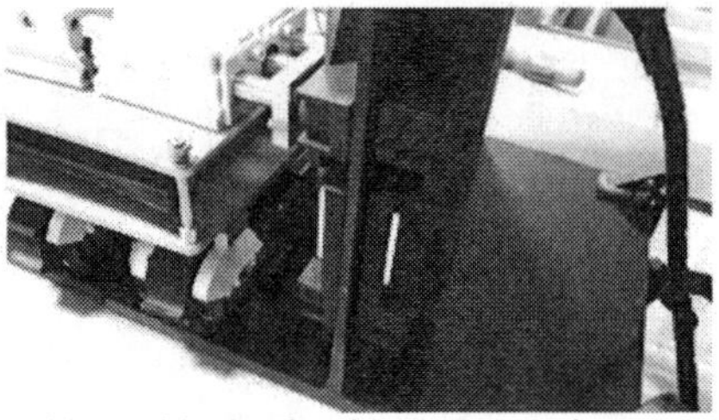

Fig. 4.UT holder with Neodymium magnets in transmitter side.

5. Development of Ultrasonic transducers

In order to cover the entire inspection area with the least number of scanning, the two-dimensional array [2 -3] was designed as shown in Figure 5.

This array is completely modular, organized in 8 linear elements and 4 rows displaced 6 mm between them. This displacement is calculated based on the diameter of the transducer which is 12x12 mm, thus allowing having a pitch of 2 mm which fits the resolution requirements of Ultrasound inspections in the Aeronautical sector [4-5].

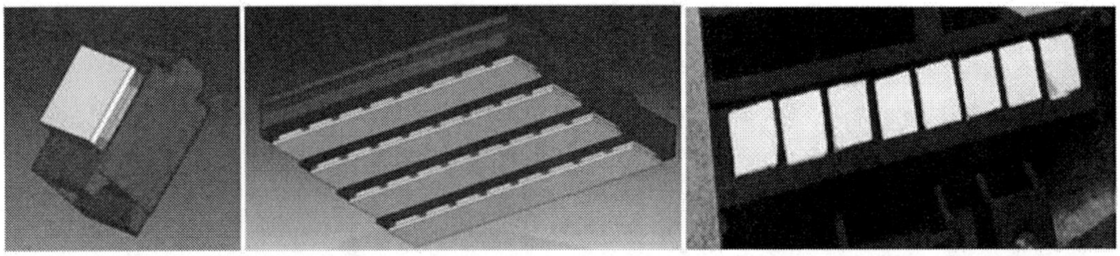

Fig. 5. a) Individual element of 12x12mm. b) Full array of 32 elements. c) Developed array.

5.1 Characteristics of the developed UT array

In order to minimize the size of the transducer without losing bandwidth, sensitivity and resolution, we opted for a 250 kHz square transducer, where the characteristics are summarized in Figure 6.

Width	Length	Pitch	Frequency	Total length	Preamp Included	Bandwidth
12mm	12 mm	2 mm	250 kHz	114 mm	Yes	70% (-20dB)

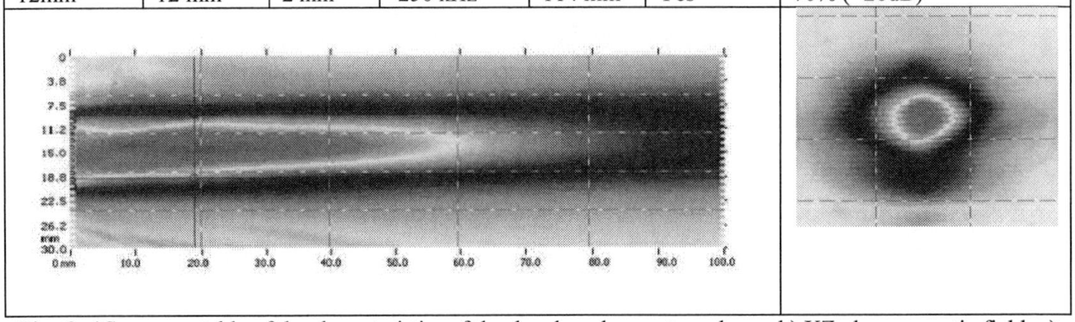

Fig. 6. a)Summary table of the characteristics of the developed array transducer. b) XZ plane acoustic field. c) XY plane acoustic field in z = 20mm.

6. Complete inspection system

The initial tests were performed with an 8x1 array, however the NUTHIC system internally can control up to 32 emission channels and 32 reception channels.

The acquisition on the real component was made at a speed of 50 mm/s, registered the advance position with an optical encoder in step of 2 mm. The divergence of the width of the component is controlled with the two axial pistons which support the UT transducers. On the other hand, a system of springs adapts in every moment the position of the same ones.

The necessary system gain was 26dB, for a solid laminate of 4 mm, with no need to average the signal. The acquisition was made simultaneously on the 8 channels (the crosstalk is below -30dB), recording the entire signal in radio frequency mode (thus allowing its further processing). The C-scan representation is made in real time taking the maximum of the signal.

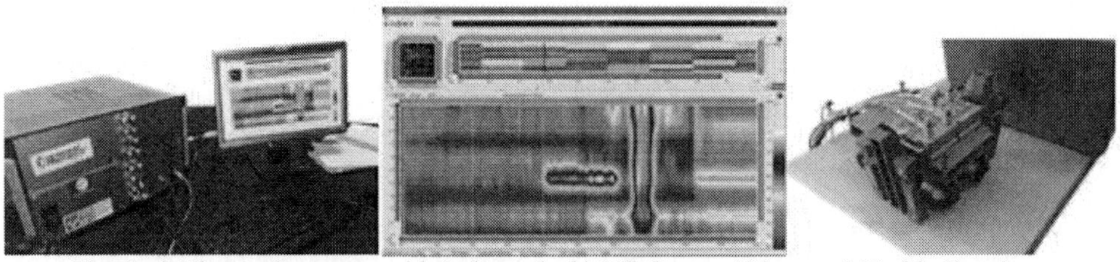

Fig. 7. a) Ultrasonic system with 8 channels. b) UT C-scan 400 mm. c) Nuthic Crawler.

7. Porosity measurements

The water column transmission method not only detects delaminations but can also measure the porosity content [6], which is reflected in the attenuation of the transmitted signal. In the case of using contactless technology there are two major limitations: 1) Reduction of the frequency of the working Ultrasound signal in an order of magnitude. 2) Phenomena of amplitude increase due to resonance within the material, since the wavelength is close to the thickness of the laminate. In Figure 8, the porosity measurements in a 270x220mm test plate are intercompared using 5 MHz in total immersion vs. non-contact measurement in air with normalization of amplitude.

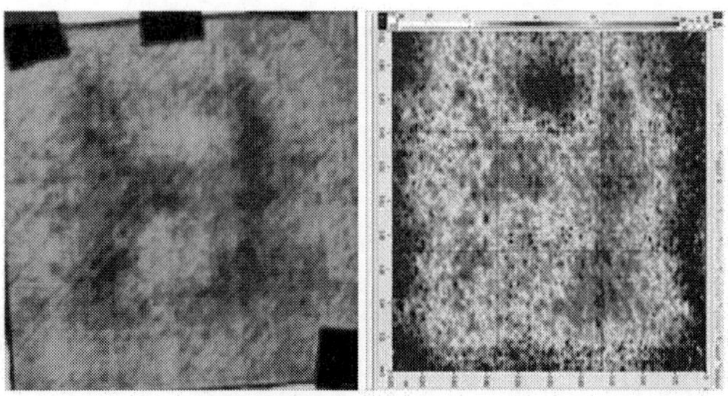

Fig. 8. a) Attenuation C-scan map 5Mhz. b) Attenuation C-scan map 400 kHz Sonojet.

8. Summary

The NUTHIC project concluded in December 2017, allowing during these three years to provide the consortium with a differentiating capacity in the development of flexible inspection systems without contact. It is important to highlight the fundamental action of the CSIC in the NUTHIC project, through the TUPACA and MICA research contracts.

The non-contact UT technology has shown that it is able to compete with traditional immersion methods.

The initial tests carried on porosity measurement are very promising, but must be corroborated with the manufacture of a set of specimens of different thicknesses and porosity levels.

Acknowledgement.

This work has been jointly funded by the Eurostar 2 programme E! 8929 and CDTI

References

[1] B T. Khuri-Yakub, J. H. Kim, C. Chou, P. Parent and G. S. Kino, "A new design for air transducers", *Proc. IEEE Int'l Ultrason. Symp.*, pp. 503–506, 1988.
[2] W. Manthey, N. Kroemer and V. Mágori, "Ultrasonic transducers and transducers arrays for applications in air", *Meas. Sci. Tech.* 3, pp. 249-261, 1992.
[3] D.W. Schindel, D.A. Hutchins and W.A. Grandia, "Capacitive and piezoelectric air-coupled transducers for resonant ultrasonic inspection," *Ultrasonics*, 34, 6, pp.621-627, 1996.
[4] T.E. Gómez Alvarez-Arenas, "Acoustic Impedance Matching of piezoelectric transducers to the air", *IEEE Trans. Ultrason. Ferroelectr. Freq. Control*, 51, 5, pp. 624-633, 2004.
[5] T. E. Gómez Alvarez-Arenas, "Wide-band air-coupled ultrasonic piezoelectric transducers".*International Symposium on Piezocoposite Applications,* Dresden(Germany), Sept.22-23, 2011.
[6] Ultrasonic pulse-echo inspection of carbon fibre plastics AITM6-4005, Issue 10 Page 1 of 28 March 2013

Crack detection on aerospace composites by means of photorefractive interferometry

Tommaso SERESINI [1], Xiong JICHUAN [2], Martine WEVERS [3], Helge PFEIFFER [4], Christ GLORIEUX [5]
[1] KU Leuven, Leuven, Belgium
[2] NJU, Nanjing, China
[3] Katholieke Universiteit Leuven Dept. of Matellurgy and Materials Engineering, Heverlee, Belgium
[4] Katholieke Universiteit Leuven, Heverlee, Belgium
[5] Laboratorium voor Akoestiek en Thermische Fysica Departement Natuurkunde en Sterrenkunde Katholieke Universiteit Leuven, Heverlee, Belgium

Contact e-mail: tommaso.seresini@kuleuven.be

Abstract

Aircraft industry requires fast, robust and reliable tools for assessing structural integrity. Linear ultrasonic techniques are among the tools of choice for Non-Destructive Testing, however they fail to sense a defect when the limbs are in contact.

The inspection tool presented in this work is based on the detection of defect-induced elastic cross-modulation phenomena using photorefractive interferometry (PRI). This non-contact method is full-field and ultimately, it will allow to inspect large areas at once and enable detection of defects at an early stage.

The novelty of the system lies in both the ultrasonic excitation and in the detection method. Two surface waves are sent along the sample: one of low amplitude and high frequency (probe wave) and one of high amplitude and low frequency (modulating wave). The modulating wave opens and closes the limbs of the defect so that the probing wave experiences a non-linear modulation, thus evidencing its existence.

To sense the surface motion and measure its spectrum accurately, we have opted for a PRI setup, which can be configured to detect vibration frequency components of interest without cross-talk from other vibrations, e.g. the high amplitude vibration that opens and closes the defect. Using standard interferometry, frequency mixing between the modulation frequency and the probe frequency resulting from the defect response would be scrambled by the presence of frequency mixing due to the nonlinear relation between the interferometer light intensity and the measured displacements. This work highlights that photorefractive interferometry allows for frequency sensitive detection and enables to identify the modulated probing signal amid an intense background.

An important aspect of this study is the aerospace applicability of this tool. Therefore, experimental observations were focused on plausible defects: surface breaking cracks in metallic plates and internal delamination in carbon fibre plates. The samples are observed both under static conditions as well as under a time dependent load.

Acknowledgment: "NDTonAIR" project (Training Network in Non-Destructive Testing and Structural Health Monitoring of Aircraft structures) H2020-MSCA-ITN-2016-GRANT 722134.

On detecting kissing bonds in adhesively bonded joints using electric time domain reflectometry

Philip Johannes STEINBILD [1], Robin HÖHNE [1], René FÜßEL [1], Niels MODLER [1]
[1] TU Dresden, Institut für Leichtbau und Kunststofftechnik,
Holbeinstr. 3, 01307 Dresden, Germany
Contact e-mail: philip_johannes.steinbild@tu-dresden.de

Abstract. The use of fibre reinforced plastics to achieve low mass aircrafts calls for suitable joining technologies. The joining technology with the highest potential in lightweight construction with fibre reinforced plastics is adhesive bonding. However, in today's aircrafts the potential of adhesive bonding is limited due to the additional use of rivets as crack arrestors for regulation and safety issues. To meet the regulations and reduce the number of joint-weakening, heavy rivets, an adhesive bond needs to be monitored during its production and throughout its lifetime in operation. State of the art are ultrasound and thermal imaging technologies for non-destructive testing of adhesive bonds. The use of such technologies in a structural health management system is not practical. Additionally, ultrasound and thermal imaging technologies are only capable of detecting gross defects like areas of uncured adhesive and voids. The detection of adhesion defects like kissing bonds still pose a serious problem. This paper describes a novel adhesive sensor principle based on the electric time domain reflectometry, which can detect differences in the deformation of the adherents in an adhesively bonded joint under load to infer that a kissing bond is taking effect. The sensor is integrated into the joint. Results of the experimental validation by shear tension testing of single lap shear specimens with the adhesive sensor integrated into the joint are presented. The deformation of the adherents is monitored by digital image correlation and compared to the data obtained by the electric time domain reflectometry. The results show that by integrating the proposed sensor into an adhesively bonded joint the joint can be monitored operando. This novel technology can be used in a structural health management system to raise the confidence in adhesive bonds and reduce the number of rivets thus reducing mass. Further developments will include the use of a finite difference time domain model to numerically test sensor configurations regarding its geometry and electrical properties.

1. Introduction

The increasing use of composite design and fibre reinforced plastics (FRP) in the aerospace industry calls for suitable lightweight joining technologies. The joining technology featuring the highest lightweight potential is the adhesively bonded joint. The adhesively bonded joint enables transferring forces through low tension over large areas of the joint without using additional high-mass joining elements like screws or rivets. Especially the use in joining FRP parts is advantageous because continuous fibres are not damages by the joining process [1].

Incorrect manufacturing can lead to defects, which lower the carrying capacity of the adhesive joint drastically. The defects can also induce cracks into the joint, causing failure. One of the challenges is the prevention of interfacial weak bonds like kissing bonds. The relatively small dimensions of interfacial weak bonds make a detection via non-destructive testing (NDT) difficult [2]. This paper gives an insight on the detection and surveillance of kissing bonds using the difference in deformation of the adhesive joint at and around the defect. A novel sensor principle based on the electric time domain reflectometry (ETDR) is presented and experimentally investigated.

2. State of the Art

2.1 Kissing Bond

A kissing bond, also known as zero volume debond, is an interfacial weakness between adhesive and adherent. The kissing bond is a region in the adhesive bond in which the adhesive could not or only to certain extend establish adhesion while curing. Thus, the adhesive and the adherent are only partly joined or adhesive and adherent are touching but no forces can be transferred [2-6]. Figure 1 shows a schematic diagram of a kissing bond.

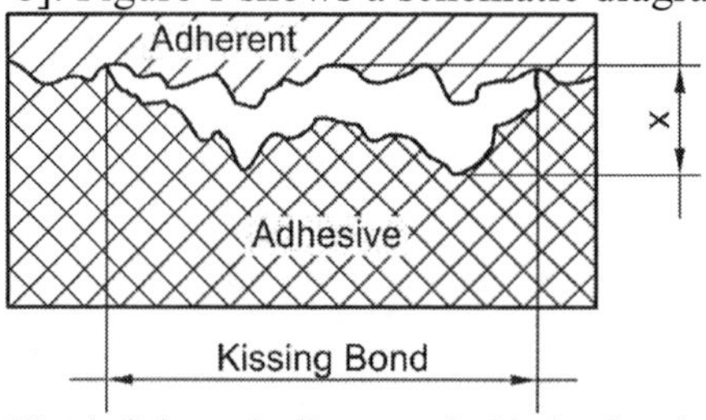

Fig. 1. Schematic diagram of a kissing bond.

A typical reason for a kissing bond is a not or insufficiently cleaned adherent surface which is still contaminated e.g. by a release agent or oil used in the manufacturing process of the adherent.

The detection of kissing bonds via conventional NDT like advanced ultrasonics or radiographic techniques are few and still subject of interest. The defects introduced in the adhesive bond are as small as nanometres (thickness x in Fig. 1). The mentioned conventional NDT detect differences in transmission properties of the bond. Since the kissing bond is a small defect, the effect on the transmission properties are too small to be detected in most cases [6].

The detectability of a kissing bond under load via digital image correlation (DIC) has been evaluated in [5]. Single lap shear samples made of glass fiber reinforced plastics (GFRP) as adherents and epoxy resin as the adhesive layer containing varied areas of kissing bond were examined under shear load. The kissing bond was introduced by placing an ethylene-tetraflourethylen-copolymer (ETFE) based release film on the adhesive surface before the actual joining process. The results showed different strain on the outer surface of an adherent in areas with kissing bond compared to those without kissing bond [5].

2.2 Electric Time Domain Reflectometry

The ETDR method is well-established in the electrical industry to analyse transmission lines. Based on run time and characteristics of reflections the ETDR method allows the space-resolved measurement of electric signals fed into transmission lines such as coaxial cables. A transmission line consist in the most basic case of two conductors with a dielectric layer in between (Fig. 2).

Maxwell's equations are the mathematical basis for describing the propagation of electromagnetic waves. Assuming only transverse electromagnetic (TEM) waves or quasi TEM waves exist in a certain transmission line, the calculation of the electromagnetic properties can be done by applying the transmission theory. Rather than calculating the field strengths ***E*** and ***H***, the transmission theory allows the calculation of the transmission properties ***I*** and ***U*** without the need to determine the electromagnetic field by use of Maxwell's equations. This allows determining the propagation of electromagnetic waves by solely analysing the geometry of the transmission line. The derivation of the transmission theory is based on the assumption that the electric and magnetic field structures in the transverse plane are identical to the stationary field structures at direct current excitation. The conformity of the field structures allow the definition of the transmission line properties ***I*** and ***U*** as well as the derivation of an equivalent circuit diagram through the method of describing stationary fields [9].

A basic transmission line and the corresponding equivalent circuit diagram are shown in Figure 2.

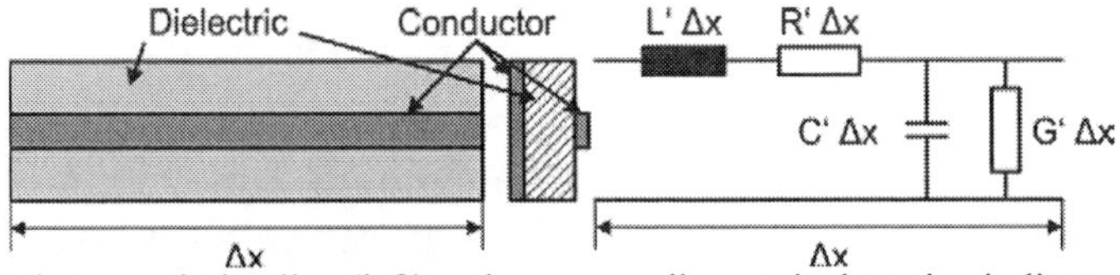

Fig. 2. Basic transmission line (left) and corresponding equivalent circuit diagram (right).

The properties of the transmission line show in Figure 2 can be described by the properties resistance, inductance, lateral conductivity, and capacity per unit length ***R'***, ***L'***, ***G'***, and ***C'*** respectively. One of the properties of a transmission line is the characteristic impedance Z_0. When excited by alternate or pulsed current the characteristic impedance is calculated by equation 2.1 [10].

$$Z_0 = \sqrt{\frac{R' + j\omega L'}{G' + j\omega C'}} \tag{2.1}$$

For high frequencies or pulsed current ***R'*** and ***G'*** are negligibly small compared to the frequency dependent quantities ***C'*** and ***L'***. Assuming small loss or lossless transmission lines, ***R' = G' = 0*** which leads to (2.2) in which ***a*** represents the width of the conductors, ***b*** the distance between the conductors, μ the magnetic permeability, and ε permittivity [10].

$$Z_0 = \sqrt{\frac{L'}{C'}} = \sqrt{\frac{b^2\mu}{a^2\varepsilon}} \tag{2.2}$$

The mentioned reflections in an ETDR measurement are induced by the change of electric properties of the transmission line. Since the characteristic impedance Z_0 depends on the basic electric properties inductance per unit length ***L'*** and capacity per unit length ***C'***, a change in ***L'*** or ***C'*** leads to a change in the characteristic impedance of the transmission line.

The ETDR method relies on a high frequency signal being fed into the transmission line. For this, an impulse generator is needed to generate the high frequency signal. To be able to measure the reflections an oscilloscope is also needed. The ETDR measurement device used in this paper is the D-TDR 3000 by Sympuls Aachen, shown in Figure 3 (right). It combines both devices needed for an ETDR measurement into one compact device. The equivalent circuit diagram in Figure 3 (left) shows the configuration of both, impulse generator and oscilloscope.

To perform measurements of the reflections in a transmission line, the investigated transmission line is connected to an ETDR measurement device via a coaxial cable. The outer and inner conductor is connected to one of the conductors of the transmission line respectively. The impulse generator sends repeating wave signals through the transmission line. At a change in impedance in the transmission, part of the wave is reflected while the remaining part keeps going through the transmission line. When the partly reflected signal reaches the oscilloscope, the location of the change in impedance can be determined by means of runtime and phase velocity. The phase velocity v_p complies with (2.3), with ε_0 electric constant, ε_r .relative permittivity, μ_0 magnetic constant, and μ_r relative permeability.

$$v_p = \frac{1}{\sqrt{L'C'}} = \frac{1}{\sqrt{\varepsilon_0 \varepsilon_r \mu_0 \mu_r}} \tag{2.3}$$

To outline the impedance over the course of the transmission line it is necessary to calculate the impedance after every change in impedance i.

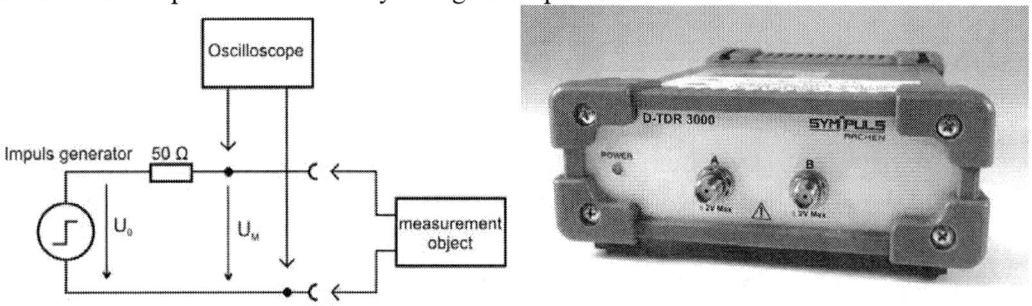

Fig. 3. ETDR measuring device D-TDR 3000 by Sympuls Aachen (right) and the corresponding equivalent circuit diagram (left) [11].

The reflection coefficient Γ decribes the ratio of the reflected (V_{ref}) to the incoming voltage amplitude (V_{inc}) and can be determined via the ratio of the impedances before (Z_{pre}) and after the change in impedance (Z_{after}) (2.4) [12].

$$\Gamma = \frac{V_{ref}}{V_{inc}} = \frac{(Z_{after} - Z_{pre})}{(Z_{after} + Z_{pre})} \tag{2.4}$$

For $i = 1$ Γ is calculated by means of (2.4). For $i > 1$ Γ corresponds to (2.5) [13].

$$\Gamma_i = \frac{\Delta V_i}{V_i \prod_{k=1}^{i-1}(1-\Gamma_k)} \tag{2.5}$$

ΔV_i represents the difference in voltage amplitude at the change of impedance i and V_i the voltage amplitude before the change of impedance i. V_i corresponds to (2.6) [13].

$$V_i = (1 + \Gamma_{i-1}) \cdot V_{i-1} \tag{2.6}$$

The impedance after the change of impedance i, Z_i, corresponds to (2.7) [13].

$$Z_i = Z_{i-1} \frac{(1+\Gamma_i)}{(1-\Gamma_i)} \tag{2.7}$$

The characteristic impedances from (2.7) plotted over the runtime give the characteristic impedance curve. In practice, the characteristic impedance curve is not calculated but measured at the undamaged object and used as a reference for further measurement.

The spatial or rather the time resolution of the ETDR measuring device depends on the rise time of the impulse signal. An infinitely high resolution is only possible in theory when the voltage can be built up in $t_R = 0$ s. Since this is not possible in practice, consecutive reflections are only distinguishable when the distance between them is at least the distance which the impulse signal travels in half of the rise time t_R [14].

Since the 1950s, ETDR is being used in energy and communication industry to detect imperfections in various cables. In the 1970s, the technology was adapted for geotechnology applications [15]. In geotechnology ETDR is mainly used for determining the relative permittivity or conductivity in a probing or for failure detection. The relative permittivity and conductivity of geotechnically relevant materials give information about their properties like moisture content while sudden changes in the electrical properties might indicate collapsed mine shafts [16].

In the 1990s, ETDR was developed further for structural health monitoring of concrete beams. Lin et. al. developed a method to detect and quantify strains and cracks in concrete beams [17]. Further developments used an elastomer as a dielectric to enhance the sensitivity of the sensor for strain [13].

Höhne et. al. developed a sensor principle for the distributed detection of strains by means of carbon fibre which shows very high sensitivity to strain in 2016 [18].

For the characterisation of crack propagation in fibre reinforced plastics Abu Obaid et. al. 2005 used an ETDR method and achieved a resolution of up to 0.3 mm [19].

ETDR and electrical frequency domain reflectometry (EFDR) measurements on adhesively bonded joints were first described in [7]. Samples using aluminium adherents and various polymer adhesives were examined. The relative permittivity of the polymers were calculated and compared using the results obtained by ETDR and electric frequency domain reflectometry (EFDR). Furthermore, the investigations showed that the time domain signals changed when the geometry was altered due to differences in joint thickness [7]. In [8] the influence of humidity and aging on adhesively bonded joints using ETDR was investigated.

These investigations confirm the applicability of ETDR for adhesively bonded joints. However, in adhesively bonded joints only the overall behaviour of the ETDR signals were considered. To allow the use of ETDR methods for structural health monitoring of adhesively bonded joints a novel sensor principle is presented and investigated in this paper.

3. Sensor Principle and Experimental Setup

Vijaya Kumar et. al. suggest that the monitoring of axial strain of the adherents can be used to detect kissing bonds in adhesively bonded joints [5]. The deformation of the adherents under load differs in damaged and undamaged areas of the joint when affected by kissing bonds. This forms the basis for the proposed sensor principle described in this paper.

The proposed sensor principle is shown in Figure 4. The sensor itself consists of an upper (red) and a lower (green) conductor. The adhesive acts as a dielectric between the upper and lower conductor. The conductors are connected to an ETDR measuring device, which measures the spatial impedance of the pair. In the unloaded condition (Fig. 4, left), the conductors cover each other completely. This delivers an impedance of $\boldsymbol{Z_{w0}}$ in the transmission line over the width of the sensor thus the adhesive bond. When the bond is loaded (Fig. 4, right), the kissing bond in area II causes the upper adherent to behave differently in area II than in area I. The difference in deformation is transferred to the upper conductor. The measurement shows higher impedances in area II ($\boldsymbol{Z_{w2}}$) than in area I ($\boldsymbol{Z_{w1}}$). Also, impedances in area I are smaller than in the unloaded state.

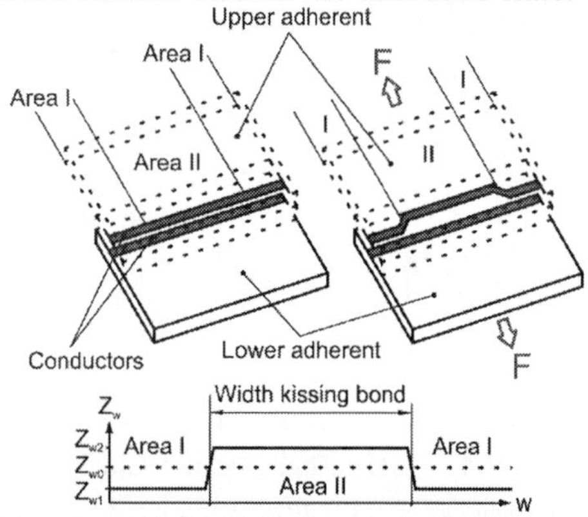

Fig. 4. Proposed sensor principle detecting kissing bonds in area II via ETDR.

This electrical behaviour correlates with the mechanical behaviour. Since there is no adhesion in area II, no lateral contraction of the adhesive occurs in this area. This causes the distance between the two conductors to increase in area II while it decreases in area I. The lateral contraction is overlaid with the difference in deformation of the adherent, respectively the conductor that leads to a characteristic distribution of impedance, Figure 4, bottom.

For the experimental validation single lap shear specimens with an integrated ETDR sensor were used. The adherents were made of the GFRP laminate Polydet® PowerStar having a thickness of 3.3 mm with bidirectional (BD) E-glass as reinforcement fibre and unsaturated polyester (UP) as a matrix. The structural adhesive DOW® Betaforce 2850 based on polyurethane (PUR) was used. The flat conductors were made of shielding 3M Scotch® 1181 (width 3 mm, thickness 0.1 mm). The overall thickness of the joint is 8.6 mm while the thickness of the adhesive is 2 mm, respectively 1.8 mm between the conductors. The adhesive surfaces of the adherents were prepared for joining by sandblasting before applying the flat conductors. The kissing bond was introduced by tape Airtech Flashbreaker® 1 with a thickness of 55 μm. The kissing bond was introduced in the middle of the adhesive surface over the full length of the adhesive surface and a defined width. The adherents were bonded in a bonding tool to maintain a uniform bonding thickness and cured at room temperature for 60 min. The tested specimen configurations are displayed in Table 1.

Table 1. Single lap shear specimens with integrated ETDR sensors used for experimental tests

Specimen ID	Adhesive surface	Kissing bond width
SLS_1	200 × 25 mm²	100 mm
SLS_2	200 × 25 mm²	50 mm

For the ETDR measurements a time domain reflectometer Sympuls D-TDR 3000 was used. The reflectometer and the sensor were connected via a coaxial cable RG-174/U. The inner and outer conductor of the coax were soldered to the upper and lower conductor respectively. A matched termination resistor connect both conductors on the other end of the sensor. A speckle pattern for DIC measurements was applied to the adherent with the induced kissing bond. A single lap shear specimen with integrated ETDR adhesive sensor, soldered coax, termination resistor, and speckle pattern is shown in Figure 5. To make the arrangement of the conductors clear, a cross section of the specimen is also given in Figure 5.

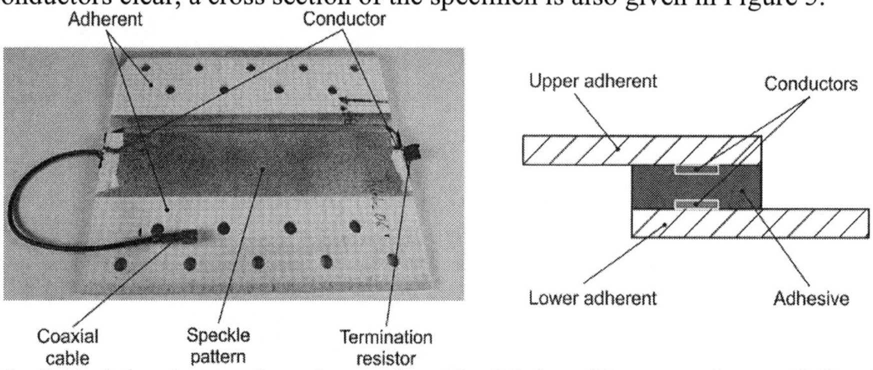

Fig. 5. Single lap shear specimen for experimental validation of the proposed sensor (left) and schematic cross section of the specimen showing the conductor arrangement.

Measurements used for the calculation of a DIC were taken to monitor strain on the outer surface of the kissing bond affected adherent to confirm that the tape introduced kissing bond lowers the carrying capacity of the adhesive bond during the tensile shear test. The data for the calculation of the DIC was measured by the 3D movement and deformation measurement system GOM ARAMIS. A single lap shear specimen with integrated ETDR adhesive sensor was mounted to the universal testing machine Zwick Roell UPM 1465 so that the ARAMIS system can observe the kissing bond affected adherent with applied speckle pattern. The test setup is shown in Figure 6.

4. Experimental Results and Discussion

The strain fields on the adherents outer surfaces obtained by DIC show that kissing bonds were successfully introduced by tape to every specimen. Figure 7 shows the calculated DIC of both specimens at a displacement of 2.2 mm. The projected area of the surface on which a kissing bond was induced by tape shows no significant strain while the undamaged areas right and left of the damaged area show strain as high as 0.3 %.

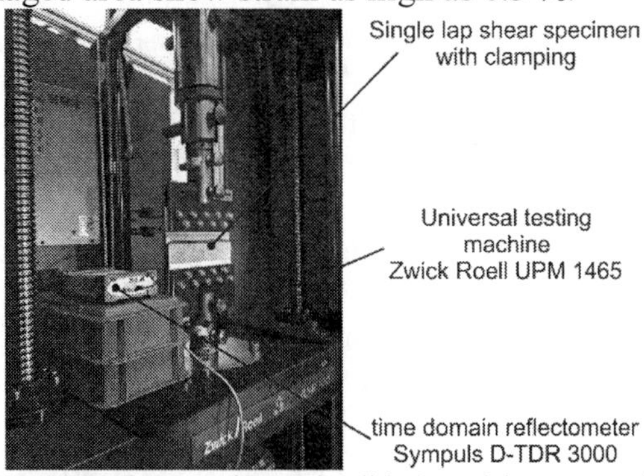

Fig. 6. Test setup for experimental validation of the proposed sensor.

The impedance is referred to the resistance of the termination resistor (50 Ω). The spatially resolved relative impedances over the width of the sensor, respectively the width of the adhesive bond, are shown in Figure 8 for both types of specimens at the unloaded state (0.0 mm) and for a displacement of 2.2 mm.

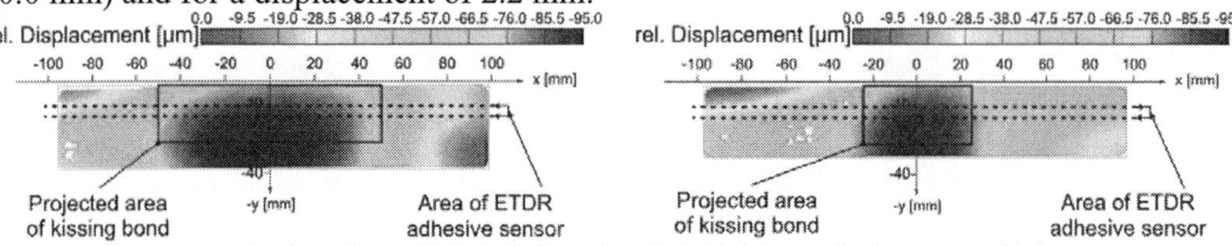

Fig. 7. DIC of specimen SLS_1 (left) and SLS_2 (right) at a displacement of 2.2 mm.

At the transition from the coaxial cable to the sensor at 0 mm a peak indicates big reflections caused by the soldering. At 200 mm the transition from the sensor to the termination resistor takes place, indicated by a smooth rise of the relative impedance. Between those two distinctive points the relative impedances of the sensing transmission line are plotted.

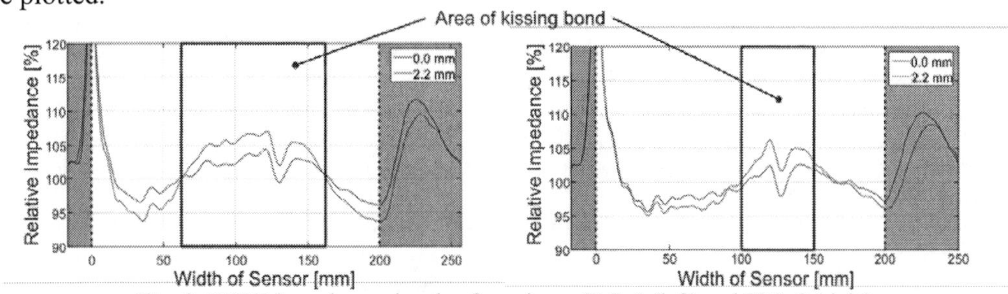

Fig. 8. ETDR impedance signals of specimen SLS_1 (left) and SLS_2 (right) at a displacement of 0.0 mm and 2.2 mm

The level of impedance in the kissing bond affected area is about 2.5 % higher under load than unloaded. In the area with no kissing bond the level of impedance is about 2.5 % lower under load than at no load caused by the lateral contraction of the adhesive. This behaviour can be observed for kissing bonds induced by tape for both widths, 100 mm and 50 mm. Since the dielectric properties of the sensors transmission line change drastically in the area of the kissing bond, the data shows a shift to the right of the sensor, indicating that the kissing bond is not located in the middle. The DIC however showed that the kissing bonds are in fact located in the middle of the specimens. This effect needs to be analysed and compensated in further investigations. The observed behaviour confirms the principle presented in Figure 4.

5. Summary and Conclusion

As shown in the state of the art, conventional NDT methods are not suitable for monitoring adhesively bonded structures operando. Since the sensor solution presented in this paper is integrated in the bonding, it can be used as part of a structural health monitoring system in the future to monitor the carrying capacity of the adhesive bond. In addition, the used measuring equipment to acquire ETDR signals is by far not as high cost as conventional NDT methods.

The adhesive sensor is able to detect adhesive defects like kissing bonds. This function has been experimentally validated. For the experimental validation tensile shear tests using single lap shear specimens with integrated transmission lines as a sensor were manufactured. The DIC calculations for outer adherent surfaces showed that kissing bonds can be prepared by an adhesive tape. ETDR measurements of the integrated transmission lines show different levels of impedances in kissing bond affected and non-kissing bond affected parts of single lap shear specimens, detecting kissing bonds under load.

The ETDR adhesive sensor has the potential to increase the confidence in adhesive bonds in particular and in lightweight structures generally.

References

[1] Schürmann, H.: Konstruieren mit Faser-Kunststoff-Verbunden. 2nd edition. Berlin, Heidelberg: Springer-Verlag, 2007. – ISBN 978-3-540-72189-5

[2] Nagy, P. B.: Ultrasonic Classification of Imperfect Interfaces. In: Journal of Nondestructive Evaluation, vol. 11 (1992), no. 3/4, pp. 127-139

[3] Najib, M. and Nobari, A.: Kissing bond detection in structural adhesive joints using nonlinear dynamic characteristics. In: International Journal of Adhesion & Adhesives, vol. 63 (2015), pp. 46-56

[4] Yan, D.; Neild, S.; Drinkwater, B.: Modelling and measurement of the nonlinear behaviour of kissing bonds in adhesive joints. In: NDT&E International, vol. 47 (2012), pp. 18-25

[5] Vijaya Kumar, R.; Bhat, M.; Murthy, C. R. L.: Evaluation of kissing bond in composite adhesive lap joints using digital image correlation: Preliminary studies. In: International Journal of Adhesion & Adhesives, vol. 42 (2013), pp. 60-68

[6] Jeenjitkaew, C. and Guild, F.: The analysis of kissing bonds in adhesive joints, vol. 75 (2017), pp. 101-107

[7] Li, Z.-C.; Joshi, S.; Hayward, D.; Gilmore, R.; Pethrick, R.: High frequency electrical measurements of adhesive bonded structures: An investigation of model parallel plate waveguide structures. In: NDT&E International, vol. 30 (1997), no. 3, pp. 151-161

[8]Affrossman, S.; Banks, W.; Hayward, D.; Pethrick, R.: Non-destructive examination of adhesively bonded structures using dielectric techniques: Review and some results. In: Proceedings of the Institution of Mechanical Engineers, vol. 214 (2000), pp. 87-102

[9] Paul, C. C.: Analysis of multiconductor transmission lines. Hoboken, New Jersey: John Wiley & Sons, Inc., 2008. – ISBN 978-0470131541

[10] IPC-2141A: Design Guide for High-Speed Controlled Impedance Circuit Boards. Northbrook, IL: IPC – Association Connecting Electronics Industries, 03.2004

[11] Sympuls – Gesellschaft für Pulstechnik und Messsysteme mbH: Wellenwiderstandsmessgerät D-TDR 3000: Gerätehandbuch. Aachen, 2014.

[12] Liao, S. Y.: Microwave Devices and Circuits: Third Edition. Englewood Cliffs, NJ: Prentice Hall, 1996

[13] Lin, M. W.; Thaduri, J.; Abatan, A.: Development of an electrical time domain reflectometry (ETDR) distributed strain sensor. In: Measurement Science and Technology, vol. 16 (2005), pp. 1495-1505

[14] Tektronix Inc.: TDR Impedance Measurements: A Foundation for Signal Integrity. 2007

[15] Chen, G.; Mu, H.; Pommerenke, D.; Drewniak, J. L.: Damage Detection of Reinforced Concrete Beams with Novel Distributed Crack/Strain Sensors. In: Structural Health Monitoring, vol. 3 (2004), no. 3, pp. 225-243

[16] Benson, C. H. and Bosscher, P. J.: Time-Domain Reflectometry (TDR) in Geotechnics: A Review. In: Nondestructive and Automated Testing for Soil and Rock Properties. ASTM STP 1350 (1999), pp. 113-136

[17] Lin, M. W.; Abatan, A.; Zhang, W.-M.: Crack damage detection of concrete structures using distributed electrical time domain reflectometry (ETDR) sensors. In: Proceedings of SPIE Conference on Smart Systems for Bridges, Structures, and Highways 3671 (1999), pp. 297-304

[18] Höhne, R.; Ehrig, T.; Kostka, P.; Modler, N.: Phenomenological investigation of a carbon fibre based strain sensor with spatial resolution by means of time domain reflectometry. In: Materialwissenschaft und Werkstofftechnik, vol. 47 (2016), no. 11, pp. 1024-1033

[19] Abu Obaid, A.; Yarlagadda, S.; Yoon, M. K.; Hager III, N. E.; Domszy, R. C.: A Time-domain Reflectometry Method for Automated Measurement of Crack Propagation in Composites during Mode I DCB Testing. In: Journal of Composite Materials, vol. 40 (2006), no. 22, pp. 2047-2066

Determination of the nonlinear elastic constants in a surface treated layer for aero-engine disk residual stress measurement

Sergey GARTSEV [1], Marek RJELKA [1], Bernd KÖHLER [1], Andreas MAYER [2]
[1] Fraunhofer-Institut für Keramische Technologien und Systeme (IKTS), Dresden, Germany
[2] Hochschule Offenburg, Offenburg, Germany
Contact e-mail: sergey.gartsev@ikts.fraunhofer.de
Fraunhofer IKTS, 01109, Dresden, Maria Reiche Straße 2

Abstract. Existing ultrasonic stress evaluation methods utilize the acoustoelastic effect for bulk waves propagating in volume, which is unsuitable for a surface treated material, possessing a significant variation in material properties with depth. With knowledge of nonlinear elastic parameters – third-order elastic constants (TOEC) close to the surface of the sample, the acoustoelastic effect might be used with surface acoustic waves. This work is focused on the development of an independent method of TOEC measurement using the effect of nonlinear surface acoustic waves scattering – i.e. the effect of elastic waves interaction in a nonlinear medium.

In this paper, the possible three wave interactions of surface guided waves and bulk waves are described and formulae for the efficiency of harmonic generation and mode mixing are derived. A comparison of the efficiency of surface waves scattering in an isotropic medium for different interaction types is carried out with the help of nonlinear perturbation theory. First results for surface and bulk wave mixing with known second- and third-order elastic constants are shown.

Introduction

Ultrasonic methods provide viable tools for non-destructive evaluation of material properties, including external or residual stress. These methods for stress evaluation utilize the acoustoelastic effect of the dependency of the phase velocity of acoustic waves on (external and residual) stress [1]. The change in the wave velocity is linked to the stress by the material nonlinearity expressed in terms of acoustoelastic constants (AEC). These constants have to either be known from literature or by calibration with a known (e.g. applied) stress level. The acoustoelastic effect is widely used for volume inspection using bulk waves, which is as such not directly applicable to the evaluation of near surface residual stress.

A common method to enhance the fatigue life of components is the creation of compressive residual stress in the components' surface which suppresses the surface crack initiation and growth under (cyclic) tensile load. One major application is the surface treatment of blade discs in jet engines, fabricated of titanium and superalloys. These discs are exposed to high temperatures and in parallel to high centrifugal tensile stress. Up to the moment, evaluation of induced compressive stress amplitudes can only be done either during production by reference samples (Almen stripes) or with destructive (borehole drilling)/laboratory (x-ray) methods. Current airworthiness regulations call for an NDE tool for in-service control of residual stress. Otherwise the advantages of the surface treatment cannot be considered in the design [2].

The availability of a method to prove that the amount of compressive stress did not decrease during operation would allow a further weight optimization of jet engine elements, along with operational costs reduction and increased performance. Published work [3, 4, 5, 6] indicates, that surface layer stress estimation by the acoustoelastic effect might work but also reveals the challenges of that concept. One of the difficulties is that the surface wave velocity is not only influenced by the stress via the acoustoelastic effect but also directly by the microstructure changes due to plastic deformation. So we have (at least) two contributions, which have to be separated. Moreover, the first contribution depends on the AEC, representing the nonlinearity of the material, which is known to change significantly by plastic deformation [7], which is caused by the surface treatment to introduce the stresses. The main task of the present work is therefore to find an independent way to determine the nonlinearity of the surface layer that underwent plastic deformation at the sample itself, as a prerequisite to capture the stress contribution to the velocity change correctly. To achieve this goal, we suggest to make use of nonlinear propagation effects as alternatives to the acoustoelastic effect, namely harmonic generation and nonlinear wave mixing, including non-collinear wave mixing processes. The concept of the present work is to use these interactions to determine the third-order elastic constants (TOEC), which are linked with the sought AEC [8]. As a first step we shall solve the forward problem of predicting the harmonic and mixed wave generation efficiency for known TOEC. In this paper the current progress on the method development is reported. The possible interactions of surface guided waves and bulk waves with low inclination angles are described and formulae for the efficiency of harmonic and mixing mode generation are derived. First results of surface acoustic waves (SAW) and bulk waves mixing modeling for known second order elastic constants (SOEC) and TOECs are presented.

1. Concept

For isotropic elastic media, the potential energy density may be expanded in powers of three strain invariants I_1, I_2, I_3 [9,11]:

$$\Phi = \frac{1}{2}\lambda I_1^2 + \mu I_2 + \frac{1}{6}\alpha I_1^3 + \beta I_1 I_2 + \frac{4}{3}\gamma I_3 + \ldots \quad (1)$$

where λ and μ are the second-order Lamé constants, α, β, γ are third-order Lamé constants (TOEC) [10], which may be expressed in terms of the Landau-Lifshitz constants A, B, C [9] or the Murnaghan constants l, m, n [11].

Within the cubic approximation of the strain energy, two input waves with frequencies ω_1 and ω_2 give rise to scattered waves of mixed frequency $n_1\omega_1 + n_2\omega_2$, where n_1 and n_2 are integers. This effect has been described theoretically [12, 13, 14] and observed experimentally [15, 16, 17].

One of the most well-known practical examples of wave scattering is the generation of the second harmonic along the propagation path in nonlinear media [18, 19, 20], which can be interpreted as interaction of two equal collinear waves. The process can be characterized by the acoustic nonlinearity parameter β (ANP), usually proportional to the ratio of the second harmonic amplitude to the square of the fundamental harmonic amplitude. ANP measurements in a nonlinear medium can also be utilized for surface layer NDE [20], as the ANP is related with the treatment intensity of the surface.

Available research results, related to practical applications of non-collinear wave scattering in NDE (i.e. works of Korneev and Demchenko [17, 22]), consider the interaction of bulk waves. For evaluation of material properties in a thin sub-surface layer one may also utilize surface acoustic waves (SAW or Rayleigh waves). This kind of waves propagates in a surface layer with effective penetration depth comparable to the wavelength [21]. To our knowledge, the use of Rayleigh wave mixing for elastic constants characterization of isotropic materials is not widely known.

The type (longitudinal, shear, SAW) and intensity of scattered wave is dependent on the types and intensities of the input waves, their frequencies, relative orientation of wave-vectors and polarization vectors, and on the material constants. Therefore, measurements of the scattering process in a surface layer can provide the information on the TOEC, necessary for stress characterization within the acoustoelastic effect.

2. Theoretical description

To get significant mixing intensity in the nonlinear interactions of longitudinal, shear and Rayleigh waves they must fulfill the phase matching (or resonance) conditions, which correspond to the energy and momentum conservation conditions in anharmonic phonon-phonon interaction [23]:

$$\omega_3 = \omega_1 \pm \omega_2 \qquad (2,a)$$

$$\overrightarrow{k_3} = \overrightarrow{k_1} \pm \overrightarrow{k_2} \qquad (2,b)$$

Analysis of allowed interactions and scattered amplitude intensities for bulk waves was introduced first by Jones and Kobett [14]. Later Korneev et al. took the polarization restrictions into account [22].

For the bulk wave - surface wave interactions in isotropic solids the only available reference is [24]. Allowed interactions are schematically shown in Fig. 1, where straight and curly lines represent bulk and Rayleigh waves, black and orange color stands for initial and scattered waves. Interaction processes of other combinations, involving surface acoustic waves, are forbidden because of the relative size of the phase velocities that enter Eq. 2, b.

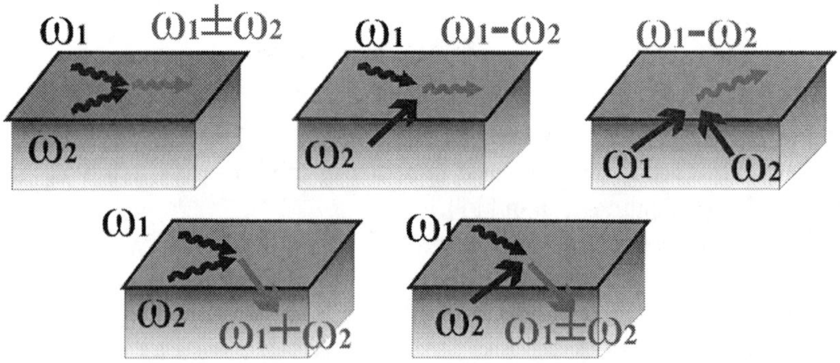

Figure 1: Available scattering modes for acoustic waves involving SAW [24]. Wavy lines: Rayleigh waves, straight lines: acoustic bulk waves, black: input waves, orange: scattered waves due to 2nd-order nonlinearity.

For determination of three TOECs in isotropic solids at least three independent measurements have to be performed. To identify the geometries with maximal scattering efficiency, the intensities of waves generated in nonlinear scattering processes for all allowed incoming and scattered wave combinations should be calculated.

As the numerical simulations of 3D wavefields in the time-domain for nonlinear isotropic media is computationally expensive, an approximate analytical solution for the scattered wave intensity [25] was implemented in Wolfram Mathematica 11.2 and solved for relevant interaction cases. This analytical solution uses perturbation theory and relies on the following assumptions:

1. The nonlinear perturbation is small compared to the linear part of the displacement.
2. The involved wavelengths are small compared to the sample thickness.
3. The medium is homogeneous and isotropic containing a uniform stress distribution.

While assumption 1 is clearly justified in practical cases and so is assumption 2 as long as the frequencies of the ultrasonic waves are sufficiently high, this does not need to be the case for assumption 3. This aspect will be further discussed below.

The equation of motion for the displacement field *u* can be presented in the following way:

$$\rho\, \ddot{u}_j(\vec{X},t) = T_{Lj,L}(\vec{X},t), \tag{3}$$

Where

$$T_{Lj} = \frac{\partial \Phi}{\partial u_{j,L}} = S_{jL} + S_{jLmM}u_{m,M} + \frac{1}{2}S_{jLmMnN}u_{m,M}u_{n,N} + \ldots \tag{4}$$

is the first Piola-Kirchhoff stress tensor, ρ is the mass density and the tensors *S* are defined in [26]. It is straightforward to express the quantities S_{jLmMnN} as linear combinations of second- and third-order elastic moduli λ, μ, α, β, γ.

Let the displacement field of the two input waves be of the form:

$$\vec{u}(\vec{X},t) = e^{i(\vec{q}\cdot\vec{X}-\omega t)}\vec{w}(Z|\vec{q}) + c.c\,, \tag{5}$$

where $\vec{X}$ - coordinate vector, $\vec{q}$ – 2D wave vector, and axis OZ is oriented normal to surface (the medium fills the halfspace Z < 0). Functions $\vec{w}(Z|\vec{q})$ are displacement profiles. For bulk waves as input waves, $\vec{q}$ is the projection of the 3D wave vector onto the surface plane (the OXY plane).

All of the analytical expressions and simulation results use the notation shown in Fig. 2. Wave vectors of incident waves ($\overrightarrow{q_1}$ and $\overrightarrow{q_2}$) are colored with black and oriented with angles α_1 and α_2 towards $\overrightarrow{OX}$ and the wave vector $\overrightarrow{q_3}$ of the scattered wave is colored with orange and points into the $\overrightarrow{OX}$ direction. For Rayleigh waves in isotropic media, the depth profile can be written as a sum of two exponentials:

$$\vec{w}(Z|\vec{q}) = \sum_{r=1}^{2} c(r)\begin{pmatrix}1\\0\\p(r)\end{pmatrix}e^{\alpha(r)Z}\,, \tag{6}$$

where coefficients *c*(*r*), *p*(*r*) and $\alpha(r)$ are defined in [25]. For longitudinal and shear waves expressions for $\vec{w}(Z|\vec{q})$ can be found in [21].

If the resonance conditions of Eq. 2 are fulfilled and the scattered wave type is a SAW, the solution for the displacement field of the scattered wave can be found in the form

$$\overrightarrow{u_3}(\vec{X},t) = e^{i(\overrightarrow{q_3}\cdot\vec{R}-\omega_3 t)}\vec{w}(Z|\overrightarrow{q_3})(-i)hX + \zeta(Z) + c.c\,, \tag{7}$$

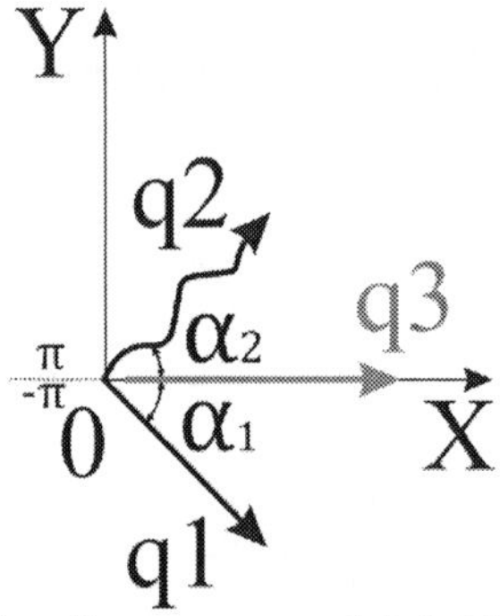

Figure 2: Coordinate system and definition of angles
(α defined in range between -π and π in OXY plane)

where $\vec{R}$ is the projection of $\vec{X}$ on the OXY plane. By inserting Eq. 7 into the equation of motion, projecting on $\vec{w}(Z|\vec{q})$, integrating by parts and making use of the nonlinear boundary conditions at the surface one obtains the following expression for the efficiency h of the Rayleigh wave generation through nonlinear interaction of two input waves:

$$h = \frac{\omega_3}{S_R} S_{jLmMnN} \int_{-\infty}^{0} [\hat{Q}_L\,(\overrightarrow{q_3}) w_l(Z|\overrightarrow{q_3})]^* [\hat{Q}_M(\overrightarrow{q_2}) w_m(Z|\overrightarrow{q_2})]\,[\hat{Q}_N\,(\overrightarrow{q_1}) w_n(Z|\overrightarrow{q_1})] dZ. \quad (8)$$

If the output wave is a bulk wave with 3D wave vector K = ($\overrightarrow{q_3}$,-k), it can be described as:

$$\overrightarrow{u_3}(\vec{X},t) = e^{i(\overrightarrow{q_3}\cdot\vec{R}-\omega_3 t)}(B\overrightarrow{W}e^{-ikZ} + \vec{\zeta}(Z)) + c.c, \quad (9)$$

where zeta represents additional partial waves to satisfy the boundary conditions at the surface, and the solution for H (component of complex amplitude B of scattered bulk wave: $B = HA_1A_2$), can be calculated as an overlap integral similar to h in eq. (8), involving the displacement fields of the three nonlinearly interacting waves [25]. S_R depends on the normalization of w and on the Rayleigh wave velocity [25], $\overrightarrow{W}$ is a polarization unit vector and $\hat{Q}_L = \delta_{L1} iq_1 + \delta_{L2} iq_2 + \delta_{L3}\frac{\partial}{\partial Z}$. Therefore, the scattered wave efficiency can be expressed through depth profiles, polarization and wave vector components of initial waves. Based on this model, the following procedure has been implemented:

- Interaction case selection
- For each point on an (ω_1, ω_2) frequency grid:
  - Search for permitted initial waves orientation on a surface for a given scattered wave direction (Eq. 2)
  - Calculation of scattering efficiency for a given parameter combination
- Data storage, visual layout

Details of simulation are described in the following section.

3. Surface wave mixing modeling results

The forward problem (modeling of scattered wave efficiency with known material parameters) has been investigated for the nickel-base superalloy In718 ($\rho = 8.91\frac{g}{cm^3}$, B = 199,94 *GPa (bulk modulus)*, v = 0.294, TOEC values [6]: l = -520 *GPa*, m = -600 *GPa*, n = -480 *GPa*). In order to comply with the boundary conditions, the bulk waves were inclined 1.5° towards –OZ. Wave scattering modeling was preceded by several consistency checks. Comparison with [27] has been made for the case of collinear Rayleigh wave mixing.

Unfortunately, no published results for non-collinear wave scattering are known to the authors, which could have served as basis for a further check of our model.

As all of the possible α_1 and α_2 combinations are symmetrical with respect to the origin, we are expecting symmetry for scattered wave amplitudes for such coordinate pairs. In Fig. 3 a typical layout for scattering is presented as 3D plots with projections. Axes 1 and 2 (α_1 and α_2) form the plan view. The vertical axis is the relative scattering efficiency normalized by the scattered wave frequency ω_3. The frequency ratio is displayed by color-coding.

Simulation results for the case of two Rayleigh waves generating a longitudinal wave with sum-frequency are shown in Fig. 3 (in the following, the interactions would be labeled as /First input wave/ *[sign distinguishing sum and difference frequency generation]* /Second input wave/ → /Scattered wave/, i.e. RW+RW→P for the given example).

Simulation results are shown on Tab. 1 with the three most intensive interactions highlighted.

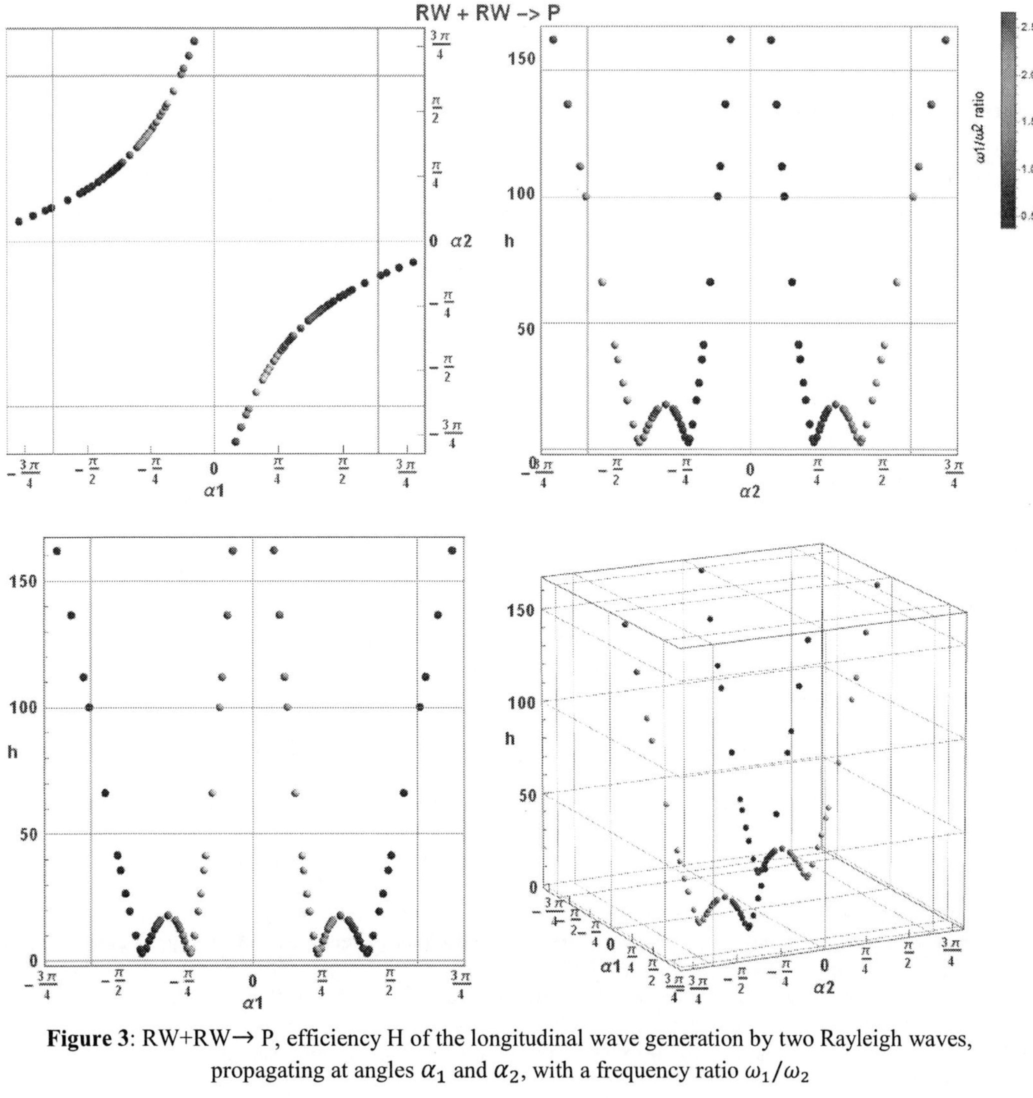

Figure 3: RW+RW→ P, efficiency H of the longitudinal wave generation by two Rayleigh waves, propagating at angles α_1 and α_2, with a frequency ratio ω_1/ω_2

Table 1: Scattering efficiency simulation results

Interaction case	α_1, rad	α_2, rad	**Scattering efficiency** $H/\omega_3 \mid h/\omega_3$	ω_1/ω_2
RW+RW→P	-2.42	0.25	162.23	0.38
RW+RW→SV	-1.34	0.1	60.86	0.1
RW+P→P	0.19	-2.4	70.38	1.8
RW+ P→SV	-0.02	0.3	6.17	6.
RW+SV→P	-2.96	0.05	180.41	0.29
RW+SV→SV	-0.09	1.33	75.71	10.
P-P→RW	0.96	1.14	24.19	1.11
P-SV→RW	2.96	3.02	72.82	1.29
SV-P→RW	-0.05	-0.88	22.99	9.
SV-SV→RW	-0.1	-1.34	227.83	10.

4. Discussion and further work

First results on anharmonic wave scattering simulations involving surface acoustic waves and bulk acoustic waves are presented. Geometrical layouts and a comparison of the relative efficiencies of the various scattering processes served as a basis for the design of experiments. Laser Doppler vibrometry and contact measurements are currently underway to verify the scattering model described above. The following step will be the investigation of the inversion problem: evaluation of third-order elastic constants by recording the intensity of the scattered wave as function of the parameters of the input waves. Subsequently, the model will be extended to allow for variations of the material constants along the coordinate axis normal to the surface, i.e. assumption 3 in section 2 will be relaxed.

Acknowledgment

One of the authors, Sergey Gartsev, gratefully acknowledges funding from the European Union's Horizon 2020 research and innovation program under the Marie Sklodowska-Curie grant agreement N. 722134 - NDTonAIR.

References

[1] D. Husson and G. S. Kino. A perturbation theory for acoustoelastic effects. In: Journal of Applied Physics 53.11 (Nov. 1982), pp. 7250-7258. doi: 10.1063/1.331623.

[2] AMC 25.571 Certification Specifications. Acceptable Means of Compliance for Large Aeroplanes. In: CS-25. Amendment 11.4 (2011).

[3] B. Koehler et al. Characterization of surface treated aero engine alloys by Rayleigh wave velocity dispersion. In: Conference. AIP, 2010. doi: 10.1063/1.3362402.

[4] M. Barth et al. Universal ultrasonic goniometer for Rayleigh and surface skimming longitudinal wave dispersion measurements. In: Conference. AIP, 2012. doi: 10.1063/1.4716439.

[5] S. Hubel et al. Ultrasonic evaluation of residual stresses in aero engine materials using bulk and Rayleigh surface waves. In: AIP Publishing LLC, 2014. doi: 10.1063/1.4864876.

[6] M. Rjelka et al. Third Order Elastic Constants and Rayleigh Wave Dispersion of Shot Peened Aero-Engine Materials. In: Materials Science Forum 768-769 (Sept. 2013), pp. 201208. doi: 10.4028/www.scientific.net/msf.768-769.201.

[7] P. B. Nagy. Opportunities and Challenges for Nondestructive Residual Stress Assessment. In: AIP Conference Proceedings. AIP, 2006. doi: 10.1063/1.2184508.

[8] D. E. Bray and R. K. Stanley. Nondestructive evaluation: a tool in design, manufacturing and service. CRC press, 1996.

[9] L. D. Landau and E. M. Lifshitz. Theory of Elasticity, 3rd. In: Pergamon Press 3 (1986).

[10] S. Tamura, Spontaneous decay rates of LA phonons in quasi-isotropic solids, Phys. Rev. B31, 2574 (1985).

[11] F. D. Murnaghan. Finite Deformation of an Elastic Solid, New York (1951). John Wiley and Sons, 1951.

[12] L. H. Taylor and F. R. Rollins. Ultrasonic Study of Three-Phonon Interactions. I. Theory. In: Physical Review 136.3A (Nov. 1964), A591-A596. doi: 10.1103/physrev.136.a591.
[13] R. Normandin and G. I. Stegeman. Surface-wave generation by nonlinear mixing of bulk waves. In: Journal of Applied Physics 50.1 (Jan. 1979), pp. 87-90. doi: 10.1063/1.325622.
[14] G. L. Jones and D. R. Kobett. Interaction of Elastic Waves in an Isotropic Solid. In: The Journal of the Acoustical Society of America 35.1 (Jan. 1963), pp. 5-10. doi: https://doi.org/10.1121/1.1918405.
[15] F. R. Rollins, L. H. Taylor, and P. H. Todd. Ultrasonic Study of Three-Phonon Interactions. II. Experimental Results. In: Physical Review 136.3A (Nov. 1964), A597-A601. doi: 10.1103/physrev.136.a597.
[16] W. J. Tanski and W. G. Mayer. Further investigation of noncollinear surface-wave interactions. In: Journal of Applied Physics 44.1 (Jan. 1973), pp. 506-507. doi: 10.1063/1.1661920.
[17] A. Demcenko, V. Koissin, and V.A. Korneev. Noncollinear wave mixing for measurement of dynamic processes in polymers: Physical ageing in thermoplastics and epoxy cure. In: Ultrasonics 54.2 (Feb. 2014), pp. 684-693.doi: 10.1016/j.ultras.2013.09.011.
[18] P. J. Vella et al. Nonlinear surface-wave interactions: Parametric mixing and harmonic generation. In: Journal of Applied Physics 45.5 (May 1974), pp. 1993-2006. doi: 10.1063/1.1663536.
[19] D. J. Shull et al. Harmonic generation in plane and cylindrical nonlinear Rayleigh waves. In: The Journal of the Acoustical Society of America 94.1 (July 1993), pp. 418-427. doi: https://doi.org/10.1121/1.407053.
[20] K. H. Matlack et al. Review of Second Harmonic Generation Measurement Techniques for Material State Determination in Metals. In: Journal of Nondestructive Evaluation 34.1 (Nov. 2014). doi: 10.1007/s10921-014-0273-5.
[21] J. Achenbach. Wave propagation in elastic solids. Vol. 16. Elsevier, 2012.
[22] V. A. Korneev, K. T. Nihei, and L R. Myer. Nonlinear interaction of plane elastic waves. Tech. rep. Berkeley Lab, June 1998. doi: 10.2172/290877.
[23] L. K. Zarembo, B. A. Krasilnikov, Introduction to Nonlinear Acoustics (Nauka, Moscow, 1966).
[24] G. I. Stegeman and F. Nizzoli. Surface vibrations. In: *Surface Excitations*, ed. by R. Loudon and V. M. Agranovich (North-Holland, Amsterdam, 1984), Chap. 2, pp. 195-378.
[25] A. P. Mayer. Surface acoustic waves in nonlinear elastic media. In: Physics Reports 256.4-5 (1995),pp. 237-366.
[26] G. Leibfried and W. Ludwig. Gleichgewichtsbedingungen in der Gittertheorie. In: Zeitschrift fur Physik 160.1 (Feb. 1960), pp. 80-92. doi: 10.1007/bf01337481.
[27] M. B. Morlock et al. Mixing of two co-directional Rayleigh surface waves in a nonlinear elastic material. In: The Journal of the Acoustical Society of America 137.1 (Jan. 2015), pp. 281-292. doi: 10.1121/1.4904535.
[28] J. J. Ditri and D. Hongerholt. Stress distribution determination in isotropic materials via inversion of ultrasonic Rayleigh wave dispersion data. In: International Journal of Solids and Structures 33.17 (July 1996), pp. 2437-2451. doi: 10.1016/0020-7683(95)00165-4.
[29] D. Husson. A perturbation theory for the acoustoelastic effect of surface waves. In: Journal of Applied Physics 57.5 (Mar. 1985), pp. 1562-1568. doi: 10.1063/1.334471.

The Use of Pulse Compression Technique in Non-Destructive Testing: A Review

Muhammad Khalid RIZWAN [2], Hamed MALEKMOHAMMADI [1], Stefano LAURETI [1], Pietro BURRASCANO [2], David A. HUTCHINS [3], Gui Yun TIAN [4], Junzhen ZHU [4], Qiuji YI [4], Marco RICCI [1]
[1] University of Perugia, Perugia, Italy
[2] Universita degli Studi di Perugia, Terni, Italy
[3] University of Warwick, Warwick Manufacturing Group (WMG), Coventry, UK
[4] Newcastle University, Newcastle upon Tyne, UK
Contact e-mail: muhammadkhalid.rizwan@unipg.it

Abstract

Non-Destructive Testing (NDT) refers to an extended group of techniques used both in research and industry, which are exploited for inspecting, characterizing and evaluating various kinds of materials and goods. Eddy Current Testing (ECT), Ultrasonic Testing (UT) and InfraRed Thermography (IRT) are among the most employed NDT methods. In the said methods, a short duration delta-like signal provided by different types of transducers is typically employed to excite the sample under test (SUT) within an extended bandwidth. Features of interest, e.g. presence of flaws and inclusions inside the SUT, can be extracted by analyzing the system's impulse response both in time and frequency domains. However, the maximum achievable Signal-to-Noise Ratio (SNR) is directly related to the source power. This limit can be overcome by using coded excitations together with Pulse Compression (PuC) technique. In fact, in PuC the frequency spectrum of the coded excitation can be tailored to suit the investigation of a given sample, while the SNR can be increased almost arbitrarily by enhancing the signal time duration. Finally, the system's impulse response is retrieved by applying the PuC algorithm on the acquired data. In this paper, the use of PuC technique in IRT, UT, ECT will be reviewed and discussed.

THE USE OF PULSE COMPRESSION TECHNIQUE IN NON-DESTRUCTIVE TESTING: A REVIEW

Muhammad Khalid Rizwan[1], *Hamed Malekmohammadi[1], Stefano Laureti[1], Pietro Burrascano[1], David A. Hutchins[2], Gui Yun Tian[3], Junzhen Zhu[3], Qiuji Yi[3], Marco Ricci[4]*

[1]Department of Engineering
University of Perugia, Italy

[2]School of Engineering
University of Warwick, UK

[3]School of Engineering
Newcastle University, UK

[4]DIMES Department
University of Calabria, Italy

Outline

→ ***Background***
→ ***Limits of PuE approach***
→ ***Pulse Compression (PuC)***
- ***Basic principle***
- ***Signals for PuC***
- ***Choice of the Signals***

→ ***Application in NDT***
- ***PuC Ultrasonic***
- ***PuC Thermography***
- ***PuC Eddy Current Testing***

→ ***Conclusions***
→ ***References***

THE USE OF PULSE COMPRESSION TECHNIQUE IN NON-DESTRUCTIVE TESTING: A REVIEW

Background

General procedures in Ultrasonic, Eddy Current, Thermography NDT inspections are based on measuring the response $h(t)$ of the sample under test (SUT) to an impulsive excitation which approximates a delta $\delta(t)$ like function (e.g. Pulse Echo)

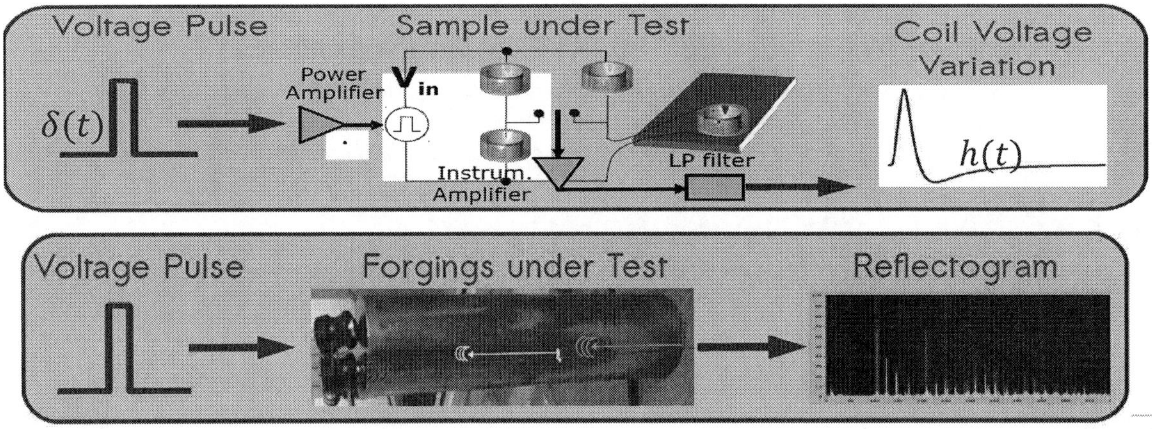

THE USE OF PULSE COMPRESSION TECHNIQUE IN NON-DESTRUCTIVE TESTING: A REVIEW

Background

- In circuit theory, the response to a delta $\delta(t)$ excitation is called impulse response h(t) and it describes the input-output relation of the system as

$$y(t) = x(t) \otimes h(t)$$

$\Rightarrow$Measuring $h(t)$ allows the inspection of the sample as well as the calculation of the response of the SUT to any other possible excitation

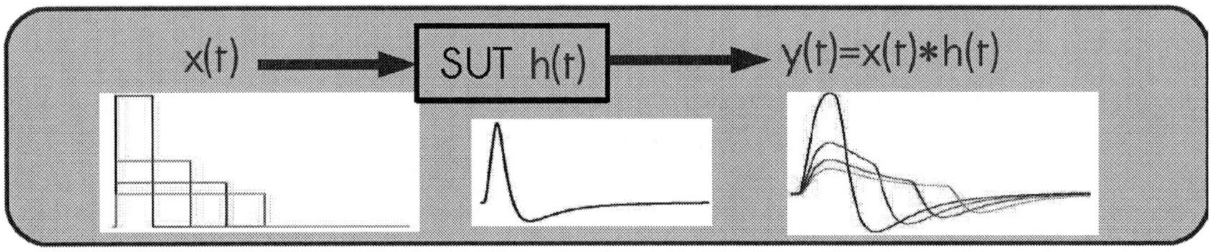

THE USE OF PULSE COMPRESSION TECHNIQUE IN NON-DESTRUCTIVE TESTING: A REVIEW

Limits of the Pulse-echo approach

→ Maximum achievable SNR is not sufficient for the defect detection and characterization in the presence of noise or high attenuation

↓

→ Aim is to measure the impulse response even in the presence of noise, to enhance sensitivity of the system

How?

→ Increase energy of the excitation pulse → Amplitude of the pulse; constrained by the system and the transducer's limitations

→ Averages → Long time of measurements

THE USE OF PULSE COMPRESSION TECHNIQUE IN NON-DESTRUCTIVE TESTING: A REVIEW

NDTonAIR

Enhance SNR → HOW?

Proposed Solution

↓

Pulse Compression

An alternative technique to increase the SNR of the measurement setup

→ increase the time bandwidth product → increase the time duration of the excitation pulse rather than the amplitude

- Flexible both in time and frequency shaping
- Widely applicable
- Various techniques to implement

THE USE OF PULSE COMPRESSION TECHNIQUE IN NON-DESTRUCTIVE TESTING: A REVIEW

NDTonAIR

Pulse Compression (PuC)- *basic principle*

- Pair of signals whose correlation is a delta $\delta(t)$ like function, i.e.,

$$g(t) * \Psi(t) = \hat{\delta}(t)$$

- Excite the Linear (and in some cases nonlinear) system with one of the signals, $g(t)$
- Impulse response of the unknown system can be estimated when the output is convolved to the second signal $\Psi(t)$ in the pair, i.e.,

$$y(t) * \Psi(t) = [g * h](t) * \Psi(t) = h(t) * \hat{\delta}(t) \simeq C\hat{h}(t)$$

- SNR is maximized when $\Psi(t)$= $g(-t)$ ⟶ *matched filter theory*

THE USE OF PULSE COMPRESSION TECHNIQUE IN NON-DESTRUCTIVE TESTING: A REVIEW

NDTonAIR

Pulse Compression- *basic principle*

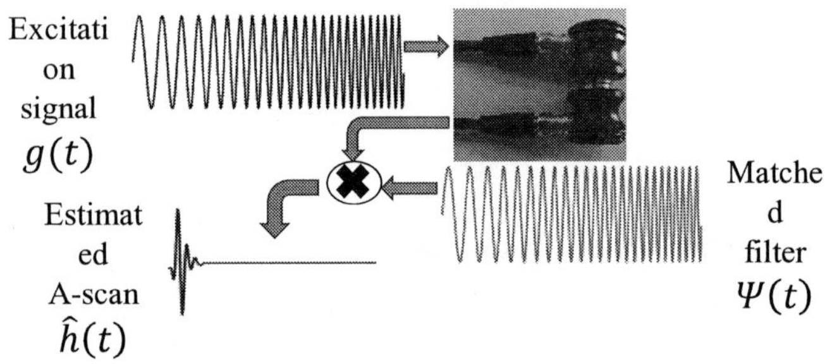

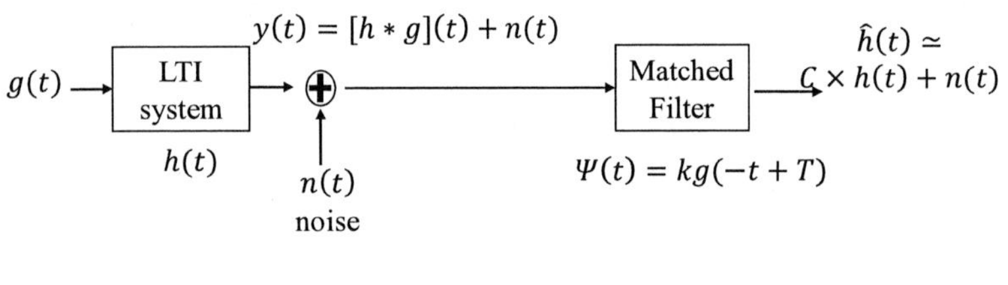

Advantages

- ☺ High SNR ⟶ Required SNR Gain can be achieved by increasing the time duration of the excitation pulse without effecting the bandwidth

↓

Bandwidth is independent of T_{PuC}

- ☺ Flexibility in time and frequency domains shaping
- ☺ No hardware complexity for implementation

THE USE OF PULSE COMPRESSION TECHNIQUE IN NON-DESTRUCTIVE TESTING: A REVIEW

NDTonAIR

Signals for PuC

- Ideally, the signal should prevail the property of autocorrelation as;

$$R_{gg} = \delta(t)$$

- In Real cases, the closer the correlation ⟶ $R_{gg} = \hat{\delta}(t) \sim \delta(t)$, the more accurate is the estimation of the impulse response, ⟶ $\hat{h}(t) \sim h(t)$
- Practically, frequency spectrum of the selected signal should be wide enough to cover the bandwidth of the sample under test

THE USE OF PULSE COMPRESSION TECHNIQUE IN NON-DESTRUCTIVE TESTING: A REVIEW

NDTonAIR

Applications of PuC in NDT

→ Wave Phenomenon   Ultrasonic NDT

- SNR_{Gain} is significantly improved ⟶ widely applicable
- h(t) consists of several reflections/paths
- Due to propagation, duration of impulse response, T_h is much longer than the required time resolution δT, i.e., $T_h >> \delta T$ ⟶ $T_h B >> 1$ ⟶ SNR_{Gain} is high

→ Diffusion Phenomenon   EDDY Current, Thermography

- SNR_{Gain} is reasonable, but other advantages exist
- h(t) is smooth and exponentially decaying
- Time duration of the impulse response T_h is greater than time resolution δT, but reasonable

THE USE OF PULSE COMPRESSION TECHNIQUE IN NON-DESTRUCTIVE TESTING: A REVIEW

NDTonAIR

Choice of the Signals

Swept frequency Signals:

Chirps (both linear and nonlinear)

- Chirps are the widely used signals in PuC NDT ⟹ Easy control over the bandwidth and almost a flat spectrum

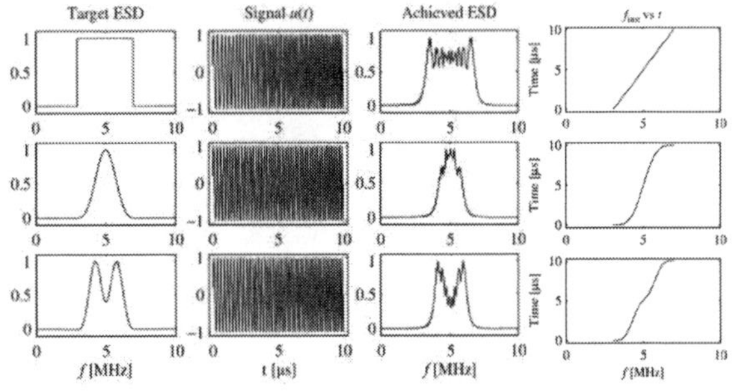

Coded Signals:

Binary codes, mapped into continuous waveforms (bits in the form of square wavelets), e.g. Multi-Length Sequences (MLS), Golay Complementary Sequences, Barker Codes, etc.

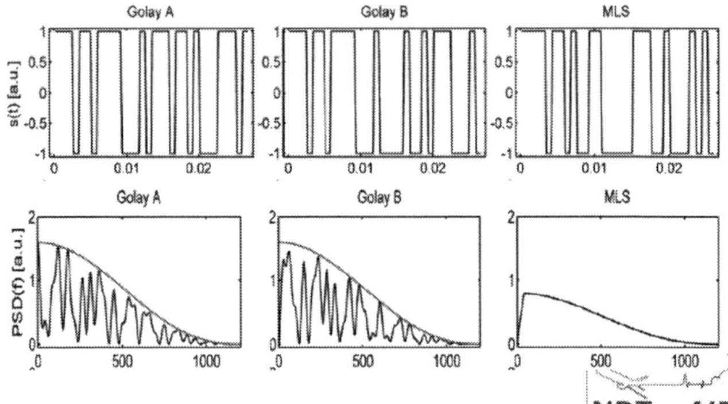

THE USE OF PULSE COMPRESSION TECHNIQUE IN NON-DESTRUCTIVE TESTING: A REVIEW

NDTonAIR

PuC Ultrasonic NDT: *Schematic of the Setup*

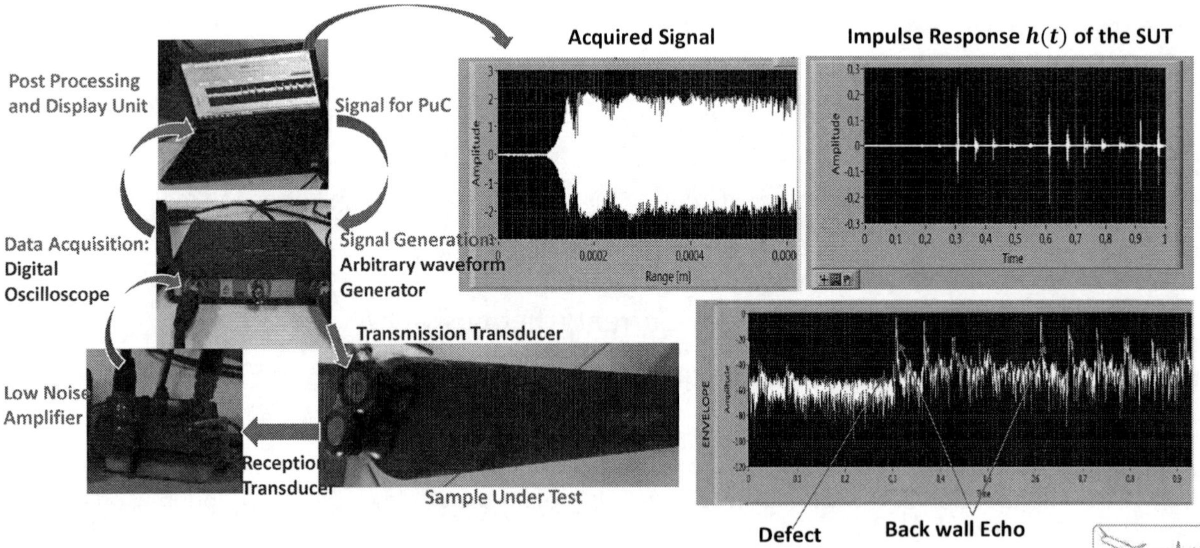

THE USE OF PULSE COMPRESSION TECHNIQUE IN NON-DESTRUCTIVE TESTING: A REVIEW

NDTonAIR

PuC Ultrasonic NDT: *Results comparison to PuE*

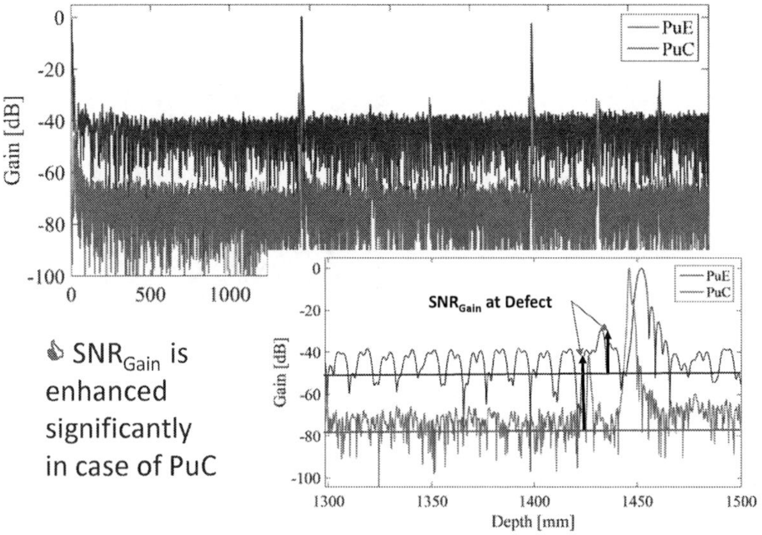

- PuC (using a Linear chirp as an excitation signal) and PuE UT inspections of a steel forging, 1.5m long and diameter of 0.6m
- Using the same transducers, V108-VEDIOSCAN from Olympus
- AWG Vp-p;
  - in case of PuE 200V
  - in case of PuC 24V

SNR$_{Gain}$ is enhanced significantly in case of PuC

NDTonAIR

THE USE OF PULSE COMPRESSION TECHNIQUE IN NON-DESTRUCTIVE TESTING: A REVIEW

PuC Ultrasonic NDT: *Results*

- Echograms of PuC UT inspection of forgings
- SNR$_{Gain}$ is significantly high, because the duration of the signal (4m) is much higher than the resolution (in mm)
- Exploit the advantage of broad bandwidth of a Linear Chirp
- Obtain the optimum frequency of Inspection

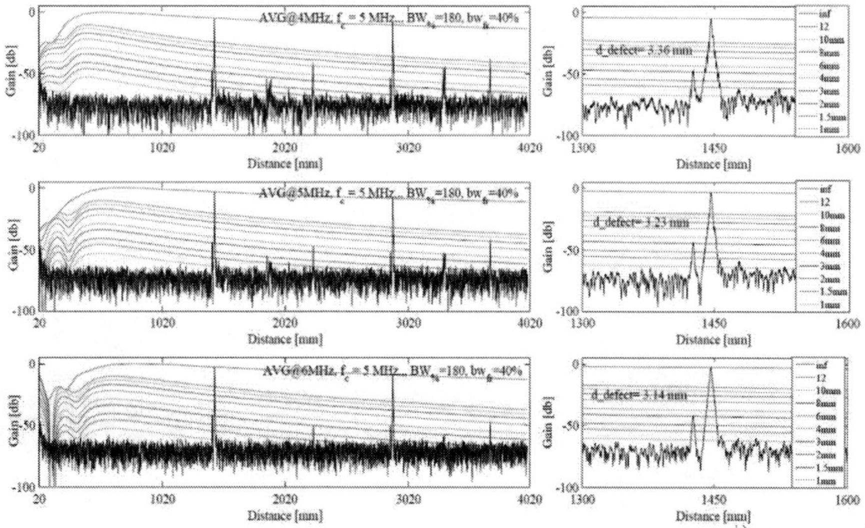

NDTonAIR

THE USE OF PULSE COMPRESSION TECHNIQUE IN NON-DESTRUCTIVE TESTING: A REVIEW

PuC Thermography, *Setup*

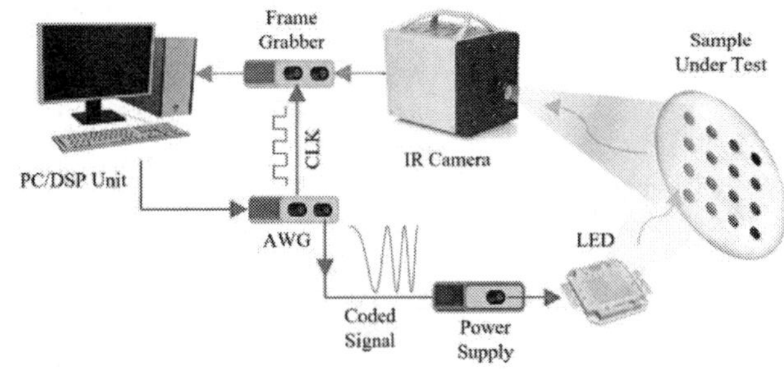

Schematic of the PuC based thermographic Experiment [1]

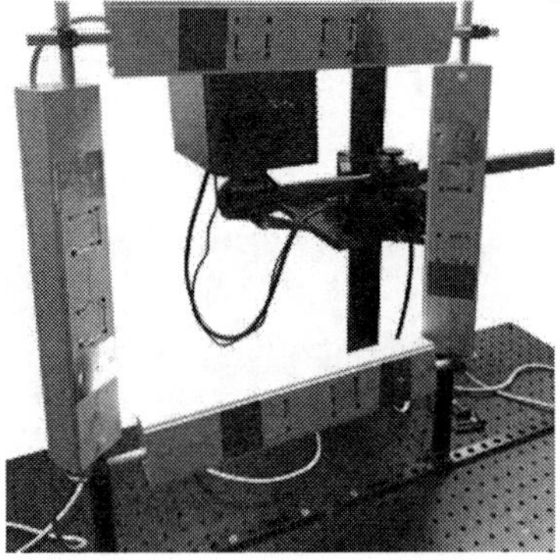

The Setup at our Lab (the NDT Laboratory, Terni) [2]

- Heating Source: 8 x 50W LED chips
- Thermal Camera: Xenics Onca-MWIR-InSb IR
- Signal Generation/Acquisition: Labview™
- Camera in Reflection mode, 40 FPS

[1],[2] Diagram and the setup from presentation of **Hamed Malekmohammadi,**
Comparison of optimization strategies for the improvement of depth detection capability of Pulse-Compression Thermography, **QIRT-2018**

NDTonAIR

PuC Thermography, *Results*

- CFRP sample with defects at different depths
- Impulse Response $h(t)$ is slow, SNR_{Gain} is not very high
- Binary Sequences, Barker Code is used for PuC

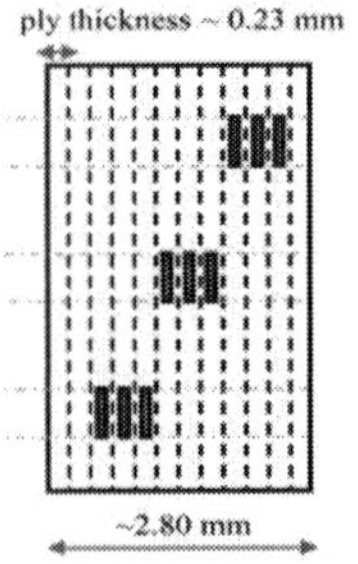

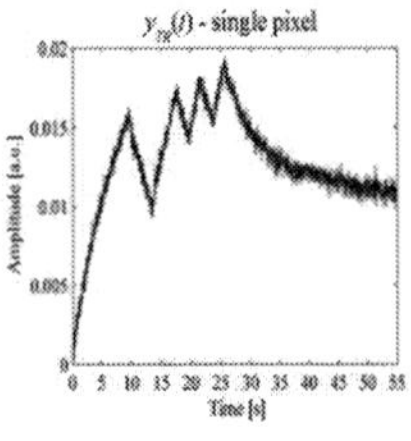

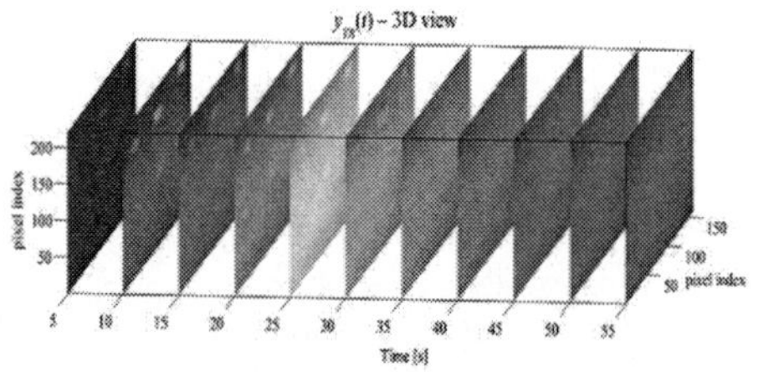

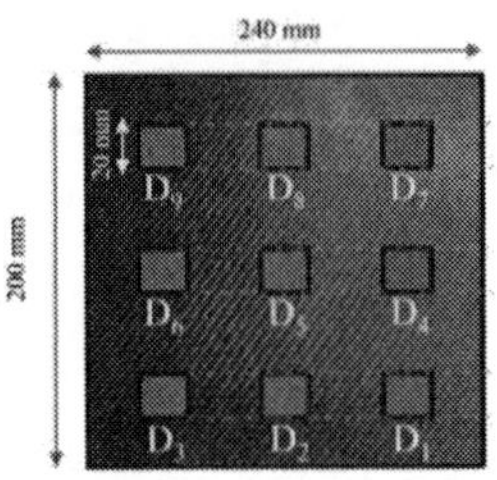

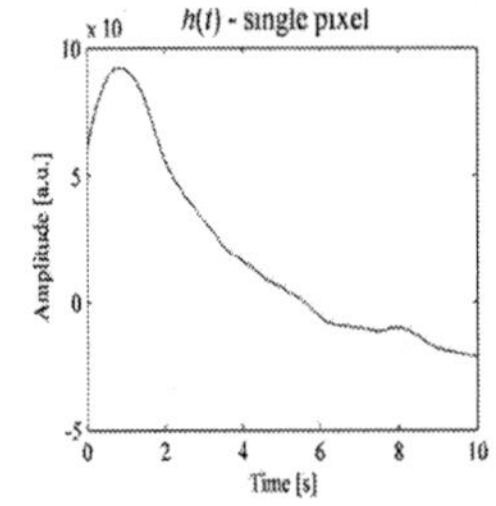

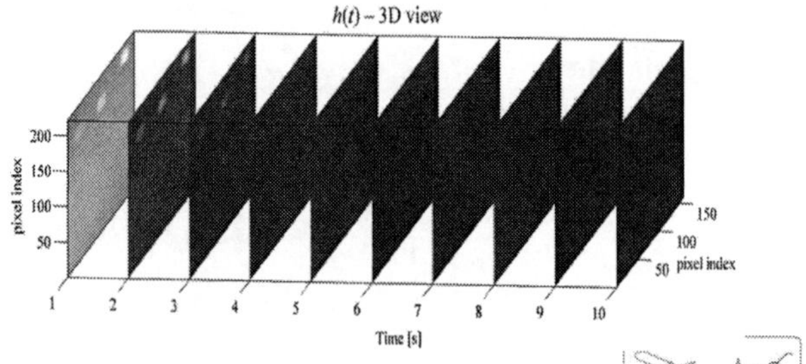

NDTonAIR

PuC Eddy Current Inspections, *Setup*

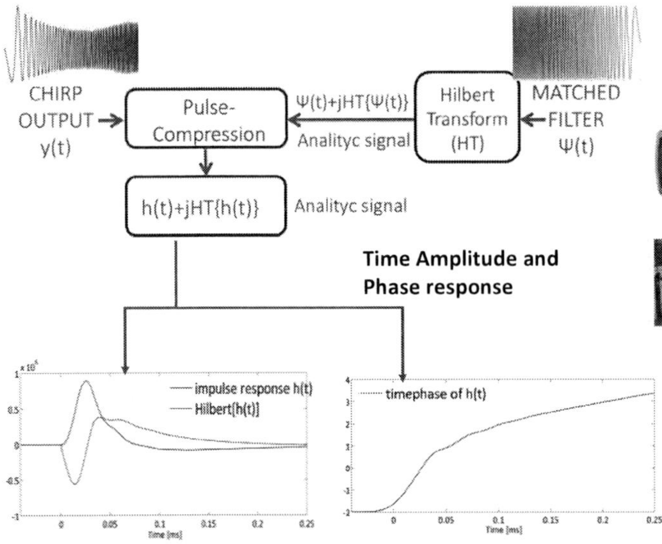

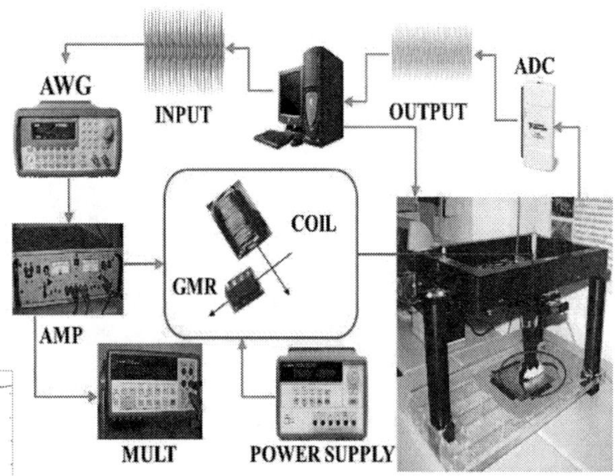

THE USE OF PULSE COMPRESSION TECHNIQUE IN NON-DESTRUCTIVE TESTING: A REVIEW

PuC Eddy Current Inspections, *Results*

- Comparison of Time and Frequency domain analysis for the defect detection, using Linear chirps (varying duration of the signal)
- A surface defect of (0.1x0.4x3 mm^3), covered with another sound plate of thickness = 2 mm
- SNR_{Gain} is better because the time resolution of impulse response is in between the UT and thermographic NDT
- In this case, time domain analysis outperforms the frequency domain analysis in terms of defect detection
- One can obtain more robustness of the system to the lift off variations, while applying time or frequency phase analysis

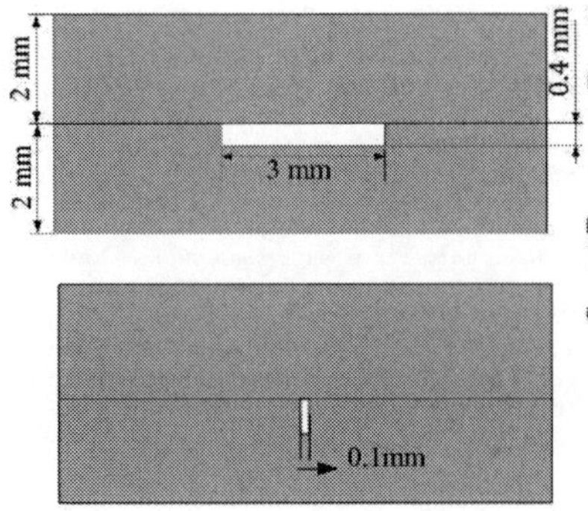

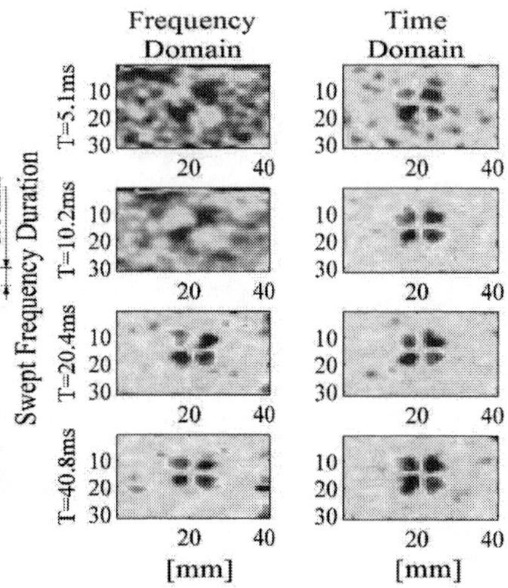

THE USE OF PULSE COMPRESSION TECHNIQUE IN NON-DESTRUCTIVE TESTING: A REVIEW

Conclusions

- *PuC procedures enhances the sensitivity of NDT techniques,*
- *As discussed, the PuC methods,*
  - *Applied in UT inspection of large structures, improved SNR has resulted in better spatial and temporal resolution*
  - *Applied in Eddy currents testing, lift off compensation can be achieved via time and frequency phase analysis*
  - *Applied in thermographic, thermograms with improved SNR and fidelity can be achieved*

- *The PuC method can easily be incorporated to the NDT systems with simple hardware*
- *Flexibility over the time and frequency domain shaping of the signal, spectral density of the excitation pulse can developed well matched with the bandwidth of the system*
- *One can exploit the broadband of the PuC procedure to obtain a multifrequency defect evaluation system, to gain the optimum frequency of inspection for a specified SUT*
- An NDT system with improved SNR gain opens further application of the method for the inspection of highly attenuating materials and structures, like CFRP and GFRP
- Performance of the PuC NDT system can further be improved combing other advanced signal and image processing techniques

THE USE OF PULSE COMPRESSION TECHNIQUE IN NON-DESTRUCTIVE TESTING: A REVIEW

References

1. Misaridis, Thanassis, and Jørgen Arendt Jensen. "Use of modulated excitation signals in medical ultrasound. Part I: Basic concepts and expected benefits." IEEE transactions on ultrasonics, ferroelectrics, and frequency control 52.2 (2005): 177-191.
2. Pollakowski, Martin, and Helmut Ermert. "Chirp signal matching and signal power optimization in pulse-echo mode ultrasonic nondestructive testing." *IEEE transactions on ultrasonics, ferroelectrics, and frequency control* 41.5 (1994): 655-659
3. Turin, George. "An introduction to matched filters." *IRE transactions on Information theory* 6.3 (1960): 311-329
4. Burrascano, Pietro, et al., eds. *Ultrasonic nondestructive evaluation systems: industrial application issues*. Springer, 2014.
5. Hutchins, David, et al. "Coded waveforms for optimized air-coupled ultrasonic nondestructive evaluation." *Ultrasonics* 54.7 (2014): 1745-1759.
6. Callegari, Sergio, et al. "From chirps to random-FM excitations in pulse compression ultrasound systems." *2012 IEEE International Ultrasonics Symposium*. IEEE, 2012.
7. Ricci, Marco, Luca Senni, and Pietro Burrascano. "Exploiting pseudorandom sequences to enhance noise immunity for air-coupled ultrasonic nondestructive testing." IEEE Transactions on Instrumentation and Measurement 61.11 (2012): 2905-2915.
8. Burrascano, Pietro, et al. "Reactance transformation to improve range resolution in pulse-compression detection systems." *Telecommunications and Signal Processing (TSP), 2017 40th International Conference on*. IEEE, 2017.
9. M. Khalid, M.Ricci, L.Senni, "Optimizing Ultrasonic Forgings Inspection by combining Pulse-Compression technique and multi-frequency AVG analysis ", ECNDT, Conference Proceedings, 2018
10. Ricci, Marco, et al. "Exploiting non-linear chirp and sparse deconvolution to enhance the performance of pulse-compression ultrasonic NDT." *Ultrasonics Symposium (IUS), 2012 IEEE International*. IEEE, 2012.
11. Oelze, Michael L. "Bandwidth and resolution enhancement through pulse compression." *IEEE transactions on ultrasonics, ferroelectrics, and frequency control* 54.4 (2007).
12. Silipigni, Giuseppe, et al. "Optimization of the pulse-compression technique applied to the infrared thermography nondestructive evaluation." *NDT & E International* 87 (2017): 100-110.
13. Laureti, S., et al. "Comparative study between linear and non-linear frequency-modulated pulse-compression thermography." *Applied Optics* 57.18 (2018): D32-D39.
14. Laureti, S., et al. "The use of pulse-compression thermography for detecting defects in paintings." *NDT & E International* 98 (2018): 147-154.
15. Betta, Giovanni, et al. "An experimental comparison of multi-frequency and chirp excitations for eddy current testing on thin defects." *Measurement* 63 (2015): 207-220.
16. Betta, Giovanni, et al. "On the use of complex excitation sequences for eddy current testing." *Proceeding of 19th IMEKO TC4 Symposium, Barcelona, Spain*. 2013.

THE USE OF PULSE COMPRESSION TECHNIQUE IN NON-DESTRUCTIVE TESTING: A REVIEW

Acknowledgements

This research work has been supported from the European Union's Horizon 2020 research and innovation programm under the Marie Skłodowska-Curie grant agreement No 722134 – NDTonAIR

www.ndtonair.eu

Non-uniform spatial sampling patterns for local wavenumber-based damage evaluation

Jakub SPYTEK [1], Przemyslaw RUSIN [1], Kajetan DZIEDZIECH [1], Lukasz PIECZONKA [1], Lukasz AMBROZINSKI [1]

[1] AGH University of Science and Technology, Krakow, Poland

Contact e-mail: jspytek@agh.edu.pl

Abstract. A crucial factor for non-destructive inspections is time required to test the structural components. Local-wavenumber estimation of guided ultrasonic waves is a promising technique that permits to balance the required spatial resolution with the inspection time. In this method a single source generates guided waves propagating in a plate. Using laser interferometry the vibration responses are collected in a number of points. Next, the data are processed in the frequency-wavenumber domain to obtain spatial distribution map of the wavenumbers. Change of the wavenumber is typically due to a local change in thickness of the sample. Therefore, this method can be effectively applied for detection of corrosion patches, delamination, disbonds.

Usually, the measurement points are spaced uniformly at a distance smaller than half of the wavelength to satisfy the Nyquist theorem for spatial sampling. However, as we show in the paper, different distributions of sensing points are also possible. We discuss the imaging performance of various spatial sampling schemes using phased arrays of limited aperture and extend this concept to laser-probing over the complete imaged surface.

1. Introduction

Local-wavenumber estimation (LWE) of guided ultrasonic waves is a technique in which a single transmitter is used to generate Lamb waves in an inspected plate. The responses are acquired using laser interferometer in a set of points spaced uniformly over the plate at a distance smaller than half of the wavelength. To observe frequency-wavenumber variations which may result from defects, e.g. corrosion patches or delaminations, the acquired data are processed using 3D Fourier transform [1], [2]. The implementation of the algorithm can benefit from equidistant location of the detectors since fast algorithms for Fourier transform calculation can be used. On the other hand, different distributions of sensing points are also possible. This problem has been extensively explored in the field of synthetic focusing of 2D arrays used to image the plate from a fixed location. A number of sensors distributions were considered [3], including circular [4] , star [5] and spiral [6], honeycomb and Fermat spiral shaped [7] arrays. Using arrays consisting of non-uniformly distributed sensors permits achieving high resolution with a reduced number of elements. Since estimation of direction of wave arrival in array processing can be understood as 2D wavenumber estimation [3], [7], the concept can be extended to LWE applied to a sensors grid distributed over the investigated plate.

2. Theoretical Background

2.1 Local Wavenumber Estimation (LWE)

The LWE is an algorithm from the family of signal processing methods which is referred to as the wavenumber spectroscopy. It takes advantage of the dispersive properties of the Lamb Waves. For a given material, the wavelength (or the wavenumber) is dependent on the frequency of the wave and the thickness of the plate. Therefore, for a monoharmonic signal it is possible to distinguish the changes of thickness of the plate by determining the local change of the wavenumber. This is attained by performing the filtration in the wavenumber – frequency representation of the wavefield measured from the surface of the examined plate. The first step is to use the 3D Fourier transform on the measured data set $V(x, y, t)$ to obtain the representation in the spectral domain:

$$V_f(k_x, k_y, f) = \iiint \overline{V}(x, y, t)e^{-j2\pi(ft+k_x x+k_y y)}dxdydt \qquad (1)$$

where k_x, k_y are the orthogonal components of the wavenumber vector and f is the frequency variable. This representation contains all the information on the distribution of the values of the local wavenumber but this information needs to be decomposed. One of the possible methods to realize this is to perform the filtration of the single wave mode and then to apply the wavenumber filter-bank, thus obtaining a set of separated values of the wavenumber. After transforming the obtained datasets back into spatial domain and choosing the local maxima a map presenting the spatial distribution of the wavenumber is obtained (Figure 1b)). For a more detailed explanation on the process of the LWE the reader is advised to look into [1].

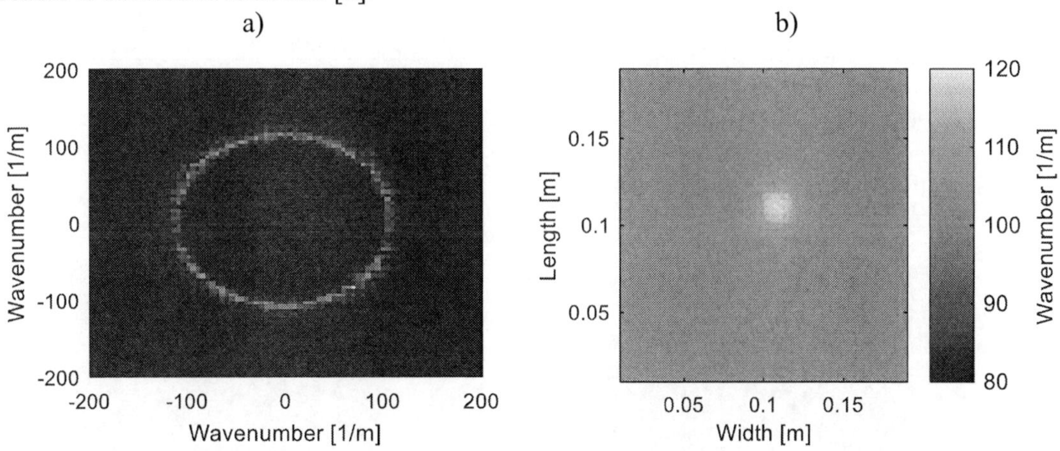

Fig. 1. (a) Wave mode for selected central frequency. (b) Local wavenumber distribution map obtained based on the filtered mode. The color of the image shows the local value of wavenumber, and the local change indicates the change of thickness

2.2 Phased Array Topologies

The subject of array geometries is a basic notion in the theory of array signal processing. The way, in which the transducers in the array are aligned influences the capability for signal sensing and beamforming in a profound manner [8]. Each array is described by an intrinsic function which is referred to as the array pattern. It can be calculated using the following expression:

$$A(\vec{k}) = \frac{1}{M}\sum_{m=0}^{M-1} w(\vec{s_m})e^{j\vec{s_m}\vec{k}} \qquad (2)$$

where $\vec{k} = [k_x, k_y]$ is the wavenumber vector, $\overrightarrow{s_m} = [x_m, y_m]$ is the position of the m-th receiver and M is the number of elements in the array. The array pattern can be understood as analogous to the 2D representation of the 1D temporal window function in the classical signal processing. This function can be used to study the directional response of the aperture to the propagating signal. By assuming the input wavefield in the form of the Dirac delta $\delta(\vec{k} - \vec{k}_0)$ defined in two dimensional space, the steered response can be expressed as:

$$A(\vec{k}) * \delta(\vec{k} - \vec{k}_0) = \frac{1}{M}\sum_{m=0}^{M-1} w(\overrightarrow{s_m}) e^{j\overrightarrow{s_m}(\vec{k}-\vec{k}_0)} \quad (3)$$

where $\vec{k}_0$ defines the wavenumber of the wavefield. With this representation the capabilities of several array geometries are presented in this section. The grating lobes of rectangular array (Figure 2) are aligned in the regular pattern with spacing equal to $\frac{2\pi}{d_x}$, where d_x is a spatial sampling interval. In the direction of the main axes, this satisfies the Nyquist sampling criterion, which states that the sampling rate of the signal must be at least equal the double of the highest frequency in the investigated signal. However, at the angle 45° from the main axes the higher possible wavenumber can be resolved without the creation of the grating lobes. This is visualised by drawing a white circle of radius $\frac{2\pi}{d_x}$, which corresponds to the maximum usable range according to the Nyquist criterion, when the k_x component of the $\overrightarrow{k_0}$ term in the equation 2 is equal $\frac{\pi}{d_x}$. It is clearly visible that in the observable region grating lobes may appear, as both the original signal and its alias are included inside the circle. This means, that wave of wavelength $\frac{d_x}{2}$ approaching the array at normal angle may create aliases. However, at the angle of 45° the same wave do not produce aliases. This property can be valuable from the point of view of LWE because different geometries may require a less total amount of measurement points to represent the wave without aliasing. Based on the literature two additional topologies are taken into consideration: hexagonal grid and Fermat's matrix [7].

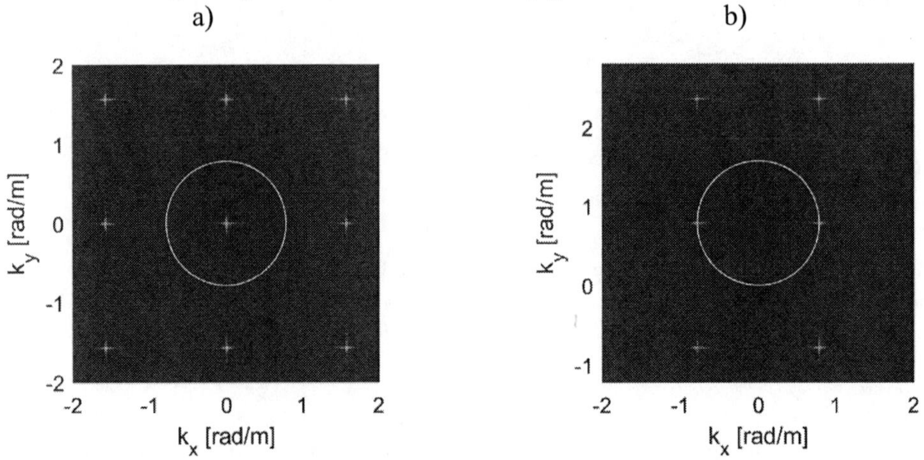

Fig. 2. Array pattern of the rectangular grid. White circle denotes the maximum usable range of wavenumber. (b) shows that at k_x equal $\frac{\pi}{d_x}$ the grating lobes may appear

Hexagonal array pattern is presented in the Figure 3 in a similar manner to the rectangular pattern. In a given example, the width of the hexagon's side is equal to the sampling length of the rectangular array and so is the white circle representing the Nyqusit criterion region. The aliases of equal amplitude are ordered in the hexagonal pattern and so are the side lobes, which are equal to roughly 50% amplitude of the main lobe. It is clearly visible that in the same window as in the rectangular array a main lobe and its alias cannot appear together. This means that the effective observable region increases according to the ratio of $\frac{2}{\sqrt{3}}$, which is around 15% [7]. This is indicated by the red circle in the Fig 3 b). Moreover, with hexagonal mesh of this spacing the 20% less points is needed to cover the area of the same size. The potential issue for this type of window is the large amplitude of the side lobes.

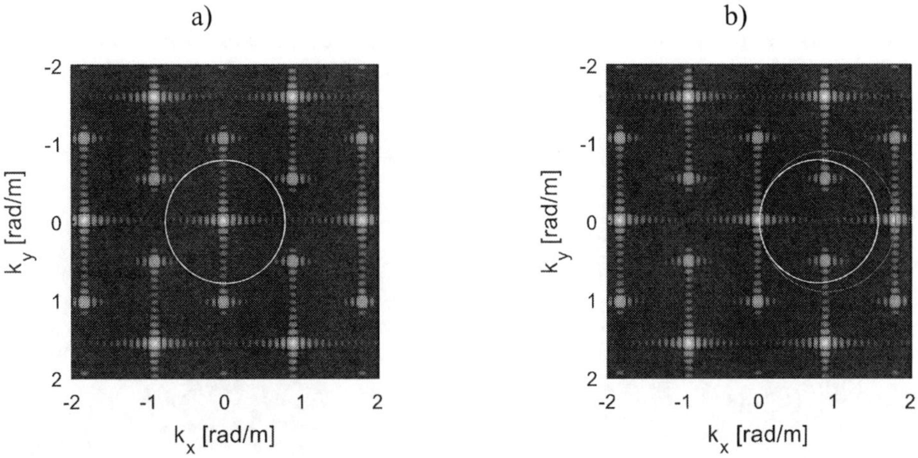

Fig 3. Array pattern of the hexagonal grid. White circle denotes the maximum usable region for the rectangular grid and the red circle for the hexagonal grid

The third array pattern considered in this work is the grid based on Fermat's matrix. This pattern, which resembles sunflower, possess several interesting properties. It is a circular pattern where each of the points possess unique angular and radial value. This property results from the following expressions:

$$\rho_n = s\sqrt{\frac{n}{\pi}} \tag{4}$$

$$\phi_n = 2\pi n\beta \tag{5}$$

where ρ_n is a distance between n-th point and the centre, s is a parameter related to the mean distance between neighboring elements, ϕ_n is an angular position of each point and β is a density-based parameter of the mesh. The exemplary mesh is presented in the Fig. 4 a). The resulting property of Fermat matrix is that it does not exhibit grating lobes, so there is no clear limit on the minimum point spacing to represent the wave of particular wavelength. On the other hand, the resulting side lobes create a noisy pattern which depends mainly on the matrix density. Due to those properties this is a very attractive choice of mesh pattern.

2.3 Non-uniform Fourier Transform

In order to use the non-rectangular array patterns in the LWE algorithm it is necessary to utilize the non-uniform representation of the Fourier Transform in the spatial dimensions. There are several cases of the transformation, depending on whether the space-time domain or the frequency domain is uniformly sampled. In the case of the LWE it is the most appropriate to assume non-uniform distribution of space-time domain and perform Fourier transform at the uniformly spaced wavenumbers or frequencies. By this assumption it is then possible to perform all the steps of LWE algorithm (mode filtration, wavenumber decomposition and choosing the maxima) as normal on the equispaced rectangular mesh. In the 2D case this non-uniform Fourier transform is given by the following expression:

$$F(k_x, k_y) = \sum_{n=1}^{N} f(n) e^{-j(k_x x_n + k_y y_n)} \tag{6}$$

where $f(n)$ is the value of the function at n-th coordinate point, x_n and y_n are the coordinates of the point, k_x and k_y are the equispaced vectors of the wavenumbers. This expression followed by the standard Fourier transform along the temporal axis (as the time measurements are assumed to be regularly sampled) gives the full 3D Fourier representation of the measurement data from the irregularly sampled grid of points.

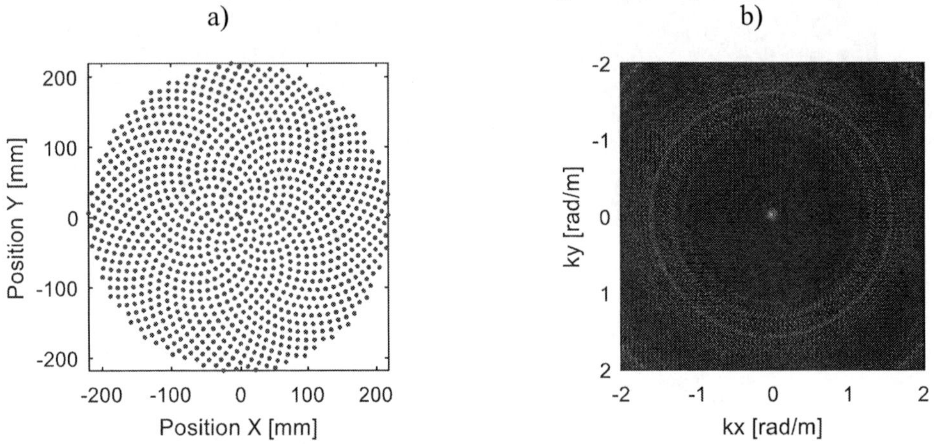

Fig. 4. An example of Fermat array (a) and its array pattern (b)

3. Experimental Setup

In order to test the effectiveness of the LWE algorithm with different array topologies an experimental setup was assembled. The test specimen is the 2 mm thick 500x500 mm plate with added damage in the form of two 20x20 mm milled surfaces. The first damage is 0.5 mm deep and the second is 1 mm deep. The sample was excited with a wavelet packet using a single piezoelectric transducer attached to the surface of the plate. The mutual positions of the transducer, damage and the general placement of the measurement grid is presented in the **Fig. 5**. The rectangular mesh is considered to be a baseline for further evaluation of the properties of non-standard hexagonal and Fermat arrays. Therefore the grid sampling and the frequency of excitation were selected in such a way, so that the spatial aliasing was visible on the spectrum of the signal. Based on the dispersion curves of the aluminum plate this condition is met if the central frequency equals 230 kHz and spatial sampling length equals 4 mm. Therefore, the 4 mm rectangular grid of 200x200 mm area and 2601 points is the frame of reference for other configurations:

- 3 mm rectangular grid, which is sufficiently sampled to obtain a good visualization of the damage, area 200x200 mm, 4489 points total
- regular hexagonal grid with 3 mm side length, encompassed by the area 199 by 203 mm, 3555 points total
- regular hexagonal grid with 4 mm side length, encompassed by the area ,200 by 201 mm, 2006 points total
- Fermat array with the radius of 100 mm and 3000 points total

As a measuring tool the Polytec PSV-400 Laser Doppler Vibrometer was used. It was synchronized with the Agilent 3351 arbitrary waveform generator using laboratory PC running Matlab script.

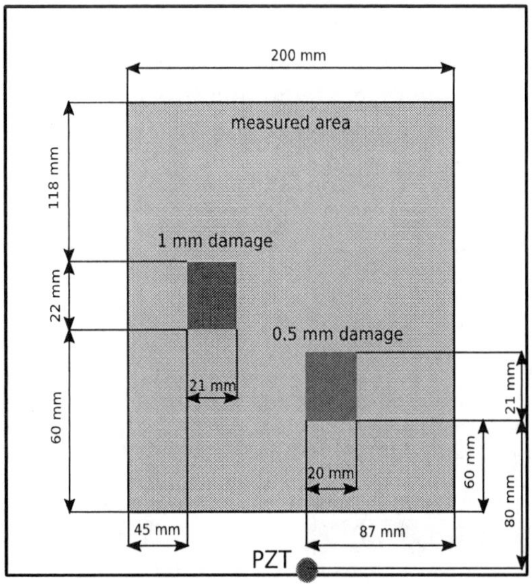

Fig. 5. The scheme representing the locations of PZT transducer, 1 mm and 0.5 mm damage with respect to the measurement area

4. Evaluation of the Array Topologies

As expected, for the well-resolved case of 3 mm rectangular array the wavenumber image presents explicit damage evaluation. The dominant color representing 122 1/m wavenumber corresponds to the intact plate, while 140 1/m and 164 1/m correspond to 1.5 and 1 mm thick plates accordingly (Fig. 6). This proves that LWE is a viable method for damage identification in the given specimen. On the other hand, as expected, the evaluation of LWE for 4mm grid for the 230 kHz frequency is almost illegible, as it lies just on the intersection of two modes (Figure 7). The 0.5 mm damage is barely visible and the 1 mm damage is jagged and below the expected value. Based on the spectral analysis, the inferior quality results from the fact that part of the data is obscured by the aliases of main lobe.

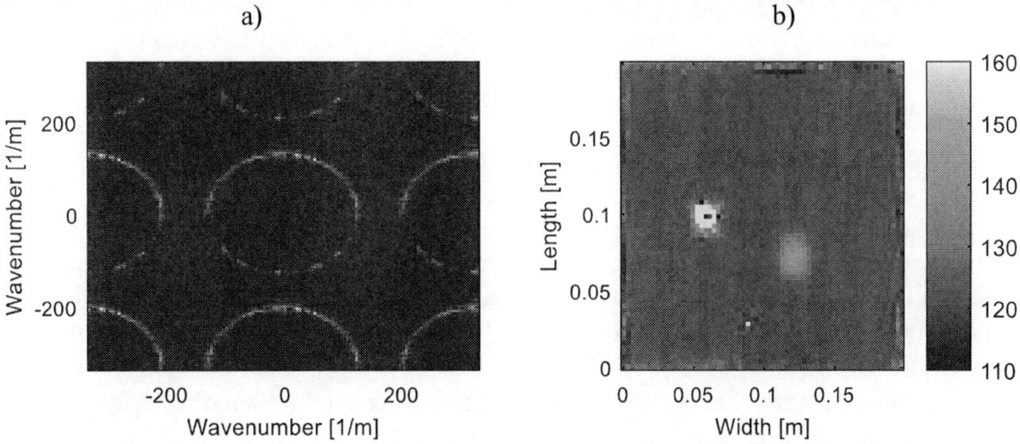

Fig. 6. Spectrum of the 3 mm rectangular grid (a) and its corresponding LWE result (b)

Taking a look at the results of the LWE based on 3 mm hexagonal grid it is noticeable, that the main lobe does not overlap with its aliases, accordingly with the predictions. However, the high-amplitude side lobes which are visible on the Fig 3 produce copies of the wavenumber spectrum close enough, that they intersect with the main lobe. The intersection points are visible on Fig. 8 a) and they remain relevant even after mode filtration. However, the two damage locations can be discerned from the intact background on the image. Both 0.5 mm and 1 mm damage locations are similarly represented as in the case of 3 mm rectangular array but the overall image is less crisp due to the influence of the intersecting side lobes. However, the reduction of the matrix size by over 21% is a noticeable improvement in the performance.

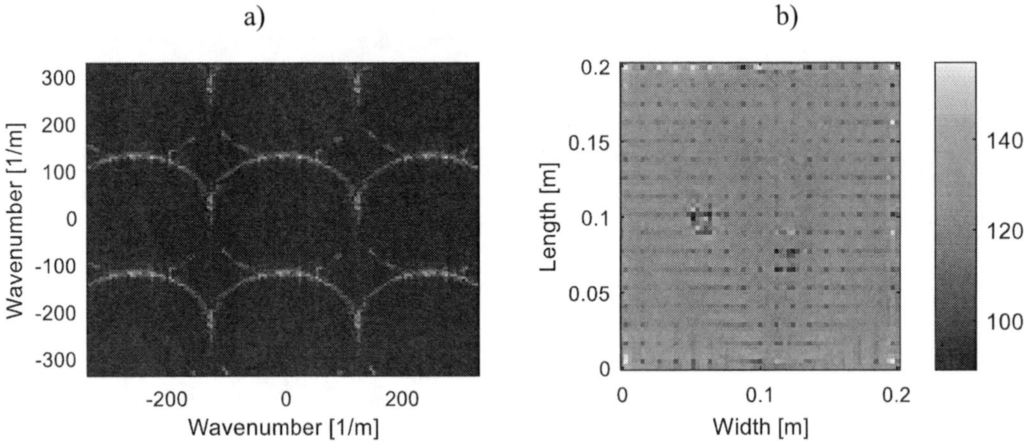

Fig. 7. Spectrum of the 4 mm rectangular grid (a) and its corresponding LWE result (b)

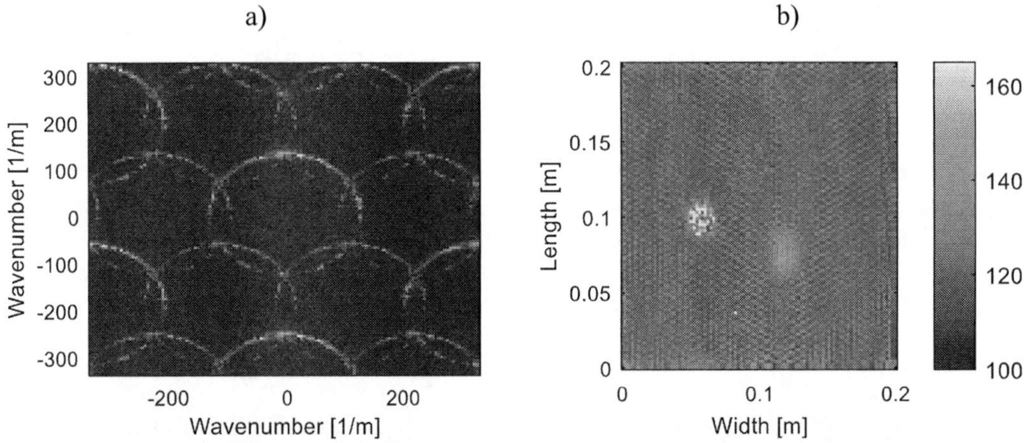

Fig. 8. Spectrum of the 3 mm hexagonal grid (a) and its corresponding LWE result (b)

The capabilities of the hexagonal grid can be further examined by taking a look at the more sparse 4 mm hexagonal grid. Due to the fact, that the grating lobes are spread further apart and useful wavenumber range should increase compared to similarly spaced rectangular grid, the problem of aliases overlapping should be reduced. The results are presented in the Fig.9. The LWE shows the localization of the damage properly and the 0.5 mm damage is clearly resolved in terms of the size and wavenumber value. However, the 1 mm damage is scattered and badly represented. This results from the influence of side lobes as well as from the fact, that the $\frac{2}{\sqrt{3}}$ gain in effective wavenumber range resulting from the grid topology is insufficient to represent the 164 1/m corresponding to the deeper damage (if the maximum usable range in terms of rectangular array equals 122 1/m, then the extended range equals 140 1/m). Overall, due to the increased influence of intersecting side lobes the 4 mm hexagonal grid presents worse quality than the 3 mm grid. However, it resolves the damage localization better than in the case of comparable rectangular 4 mm array. The overall reduction in the number of points is also considerable (from 2601 to 2006 points).

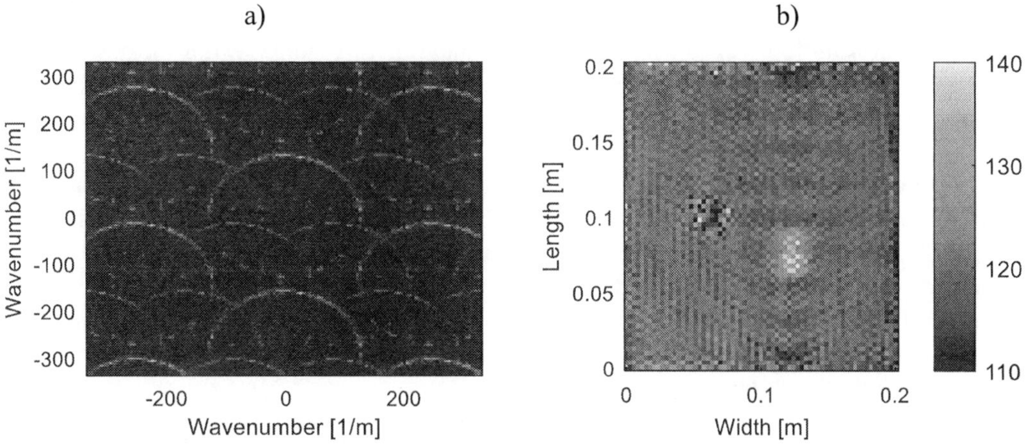

Fig.9. Spectrum of the 4 mm hexagonal grid (a) and its corresponding LWE result (b)

When it comes to the Fermat matrix, the spectra of the results are in good agreement with the theoretical predictions and there are no aliases in the spectrum. However, the noisy side lobes are clearly visible and greatly distort the mode shape (Fig. **10**). As a result, the damage representation on the obtained image shows inferior quality to the results from the 3 mm rectangular mesh. The locations of the damage are visible, the distribution of the wavenumber values is similar, however the damage boundaries are distorted and unclear. Due to the lack of aliases and intersecting modes the noise in the LWE image can be attributed solely to the effect of the side lobes.

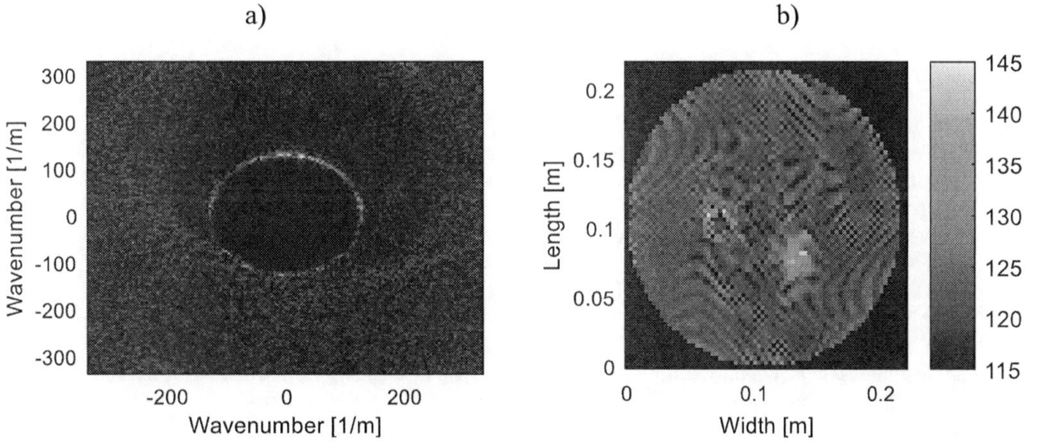

Fig. 10. Spectrum of the 3000 point Fermat array (a) and its corresponding LWE result (b). The wavenumber map is limited only to the circular area corresponding to the actual area of the grid.

5. Conclusions

The effects of choosing different array topologies on the results of the LWE algorithm were studied in this work. Rectangular, hexagonal and Fermat topologies possess different characteristics in terms of the array patterns, but come with their own limitations. While the rectangular mesh requires more points to avoid aliasing it ensures clear and well-resolved images when the Nyquist conditions is met with some necessary surplus to include the higher wavenumbers related to the damage locations. The hexagonal mesh provides possibility to observe higher maximum wavenumber values and allows for considerable reduction in the number of measurement points. On the other hand, it exhibits the influence of side lobes which result in mode intersection and make the wavenumber map less clear. Finally, the Fermat array takes advantage from the complete lack of grating lobes, but the effect of noisy side lobes is detrimental to the image quality, making the locations of the damage fuzzy and the wavenumber values non-uniform. Additionally, it is difficult to assess what amount of measurement points is necessary to image particular size of damage.

Although the advantages of the hexagonal and Fermat matrices over rectangular ones are evident, their shortcomings are critical to the damage detection quality and need to be addressed in the further work. Reduction of the side lobes is the common challenge in both of the cases and can perhaps be resolved using windowing, tapering or thinning techniques, which are widely used in the telecommunication industry. The improvement in the spectral quality of the non-rectangular arrays is potentially beneficial for the practical application of LWE algorithm, as the duration of the examination time is a crucial factor for NDT of the large structures.

Acknowledgement

J.S., P.R. and L.A. would like to acknowledge Foundation for Polish Science for project "Nondestructive evaluation of inhomogeneous layered materials using synthetic and self-focusing techniques" carried out within the Homing programme, co-financed by the European Union under the European Regional Development Fund.
K. D. and L.P. would like to acknowledge National Centre for Research and Development for financing within the scope of the project no. LIDER/2/0117/L-7/15/NCBR/2016.

References

[1] E. B. Flynn, S. Y. Chong, G. J. Jarmer, and J.-R. Lee, "Structural imaging through local wavenumber estimation of guided waves," *NDT E Int.*, vol. 59, pp. 1–10, Oct. 2013.

[2] L. Yu, Z. Tian, and C. A. C. Leckey, "Crack imaging and quantification in aluminum plates with guided wave wavenumber analysis methods," *Ultrasonics*, vol. 62, pp. 203–212, Sep. 2015.

[3] Ł. Ambroziński, T. Stepinski, and T. Uhl, "Efficient tool for designing 2D phased arrays in lamb waves imaging of isotropic structures," *J. Intell. Mater. Syst. Struct.*, vol. 26, no. 17, pp. 2283–2294, 2015.

[4] P. D. Wilcox, "Omni-directional guided wave transducer arrays for the rapid inspection of large areas of plate structures," *IEEE Trans. Ultrason. Ferroelectr. Freq. Control*, vol. 50, no. 6, pp. 699–709, Jun. 2003.

[5] P. Malinowski, T. Wandowski, I. Trendafilova, and W. Ostachowicz, "A Phased Array-based Method for Damage Detection and Localization in Thin Plates," *Struct. Heal. Monit. An Int. J.*, vol. 8, no. 1, pp. 5–15, Jan. 2009.

[6] B. Yoo, A. S. Purekar, Y. Zhang, and D. J. Pines, "Piezoelectric-paint-based two-dimensional phased sensor arrays for structural health monitoring of thin panels," *Smart Mater. Struct.*, vol. 19, no. 7, p. 075017, Jul. 2010.

[7] P. Rusin, "Synthetic focusing of Lamb waves using two dimensional arrays," AGH University of Science and Technology, 2018.

[8] D. H. Johnson and D. E. Dudgeon, *Array Signal Processing: Concepts and Techniques*. New York, NY: Simon & Schuster, Inc, 1992.

Monitoring of structures and systems of aircraft by highly non-linear sensing devices

Helge PFEIFFER [1], Sevilia SUNECHIIEVA [2], Hans SEKLER [3], Daniel BACKE [4], Martine WEVERS [2]
[1] Katholieke Universiteit Leuven, Heverlee, Belgium
[2] Katholieke Universiteit Leuven Dept. of Metallurgy and Materials Engineering, Heverlee, Belgium
[3] Lufthansa Technik AG FRA T/FG 853, Frankfurt am Main, Germany
[4] PFW Aerospace GmbH, Speyer, Germany
Contact e-mail: helge.pfeiffer@mtm.kuleuven.be

Abstract. Sensing systems, especially when used for interrogating the structural integrity of aircraft in traditional inspections, are frequently working in a quasi-linear mode. This approach however creates difficulties in some cases, especially when applications for structural health monitoring (SHM) are concerned. One of the major problems are base-line deviations interfering with damage-related signals that are in some cases moreover complicated due to interferences at complex aircraft structures creating major obstacles for a broad-scale implementation of SHM in aircraft. Advanced data processing and dedicated high-end hardware components are certainly capable of tackling part of these problems, but additional hardware requirements require extra power supply, together with advanced data transmission and processing facilities, also additional maintenance needs emerge.

In some cases, an interesting alternative is offered by highly non-linear sensing devices. They produce a sharp sensor response that ideally only depends on a certain damage-related outer parameter representing in this way a material-based threshold. Moreover, this sharp sensor response ideally ranges over many orders of magnitude above the baseline variations and in this manner, an ideal tool would be established to finally enormously increase the probability of detection, at least within the range of damage to detect.

After a short review on examples taken from the literature, that are in some cases successfully applied in operations, new developments are presented, such as water leakage sensors based on the percolation effect, dedicated optical sensors for reliably detecting damage in hydraulic pipes as well as defects in bleed air ducts. Most of them are able to cover bigger surfaces and are, when appropriate, also partially equipped with dedicated facilities for reliable damage localisation. Finally, a couple of examples are given that were already implemented in operational airliners, such as devices for the detection of corrosive liquids in aircraft (Boeing 737-500, Boeing 747-400).

1 Introduction

Highly non-linear sensing devices (HNSD) are here defined as sensing systems that show a clear sigmoidal, thus S-shaped response curve that is tailored to a pre-set damage range, or even only at a certain damage size, but with a large change in magnitude in contrast to diverse baseline variations [1]. In this way, they are in a certain sense an analogue to e.g. electrical fuses only providing information on a certain threshold-specific overcurrent without providing specifications on position or the actual amount of overcurrent.

An apparent limitation is the limited range of damage that is measured, but this drawback can be highly compensated by the partially high signal-to-noise ratio (SNR) that can even approach infinity providing new applications and options.

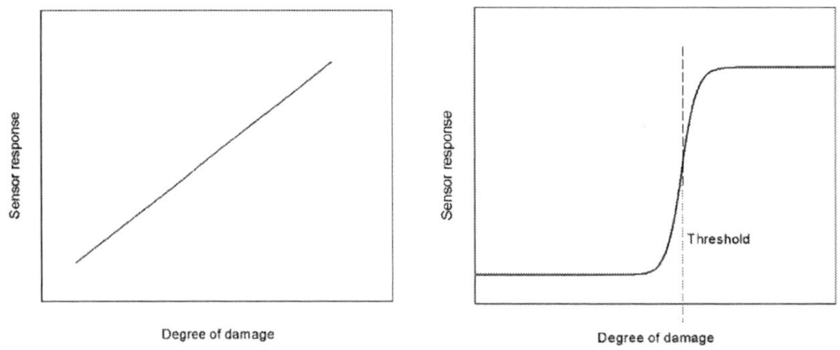

Fig. 1: Principal response of a quasi-linear and a highly non-linear sensing material [1].

Appropriate physical phenomena need to be selected show such S-shaped in essence, discontinuous behaviour as a function of the intended damage-related parameter to obtain a response curve such in Fig. 1. Those useful "large-effect" phenomena can be categorised into: i) mechanically-driven fracture, ii) phase transitions, iii) absorption and iv) chemical degradation. Related to that, the variation of the sensor data, i.e. the sensor response can ideally proceed over a big range of magnitude and therefore, the y-axis in the above figure, right can in some cases even be understood as a logarithmic scale.

A well-known, already commercialised example are crack propagation gauges that provide stepwise changes of an electrical signal as a function of a progressing fracture and their use for structural health monitoring in aircraft was also already proposed and reported.

Comparative vacuum monitoring (CVM) gauges [2] are a further typical example. Here, a propagating crack causes a loss of vacuum in fine capillaries that is monitored by appropriate flow gauges. Very recently, a similar approach was proposed when using pressure-medium-filled capillaries inside materials embedded via additive manufacturing [3]. In this case, not a flow-change, but a pressure drop will indicate the presence of a crack.

Highly interesting options are provided by phase transitions, such as the melting processes, solid state modifications or glass transitions. Already since many years the so-called Fenwal elements [4] are used in commercial aircraft applied for bleed air systems. These are "alarm wires" containing a salt mixture that melts at a certain target temperature range, establishing in this way an electrical conductivity leading to an intended shortcut between electrodes. Another example are appropriate solid state modifications in graphite material that is applied in PTC (positive temperature coefficient) [5] elements in order to protect electric circuits against overheat. A more recent application also implemented in aircraft refers to water absorption in hydrophilic polymers that trigger glass transitions to detect water leakage [6]. A similar concept is applied with for leakage detection of kerosene, in this case oil-absorbing elastomers are used to disrupt electrical percolation conductivity [6].

Chemical degradation offers further opportunities for monitoring, for instance when filaments or metallic wires are applied to detect corrosive conditions in engineering structures [7]. Also the detection of ester-type hydraulic liquids frequently applied in aviation is possible via the solvation of electrically conductive composites made of acrylics [6].

2 Probability of Detection (POD) of HNSD

It is a question of definition whether a distinct material feature is considered as a damage or just a particular variation of a material property, and this is also depending on the specific use of that material and its specifications. Nowadays, the performance of a certain NDT technology to find defects should ideally be characterized by the "probability of detection (POD)" where statistical functions modelling for given damage sizes the statistical probability to find this particular defect. To determine POD graphs, sufficient measurements of pre-defined damages need to be done and under ideal circumstances, related series of tests are performed on various test samples, by different operators that use different devices with the same specifications. Also the "background noise" needs to be taken into account when determining POD plots as they are part of the "false positives". In most of the cases, parallel to the POD value, an appropriate confidence interval is provided.

Two major approaches exist. In the case of the established hit/miss analysis, for every defect, tests are performed and a statistics is established for all "existing defects detected" versus "existing defect not detected". The other method is using real sensor data, called â (a-hat) that is analysed versus the real damage sizes, which is annotated by "a". Using the slope and the scattering of these data points, a non-binary POD curve is obtained using generalized linear regression methods (GLM) [8] that usually includes corresponding confidence intervals.

The POD analysis is well-developed for traditional non-destructive testing but still in an earlier phase for structural health monitoring [9]. Moreover, for obvious reasons, the highly non-linear sensing devices require a different approach for the "Probability of Detection" analysis as the most basic assumption for traditional NDT analysis is not fulfilled, which is the quasi-linearity of sensor response (in most cases log-linear sensor behaviour). Some thoughts about alternatives are presented in Pfeiffer et al. [10]. In essence, the probability of detection of a highly non-linear sensing device should be a normalised delta-function at the targeted damage size, thus it is approximately POD = 1 at the pre-set threshold and POD = 0 outside that range.

3 Crack detection via crack gauges

A perfect example for a HNSD are crack propagation gauges that are already widely commercially available (e.g. those provided by Vishay). Usually, they are adhesively attached to the potential crack area which might e.g. be situated around the crack initiation point during fatigue testing to follow the crack propagation [11]. They are made of a series of metallic wires leading to stepwise increase of the overall electrical resistance during successive wire breakage. The POD for crack propagation gauges is in this logic a series of small ranges with respective POD=1.

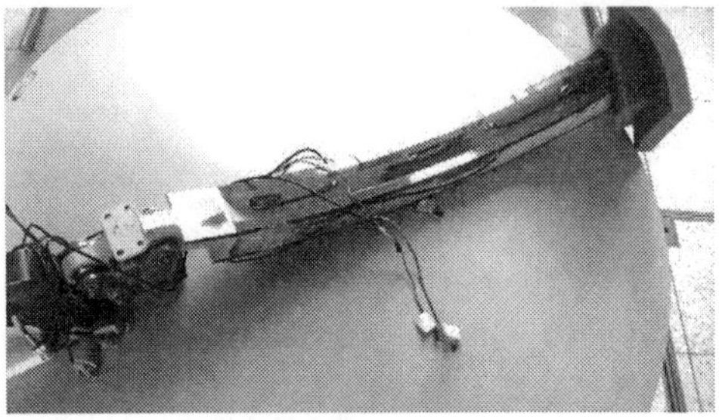

Fig. 2. Slat Track from Airbus A 320 with installed crack propagation of Vishay, partially overpainted.

A typical curve showing voltage development at the recording device obtained from crack gauges mounted on slat tracks of an Airbus A 320, Fig. 2 and Fig. 3.

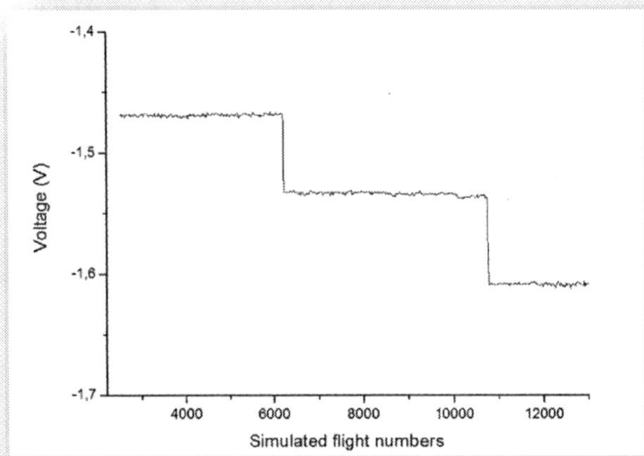

Fig. 3. Readings from crack propagation gauges, the distances between the "steps" represent 0,25 mm.

Between the periodic "steps", crack gauges are more or less insensitive to crack propagation, thus, one can state that those systems establish a kind of "digitization" of material testing as only at the pre-set crack points, we have POD=1, in the other ranges it is POD=0.

4 Moisture monitoring for leakage detection

The leakage of water after structural damage/malfunction of sealings is reliably detectable by a change of the electrical conductivity of appropriate hygroscopic sensing materials. Also in this case, the sensing material tailored as HNSD is in a large humidity ranges quite insensitive to the liquid to be detected, but at the so-called percolation point the conductivity changes over many orders of magnitude [12].

In the example shortly presented here Fig. 4, the hygroscopic matrix material is made of polyvinyl alcohol (PVA) and the targeted solvent is water. The electrically conductive dispersed material is titanium carbo-nitride (TiCN). The percolation threshold related to humidity is approximately in the range of RH=80% and only at this point, a reliable detection of liquid is possible, but that signal is many orders of magnitude above baseline variations making this measuring principle very useful for practical applications.

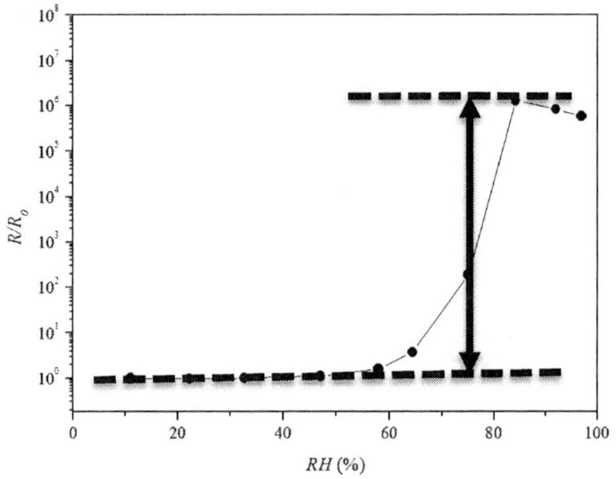

Fig. 4. Huge jump of electrical resistance of a water sensor at the percolation point RH=80% [12], note the logarithmic scale.

5 Monitoring of hydraulic liquids (Skydrol®) using conductive composites

Leakage of hydraulic liquids in aircraft does not only cause problems with potential loss of hydraulic performance, the leakage as such is harmful to different kinds of materials, including all acrylics, such as perplex glass [6]. Moreover, it is irritant for human beings creating in this way a hazard for mechanics and even passengers.

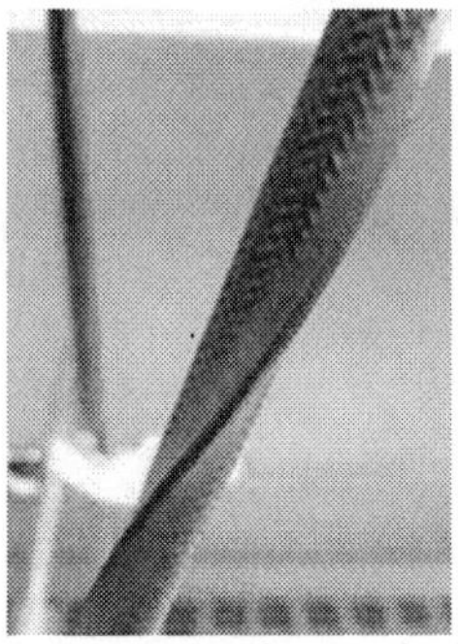

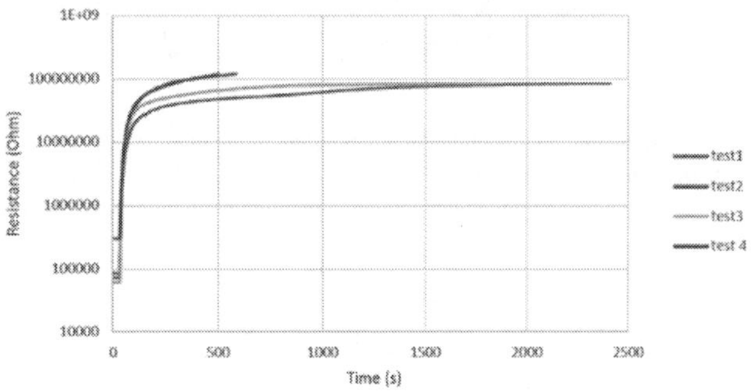

Fig. 5: Sensor wrapped around a hydraulic tube and sensor response recorded when exposed to tiny amounts of hydraulic liquids.

To detect early loss of hydraulic liquids in the corresponding tubes, a shroud system has been developed containing a wire sensor with a sensitive coating that is only detecting those kinds of phosphate-ester-type liquids. It can easily be installed and not only leakage due to cracks, but also caused by lose fittings will be detected. The implemented wire sensor and a typical response curve is shown in Fig. 5.

6 Monitoring of a bleed air system by optical fibre technology

Hot air escaping from bleed air systems is one of the most dangerous damage modes in aircraft as hot air is capable of causing severe thermal damage to cables and other components. To monitor such leakage, monitoring systems already exist that determine the conductivity of eutectic salt mixtures (Fenwal elements, see introduction). In the case of overheat, the salt mixtures melt and an electrical shortcut is established enabling the pilot to show-down the system early enough. Newer developments provide even functionality towards localisation of the leakage reducing inspection time during repair operations.

An alternative approach proposed here is the use of polymer optical fibres (acrylic POF) that have an intrinsic glass transition at the target temperature (Fig. 6 and Fig. 7).

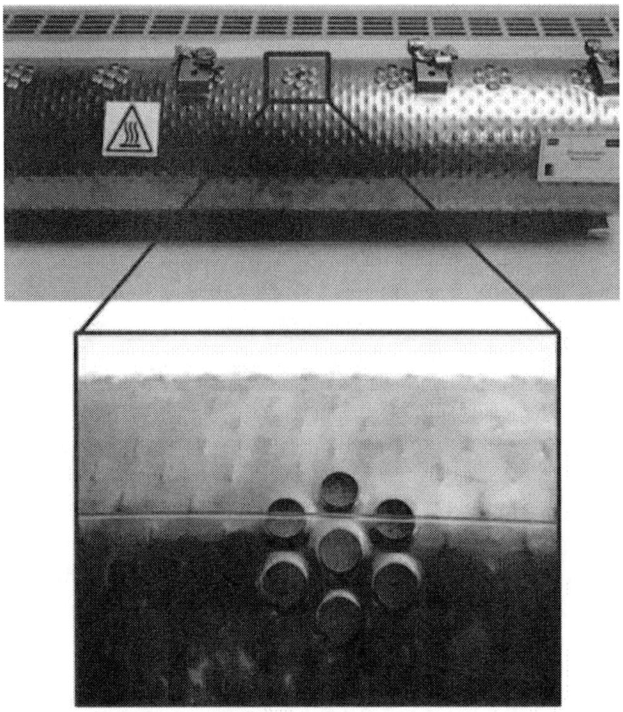

Fig. 6. Demonstrator for the detection of bleed air leakage. Above the pre-set "escape holes" for hot air, a robust polymer optical fibre is installed undergoing glass transition at the target temperature, easily detectable by state-of-the-art optical time domain reflectometry (TDR).

a)

b)

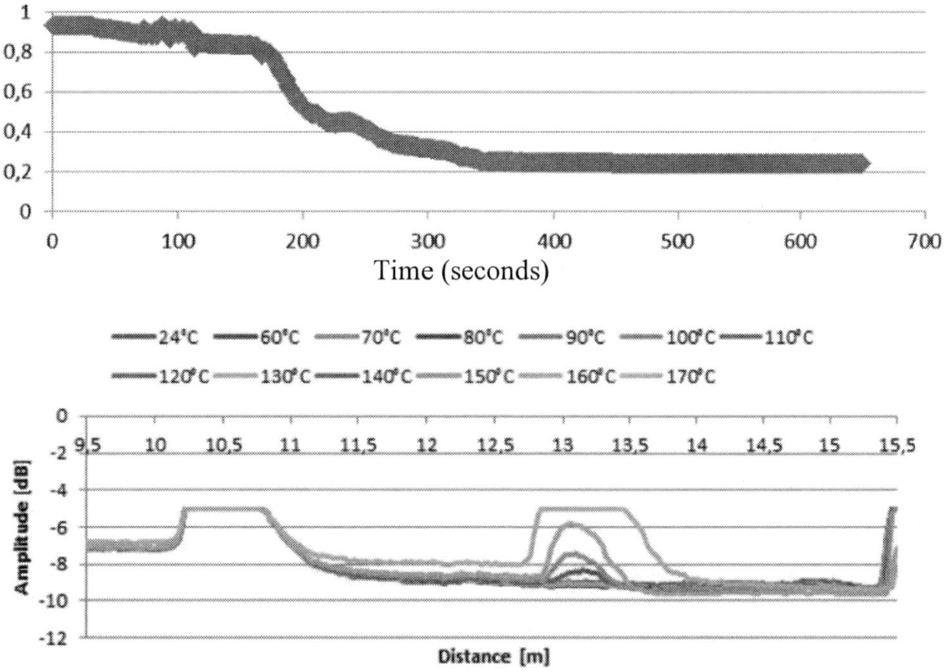

Fig. 7. Sigmoidal trend of light transmission through a polymer optical fibre a) when a target temperature of 135°C is reached, here after 180 seconds. The light transmission that has started at a plateau of almost 1 (100%) and falls down to a steady value of approximately 20%. The plot in time domain reflectometry is given by b), here the sensor starts at 10,5 meters and the defect is applied at 13 meters.

In this way, the optimised total reflection is lost at that area when heat damage occurs and this clear signal can even be determine by optical time domain reflectometry Fig. 7, b). Although the sensor is lost after melting, this is not a critical issue because the defect systems have to be repaired anyway and the cost for new optical fibres is negligible.

7 Example for moisture monitoring in operational aircraft

A working example developed by the authors is monitoring of corrosive liquids present in diverse floor structures of operational aircraft (Boeing 737-500, Boeing 747-400). The detection of those liquids provides interesting options for corrosion prevention already at very early stages [13]. Here, wire-sensors were embedded into between floor panels under galleys, lavatories and entrance areas, and the system are partially in service since 7 years now. The read-out of sensor data is performed after approximately 100 flight hours by use of simple ohmmeters and it has been proven that this sampling interval is highly sufficient.

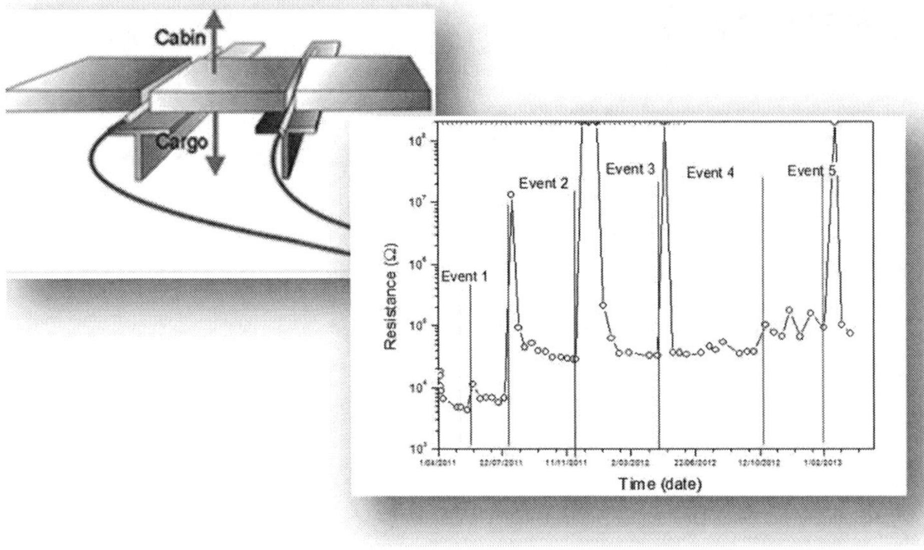

Fig. 8 Wire sensor for detecting "wetness events" in floor structures of a Boeing 737-530.

A characteristic response curve is shown in Fig. 8 that gives clear indications on "wetting events" in the floor under the galley area of a Boeing 737-530, most probably enabled by faulty isolations, please note the logarithmic scale suppressing non-relevant baseline variations. A typical feature is that wetness in the beginning appears just temporally as "peaks"; the area partially dries afterwards. The whole sensor is hygroscopic as such and works in this phase partially as a kind of water buffer, also visible by the fact that the baseline of the sensor as such increases gradually. These plots with this typical "peaks" give a clear information to the maintenance service teams to repair the respective sealings at the next occasion. Finally, currently, three operational airliners are equipped with these HNSD with an integral sensor length of app. 50 meters.

8 CONCLUSIONS

The intention of this paper is to present and encourage the implementation of "highly non-linear sensing devices" for health monitoring of structures and systems in aircraft to essentially avoiding problems arising from a weak "contrast" between damage data and background noise. Although, the application of this principle is still limited to certain niche applications, in some cases they enable easy to implement, straightforward solutions to practical problems in maintenance operations of engineering structures towards Industry 4.0. A few working examples are presented and further ideas will be developed to establish this concept as an alternative for traditional SHM techniques.

Acknowledgements

Part of the research leading to these results has received funding from the European Community's Seventh Framework Programme [FP7/2007-2013] under grant agreement n°212912 and the "NDTonAIR" project (Training Network in Non-Destructive Testing and Structural Health Monitoring of Aircraft structures) under the action: H2020-MSCA-ITN-2016- GRANT 722134.

9 References:

[1] H. Pfeiffer, S. Sunechiieva, H. Sekler, M. Wevers, Highly non-linear sensing devices in structural health monitoring, 12th European Conference on Non-Destructive Testing (ECNDT 2018)Gothenburg, 2018.

[2] D. Roach, Real time crack detection using mountable comparative vacuum monitoring sensors, Smart Struct Syst 5 (2009) 317-328.

[3] D. De Baere, M. Strantza, M. Hinderdael, W. Devesse, P. Guillaume, Effective Structural Health Monitoring with Additive Manufacturing, in: V.a.M. Le Cam, Laurent and Schoefs, Franck (Ed.) 7th European Workshop on Structural Health MonitoringNantes, France, 2014.

[4] MaintenanceHandbook, Aviation Maintenance Technician Handbook Federal Aviation Administration2012.

[5] TycoElectronics, Fundamentals of PolySwitch Overcurrent and Overtemperature Devices - Polymeric PTC Technology, Advertisement Tyco Electronics (2015).

[6] H. Pfeiffer, P. Heer, M. Winkelmans, W. Taza, I. Pitropakis, M. Wevers, Leakage monitoring using percolation sensors for revealing structural damage in engineering structures, Struct Control Hlth 21 (2014) 1030-1042.

[7] H. Budelmann, A. Holst, New Sensors for Rebar Corrosion Monitoring, Proceedings of the Fourth European Workshop on Structural Health Monitoring 2008 (2008) 211-218.

[8] Department of Defense - Handbook Nondestructive Evaluation System Reliability Assessment, MIL-HDBK-1823, USA, 2004.

[9] L.E. Pado, J.B. Ihn, J.P. Dunne, Understanding Probability of Detection (POD) in Structure Health Monitoring Systems, in: F.-K. Chang (Ed.) 9"‘ International Workshop on SHMUniversity of Stanford, 2013.

[10] H. Pfeiffer, J. Perremans, H. Sekler, M. Schoonacker, M. Wevers, The potential of highly non-linear sensing systems in engineering structures – operating applications in civil aircraft and chemical installations, 8th European Workshop on Structural Health MonitoringBilbao, 2016.

[11] H. Pfeiffer, D. De Baere, F. Fransens, G. Van der Linden, M. Wevers, Structural Health Monitoring of Slat Tracks using transient ultrasonic waves, EU Project Meeting on Aircraft Integrated Structural Health Assessment (AISHA), www.ndt.net, Leuven, Belgium, 2008.

[12] H. Pfeiffer, P. Heer, I. Pitropakis, G. Pyka, G. Kerckhofs, M. Patitsa, M. Wevers, Liquid detection in confined aircraft structures based on lyotropic percolation thresholds, Sensor Actuat B-Chem 161 (2012) 791-798.

[13] H. Pfeiffer, Structural Health Monitoring makes sense, LHT Connection - The Lufthansa Technik Group Magazine (2012).

A Bayesian Probabilistic Approach for Damage Imaging Utilizing Responses at Vibration Nodes

Tianxiang HUANG [1], Bo HAO [1], Kai-Uwe SCHRÖDER [1]
[1] Institute of Structural Mechanics and Lightweight Design, RWTH Aachen University, Aachen, Germany

Contact e-mail: tianxiang.huang@sla.rwth-aachen.de

Abstract. Most of the Structural Health Monitoring (SHM) methods are struggling between the number of sensors and the accuracy in damage detection. This paper explores the possibility of using a Bayesian probabilistic approach for damage imaging utilizing dynamic response at a few vibration nodes. The vibration amplitude at the nodal points can be considered as an efficient structural damage indicator (SDI). This SDI serves as not only a global but also a local indicator, as nodal points are part of vibration mode shapes. Instead of providing a rough damage detection, the posterior probability density function of the damage parameters is calculated under the Bayesian statistical system identification framework to quantify the confidence level of the identified results, which can be further interpreted as damage imaging results. In addition, the uncertainty in measurement due to the problems of measurement noise and time-varying environment can also be revealed by the framework. In the present paper, several case studies demonstrate the advantage and feasibility of the proposed method.

Introduction

Conventional NDT techniques use external equipment like ultrasound, X-ray or other systems to assess the health stage of the target structure. Unlike Conventional NDT, in SHM, sensors and actuators are permanently attached to the structure, continuously collecting data during the entire service life of the structure [1]. By adopting an SHM system, the schedule-driven maintenance can be gradually shifted to the condition-based one, thus the cost of maintenance inspection can be reduced. In addition, the safety of the structure can be further guaranteed.

Level-1 SHM [2] which only gives the existence of damage is far from reaching the aforementioned benefits. Thus, further quantitative damage detection, like the location and the severity, should be considered [3]. Due to the influence of noise and the environmental effect (e.g. temperature), the damage detection results are contaminated. In that case, rather than pointing a specific damage parameter (i.e. the location and the severity) which might be a false alarm, the probability distribution of the damage parameter can be used to quantify the confidence level of the identified results. The Bayesian statistical framework is frequently used to handle and describe the uncertainty problems in model updating and further in damage detection.

The Bayesian statistical framework was firstly applied in updating structural models by utilizing dynamic response [4]. The adjustable model parameters are considered as random variables. A prior can be assigned to the random variables based on engineering knowledge. Then the posterior can be updated with measurement by adopting the Bayes' rule. Some researchers [5] set the damage parameters as the random variables and a similar framework is adopted to update these damage parameters. Recently, a lot of researchers [6, 7] adopt the Bayesian statistical framework for real-time damage detection.

The vibration amplitude at the nodal points can be considered as an efficient structural damage indicator (SDI) [8]. This SDI is identified as NOde DISplacement (NODIS). As nodal points are part of vibration mode shapes, the NODIS serves as not only a global but also a local indicator. In this paper, the NODIS is used as the dynamic measurement for Bayesian probabilistic damage detection.

This paper is organized as follows. Section 1 explains the principle of the NODIS and the theoretical background of the Bayesian statistical identification framework. Section 2 will validate the proposed damage imaging method based on the analytical and numerical data. The effect of uncertainty, such as noise and temperature, is studied in Section 3. At last, the conclusion is drawn in Section 4.

1. Theoretical background

In this section, the idea of the NODIS is explained with a double clamped beam. Then, the Bayesian statistical identification framework for damage detection is derived. Furthermore, a flowchart of the framework is given.

1.1 Principle of the NODIS

The NODIS is the responses at vibration nodes. When damage occurs on a structure, the mode shape of the structure will be changed. The NODIS is therefore changed. In order to investigate the influence of the location and the severity on NODIS, a double clamped Euler-Bernoulli beam with an open crack is considered.

According to [9], a beam with a crack can be modeled as two beams connected by a rotational spring as illustrated in Figure 1. If the crack is located at $\xi_s = s/L$, the mode shapes of the two beams can be expressed as

$$w(\xi) = \begin{cases} c_1\cos(\lambda\xi) + c_2\sin(\lambda\xi) + c_3\cosh(\lambda\xi) + c_4\sinh(\lambda\xi) & \xi \in (0, \xi_s) \\ c_5\cos(\lambda\xi) + c_6\sin(\lambda\xi) + c_7\cosh(\lambda\xi) + c_8\sinh(\lambda\xi) & \xi \in (\xi_s, 1) \end{cases} \tag{1}$$
$$\xi = \frac{x}{L}, \quad x \in (0, L), \quad \lambda^4 = \frac{\rho A}{EI}\omega^2 L^4$$

where w represents the lateral deflection of the beam, ξ is the coordinate in the longitudinal direction, L is the beam length, ρ is the density, ω is the natural frequency, E represents the Young's modulus and I is the moment of inertia of the cross-section.

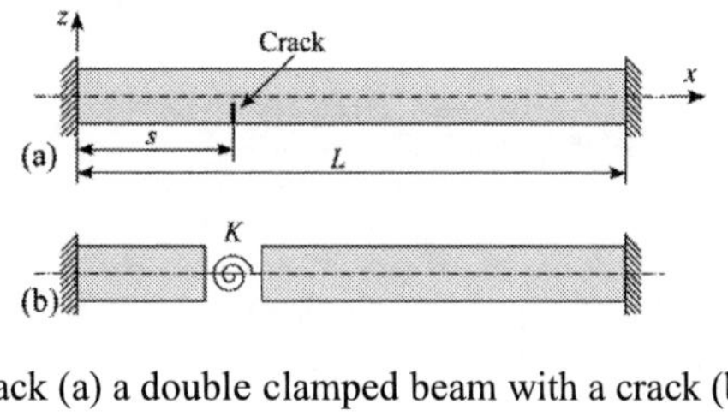

Fig. 1. Sketch of a beam with crack (a) a double clamped beam with a crack (b) model of the beam with crack.

The severity of the crack $\mu = \frac{EI}{KL}$ is the dimensionless cracked section flexibility, in which K represents the stiffness of the spring. After solving the natural frequency equation under the boundary conditions, the coefficients in mode shape Eq. (1) can be acquired by the solution set of the homogeneous linear system. The mode shape in Eq. (1) only related to the damage location and severity. And the position of the node does not depend on the material properties of the structure. A more detailed explanation can be found in [8].

Mode 2 has one node ($\xi_2 = 0.5$), while mode 3 has two nodes ($\xi_1 \approx 0.36$, $\xi_3 \approx 0.64$). Therefore, the mode shapes of mode 2 and 3 are taken into consideration, as shown in Figure 2 - (a) and (b). The displacements of the nodes in the intact state are equal to zero by definition. For a certain crack location and severity, the value of the node displacement at the chosen mode can be acquired. The effect of the crack location and severity on the node displacement is quantitatively studied. The location of the crack ξ_s is studied from 0 to 1 with an interval of 0.01, while the severity μ is investigated from 0.01 to 1 with an interval of 0.01. As shown in Figure 2 - (c), (d) and (e), the abscissa represents the location of the crack; the ordinate represents the severity of the crack. The color represents the value of the lateral node displacement of N_1, N_2 and N_3. Figure 2 - (c) is the node displacement of N_1, while Figure 2 - (d) and (e) are the node displacements of N_2 and N_3 respectively.

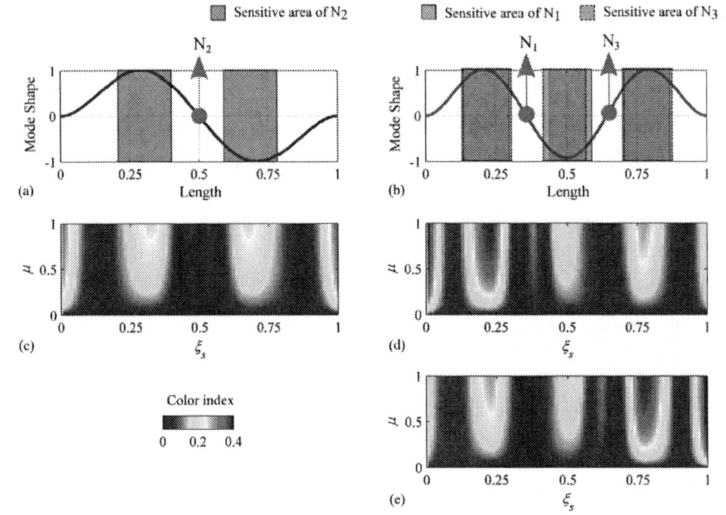

Fig. 2. Mode Shape of a double clamped beam (a) 2nd mode (b) 3rd mode and the value of node displacement affected by crack location and severity (c) N2 (d) N1 (e) N3.

1.2 Bayesian statistical identification framework

In the proposed method, the damage parameter $\boldsymbol{\theta} = \{\xi_s, \mu\}$ is identified with the Bayesian statistical system identification framework. The damage parameter $\boldsymbol{\theta}$ is represented by a prior probability distribution $p(\boldsymbol{\theta}|I)$ conditioned upon the background information such as expert opinion or practical experience. A proper prior distribution can be a uniform distribution or a Gaussian distribution when the mean and the relative error of the parameter are known. The posterior probability density function (PDF) $p(D|\boldsymbol{\theta}, I)$, can be given by the Bayes' rule:

$$p(\boldsymbol{\theta}|D, I) = \frac{p(D|\boldsymbol{\theta}, I)p(\boldsymbol{\theta}|I)}{p(D|I)} \tag{6}$$

where the likelihood function $p(D|\boldsymbol{\theta}, I)$, represents the contribution of the measured data to the posterior PDF. The denominator term, known as the marginal likelihood or the evidence, can be considered as a normalization factor ensuring that the integration of the posterior PDF is equal to 1. According to the total probability theorem, it can be written as:

$$p(D|I) = \int_{\boldsymbol{\Theta}} p(D|\boldsymbol{\theta}, I)p(\boldsymbol{\theta}|I)\boldsymbol{d\theta} \tag{7}$$

Different from the deterministic approach, the Bayesian framework which is a probabilistic approach aims to calculate the posterior PDF of the damage parameter for a given set of measurement data *D*.

The likelihood function is often set to be a multivariate normal distribution,

$$p(D|\boldsymbol{\theta}, I) = \prod_{i=1}^{N} \frac{1}{\sqrt{2\pi}\sigma_i} exp\left(-\frac{\left(z_i^e - z_i(\boldsymbol{\theta})\right)^2}{2\sigma_i{}^2}\right) \tag{8}$$

where z_i^e is the mean of the measured i th modal parameter, σ_i is the standard deviation of the measured i th modal parameter, $z_i(\boldsymbol{\theta})$ is i th modal parameter from the model with different damage parameter $\boldsymbol{\theta}$.

The flowchart of the Bayesian statistical identification framework is illustrated in Figure 3.

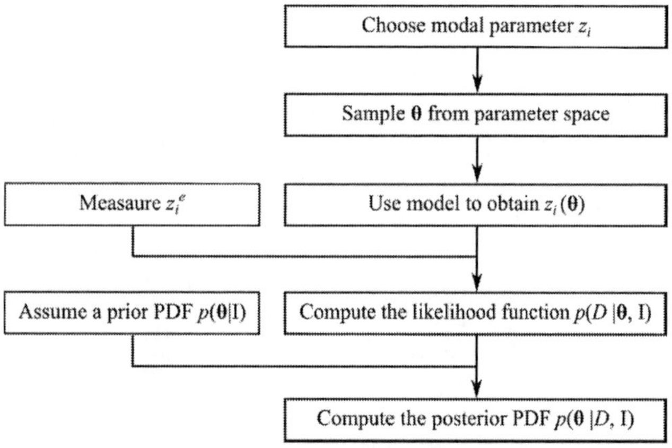

Fig. 3. Flowchart of Bayesian statistical identification framework

2. Analytical and experimental validation

2.1 Analytical validation

Continue with the example of the double clamped beam. The modal parameter z_i are the three NODIS of mode 2 and 3. The damage parameter $\boldsymbol{\theta} = \{\xi_s, \mu\}$ represents the location of the damage and the severity of the damage. ξ_s is sampled from 0 to 1 with an interval of 0.01, and μ is sampled from 0.01 to 1 with an interval of 0.01.

With the analytical model, the $z_i(\boldsymbol{\theta})$ at each sampling points can be obtained as illustrated in Figure 2. If the damage is located at $\xi_s = 0.25$ with a severity of $\mu = 0.5$, we can have the measurement z_i^e. The likelihood function of each z_i can be calculated according to Eq. (8) and they are illustrated in Figure 4. In this test, as there is no information about the damage, the prior is set to be a uniform distribution. By applying Eq. (6), the total likelihood function can be calculated. Then by introducing the prior PDF and the Bayesian rule, the posterior PDF can be acquired. As illustrated in Figure 5, the posterior PDF has a peak at the damage location and severity.

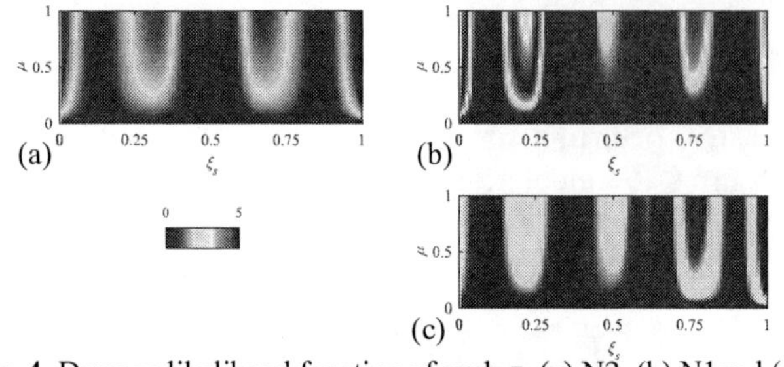

Fig. 4. Damage likelihood function of each z_i (a) N2, (b) N1and (c) N3

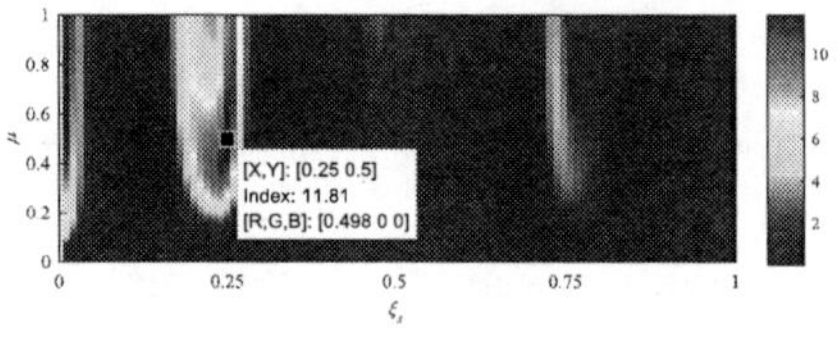

Fig. 5. Posterior damage PDF

Figure 6 illustrates the posterior PDF under different damage location, the severity is still set to be $\mu = 0.5$. It can be seen that when the structure is intact, the damage PDF has an output in low severity area and the area close to $\xi_s = 0.1, 0.4, 0.6, 0.9$, as shown in Figure 6 (a). This shows the minimum severity which can be detected by the proposed method and the possible false alarm locations. When the damage locates at other areas, the detection results are also acceptable, as illustrated in Figure 6 (b), (c) and (d).

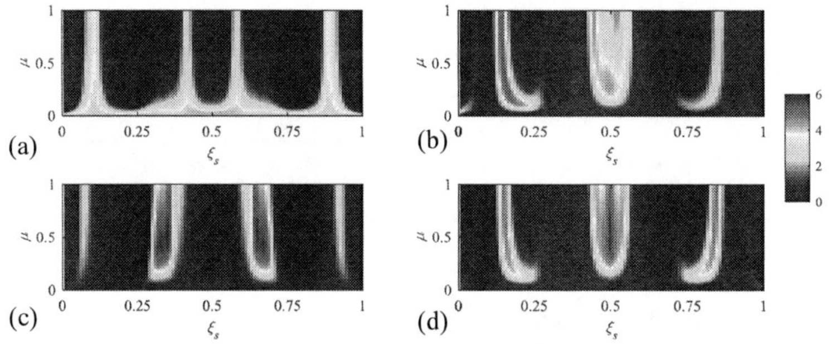

Fig. 6. Posterior PDF under different damage location (a) Health, (b) $\xi_s = 0.15$, (c) $\xi_s = 0.35$ and (d) $\xi_s = 0.5$

2.2 Experimental validation

2.2.1 Experimental settings

A steel beam with the dimension of 980mm×45mm×10mm is chosen. The beam is double clamped on a steel workbench as presented in Figure 7. The effective length of the beam between two clamps is 800 mm. The material properties are: $\rho = 7850$ kg/m^3, E = 200 GPa, $\upsilon = 0.26$.

Different from common mode-shape-based methods which need a large number of sensors to capture the mode shape of the beam, only three accelerometers (M352C15, PCB Piezotronics, Inc), identified as Acc1, Acc2 and Acc3, are adopted in this experiment. A manual tap hammer without force measurement is used for the impact testing. The unrecorded hammer impact is considered as an unknown impulse excitation from the environment. The damage information is acquired directly by the real-time signal. The sensors measured the acceleration in the transverse direction (vertically up).

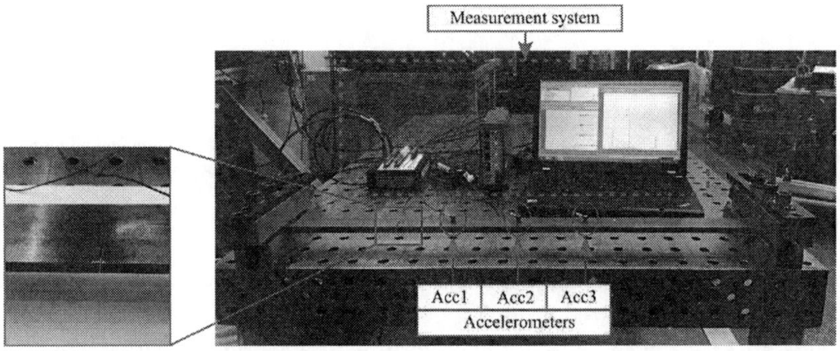

Fig. 7. The structure and the measurement system

The accurate position of each node is first estimated by the numerical model. Then the impulse force hammer is used to excite the specimen at a grid of points close to the estimated position to locate the accurate position of nodes. The accelerometers are then placed in the confirmed node locations. Based on that, the node of the second bending mode (217.6 Hz) and the two nodes in the third bending mode (430.3 Hz) are identified. Thus, the locations of the nodes are identified at 288 mm, 400 mm and 512 mm from the left end.

Saw cuts with different length (5 mm, 10 mm and 15 mm) are introduced to simulate the crack. The saw cut is introduced spanning the height. And the length of the cut is increased in the lateral direction. The saw cut is located at 200 mm from the left end. To validate the damage location procedure a magnet block with the weight of 40 g is used.

2.2.2 Experiment result

Fast Fourier Transform (FFT) is implemented to transform the time domain signal into frequency domain. The single-sided amplitude spectrums of the three accelerometers are presented in Figure 8 - (a) (Ordinate is on a logarithmic scale). All three accelerometers have a response peak at 1st mode (78 Hz). In the 2nd mode (218 Hz), Acc2 has a small amplitude (3×10^{-3}), as Acc2 locates at the node of 2nd mode. In the 3rd mode (430 Hz), the amplitudes of Acc1 (12×10^{-3}) and Acc3 (13×10^{-3}) are also very small, as Acc1 and Acc3 locate at the nodes of 3rd mode.

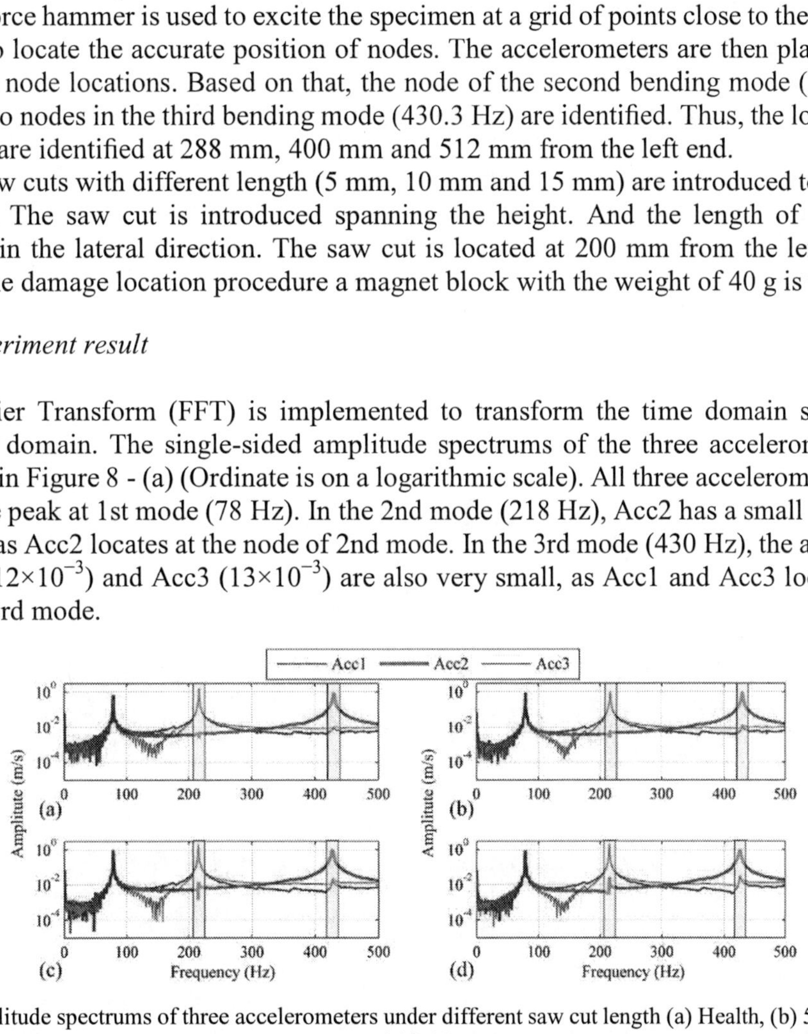

Fig. 8. Amplitude spectrums of three accelerometers under different saw cut length (a) Health, (b) 5 mm, (c) 10 mm and (d) 15 mm.

The Fourier transform of Acc1 and Acc2 are defined as $U_1(\omega_k)$ and $U_2(\omega_k)$. $U_1(\omega_k)$ and $U_2(\omega_k)$ are related to $F_0(\omega_k)$. The transfer-function-based damage indicator (TFDI) [8] is adopted:

$$\iota_1 = \frac{UR_1(\omega_3)}{UM_2(\omega_3)}, \quad \iota_2 = \frac{UR_2(\omega_2)}{UM_1(\omega_2)}, \quad \iota_3 = \frac{UR_3(\omega_3)}{UM_2(\omega_3)} \tag{9}$$

where

$$UR_i(\omega_k) = \text{range}[U_i(\tilde{\omega}_k)]$$
$$UM_i(\omega_k) = \max[U_i(\tilde{\omega}_k)] \tag{9}$$

in which, i represents the number of the node, $\overline{\omega}_2 - \Delta\omega \le \tilde{\boldsymbol{\omega}}_2 \le \overline{\omega}_2 + \Delta\omega$, $\overline{\omega}_3 - \Delta\omega \le \tilde{\boldsymbol{\omega}}_3 \le \overline{\omega}_3 + \Delta\omega$, $\overline{\omega}_k$ represent the natural frequency of the intact structure, the frequency range $\Delta\omega$ is set to be 10 Hz in this case.

The TFDI results show that for ι_1 and ι_3, a 15 mm saw cut will lead to an approximate 0.017 increase in DI as illustrated in Figure 9. And the increase for ι_2 is about 0.01 under the case of a 15 mm saw cut. This demonstrates that the NODIS method has good damage quantification and location results. Thus, it can be used as a promising modal parameter for Bayesian damage imaging.

Figure 10 demonstrates that the Bayesian damage imaging method identified the saw cut at $\xi_s = 0.25$. With the increase of the saw cut length, the method has a more concentrated imaging result. When the saw cut length below 10mm, the Bayesian damage imaging method gives some false alarm in the health area. It might because the limitation of the NODIS indicator. A damage detection method without Bayesian framework might directly give a result at the wrong location or severity. While the proposed Bayesian damage imaging method can provide the probability distribution of the entire location and severity, which is more accurate and reliable.

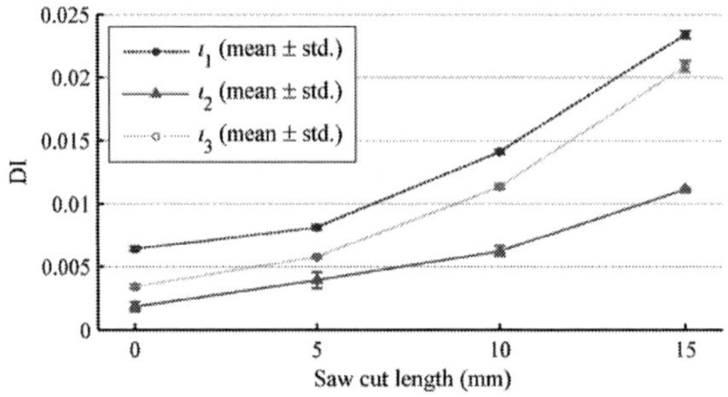

Fig. 9. DI with different saw cut length

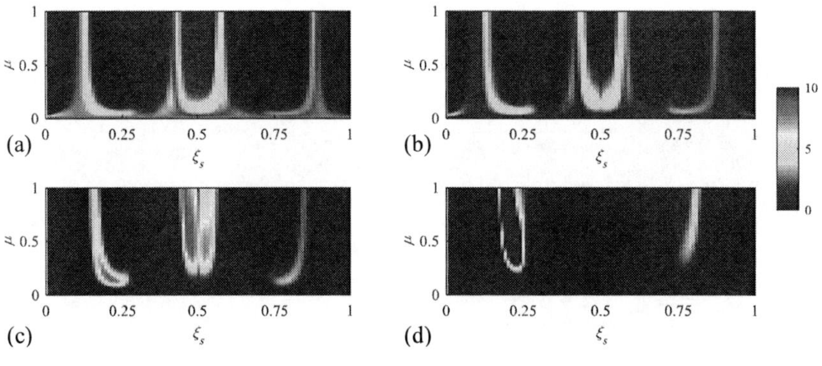

Fig. 10. Posterior PDF under different saw cut length: (a) 0 mm, (b) 5 mm, (c) 10 mm and (d) 15 mm

3. The effect of uncertainty

In this section, the performance of the proposed method against uncertainty, such as measurement noise and temperature, is studied.

3.1 The effect of measurement noise

The measurement noise of a monitoring system is usually a wideband noise. Therefore, the Gaussian white noise of SNR 15 and 20 dB are added to the time domain signals to evaluate the performance of the method.

$$\mathrm{SNR_{dB}} = 10\log_{10}\left(\frac{\text{Signal Power}}{\text{Noise Power}}\right) \tag{10}$$

The same experimental data from section 2.2 with saw cut is adopted. Figure 11 shows the original signal and the signal with different noise and their spectrum of Acc1. After adopting the TFDI, the DIs under different saw cut length and noise levels are illustrated in Figure 12. It shows that the noise will increase the DI in each case. This makes the damage detection fail or cause false alarms.

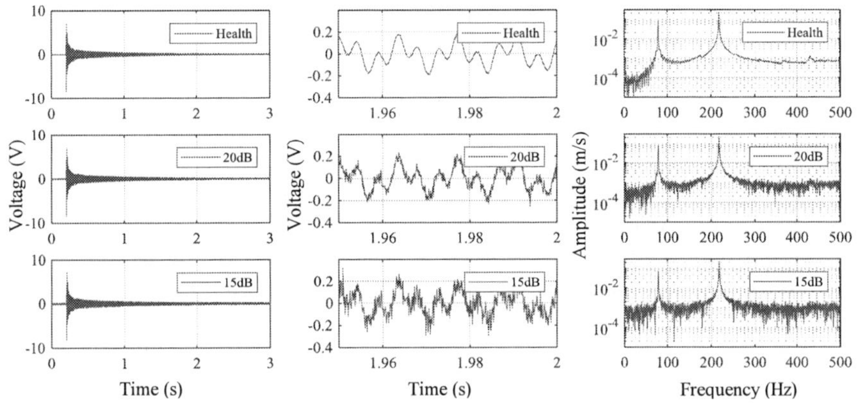

Fig. 11. Time domain and frequency domain signal with different level of noise

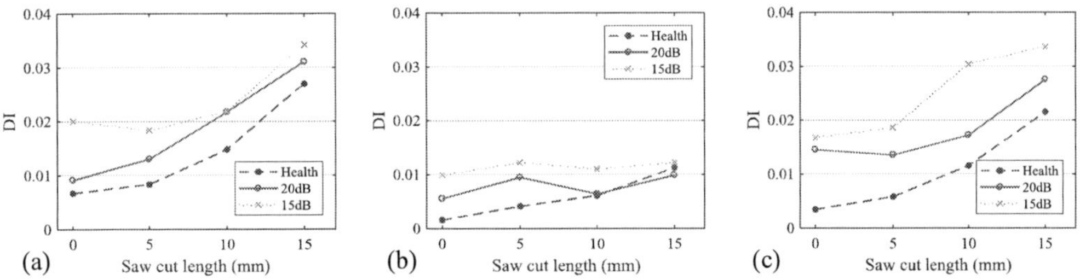

Fig. 12. The influence of noise on DI under different saw cut length (a) N1, (b) N2 and (c) N3.

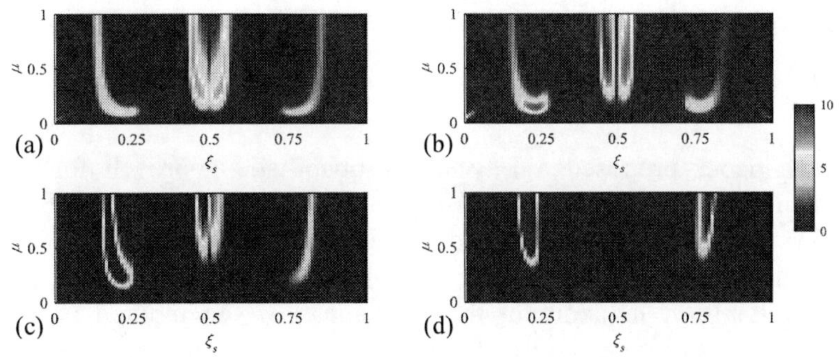

Fig. 13. Posterior PDF under different saw cut length under noise (SNR=20dB): (a) 0 mm, (b) 5 mm, (c) 10 mm and (d) 15 mm

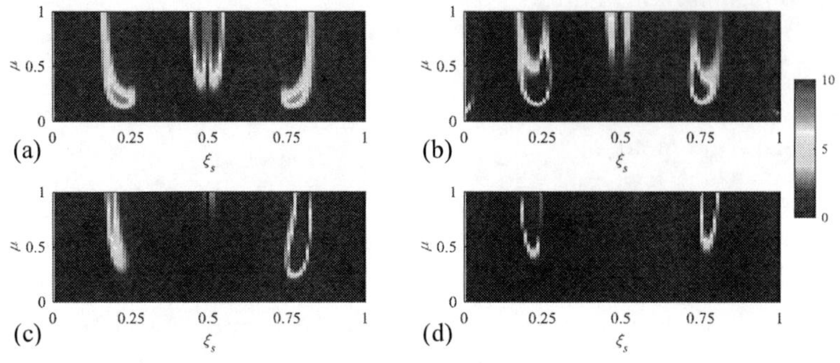

Fig. 14. Posterior PDF under different saw cut length under noise (SNR=15dB): (a) 0 mm, (b) 5 mm, (c) 10 mm and (d) 15 mm

The Bayesian imaging method can still provide damage PDFs under different level of noises as shown in Figure 13 and 14. Although the noise will increase the probability of false alarm, the right parameters of the damage are shown by the Bayesian method.

3.2 Temperature effect

A similar experiment setting as the last section is adopted, except that the effective beam dimension is 800mm×40mm×10mm. A heat gun is used to heat the beam at the location of 200 mm from the left end. The beam is heated to 40°C, 50°C and 60°C at this location (The room temperature is about 25°C). Later, two magnets blocks with the weight of 7g and 40g respectively are introduced at the same location to simulate damage.

As listed in Table 1, temperature and damage will lead to a similar change to the frequency, while the DI has a much smaller difference under changing temperature comparing to the damage. Figure 15 (a-d) shows that under different temperatures, the damage imaging result does not focus. When the damage is introduced, the Bayesian imaging method can identify the damage.

Table 1. DI and natural frequency with different temperature and damage

Stage	DI ($\times 10^{-2}$)			Frequency (Hz)		
	ι_1	ι_2	ι_3	Mode 1	Mode 2	Mode 3
Health (25 °C)	1.33	0.09	1.67	80.0	216.1	417.9
T1 (40 °C)	2.86	0.20	3.26	79.3	215.1	417.2
T2 (50 °C)	3.13	0.29	3.32	78.2	213.3	416.0
T3 (60 °C)	3.43	0.53	3.51	76.6	211.0	414.1
D1 (7g)	4.11	0.62	3.10	80.0	215.5	417.4
D2 (40g)	6.43	4.07	4.52	79.7	212.7	414.1

4. Conclusion

The presented paper proposed a Bayesian probabilistic approach for damage imaging utilizing dynamic response at three vibration nodes. The flowchart of the Bayesian statistical identification framework is illustrated. The analytical and experimental validation demonstrates that the proposed method can give a PDF of the damage location and severity with a high reliability. In addition, the robustness of the method is investigated with measurement noise and temperature. The results show that the proposed method can still have an effective damage detection result at a noise level of 15dB or a local temperature change of 35°C.

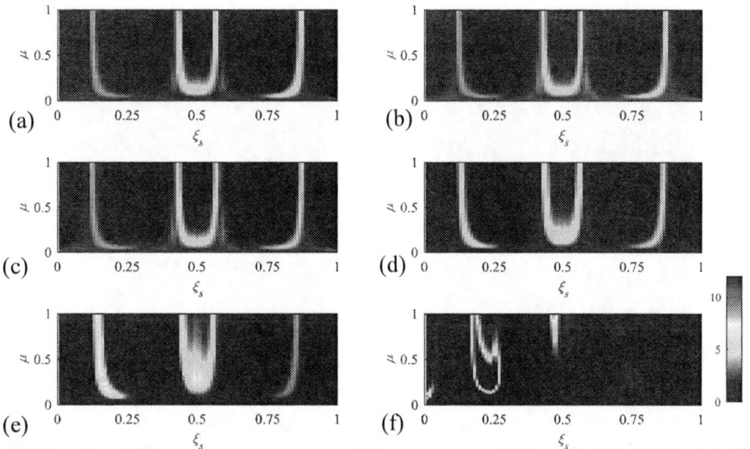

Fig. 15. Posterior PDF under different temperatures and damage case (a) Health, (b) 40 °C, (c) 50 °C, (d) 60 °C, (e) D1 and (f) D2

References

[1] W. Staszewski, C. Boller, G.R. Tomlinson, Health monitoring of aerospace structures: smart sensor technologies and signal processing, John Wiley & Sons, 2004.

[2] C.R. Farrar, K. Worden, An introduction to structural health monitoring, Philosophical Transactions of the Royal Society A: Mathematical, Physical and Engineering Sciences 365 (1851) (2007) 303–315.

[3] L. Qiu, S. Yuan, F.-K. Chang, Q. Bao, H. Mei, On-line updating Gaussian mixture model for aircraft wing spar damage evaluation under time-varying boundary condition, Smart materials and structures 23 (12) (2014) 125001.

[4] J.L. Beck, L.S. Katafygiotis, Updating Models and Their Uncertainties. I: Bayesian Statistical Framework, Journal of Engineering Mechanics 124 (4) (1998) 455–461.

[5] H. Sohn, K.H. Law, A Bayesian probabilistic approach for structure damage detection, Earthquake Engineering & Structural Dynamics 26 (12) (1997) 1259–1281.

[6] D. Giagopoulos, A. Arailopoulos, V. Dertimanis, C. Papadimitriou, E. Chatzi, K. Grompanopoulos, Structural health monitoring and fatigue damage estimation using vibration measurements and finite element model updating, Structural Health Monitoring (2018) 1475921718790188.

[7] R. Rocchetta, M. Broggi, Q. Huchet, E. Patelli, On-line Bayesian model updating for structural health monitoring, Mechanical Systems and Signal Processing 103 (2018) 174–195.

[8] T. Huang, M. Chaves-Vargas, J. Yang, K.-U. Schröder, A baseline-free structural damage indicator based on node displacement of structural mode shapes, Journal of sound and vibration 433 (2018) 366–384.

[9] T.G. Chondros, A.D. Dimarogonas, J. Yao, A continuous cracked beam vibration theory, Journal of sound and vibration 215 (1) (1998) 17–34.

Tailored Embeddable Sensor-Actuator Layers for CFRP Aerospace Structures

Sandro NITSCHKE [1], Andreas HORNIG [1], Anja WINKLER [1], Niels MODLER [1]
[1] TU Dresden Institut für Leichtbau und Kunststofftechnik, Dresden, Germany
Contact e-mail: sandro.nitschke@tu-dresden.de

Abstract. The increasing structural integration of fibre-reinforced composites and the growing demand for more electrical and intelligent structural components are one of the future markets for aerospace and aviation industry to master growing economic and ecological challenges.

Fibre-reinforced composites offer the opportunity to embed sensor and actuator networks in the material, which is of particular relevance for example for structural health monitoring (SHM) applications. In order to maintain the structural integrity and necessary safety margins of the carbon fibre-reinforced plastic (CFRP) components, the elements (e.g. sensors, conducting paths, carrier layer, etc.) which will be integrated will have to be selected by extensive experimental characterization. To realize this, a methodology has been developed to embed tailored functional layers, assembled with piezo ceramic elements, into structural components made of CFRP. Due to the high sensitivity, high temperature stability, commercially availability as well as the variable small element sizes of the piezo ceramic elements, they are predestined for the integration into CFRP.

The resulting smart structures provide both, sensoric and actoric functionalities. Thus, the developed Tailored Embeddable Sensor-Actuator Layers (TEmSAL) are particularly characterized by „off-the-shelf" manufacturing, cost-efficiency, a high automation capability, tailored sensing capabilities for customized applications and parts, and a low impact on mechanical CFRP properties. The presented investigations show the aspects of the characterization and manufacturing strategy on specimen level, the resulting structural properties of the CFRP material and sensor as well as actuator performance. Additionally, TEmSAL CFRP engine components will be covered as an outlook for further investigations and applications.

Keywords: function integration, piezo ceramic module, active composite structures, screen printing conductive paths, impact localisation, structural health monitoring

1 Introduction

Sensor and actuator networks are of rising interest for aerospace applications. An embedding in carbon fibre-reinforced plastic (CFRP) components offers the possibility for in-situ structural health monitoring (SHM) of the component, by detecting structural response changes like eigenfrequency shifts or varying damping behaviour due to damage or fatigue [1], [2]. Additionally, these systems can be used for the detection of external disturbances during operation time, like dynamic and high dynamic loading scenarios caused by tool drop, hail, stone, ice and bird impact or foreign object damage (FOD). With such information one can either adapt or individually plan the service and maintenance interval as well as gather more realistic field data for the design and analysis process of future components. This results in a better understanding of the material behaviour and consequently in improved material damage, failure and deterioration models.

The high specific mechanical properties of epoxy resin based thermoset CFRP result in a broad applicability for high performance structural components. However, the integration of electric components are one of the biggest challenges due to the inherent electrical conductivity of the CFRP material.

In this study a method is presented to design and manufacture sensor and actuator networks, which are adaptable to the needs of the component and their functional intention. One key electric element for this purpose are piezo ceramic elements. They are predestined for an embedded SHM system and impact detection due to their high sensitivity, high temperature stability, relatively low price, high availability, adaptable size and their simultaneous utilization as sensor and as actuator, see Table 1.

Table 1 Advantages of piezo ceramic elements

Advantages	**Disadvantages**
+ High sensitivity → high output and low noise voltage even for small deformations	- Bridle ceramic material (but also available as flexible PVDF polymer foil)
+ Can be used as sensor and actuator elements	- Solderability sometimes difficult
+ Relatively cheap	- Stress over xx MPa lead to depolarisation
+ Needs not necessarily a conductive connection to signals paths (high fault tolerance contacting)	- Voltage measurement equipment needs protection circuitry for high piezo loads (up to kV)
+ High temperature capabilities	
+ Available as very thin elements	

2 Material and specimen preparation

2.1 Manufacturing and sensor positioning

Starting point for the development of the Tailored Embeddable Sensor-Actuator Layers (TEmSAL) for CFRP structures is the integration and experimental investigation of so called "thermoplastic composite compatible piezo ceramic modules (TPM)" [3-6] within a prepreg (Hexel HexPly 8552) panel specimen (Fig. 1c) of 12 plies ($[0/90]_{6S}$).

TPMs are thin (0.2 mm) piezo ceramic modules with a quadratic (10 mm) shape, enclosed in a polyamide matrix with special designed electrode structure, mostly used in glass thermoplastic material (Fig. 1a). To electrically insulate the TPMs within the CFRP panel the copper contacts of the TPMs as well as the attached soldering terminal were wrapped in polyimide (PI) tape. Within the plate specimen 16 TPMs (Fig. 1b) were placed between two very thin glass fleeces (30 g/m^2). The electric contacts of the TPMs in the middle of the plate (#6, #7, #10 and #11) were extended with enamelled copper wire to the left and right edges of the plate. This TPM layer was then positioned between ply 10 and 11 of the prepreg plate specimen. To ensure the accessibility of the soldering terminals after the consolidation process, the prepreg was cut out according to the position of the soldering terminals of each sensor.

After the consolidation process at 180°C and 7 bar pressure for 180 min. within a vacuum press, the positioning of the individual TPMs was checked with a computed tomography (CT) scan (Fig. 1c) and the insulation polyimide tape of the soldering terminal was removed to contact the sensors to the measuring system.

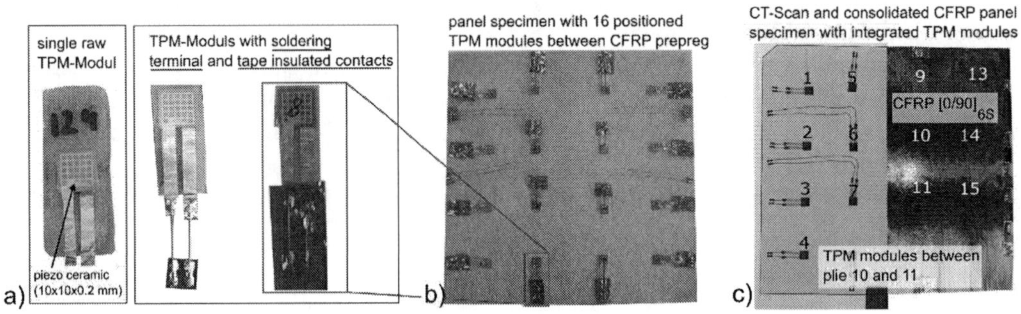

Fig. 1 a) Raw TPM with simplified contacting (soldering terminal) and isolation (polyimide tape), b) placement of 16 TPM on a 12 layer CFRP prepreg ($[0/90]_{6S}$) on additional glass fleece, c) consolidated CFRP prepreg plate specimen with integrated TPMs and additional computed tomography scan to check sensor position

2.2 *Impact detection and test results*

To examine the dynamic sensor capabilities of the piezo ceramic elements, impact tests with a modal hammer (Fig. 2a) were performed. In Fig. 2b an exemplary test result of a direct impact on TPM #2 of around 18 N with the corresponding sensor excitations of eight TPMs is shown. The voltage signals of the eight TPMs were recorded simultaneously with a sampling rate of 2 MS/s (mega samples per second).

On account of the high sensitivity of the TPM a voltage peak of around 7 V is gained. Due to mechanical wave propagation a delayed excitation of the other seven sensors can be observed. This results show the capabilities of the used piezo elements for detecting the impact location as well as the intensity of an impact event.

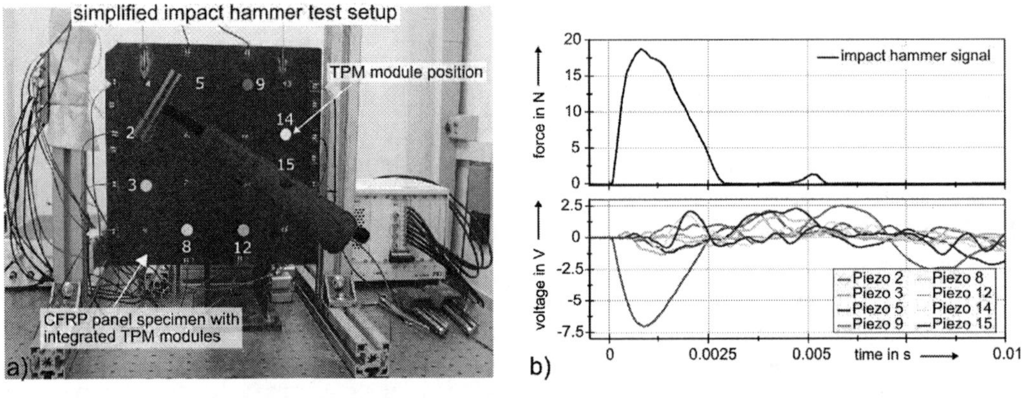

Fig. 2 a) Test setup for the impact detection experiment with a modal hammer, b) comparative diagram with the modal hammer force time signal and the voltage time response of the integrated TPMs

2.3 *Structural health monitoring and test results*

The ability to use integrated piezo ceramic elements either as sensor or as actuator are an essential advantage for their use in structural health monitoring systems. After detecting e.g. an impact event, a modal analysis with the same piezo elements as an excitation source can be performed in order to determine shifts in the eigenfrequency of the structural component, indicating delaminations. An alternative possibility for detecting changes in the integrity of the component is the analysis of the mechanical waves and their propagation und reflection behaviour. [7-10]
In Fig. 3a the test setup for actuator capability investigations is shown. TPM #1 is excited once with a rectangular voltage pulse of 250 V and a frequency of 0.2 Hz. To visualize the propagating mechanical waves (Fig. 3a top and bottom), induced by TPM #1, a scanning laser vibrometer is used. Fig. 3b shows the sensoric voltage response of the marked eight TPMs induced by the propagating wave. By comparing the response signals of the TPMs after an external loading event with a reference state it is possible to identify structural changes.

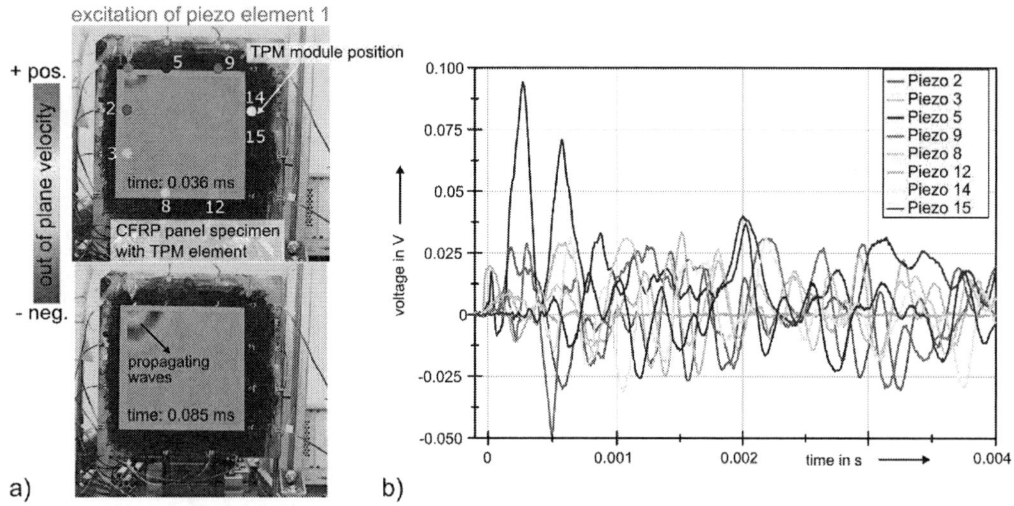

Fig. 3 a) Test setup for the structural health monitoring system, excitation of TPM #1 with half sine and corresponding wave propagation and out of plane velocity recorded with a laser vibrometer, b) voltage time response of the integrated TPMs after excitation of TPM #1

Within the impact detection and SHM tests setups the concept for the sensoric and actoric usage of the chosen piezo element has been demonstrated. However, the complicated handling of the individual TPMs, such as the addition of soldering terminals, the positioning of enamelled copper wires, the insulation of the individual modules and the TPM material (polyamide), which is unfavorable for epoxy resin-based prepreg systems, requires further development of the methodology. Furthermore, the danger of blurring of the sensors (incorrect positioning) during the manufacturing process leads to the consideration of a designated sensor-actuator layer.

3 Tailored Embeddable Sensor-Actuator Layers (TEmSAL)

The basic idea is the development of a highly adaptable to arbitrary geometry and measurement requirements. Also, it should have the capability of an automatable manufacturing process and off-the-shelf usage. The interlaminar properties of the composite component should not be weakened. The most important demands are on the one hand the electrical insulation of the layer so that it can be used in any CFRP material. On the other hand such tailored sensor-actuator layers (TEmSAL) must not affect the mechanical properties of the structural component itself, specifically the interlaminar properties should not be reduced.

3.1 TEmSAL carrier material - polymer foil characterisation

To identify suitable polymer carrier foils for TEmSAL to be used in epoxy resin based CFRP components, one of the most important design criteria are the interlaminar material properties under mode I (tension) and mode II (shear) as an indication of delamination resistance. Table 2 shows an excerpt from the experimental investigations of different polymer foils under mode I (double cantilever beam – DCB – experiment) and mode II (end loaded split – ELS – experiment) loading conditions.
The reference material configuration is a 30 ply unidirectional ($[0]_{30}$) epoxy CFRP prepreg (Hexel HexPly 8552). In the mid surface of the CFRP prepreg different polymer foils were embedded to replicate a full-surface TEmSAL. Polyamide (PA) as well as polyimide (PI), which is a widely used foil for flexible circuit boards, was characterized. Especially the energy release rate under shear loading G_{IIC} (mode II) was significantly higher than the pure CFRP reference. Under tension loading (mode I) the energy release rate G_{IC} for the pure polymer foils were partially much lower.
The most promising configuration for the TEmSAL is a combination of 2 PI. In order to connect the two PI foils an adhesive agent (AA) – Nolax cox 391 – is used. This increases the energy release rates G_{IC} (+109 %) and G_{IIC} (+346%) significantly and improves the delamination resistance substantially compared to the CFRP reference.

Table 2 Exemplary test results from the experimental investigations of different polymer foils under mode I and mode II load for the TEmSAL selection process

Configuration (CFRP material: Hexcel HexPly 8552)	**DCB experiment - mode I load** Energy release rate G_{IC} in kJ/m²	**ELS experiment - mode II load** Energy release rate G_{IIC} in kJ/m²
Pure CFRP prepreg **$[0]_{30}$**	0.280 ± 0.035 (reference)	0.73 ± 0.08 (reference)
CFRP prepreg with 100µm polyamide (PA) foil **$[0]_{15}$/PA/$[0]_{15}$**	0.052 ± 0.036 (-81.4 %)	2.16 ± 2.41 (+195.8 %)
CFRP prepreg with 125µm polyimide (PI) foil **$[0]_{15}$/PI/$[0]_{15}$**	0.192 ± 0.032 (-31.4 %)	2.78 ± 0.62 (+280.0 %)
CFRP prepreg with 2x125µm polyimide (PI) foils and 100µm adhesive agent* (AA) foil in between PI **$[0]_{15}$/PI/AA/PI/$[0]_{15}$**	**0.579 ± 0.035 (+109.7 %)**	**2.53 ± 1.46 (+346.5 %)**

*adhesive agent Nolax cox 391

In order to create a fully functional and insulated TEmSAL, two covering foils are required to position the electrical contacting surface to the piezo ceramic element and to build the conducting paths for the electrical signal. Therefore the presented combination of 2 PI polymer foils with an additional adhesive agent foil will be further investigated in detail.

3.2 Designing and manufacturing of TEmSAL

Flexible circuit boards made of PI foils, deposited and etched copper conducting paths are widely used in the electronic industry and was also a starting point for integrated sensor systems for structural composite components [11], [12]. Due to their limited size, very high price and more extensive use of chemicals (acid) an alternative process for adding conductive paths onto surfaces was chosen. By using epoxy based conductive ink, the screen printing is a very promising technology for this kind of application. One of its advantages (Table 3) is the wide industry usage with the subsequent availability and high automation capability and a fast and cheap manufacturing process. Additionally, a straight forward conductive layout design process is available, using standard CAD and vector drawing programs only.

Table 3 Advantages and disadvantages of conductive screen printing

Advantages	**Disadvantages**
+ Highly automatable, cheap, fast and widely used process	- no soldering possible, however wires could be bonded to conductive adhesive or soldered onto a hollow copper rivet (Fig. 5a)
+ Printing even with simple tools	
+ Printable on nearly any surface and size	
+ For printing no harming chemicals (acid) needed	
+ Easy designing process of electrical paths with standard CAD and vector programs	

A PI cover foil for 16 piezo ceramic elements was designed for the development of the manufacturing process. In Fig. 4a the workflow from designing the conductive paths and contacting surfaces in a standard CAD program (#1), over creating the vectorised screen print templates (#2) for the screen print mask (#3) to a one side printed PI cover foil (#4) is presented. Alternatively, a fully automated roll to roll screen print processing machine can be used, Fig. 4b which is necessary for a series production of the TEmSAL PI cover foils.

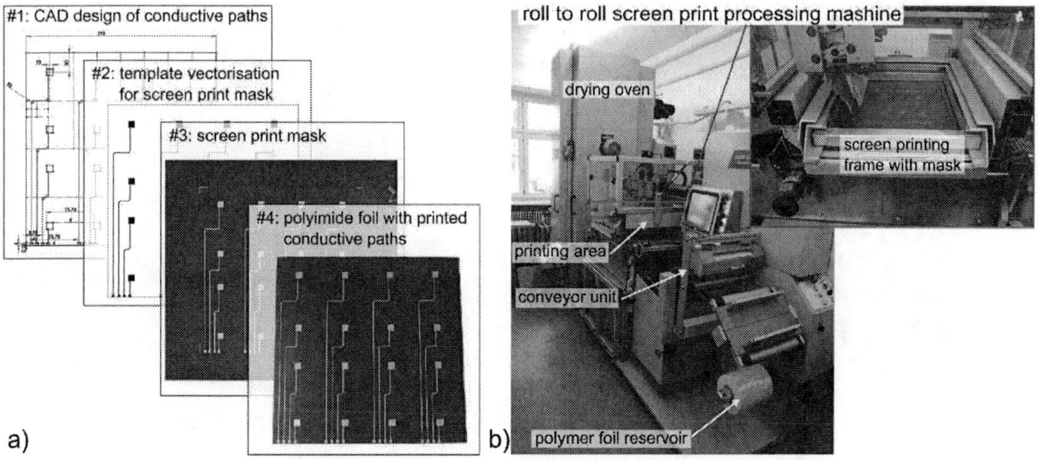

Fig. 4 a) Workflow from designing to finished conductive layout of an exemplary TEmSAL, b) roll to roll screen printing processing machine for the automated manufacturing of conductive PI cover foils for TEmSAL

In order to manufacture fully functional off-the-shelf TEmSALs (Fig. 5b) additional steps need to be done. In Fig. 5a one can see two PI cover foils with the screen printed conducting and contacting paths. On the left side the AA agent layer is placed on top of the PI cover foil. Depending on the number of measuring points, the AA layer must to be cut out on the position of the piezo contacting surface. At these points the piezo ceramic elements are fixated with additional epoxy resin. In order to be able to solder electrical wires to the screen printed conductive wire connection points, hollow copper rivets are placed. After soldering the electrical wires to the connection points, the right PI cover foil is flipped over. The package is then placed in a vacuum bag and heat treated at 180°C to activate the AA layer and consolidate the TEmSAL. Fig. 5b shows a fully functionised TEmSAL with eight piezo ceramic elements and electrical wiring.

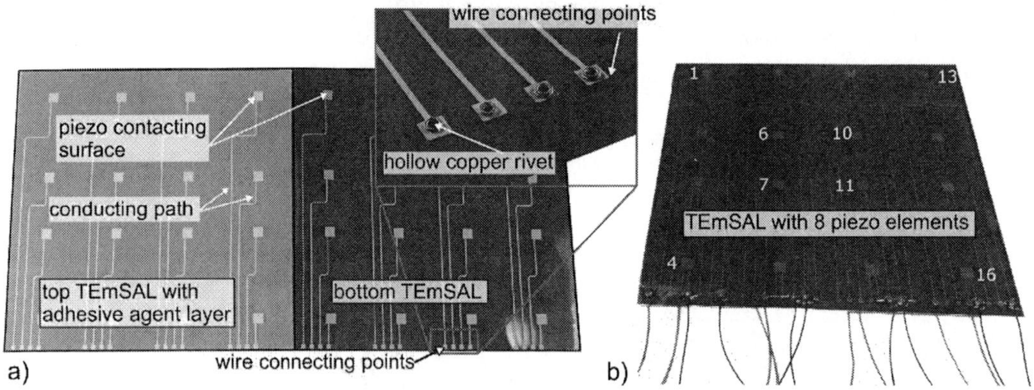

Fig. 5 a) PI cover foil with adhesive agent (left) and inserted hollow copper rivet as wire connecting points (right) before consolidation process of a TEmSAL, b) functionised TEmSAL with eight piezo elements and soldered contact wires

Consequently, the TEmSAL is positioned between ply 14 and 15 of a CFRP panel prepreg specimen $[0/90]_{8S}$ (Fig. 6a). To verify the correct positioning of the individual piezo ceramic elements a CT scan is performed (Fig. 6b). The flexibility and the small piezoelectric elements used, the integration of TEmSAL also allows an adaption to complex shaped geometries.

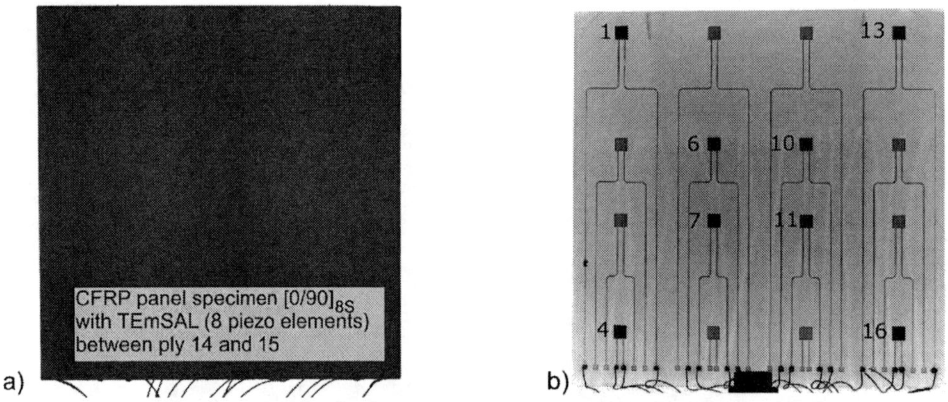

Fig. 6 a) CFRP panel specimen $[0/90]_{8S}$ with integrated TEmSAL, b) CT scan of the same panel specimen to check correct positioning of the eight integrated piezo elements

4 Structural application

To explore the potential of the developed technology on component level, the integration of the TEmSAL into a generic fan blade like structure (Fig. 7a) is presented. Due to its axially tapered (Fig. 7b, cut 1) and curved cross section (Fig. 7b, cut 2) this geometry is a suitable starting point for the validation of structural sensing and actuator capabilities. To measure the time signal of the induced piezo voltage during impact events as well as the oscillation behaviour and eigenfrequencies, the piezo ceramic elements are positioned as shown in Fig. 7c. Furthermore a modal analysis of the generic fan blade could be done in advance.

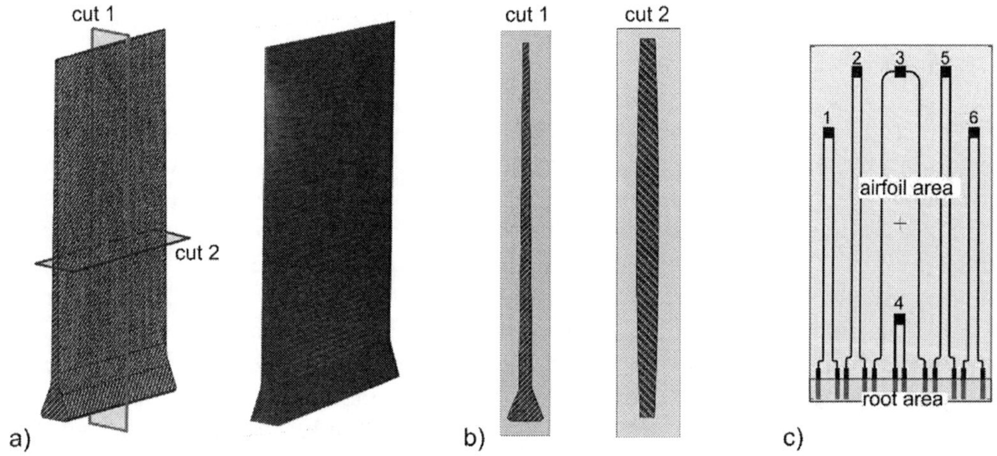

Fig. 7 a) CAD design (left) and manufacturing example (right) of a generic fan blade specimen with integrated TEmSAL, b) cross sectional cuts of the generic blade, c) sensor layer design and positioning of the piezo elements within the generic fan blade specimen

To ensure functionality of the integrated TEmSAL an impact hammer setup as well as a test setup for the SHM configuration analogously to the setups in Fig. 2 and Fig. 3 were performed. In Fig. 8a the manufactured generic blade specimen with the corresponding piezo element positons (with overlaid CT-scan), the location of the impact hammer (red circle) and the active excited piezo element #2 (orange) for the SHM test are shown.

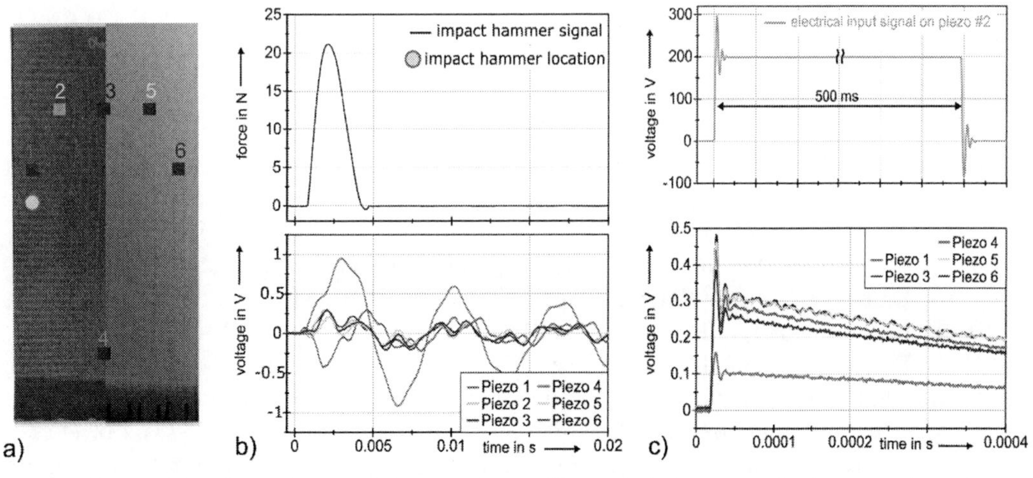

Fig. 8 a) Manufactured generic blade specimen with overlaid CT-scan to check correct piezo element positioning; location of the impact hammer (red circle) and excited piezo element #2 (orange) for the SHM test, b) force and piezo voltage time signals of the impact hammer test, c) voltage input signal of excited piezo element #2 and corresponding response of the other piezo elements

In Fig. 8b the induced impact of approx. 20 N results in a very high quality low noise response (4 MS/s sampling rate) of the integrated piezo elements. With this setup an eigenfrequency analysis can be performed to check on the one hand the manufacturing quality of the blade and on the other hand set a reference for a pre and post impact comparison to evaluate structural damage and failure.
In Fig. 8c an exemplary SHM test is presented. A voltage pulse with a nominal amplitude of 200 V and 500 ms duration is send to piezo element #2 and the corresponding voltage over time response (4 MS/s sampling rate) of the other piezo elements are plotted. This could also be repeated with the excitation of the remaining sensors in order to do an in-situ structural health comparison of the blade after an impact event.

5 Conclusions

In the presented study a methodology for designing and manufacturing embedded sensor-actuator layers for CFRP aerospace components has been developed. With Tailored Embeddable Sensor-Actuator Layers (TEmSAL) the sensor network can be individually customized for the loading and geometrical requirements of the respective component. It also features off-the-shelf capabilities for series production purposes.
It has been shown that the used PI carrier foil in combination with the adhesive agent has not only a negligible influence on the layer interface properties, but rather strengthens it and prevents delaminations. A proof of concept has been performed using impact tests to investigate sensing capabilities and excitation by the embedded piezo ceramic elements using laser vibrometer for SHM capabilities. Subsequently, the applicability of the developed concept for aerospace applications has been demonstrated by using TEmSAL in a generic fan blade like structure. Both sensor and actuator capabilities where examined. In a next step, soft body impact experiments at 120 m/s will be performed to explore TEmSAL capabilities under extreme loading conditions.

Acknowledgements

The project leading to this publication has received funding from the European Union's Horizon 2020 research and innovation programme under grant agreement No 636549, project EXTREME.

References

[1] Kostka, P.; Holeczek, K.; Hufenbach, W.: A new methodology for the determination of material damping distribution based on tuning the interference of solid waves. Engineering Structures, Volume 83, 15 January 2015, Pages 1-6
[2] Praisach, Z. I.; Micliuc, D.M.; Gillich, G.R.; Korka, Z.I.: Natural Frequency Shift of Damaged Circular Plate Clamped all around. ANNALS of Faculty Engineering Hunedoara– International Journal of Engineering, Tome XIV [2016] – Fascicule 4
[3] Hufenbach, W; Gude, M; Heber, T.: Embedding versus adhesive bonding of adapted piezoceramic modules for function-integrative thermoplastic composite structures. Composites Science and Technology, Elsevier, 2011, 71 (8), pp. 1132
[4] Winkler, A.; Modler, N.; Dannemann, M.; Starke, E.; Holeczek, K.: Aktive faserverstärkte Thermoplastverbunde mit materialhomogen integrierten Piezokeramikmodulen – ein Ausblick. Smarte Strukturen und Systeme: Tagungsband des 4SMARTS Symposiums vom 6. - 7. April 2016 in Darmstadt

[5] Hufenbach W.; Modler, N.; Winkler, A.; Ilg, J.; Rupitsch, J.: Fibre reinforced composite structures based on thermoplastic matrices with embedded piezoceramic modules. Smart Materials and Structures, 23 (2014) 025011 (10pp)
[6] Hufenbach, W.; Gude, M.; Modler, N.; Heber, T.; Winkler, A.; Weber, T.: Process chain modelling and analysis for the high volume production of thermoplastic composites with embedded piezoceramic modules. Smart Materials Research (2013), article ID 201631, S. 1-13
[7] Meo, M., Zumpano, G., Piggott, M., & Marengo, G. (2005). Impact identification on a sandwich plate from wave propagation responses. Composite structures, 71 (3-4), 302-306.
[8] De Simone, M. E., Ciampa, F., Boccardi, S., & Meo, M. (2017). Impact source localisation in aerospace composite structures. Smart Materials and Structures, 26(12), 125026.
[9] De Simone, M. E., Ciampa, F., & Meo, M. (2017). A Structural Health Monitoring Technique for the Reconstruction of Impact Forces in Aerospace Components. Structural Health Monitoring 2017.
[10] Ciampa, F., & Meo, M. (2014). Impact localization on a composite tail rotor blade using an inverse filtering approach. Journal of Intelligent Material Systems and Structures, 25(15), 1950-1958.
[11] Lin, M.; Chang, F.-K.: The manufacture of composite structures with a built-in network of piezoceramics. Composites Science and Technology 62 (2002), Nr. 7-8, page 919-939
[12] Ye, L.; Lu, Y.; Su, Z.; Meng, G.: Functionalized composite structures for new generation airframes: a review. *Composites Science and Technology* 65 (2005), page 1436–1446

A new non-contact measuring method for the evaluation of the curing status of glued lightweight components based on ultrasound

Giovanni SCHOBER[1], Pierre PFEFFER[1]
SKZ - Das Kunststoff-Zentrum, Würzburg, Germany
Contact e-mail: g.schober@skz.de

Abstract. The curing of adhesives depends mainly on the application conditions such as temperature or humidity. Manufacturers' information on pot life and processing time are therefore not always directly transferable to the real application, so that large safety factors in the curing time are observed. This unnecessarily ties up resources such as production and storage areas and delays subsequent process steps, which is accompanied by a significantly reduced added value.
There are only a few non-destructive test methods that can be used to monitor the curing processes of adhesives in laboratory. Their main disadvantages are the lack of process capability, the low penetration depth, the high system costs and finally the lack of applicability for real component structures with complex geometries. On the other hand, contact-based ultrasound methods enable monitoring of the curing process with high temporal and spatial resolution and thus offer decisive advantages over laboratory technologies. The sound attenuation and the sound velocity provide material-dependent information that can be correlated directly with the degree of curing. However, disadvantages are a lack of process capability, as between the test head and the component a coupling medium must be present, and the acquisition of locally limited point information. Guided waves by using non-contact air ultrasonic technology are an alternative approach that overcomes these disadvantages. A corresponding test set-up and an evaluation method for determination of ultrasonic characteristics, which allow conclusions to be drawn on the degree of curing, are presented. The referencing is done by rheological investigations.

1. Introduction

The increasing use of lightweight components and composite materials has been gaining in importance in many industrial sectors, such as the aerospace industry, the automotive, maritime and power plant industries. This trend is often based on the aim of weight savings, e.g. to allow a reduction of pollutant emissions while increasing the performance of the entire component [1]. Bonding is often preferred due to its economic and easily automatable application and its avoidance of stress concentrations compared to mechanical joining methods such as rivets or screws [2]. The need for new joining techniques is also accompanied by suitable methods for process monitoring. The cost-efficient use of adhesive bonds therefore requires a reliable and non-destructive testing method both for assessing the final seam quality, as well as for monitoring the actual curing status depending on varying process conditions, respectively [3,4]. While contact ultrasound for the monitoring of curing processes is described in literature, air-coupled and therefore non-contact ultrasonic systems are primarily used for the detection of defects in aerospace applications [5]. A test approach is presented below to cost-effectively monitor the curing of adhesives in a non-contact, real-time, and integrated manner across a larger component area. In this case, so-called guided waves are used, which are a constructive interference of different mode converted longitudinal and transversal waves, reflected within the plate-shaped component [6].

2. State of the art

The available test methods for monitoring the curing of adhesives are essentially limited to laboratory measurement methods such as DSC [7], DMA [8,9], NMR [10], terahertz [11, 12] or rheological investigations [9]. On the other hand, ultrasonic methods offer the advantages of greater penetration depth, applicability in final component structures, and robustness to production conditions. It has already been shown that the characterization of hardening processes is possible by investigating the longitudinal propagation velocity of the sound waves and their attenuation [7]. Furthermore, investigations by means of contact ultrasound showed a consistent correlation of the curing of a pure polyester resin with corresponding ultrasound characteristics [8,9].

3. Setup and Experiment

3.1 Investigations with air-coupled ultrasound

To measure the curing process, a system consisting of the device *Sonoinspect, Forschungszentrum-Ultraschall gGmbH,* with transducers *SONOSCAN CF 200, SONOTEC Ultraschallsensik GmbH* was used. The bonding to be monitored was excited with six square-wave bursts each with a center frequency of 218 kHz, a duty cycle of 50 % and a transmission voltage of 200 V. The data acquisition of the 70 dB amplified received signal was carried out with a sampling rate of 3.9 MHz. At a time interval of 30 s, an A-image averaged over 128 pulses with a pulse repetition rate of 50 ms was recorded in each case. The measuring setup consisted of a transmitter and a receiver, which were each directed at a 22° angle and a distance of 20 mm to a 2 mm thick polystyrene plate. The angle α was chosen according to the optimum coupling condition for generating guided waves as a function of the phase velocity in air v_{air} and in the polystyrene plate $v_{polystyrene}$:

$$\alpha = \arcsin\left(\frac{v_{\text{air}}}{v_{\text{polystyrene}}}\right) \quad (1)$$

The respective phase velocities were determined experimentally. For this purpose, the distance between the transmitter and receiver was varied continuously, the captured A-images (see Figure 1, left) were converted into corresponding B-images [10] and plotted over the transmitter-receiver distance (see Figure 1, right).

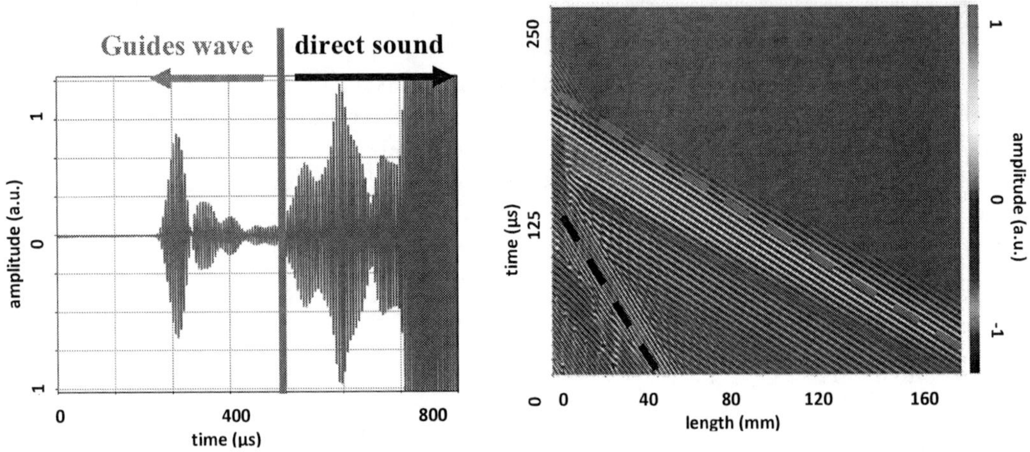

Figure 1. Exemplary A-image showing the guided wave separated from the direct sound (left). The guided wave is between approx. 280 µs and 500 µs and the direct sound can be recognized from approx. 500 µs. By continuously increasing the distance between transmitter and receiver, numerous A-images could be transferred to a B-image (right), which is applied over the running distance. The visible slopes correspond with the phase velocity in air (dashed black line) and the guided waves in polystyrene (dashed red line).

The transmitter-receiver distance and thus the running distance of the guided waves was 140 mm. The 2-component adhesives *Scotch-Weld DP460, 3M Germany GmbH,* based on epoxy resin were coated manually with a thickness of about 2 mm and a length of about 50 mm in the form of an overlap bond between two 2 mm thick polystyrene sheets (see Figure 2). According to the manufacturer, the temperature-dependent "processing time" is 60 minutes and the "handling time" is 4 to 6 hours.

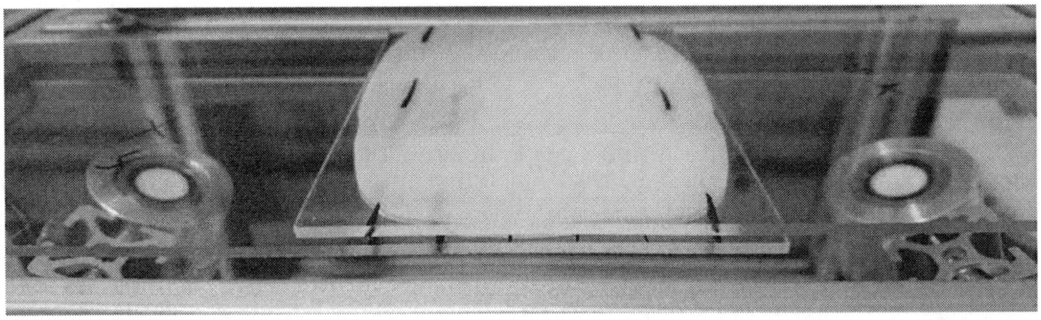

Figure 2. Photograph of the measurement setup used to monitor the curing of the Scotch-Weld DP460 adhesive, which overlaps between two polystyrene sheets of thickness 2 mm. The transducers were located at a distance of 140 mm below the polystyrene plates at an angle of 22 ° (circular, white).

The recorded measurement data were further processed with an evaluation software developed at SKZ. The signal intensity I and the group delay t_g were calculated by considering the amplitude A of the guided waves in an evaluation interval between the times t_1 and t_2 according to equations 2 and 3:

$$I = \int_{t_1}^{t_2} A(t)^2 \cdot dt \tag{2}$$

$$t_g = \frac{\int A(t)^2 \cdot t \cdot dt}{\int A(t)^2 \cdot dt} \tag{3}$$

3.2 Rheological investigations

To verify the ultrasound results, additional rheological examinations were carried out. Both adhesives were investigated with the rheometer *Haake Mars III, Thermo Fisher Scientific Inc.*, in plate/plate setup and oscillation mode. The plate diameter was 20 mm, the measuring gap 0.5 mm, the frequency 1 Hz, the shear 1 % and the temperature 25 °C.

4. Results

Figure 3 shows the temporal curves of the rheological investigations (top), the ultrasound intensity (center) and the the group velocity (bottom). For a simplified interpretation, the corresponding manufacturer's instructions are shown shaded in grey (see section 3.1).

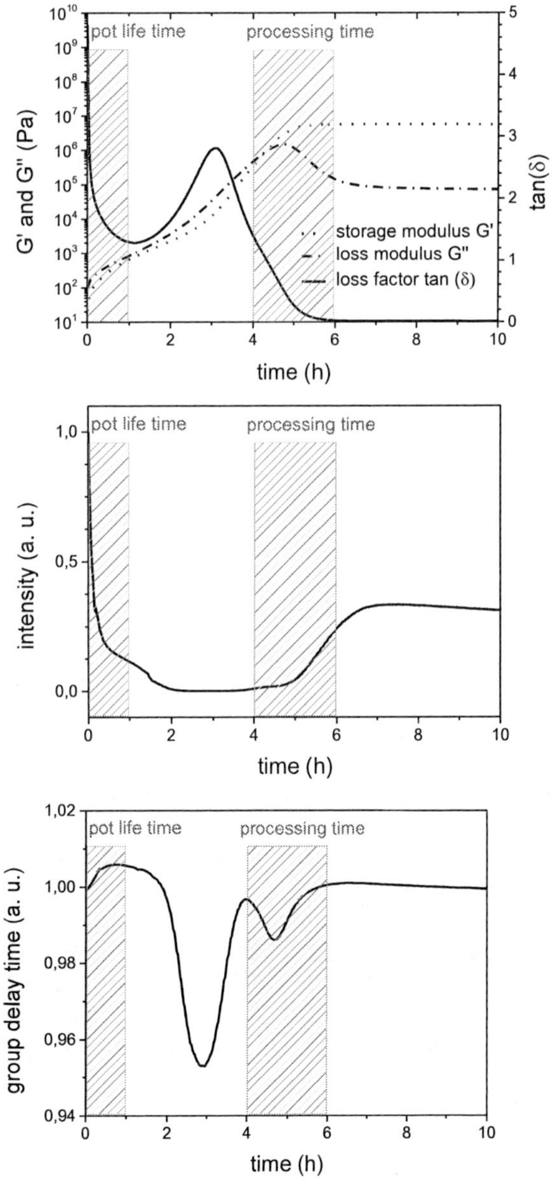

Figure 3. Rheological results (top), signal intensity (center) and group delay (bottom) measured during the curing process of the adhesive Scotch-Weld DP460.

5. Interpretation and outlook

It can be seen that the signal intensity of the guided waves initially decreases continuously until the maximum loss factor is reached. This behaviour has already been described in the curing of epoxy resins on the literature side [9,10]. The reason lies in the increasing sound attenuation due to the increasing occurring of viscoelastic effects that can be also seen in the loss factor. During the further processing time, the signal intensity increases continuously, which can be justified due to the temporally changing state of matter from liquid to solid with a physically induced decrease in damping, that can be also seen in the decrease of the loss factor. At the end of the further processing time, no significant change in the signal intensity can be observed, which is accompanied by the progressive cross-linking and the manifestation of mechanical end properties. Furthermore, a change in the group delay during the curing reaction could be detected. The course of its curve does not correspond to what was expected in the curing of epoxy resins from the literature [9]. However, it is noticeable, that the shape of the group delays' curve corresponds approximately to the reciprocal of the course of the loss factor. For example, the group delay reaches its minimum at the same time as the loss factor reaches its maximum, and after about 5.5 hours a nearly constant value can be detected. With a view to the literature, this is justified by the increasingly diffusion-controlled reaction and the associated lower mobility of the chain segments [13-15]. In addition to extended signal processing with algorithms to be developed, further measurements using DMA, NMR, and rheological investigations are planned in order to verify the measurement approach presented here. The presented measurement approach using non-contact guided ultrasound waves allows significantly greater penetration depths as well as applicability to real component structures and not only to the adhesive itself, perspectively.

6. References

[1] J. Prager, D. Brackrock, E. Dohse, M. Gaal, T. Homann und M. Grzeszkowski, „Anwendung geführter Ultraschallwellen für die Prüfung von Klebeverbindungen", *DGZfP-Jahrestagung 2014 - Mi.1.A.3.*

[2] W. Na und J. Baek, „Adhesive Defect Monitoring of Glass Fiber Epoxy Plate Using an Impedance-Based Non-Destructive Testing Method for Multiple Structures", *Sensors,* Nr. 17, p. 1439, 2017.

[3] E. Kraus, B. Baudrit, P. Heidemeyer, M. Bastian, O. V. Stoyanov und I. A. Starostina, „Limits and Possibilities in Destructive Testing of Bonded Polymer Connections", *Polymers Research Journal*, 9/3, pp. 337-344, 2015.

[4] A. Groß, H. Lohse, H. Gleich, J. Holtmannspötter, M. Wetzel, J. von Czarnecki, R. Brucksch, E. Kraus, B. Baudrit, P. Heidemeyer, M. Bastian und F. Thomsen, „Qualitätssicherung und Prüfverfahren", In: Doobe M. (Hrsg.) *Kunststoffe erfolgreich kleben*. Springer Vieweg, Wiesbaden, 2017.

[5] W. Hillger, L. Bühling und D. Ilse, „Berührungslose Ultraschallprüfung mit Ankopplung über Luft - Neuentwicklung für die Praxis", *DGZfP-Jahrestagung 2017.*

[6] D. Döring, „Luftgekoppelter Ultraschall und geführte Wellen für die Anwendung in der zerstörungsfreien Werkstoffprüfung", Universität Stuttgart, Dissertation, 2011.

[7] A. Maffezzoli, E. Quarta, V. A. M. Luprano, G. Montagna und L. Nicolais, „Cure Monitoring of Epoxy Matrices for Composites by Ultrasonic Wave Propagation", *Journal of Applied Polymer Science,* Nr. 73, pp. 1969 - 1977, 1999.

[8] N. T. Nguyen, M. Lethiecq und J. F. Gerard, „Glass Transition Characterization of Homogeneous and Heterogeneous Polymers by an Ultrasonic Method". *Ultrasonics,* Nr. 33, pp. 323 - 329, 1995.

[9] F. Lionetto und A. Maffezzoli, „Relaxations during the Postcure of Unsaturated Polyester Networks by Ultrasonic Wave Propagation, Dynamic Mechanical Analysis, and Dielectric Analysis", *Journal of Polymer Science,* Nr. 43, pp. 596 - 602, 2005.

[10] R. E. Challis, M. E. Unwin, D. L. Chadwick, R. J. Freemantle, I. K. Partridge, D. J. Dare und P. I. Karkanas, „Following Network Formation in an Epoxy/Amine System by Ultrasound, Dielectric, and Nuclear Magnetic Resonance Measurements: A Comparative Study", *Journal of Applied Polymer Science,* Nr. 88, pp. 1665 - 1675, 2003.

[11] T. Probst, S. Sommer, A. Soltani, E. Kraus, B. Baudrit, G.E. Town und M. Koch, „Monitoring the Polymerization of Two-Component Epoxy Adhesives Using a Terahertz Time Domain Reflection System", *Journal of Infrared, Millimeter and Terahertz Waves*, 36/6, pp.569–577, 2015.

[12] S. Sommer, T. Probst, E. Kraus, B. Baudrit, E. T. Graham und M. Koch, „Cure monitoring of two-component epoxy adhesives by terahertz time-domain spectroscopy", *Polymer Science Series B*, 58/6, pp. 769-776, 2016.

[13] M. Hübner, W. Lang und G. Dumstorff, „Surface Integrated Printed Interdigital Structure for Process Monitoring the Cure of an Adhesive Joint", *Proceedings,* Nr. 1, p. 631, 2017.

[14] F. Lionetto und A. Maffezzoli, „Polymer Characterization by Ultrasonic Wave Propagation", *Advances in Polymer Technology,* Nr. 27, pp. 1 - 11, 2009.

[15] A. M. Lindrose, „Ultrasonic Wave and Moduli Changes in a Curing Epoxy Resin", *Experimental Mechanics,* pp. 227 - 232, 1978.

Multiple-Flash Shearography – a New NDT Method for Reducing Thermal Stresses During the Inspection Process

Pierre PFEFFER[1], Luis WACHTER[1], Daniel Hoffmann[1], Martin BASTIAN[1], Giovanni SCHOBER[1]
[1] SKZ-KFE gGmbH, Würzburg, Germany
Contact e-mail: p.pfeffer@skz.de

Abstract. For the first time we present results of shearography measurements on fiber reinforced plastics samples being thermally excited using a flash sequence. Similar to lock-in shearography, flash shearography permits yielding higher contrasts and signal-to-noise ratios than conventional shearography and obtaining depth information of defects. For this, shearographic measurements are continuously taken during cooling of the samples. The resulting stack of images is afterwards analyzed with regard to selected frequencies using Fourier-transformations. Depending on the chosen frequencies and on the sample's material, defects can be detected up to different depths. We find that flash shearography yields results of a quality similar to that of lock-in shearography while enabling much shorter measurement times. Utilizing the novel multiple-flash shearography which excites the device under test with a burst of flashes, it is for the first time possible to heat the structure with a sequence of flashes with adjustable flash duration, voltage and energy. This creates the possibility to use lower energy pulses, thereby reducing maximum surface temperatures and thermal stresses while leaving the detectability of defects unimpaired. Corresponding references measurements were carried out with the help of high speed thermography.

1. Introduction

Shearography is a nondestructive testing method based on interferences of sheared speckle patterns [1]. Due to conventional shearography being a fast full field inspection technique, it has become a widely accepted method for the testing of tires, especially for aerospace applications [2, 3]. Moreover, in recent years inspection of fiber-reinforced plastics has become more and more established [4].

In contrast to conventional shearography, which observes the differences between phase images of a test object under load and in an unloaded state, flash shearography captures multiple images during the test object's cooling after contactless excitation with a thermal pulse. Since a pulse can be represented as a continuous range in the frequency domain, flash shearography, similar to lock-in shearography, allows to observe the sample's response at frequencies lying in the pulse's frequency band. The thermal depth of penetration being frequency dependent, flash shearography offers the potential to gain access to depth information of defects. Besides, owing to the multitude of captures higher resolutions and signal to noise ratios compared to conventional shearography can be achieved, too. In our work, we explore the possibilities and limitations of flash shearography and analyze the practicability of multiple-flash shearography, dividing a strong single pulse into several lower-energy pulses, thereby reducing surface temperatures and thus thermal stress. Similar studies have already been carried out with thermographic systems [5].

2. Methods

For conventional shearography, we use a 1200 W HEDLER C12 halogen spotlight to thermally excite the test sample. Flash shearography is done with the help of two different systems: 1) The flash lamp VH1-6000 and the flash generator Tria 6000 S by HENSEL-Visit GmbH & Co. KG, Würzburg. This combination enables single pulses with a duration of 1.3 ms to 2.2 ms and energies up to 6 kJ. 2) The flash lamp VH3-6000 and a customized flash generator, which were developed in a research project between Hensel Visit and SKZ. This allows producing a burst of up to six flashes with adjustable flash duration, temporal distance between the flashes and flash energy. Here the maximum energy depends on the number of flashes and ranges from 1.9 kJ for six flashes to 6.0 kJ for a single flash. The developed pulse generator will be commercially available soon.

We used two different test objects as measurement samples: Sample A, a carbon fiber reinforced polymer (CFRP) in which cylindrical millings with different diameter and depth were inserted. The diameter decreases from top to bottom and the depth increases from left to right. The dimensions of the sample, which is shown in Figure 1 a) and b) are 11.50 cm x 25.50 cm x 0.64 cm. Sample B is a glass fiber reinforced polymer (GFRP) with millings of varying geometries and depths and overall dimensions of 16.50 cm x 20.00 cm x 0.60 cm (Figure 1 c) and d)).

Using a commercially available shearography sensor, provided by isi-sys GmbH, Kassel, and an own customized software, we perform four kinds of shearographic measurements:

1. conventional shearographic measurements whose results stem from the phase difference images of measurements of loaded and unloaded sample-states,
2. lock-in shearographic measurements which use periodical excitations of the test object and frequency selective evaluations of the measurement data,
3. flash shearographic measurements: Here the sample is excited with one single flash pulse, shearographic images are captured during the cooling of the sample and the resulting image is obtained by pixelwisely Fourier-transforming the stack of acquired images and subsequently assigning the resulting amplitude at a chosen frequency to the corresponding pixel,
4. multiple-flash shearographic measurements: Instead of a single pulse, here we excite the sample with a series of flash pulses and continuously take shearographic measurements during the sample's cooling period. Apart from this, data processing and measurement evaluation is identical to single-flash shearography.

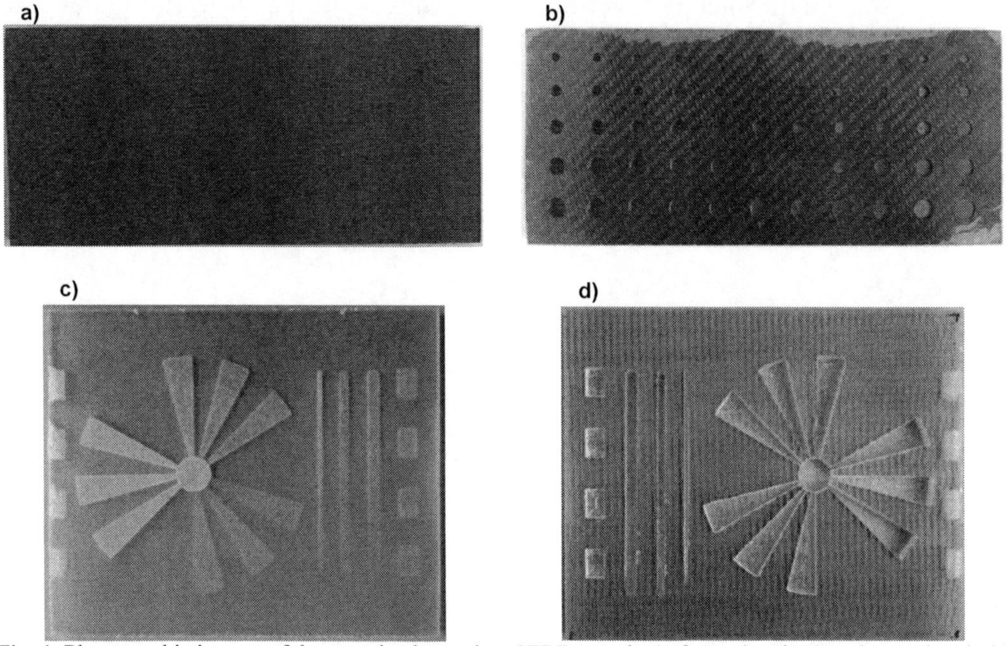

Fig. 1. Photographic images of the examined samples. CFRP sample A, front view in a) and rear view in b). GFRP sample B, front view in c) and rear view in d). During the measurements the samples were thermally excited and the shearographic images taken facing the flat front sides. Moreover, both samples' front sides were coated with a graphite layer in order to maximize light absorption and to avoid reflections in the case of sample A and a direct view of the defect structures in the case of sample B. The graphite coating is only visible in the picture of sample A. Sample B was not coated when taking the photos.

3. Results

Figure 2 shows a comparison of a conventional and three flash lamp excited shearographic measurements of sample A. While the results of the conventional measurement allow a relatively unclear and blurred detection of three full rows of defects and of the two deepest millings of the fourth and fifth row, the flash measurement at 6000 J excitation energy enables a much clearer and rich in contrast observation of three full rows. Flash energies of 3000 J and 1500 J only lead to very noisy and blurry images. A high energy of the exciting flash pulse is thus essential for the success of a measurement (N.B.: Short sample – flash lamp distances can compensate for lower lamp energies. Our setup featured a distance of about 50 cm.).

Similar to lock-in shearography, flash shearography allows to evaluate the measurement data at different frequencies in order to obtain depth information. In contrast to lock in shearography, where several time-consuming measurements at different frequencies have to be carried out, flash excitations allow to obtain the necessary data with a single measurement series in much shorter periods of time. Figure 3 demonstrates the visibility of defects up to different depths depending on the excitation/evaluation frequency using lock-in and flash shearography. It is moreover observable that flash excitation results in images of a similar quality to those obtained by lock-in measurements.

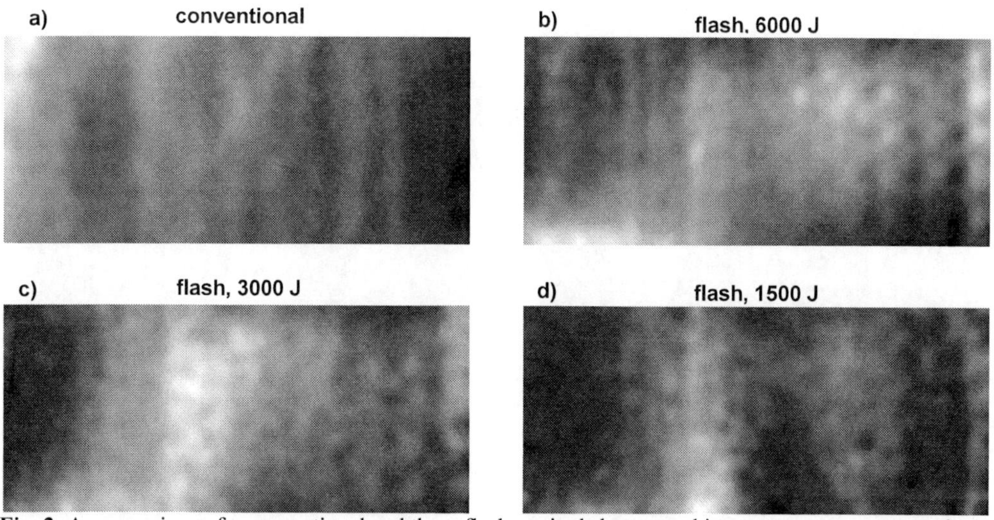

Fig. 2. A comparison of a conventional and three flash-excited shearographic measurements on sample A. a): Conventional, halogen spotlight-excited amplitude image. b), c) and d): Flash lamp (VH1-6000) excited measurement results with 6000 J, 3000 J and 1500 J pulse energy, respectively. Pulse durations were 2200 μs. Data of the flash measurements were evaluated at a frequency of 30 mHz.

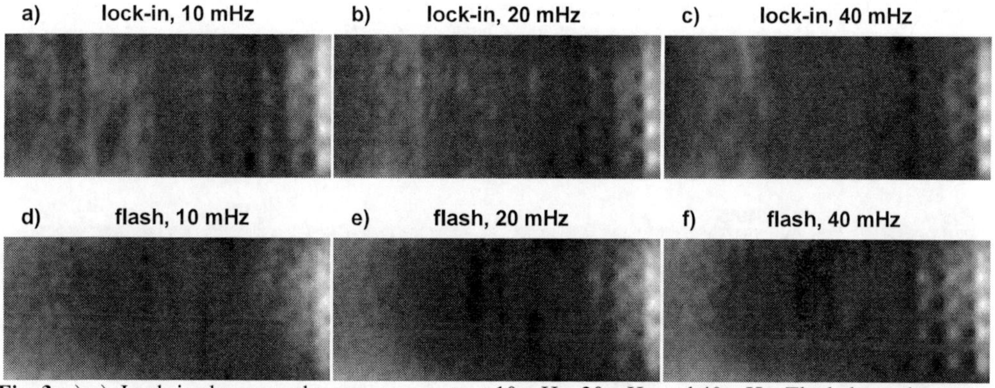

Fig. 3. a)-c): Lock-in shearography measurements at 10 mHz, 20 mHz and 40 mHz. The halogen lamp was operated at 840 W and a distance of approx. 1 m. d)-f): Flash shearography measurements using flash lamp VH1-6000 with a single pulse of 6000 J, pulse durations of 2200 μs and evaluations at 10 mHz, 20 mHz and 40 mHz.

In order to quantize these findings, we define a "defect contrast" by manually constructing a rectangle around each defect and by calculating the difference between the highest and the lowest amplitude each of these rectangles contain. Figure 4 presents these defect contrasts, averaged over the rightmost row of columns, for conventional, lock-in and flash shearography. As the results reveal, lock-in and flash offer defect contrasts of a similar order of magnitude, with lock-in results being slightly better. Contrast values for conventional shearography in contrast lie at about half of those of flash measurements.

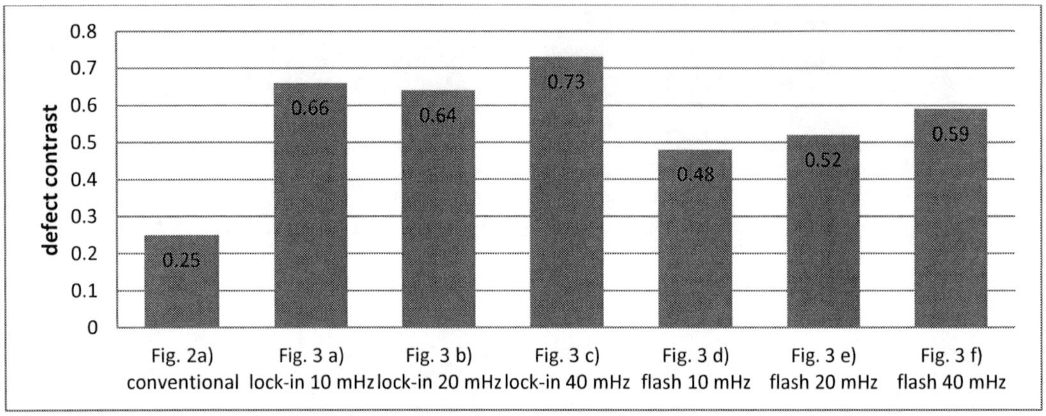

Figure 4. Mean defect contrast values of the defects in the rightmost columns of the subfigures in Figure 3 and of the rightmost column of the conventional measurement in Figure 2 a).

Since high energy flash pulses can lead to momentary heating-ups of several hundred °C for a few ms (cf. Figure 7), flash shearography can irreversibly damage the test object [6]. The small defect on the lower border of all images in Figure 3 for instance was caused by a 6000 J flash pulse which led to melting and leakage of the CFRP's resin and resulted in a small delamination near the sample's surface. A compromise between the short measurement times of flash shearography and the low excitation energies of conventional, continuously excited shearography can be reached by dividing the energy of a single pulse into a burst of several pulses. Results of this multi-flash shearography are depicted in Figure 5. As can be seen, the overall image quality of the single- and multi-pulse measurements is similar, the multi-flash results even allowing a slightly better detection of the small millings than the single-flash image.

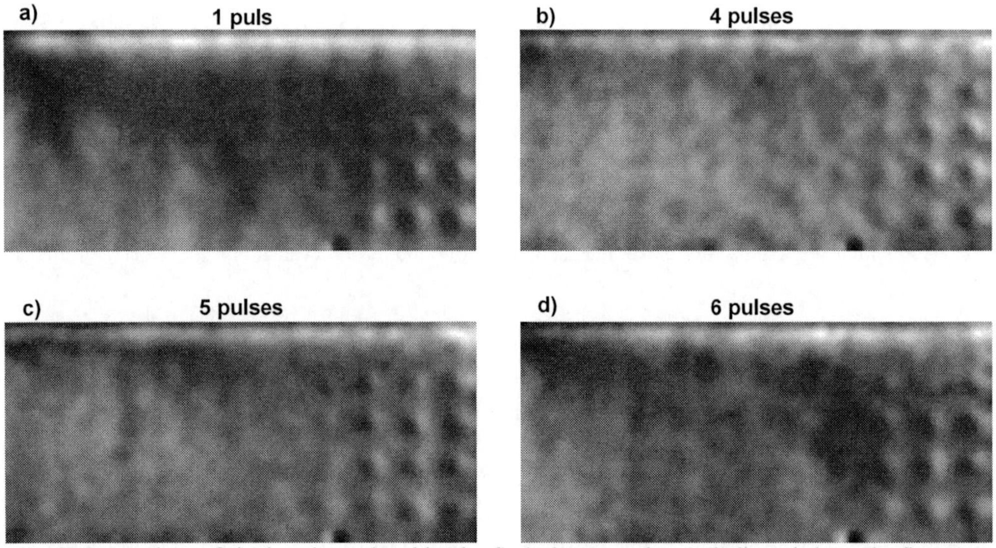

Fig. 5. Comparison of single pulse and multi pulse flash shearography. As indicated above the figures, a single pulse was used for subfigure a), four pulses for b), five for c) and 6 flash pulses for subfigure d).The overall energy input was in each case 6000 J, the evaluation frequency 45 mHz. The pulse durations were 2100 µs for 4 pulses, 2200 µs for 5 pulses and 1500 µs for 6 pulses. The time interval between the pulses was 200µs. The measurements took 18 s during which 100 phase images were captured. Flash lamp VH3-6000 was used. Due to the differences in light focussing between VH1-6000 and VH3-6000, the results of the single flash measurements here and in Figure 2 differ.

Similar results were obtained for sample B and are illustrated in Figure 6. Comparable to the CFRP sample, only little differences are discernible between the measurements taken with different numbers of flash pulses as all result images distinctly show the six deepest wedges and the uppermost rectangular milling on the left side of the sample. Furthermore they all rudimentarily allow the detection of the three vertical millings and the three deepest wedges and the three lower millings on the sample's left end.

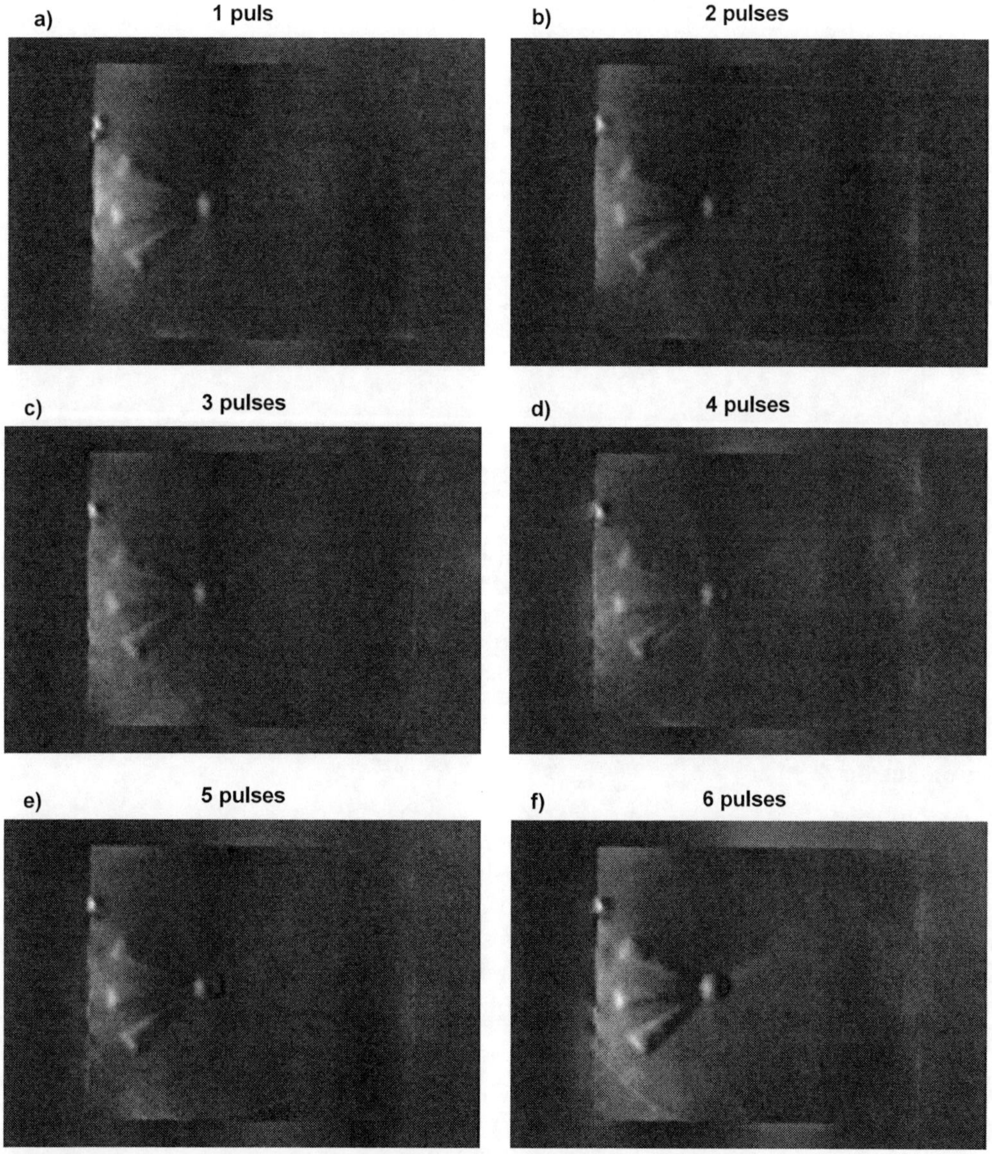

Fig. 6. Shearographic images of sample B after excitation with flash lamp VH3-6000. The total input energies were in every case 1650 J. Pulse durations were 3000 μs for the single pulse, 700 μs for 2 pulses, 400 μs for 3 pulses, 300 μs for 4 pulses, 250 μs for 5 pulses and 200 μs for 6 pulses. The temporal distance between the individual flashes was 100 ms, the evaluation frequency 20 mHz. During the measurement, 200 phase images were captured in 35 s.

The maximum surface temperature of sample A, measured with a high-speed thermography camera during a separate measurement, is plotted in dependence of the number of pulses and the excitation energy in Figure 7 [6]. As expected, lower overall pulse energies and higher numbers of pulses lead to a decrease of the maximum surface temperature. Compared to a single pulse, a burst of six pulses reduces the maximum surface temperature by approximately 35 %. The temperature values indicated in Figure 6, however, cannot be considered as representative surface temperature values, since the radiation emitted by the flash lamp's plasma is reflected by the sample surface, partially scattered into the camera and thus leads to measurement values which tend to be higher than the actual surface temperature [6].

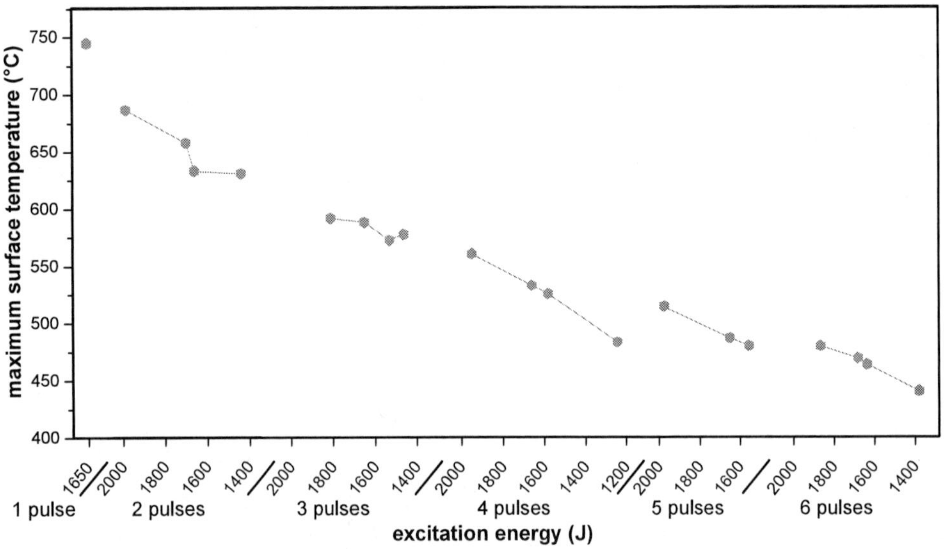

Fig. 7. Measured maximum surface temperature values in dependence of the total excitation energies for various numbers flash pulses. Excitation was done via flash lamp VH3-6000 and temporal pulse distances of 150 ms [6].

4. Conclusion

We examined flash shearographic measurements and compared them to conventional and lock-in shearographic methods. Flash shearography allows quick depth-sensitive measurements of a quality similar to those of lock-in shearography, although exerting high thermal stresses on test samples. In order to sidestep this drawback, we used multi-pulse flash shearography, dividing high excitation energies into several lower-energy pulses and demonstrated the consequential reduction of sample surface temperatures at consistently good image qualities.

Acknowledgement

The proposal project 19063 N by the research association "Fördergemeinschaft für das SKZ" was funded through the "Arbeitsgemeinschaft industrielle Forschungsvereinigungen e.V." (AiF) within the scope of the program for the Promotion of Cooperative Industrial Research and Development (IGF) by the Federal Ministry for Economic Affairs and Energy (BMWi) due to a decree by the German Federal Parliament. We would like to thank them for their financial support. Furthermore we would like to thank the members of the project-related committee for the discussions. SKZ is part of and gratefully acknowledges support of the "Deutsche Industrieforschungsgemeinschaft Konrad Zuse".

References

[1] Francis, D., Tatam, R. P., & Groves, R. M. (2010). Shearography technology and applications: a review. *Measurement science and technology*, *21*(10), 102001.

[2] Hung, Y. Y., & Ho, H. P. (2005). Shearography: An optical measurement technique and applications. *Materials science and engineering: R: Reports*, *49*(3), 61-87.

[3] Yang, L., & Xie, X. (2016). *Digital shearography: New developments and applications* (Vol. 31). Bellingham: SPIE Press.

[4] Gholizadeh, S. (2016). A review of non-destructive testing methods of composite materials. *Procedia Structural Integrity*, *1*, 50-57.

[5] Hoffmann, D., Pfeffer, P., Kolb, C., Mayr, M., Stoehr, A., Schober, G., Bastian, M. (2018), Advantages of Multi-Pulse Thermography, *Proceedings of the SHM-NDT Conference 2018*, (in print).

[6] Schober, G., Kremling, S., Hochrein T. (2017). Multiple flash thermography for non-destructive testing of lightweight components. *AMA Conferences, Proceedings IRS² 2017*, 813-817.

3D visualization of ultrasonic scanned data in the context of test methods for the approval of civil aeronautical components

Andreas HEINZE [1]
[1] IMA Materialforschung und Anwendungstechnik GmbH, Dresden, Germany
Contact e-mail: andreas.heinze@ima-dresden.de

Abstract

The aviation of the future is a driving force for the further development of the global, highly interconnected civilization. By 2050, the number of flights in Europe alone will increase from currently nearly 10 million aircraft movements to about 25 million per year. Megatrends such as globalization, logistics and mobility as driving forces for it have to be balanced with the global challenges of resource conservation, environmental and climate protection as well as sustainability. Energy efficiency will be the key to create the necessary balance. These development trends are related to the use of new energy efficient and resource-conserving materials in the context of new comfortable and interconnected airframe and cabin concepts. Furthermore, they partly demand completely new standards for aircraft construction and maintenance of the future with regard to aspects such as test, inspection and maintenance. This results in an increase of the requirements on engineering, particularly on proving the lifetime of new materials and structures. A variety of development-related tests are needed to finally prove the resistance of the entire aviation structure and to guarantee its integrity.

IMA Materialforschung und Anwendungstechnik GmbH (IMA Dresden) offers all opportunities to compare material usage and construction principles, to verify calculation methods and to experimentally study different influences on strength combined with non-destructive testing methods. In lightweight construction in aviation, minor damages caused by fatigue or external impacts are no taboo but rather are a part of the security concept ´damage tolerance´ - the largest area in the spectrum of our experiments. IMA Dresden examines, when and where damages occur, how they increase, which residual strength remains and how a structure reacts to e. g. cyclic loads after an impact. There, the specialists for non-destructive testing of IMA Dresden assist with the most appropriate process in each test and inspection phase. The concept of non-destructive testing aims at the quantitative determination of all arising damages and at their earliest recognizability.

An overview of materials and test methods for aerospace components is given in the poster contribution and a 3D visualization of ultrasonic scanned data is presented there.

IMA DRESDEN 25 since 1993

MATERIALS AND TEST METHODS FOR THE APPROVAL OF CIVIL AERONAUTICAL COMPONENTS

IMA Dresden examines, when and where damages occur, how they increase, which residual strength remains and how a structure reacts to e.g. cyclic loads after an impact.

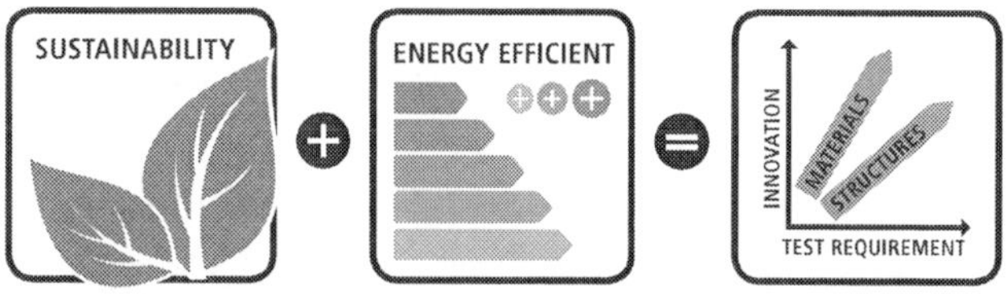

The development trends in aviation are related to the use of new energy efficient and resource-conserving materials in the context of new comfortable and interconnected airframe and cabin concepts. Furthermore, they partly demand completely new standards for aircraft construction and maintenance of the future with regard to aspects such as test, inspection and maintenance. A variety of development-related tests are needed to finally prove the resistance of the entire aviation structure and to guarantee its security.

IMA Materialforschung und Anwendungstechnik GmbH (IMA Dresden) offers all opportunities to compare material usage and construction principles, to verify calculation methods and to experimentally study different influences on strength combined with non-destructive testing methods. In lightweight construction in aviation, minor damages caused by fatigue or external impacts are no taboo but rather are a part of the security concept ´damage tolerance´ - the largest area in the spectrum of our experiments.

There, the specialists for non-destructive testing of IMA Dresden assist with the most appropriate process in each test and inspection phase. The concept of non-destructive testing aims at the quantitative determination of all arising damages and at their earliest recognizability.

3D VISUALIZATION OF ULTRASONIC SCANNED DATA EXAMPLE: CFK-STEP WEDGE

The standard presentations of ultrasonic testing or rather their results in A, B, C and D pictures are very complex. During the further processing of the results through NDT-distant departments or companies, difficulties in interpretation or imagination are likely to occur. Additional 3D presentation might not only lead to better understanding but also further application areas are conceivable:

- integration in CAD environment
- FEM examinations of impacts

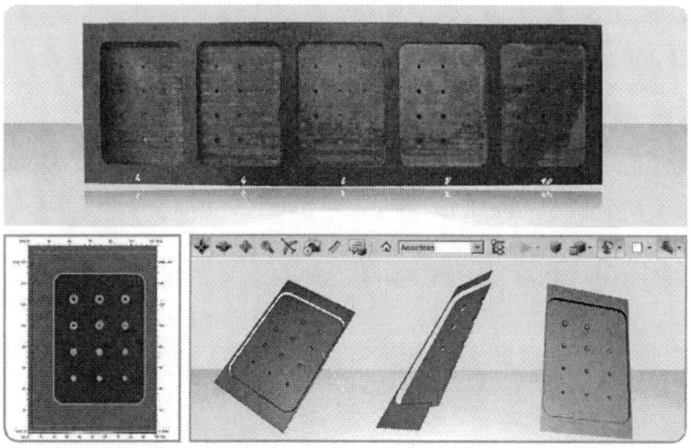

CFK STEP WEDGE - FROM THE C-SCAN FOR DISPLAY IN THE 3D-PDF AND CAD ENVIRONMENT.

AVIATION-SPECIFIC NDT-TRAININGS ACCORDING TO DIN EN 4179

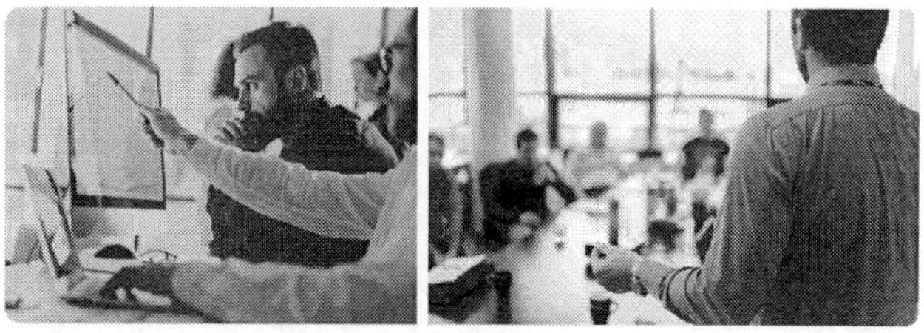

The Dresden NDT training center conducts vocational trainings and continuing education for non-destructive testing at the technology site Dresden and is operated by a cooperation of EFW Elbe Flugzeugwerke GmbH, Testia GmbH and IMA Materialforschung und Anwendungstechnik GmbH.

Course participants benefit from the long-standing material and test competence of IMA Dresden and the services of EFW such as aircraft conversion, maintenance and development of lightweight construction. In addition to exciting insights into aircraft conversion and the work with current test technologies, recent test methods are taught in the training.

The training center offers the following courses:

- ET1 Eddy Current Testing level 1
- ET2 Eddy Current Testing level 2
- UT1 Ultrasonic Testing level 1
- UT2 Ultrasonic Testing level 2

Beside fixed dates, individual courses can be arranged.

Author: Andreas Heinze

IMA Materialforschung und Anwendungstechnik GmbH
Wilhelmine-Reichard-Ring 4
01109 Dresden
+49 (0)351/8837-0
ima@ima-dresden.de
www.ima-dresden.de
Visit us:

The thermoacoustic effect and its application in air-coupled testing of composite structures

Klaas BENTE [1], Daniel KOTSCHATE [1], Saskia WENDLAND [1,2], Mate GAAL [1]
[1] Bundesanstalt für Materialforschung und -prüfung (BAM), Berlin, Germany
[2] Technical University of Ilmenau, Germany
Contact e-mail: klaas.bente@bam.de

Abstract. Airborne ultrasonic testing of lightweight, structured composite materials enables fast and contact-free non-destructive testing in aerospace and avoids material degradation due to contact with a coupling liquid. Established resonant air-coupled transducers consist of piezocomposite materials and several matching layers or more advanced materials like charged cellular polypropylene. The relaxation time and the specific frequency of such mechanical ultrasound emitters limit the spectrum of applications for each device. A short pulse length is key for reliable defect detection and each component at test can be best characterized at material- and geometry-specific frequencies. Here we show that focused thermoacoustic transducers are suited for testing lightweight, structured composite plates. Since the ultrasound is generated in air, these transducers show no resonance behavior and emit a broadband acoustic spectrum between 1.2 kHz and 1 MHz. Composite specimens of 3 mm to 9 mm thickness made of polylactide with a honeycomb structure were tested. Flat bottom holes were introduced to quantify the spatial resolution of the imaging method inside the strongly anisotropic specimen. As no broadband receivers are available yet, cellular polypropylene transducers were used as receivers, which limits the bandwidth of the method towards the bandwidth of the receiver. Nevertheless, we demonstrate the competitiveness of the thermoacoustic transducer compared to mechanical emitters at their respective resonance frequencies. Because a thermoacoustic transmitter features a nearly ideal pulse width, a single transmitter can be coupled with receivers with different resonance frequencies. With the development of broadband ultrasound receivers, air-coupled ultrasound spectroscopy will likely be possible in the near future. The analysed transducer holds the potential to speed up testing during production and maintenance in aerospace and automotives. Its combination with a broadband receiver could also expand the application field of air-coupled ultrasonic testing from a qualitative error detection towards a quantitative, spatially resolved analysis of mechanical material properties.

1. Introduction

Air-coupled ultrasonic material testing has distinct advantages when compared to other non-destructive imaging techniques. These include contactless coupling, no ionizing radiation, manageability and costs [1].

The main limitation of the technique is a high difference in acoustic impedance between air and the imaged object's material [2]. This leads to a high reflectivity and all further signals inside the object are of relatively low amplitude. Besides this fundamental physical limitation, we would like to point out three more restrictions that prevent this technique from becoming a standard in non-destructive testing in aerospace.

Most air-coupled transducers currently used in industry and developed in science work at a specific resonance frequency due to the thickness mode of vibration of the transducer material. One of the resonant transducer's main disadvantage is the time needed to find a suitable frequency and hence a suitable combination of sending and receiving devices for each set of material parameters and object geometries. The disadvantage of needing many transducer combinations to account for each new material-geometry combination can outweigh the advantages of the technique. Secondly, resonant transducers only enable a qualitative detection of interfaces and no quantification of material properties. Finally, the spatial resolution along the acoustic axis for pulse-echo sequences is low for resonant air-coupled transducers due to the limitation in frequencies that can practically be used in air at practical distances. Sound wave with frequencies above 2 MHz experience heavy damping in air and the resulting width of an acoustic pulse can often be too large to investigate relevant errors.

Broadband, air-coupled ultrasound emitters hold potential to overcome the listed limitations. Broadband pulses feature short pulse widths, which enable more precise echo-based imaging. Additionally, broadband signals might enable airborne ultrasound spectroscopy, facilitating spatially resolved material characterization when combined with broadband receivers [2]–[4]. Further, an ideal transducer would be usable at different frequencies that can be adjusted to the case needed for a certain material and object geometry.

Here, we show for transmission imaging that our broadband emitter [5], [6] can compete with a narrowband transducer that has been chosen for a certain material and object geometry. In the context of aerospace applications, we chose a hardware specimen made of a composite material with an internal honeycomb structure and flat bottom holes. We would like to emphasize that several highly sensitive narrowband transducer combinations were tested and the best performing combination was used for all further experiments. Hence, the broadband transducer competed with the best available solution that our narrowband transducer collection could offer. Thereby, we used a receiver, matched to the narrowband emitter.

We first outline the generation of broadband ultrasound signals with the thermoacoustic (TA) effect, describe how our TA emitter was built, describe the transmission imaging setup and the competing narrowband transducers and finally show and discuss the results.

2. Transducers

2.1 Broadband Acoustic Pulses Using the Thermoacoustic Effect

When heating up a closed volume V of air, the local increase of internal energy ΔU leads to an increase in pressure $p = \Delta U/V$. The resulting peak in pressure travels through air as a longitudinal sound wave [5]. The advantages of this kind of sound generation are self-evident, as no resonant mass has to perform the mechanical volume-work and no resonance frequency is related to the sound generation process. Hence, only the time dependency of the heat transfer on the surface of the heat generator determines the shape of the sound wave in air.

In order to predict an acoustic pressure, ΔU and V have to be determined. For each surface element of a transducer, one can assume V to be a half-sphere with a radius depending on the heating time and the speed of sound in air. ΔU is the internal energy that is generated inside the transducer and transfered to air. A reliably controllable heat source is a thin ohmic conductor mounted onto a substrate material, such that the available effective power is the electrical power loss inside the resistor. The ratio between the heat transfer towards the substrate, holding the thin ohmic film and air can be determined by considering the thermal properties of the substrate and air, as summarized in the next paragraph.

2.2 Realization of a Focused Thermoacoustic Emitter

A fused silica glass substrate was coated with indium tin oxide (ITO) using physical vapor deposition [5]. The substrate was chosen to maximize chemical, thermal and mechanical stability to achieve a long operating time, while at the same time providing a low thermal effusivity $e = \sqrt{k\rho c}$, where k is the thermal conductivity, ρ is the density and c_p is the specific heat capacity of the substrate material. The effusivity is the material parameter that primarily determines what proportion of the generated heat in the thin film on the material's surface is transferred into the substrate and how much heat reaches the air and causes sound wave generation.

The implemented transducer is displayed in Figure 1. ITO was chosen as a film material due to its relatively large resistance and its chemical and thermal stability, small density and heat capacity. The film thickness was 30 nm to minimize the influence of the time-dependent heat and current distribution inside the conductor.

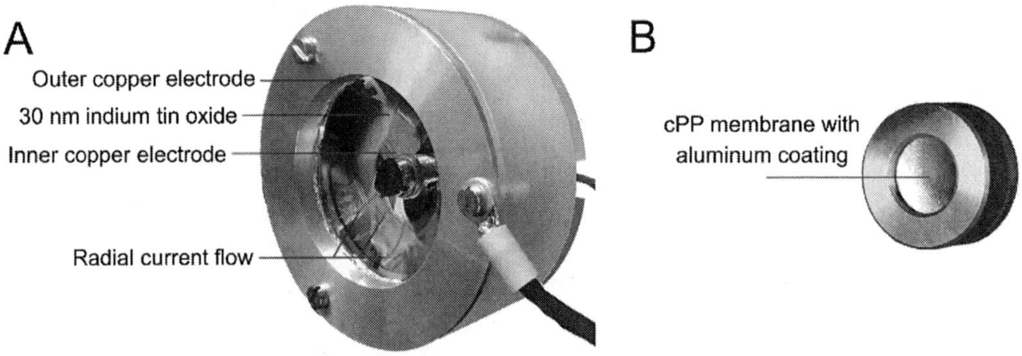

Fig. 1. Labelled photographs of the utilized transducers. **A** Broadband TA transducer. The ITO layer heats the air and the fused silica glass it is deposited on. The locally stored heat in air locally expands the air volume and a wave acoustic pressure wave propagation is initiated at each point on the ITO curved surface. This permits the generation of a focused acoustic field. **B** Cellular polypropylene transducer. The membrane is mounted onto a spherically curved, electrically conductive surface and coated with a 100 nm aluminium film and vibrates at a resonance frequency of 283 kHz.

Copper contacts were locally deposited onto the ITO surface, again using physical vapor deposition. A high voltage pulse generator was connected to the copper contacts. A concentric copper contact arrangement that led to a radial current flow was chosen. The glass substrate featured a spherically curved surface such that a focused acoustic field was generated. The diameter of the outer ITO boundary was 44 mm and the curvature was -92.5 mm.

2.3 Narrowband Reference Emitters

A focused cellular polypropylene (cPP) transducer was used as a reference emitter. These ferroelectret foil transducers feature a resonance frequency that can be adjusted during construction based on their foil thickness and diameter [7], [8]. A focusing shape was achieved by mounting the foil onto an electrically conductive, curved surface [9]. The outer surface was coated with a 100 nm aluminium film. Both surfaces were connected to an electric pulser. A relatively small aperture of 11 mm together with a high resonance frequency of 282 kHz provided the best lateral resolution, such that all relevant features of the hardware specimen were detectable. Due to its resonance behavior the emitter had to be combined with a receiver featuring the same resonance frequency.

2.4 Narrowband Receiver

The reference cPP emitter dictated the choice of the receiving transducer. A twin of the emitting transducer was used with a measured resonance frequency at 283 kHz. The same receiver was used with the TA transducer to directly compare the performance of the broadband emitter.

3. Transmission Imaging Setup

3.1 Hardware Specimen

A 3D-printed, polylactide (PLA) step wedge with flat bottom holes and an internal honeycomb structure and a fill factor of 15 % was imaged (see Figure 2). Further details on such specimen can be found in [10] The specimen was 125 mm wide and high, while the step thicknesses S were 3 mm, 6 mm and 9 mm. Each step included four flat bottom holes with a depth of 1.5 mm, 4.5 mm and 6 mm, respectively, and a width of 2 mm, 4 mm, 6 mm and 10 mm for each depth. The honeycomb parameters are depicted in in Figure 2 and were set to $R = 3$ mm, $t = 1$ mm and $h = S - 2t$. No honeycomb infill was produced below the flat bottom holes of the 6 mm and 3 mm thick steps due to the small thickness of the remaining material.

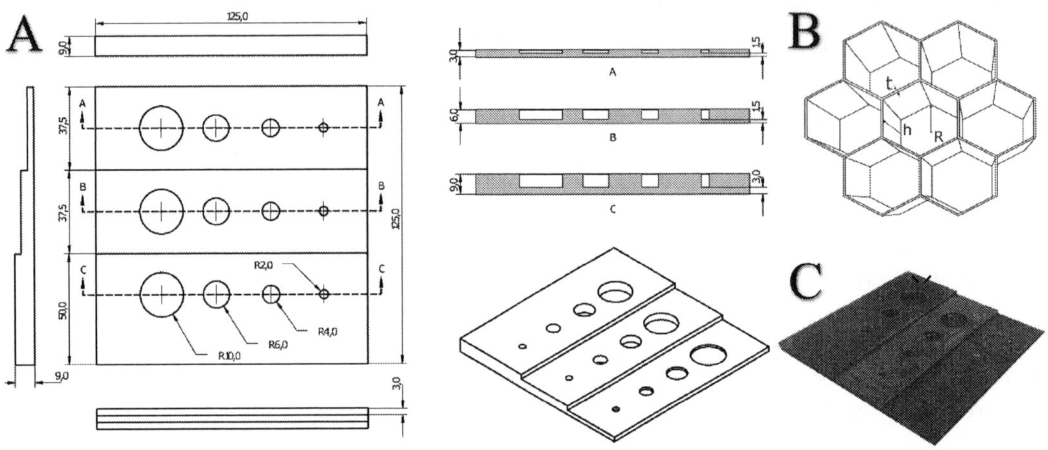

Fig. 2. The hardware specimen was designed to investigate the resolution of both the narrowband-narrowband and the broadband-narrowband transmission imaging technique. **A** Different material thicknesses, flat bottom hole diameters and depths were introduced. **B** A honeycomb infill with a fill-factor of 0.15 was incorporated. **C** The specimen was 3D printed using PLA.

3.2 Imaging Setup and Parameters

Acoustic emitter and receiver were mounted onto an automated positioning system and positioned on both sides of the specimen, facing the specimen. The distance for each transducer was adjusted such that its near field ended at the centre of the middle step's thickness. These were 24 mm for both cPP transducers and 50 mm for the TA transducer. The plate was mounted onto aluminium stacks and the edges were laminated with ultrasound-proof modelling clay to minimize deflection artefacts. The data acquisition and positioning system were part of a commercially available system (USPC 4000 AirTech; Ingenieurbüro Dr. Hillger, Ultrasonic-Techniques). The manipulator was used to scan the area of interest with a 0.3 mm step size while pulsing the emitters with pulse widths of 1.76 µs for the cPP transducer and 2 µs for the broadband transducer. No averaging was necessary while the cPP emitter was used and 3 averages were acquired for each manipulator position for the TA transducer. One and three pulses and 60 dB and 85 dB pre-amplification were used for cPP and TA emitters, respectively.

3.3 Signal and Image Processing

C-scans and D-scans were generated using the Hillgus software (Ingenieurbüro Dr. Hillger, Ultrasonic-Techniques). Thereby, only the first emerging time signal was analysed to avoid influences of reflections and diffractions. While no further processing was necessary for the cPP emitter experiment, an additional IIR bandpass (Butterworth, 4th order, 200 kHz – 350 kHz) filter was needed for the TA experiment. This was necessary to reduce the influence out the remaining acoustic transfer function of the receiving cPP transducer and to damp signals that did not belong to the resonance behavior of around 280 kHz of the receiver.

After image reconstruction, the TA images showed a strong Gaussian noise, which could be removed by a total variation (TV) regularization. It was performed with a μ-parameter of 0.4 and 0.8 for the C-scan and D-scan respectively. A standardised normal noise distribution was assumed and discretised into 9x9 elements and all further TV-parameters were set to standard values according to [11].

4. Comparison of Narrowband and Broadband Imaging Results

The reconstruction results are displayed in Figure 3. C-scans and D-scans were chosen due to best visualization of the honeycomb structure and the flat bottom holes.

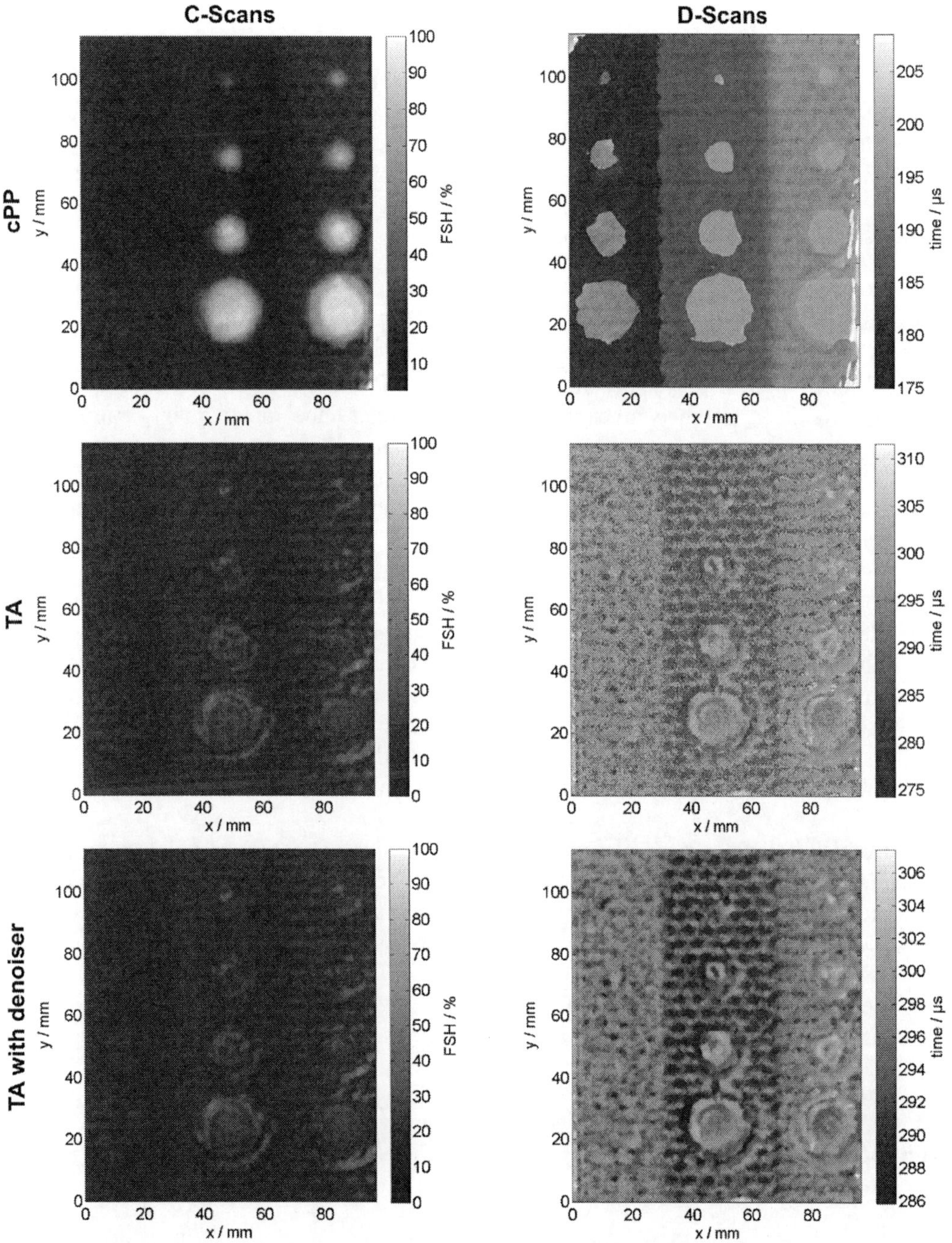

Fig. 3. Ultrasound transmission imaging results of a PLA hardware specimen with different thicknesses, differently sized flat bottom holes and a honeycomb infill. A narrowband cPP emitter and a frequency-matched cPP receiver were used and the performance of a broadband TA emitter with the same receiver compared.

4.1 cPP Performance

The C-scan clearly shows the honeycomb infill and all flat bottom holes with exception of the 1.5 mm deep and 2 mm wide hole. The D-scan shows all holes while still displaying the internal structure.

4.2 TA Performance

The TA scans display strong noise. It was reduced by using the described TV regularization. The denoised C-scan shows the honeycomb structure at 6 mm and 3 mm sample thickness (3 mm is also the thickness of the remaining material below the flat bottom holes at 9 mm step thickness). The amplitudes are inverted when compared to the cPP images. The best honeycomb images were produced at the central 6 mm thick step in the D-scan. Note, that the TA distance to the sample was adjusted at this thickness.

5. Discussion and Outlook

The difference sign in grayscale amplitude in the cPP and TA C-scans could most likely occur due to the difference in excitation pulse forms, since most other parameters are similar between the two setups - including the beam diameter. For both emitters, the receiver was an 11 mm cPP transducer, such that a comparable spatial resolution could be achieved, but the TA transducer is likely to have introduced material-dependent interferences.

The experiment was set up in a way that the cPP transducers would perform well. Thereafter it was tested whether the cPP emitter could be replaced by a TA transducer. The results show that we succeeded up to a material thickness of 6 mm. Thicker material could not be imaged since we filtered the transmitted signal using the transfer function of the receiver and by an IIR bandpass filter. We hence neglected most of the transmission signal's spectrum and energy. It is hence prospective that with more development, a single TA transducer could in future be used to replace several narrowband emitters.

Most promising development include a reduction in size of the TA transducer to increase the spatial resolution of the technique. This will, however, be accompanied by a reduction in material penetration depth due to the reduced output power, which scales linearly with the transducer's surface area.

With the advancement of broadband acoustic receivers it might be possible to locally characterize mechanical material properties using transmission ultrasound spectroscopy.

References

[1] M. C. Remillieux, B. E. Anderson, T. J. Ulrich, P.-Y. Le Bas, M. R. Haberman, and J. Zhu, "Reiview of air-coupled transduction for nondestructive testing and evaluation," *Acoust. Today*, vol. 10, no. 3, pp. 36–45, 2011.

[2] M. Gaal and D. Kotschate, "New technologies for air-coupled ultrasonic transducers," in *Proceeding of the 12th European Conference of Non-destructive Testing*, 2018.

[3] J. Grager, D. Kotschate, J. Gamper, M. Gaal, and K. Pinkert, “Advances in air-coupled ultrasonic testing combining an optical microphone with novel transmitter concepts,” in *Proceeding of the 12th European Conference of Non-desstructive Testing*, 2018, pp. 1–10.

[4] M. Daschewski, M. Kreutzbruck, and J. Prager, “Influence of thermodynamic properties of a thermo-acoustic emitter on the efficiency of thermal airborne ultrasound generation,” *Ultrasonics*, vol. 63, pp. 16–22, 2015.

[5] M. Daschewski, R. Boehm, J. Prager, M. Kreutzbruck, and A. Harrer, “Physics of thermo-acoustic sound generation,” *J. Appl. Phys.*, vol. 114, no. 11, 2013.

[6] M. Daschewski, M. Kreutzbruck, J. Prager, E. Dohse, M. Gaal, and A. Harrer, “Resonanzfreie Messung und Anregung von Ultraschall,” *Tech. Mess.*, vol. 82, no. 3, pp. 156–166, 2015.

[7] M. Paajanen, J. Lekkala, and K. Kirjavainen, “ElectroMechanical Film (EMFi) - a new multipurpose electret material,” *Sensors Actuators, A Phys.*, vol. 84, no. 1, pp. 95–102, 2000.

[8] G. M. Sessler and J. Hillenbrand, “Electromechanical response of cellular electret films,” *Appl. Phys. Lett.*, vol. 75, no. 21, pp. 3405–3407, 1999.

[9] M. Gaal, J. Bartusch, E. Dohse, F. Schadow, and E. Köppe, “Focusing of ferroelectret air-coupled ultrasound transducers,” in *AIP Conference Proceedings*, 2016, vol. 1706.

[10] D. Kotschate, S. Wendland, M. Gaal, “Airborne testing of moulded polymer compounds,” in Proceedings of 10th International Symposium on NDT in aerospace, accepted

[11] S.H. Chan, R. Khoshabeh, K.B. Gibson, P.E. Gill, and T.Q. Nguyen, “An augmented Lagrangian method for total variation image restoration,“ *IEEE Trans. Image Process.*, vol. 20, no. 11, pp. 3097–3111, 2011

AUTHOR INDEX

AUTHOR INDEX